R. Whitechurch

DOM PEDRO II,

EMPEROR OF BRAZIL

Born Dec. 2.ᵈ 1825.

After Photog. 1853.

BRAZIL

AND

THE BRAZILIANS

BY
REV. JAMES FLETCHER
AND
REV. D. P. KIDDER

Routledge
Taylor & Francis Group

LONDON AND NEW YORK

Originally published in 2005 by Kegan Paul Limited.

This edition first published in 2010 by
Routledge
2 Park Square, Milton Park, Abingdon, Oxon, OX14 4RN

Simultaneously published in the USA and Canada
by Routledge
711 Third Avenue, New York, NY 10017, USA

First issued in paperback 2018

Routledge is an imprint of the Taylor & Francis Group, an informa business

British Library Cataloguing in Publication Data
A catalogue record for this book is available from the British Library

ISBN 13: 978-1-138-86204-3 (pbk)
ISBN 13: 978-0-7103-1146-7 (hbk)

Publisher's Note
The publisher has gone to great lengths to ensure the quality of this reprint
but points out that some imperfections in the original copies may be
apparent. The publisher has made every effort to contact original copyright
holders and would welcome correspondence from those they have been
unable to trace.

PREFACE

TO THE EIGHTH EDITION.

WITHIN the last fifteen months two new editions (the sixth and seventh) of "BRAZIL AND THE BRAZILIANS" have been exhausted. While translations of portions have been made in various languages, and while an author in England has almost wholly "made up" a general book on Brazil from this work, nothing has shown a more flattering appreciation of it than the offer of Professor Laboulaye — the firm friend of America — to write an introduction for a French translation of "BRAZIL AND THE BRAZILIANS."

Since the publication of the Sixth Edition (to the Preface of which the reader is referred) several very important events have occurred in Brazil, which the authors have thought best to mention in this place, although they have noted them in the proper chapters.

THE OPENING OF THE AMAZON, which occurred on the 7th of September, 1867, and by which the Great River is free to the flags of all nations from the Atlantic to Peru, and THE ABROGATION OF THE MONOPOLY OF THE COAST TRADE from the Amazon to the Rio Grande do Sul (see page 589), whereby four thousand miles of Brazilian sea-coast are open to the vessels of every country, cannot fail not only to develop the resources of Brazil, but these measures will prove a great benefit to the bordering Hispano-American Republics and to the maritime nations of the earth. The opening of the Amazon is the most significant indication that the leaven of old narrow, monopolistic Portuguese conservatism has at last worked out. Portugal would not allow Humboldt to enter the Amazon valley in Brazil. The result of the new policy is beyond the most sanguine expectation. The exports and imports of Pará for October and November, 1867, were double those of 1866. This is but the beginning. Soon it will be found that it is cheaper for all Bolivia, Peru, Equador, and New Grenada east of the Andes to receive their goods from, and to export their India-rubber, cinchona, &c., &c., to the United States and Europe *via* the great water highway which discharges into the Atlantic, than by the long, circuitous route of Cape Horn, or the Trans-Isthmian route of Panama. The Purus and the Madeira are hereafter to be navigated by steamers. The valley of the Amazon in Brazil is as large as the area of the United States east of Colorado, while the whole valley of the Amazon, in and out of Brazil, is equal to all the United States east of California, Oregon, and Washington Territory; and yet the population is not equal to the single city of Rio de Janeiro, or the combined inhabitants of Boston and Chicago. It is estimated that a larger population can be sustained in the valley of the Amazon than elsewhere on the globe; but it will never be peopled until there is as complete freedom for emigrants, and as entire absence of *red-tapeism* in Brazil as exist in the United States.

THE SYSTEM OF EMIGRATION is improving. In 1866 there were mistakes on the part of the agents for Brazil at New York. They were not careful enough. They accepted any one and every one that applied for passage under the liberal offers (which still hold good) of the Brazilian government, and there were mistakes on the part of many well-meaning, almost penniless adventurers from our cities and from our own South, who supposed that there was a royal road to prosperity in the tropics without labor, and that slavery was a permanent institution in Brazil. But, notwithstanding the croakers who have returned, many Southerners have succeeded and are succeeding in Brazil.

SLAVERY has decreased with great rapidity during 1866–67, and the best estimates make the present number of slaves 1,400,000, — a reduction by the mild process of law and custom of 1,600,000 since 1853. The Emperor took the initiative at the last session of Parliament, and invoked legislation upon this most important subject. Dr. A. M. Perdigaõ Malheiro, an eminent advocate at Rio, has published a most important and convincing pamphlet on this question, entitled *A escravidaõ no Brazil* (Slavery in Brazil).

DIRECT TAXATION for the first time in Brazil has been brought about by the exigencies of the Paraguayan war, — a conflict which has done more to give Brazil a national feeling than any event since 1822.

THE PARAGUAYAN WAR. — The history and the aims of this contest, now waging, have been more persistently misrepresented than those of any other war of modern times, with the single exception of the misrepresentation in England of the late internal struggle in the United States. From November, 1864 (the beginning of the war), to November, 1865, the various battles and victories were impartially described in the English journals, from which source other countries, not South American, have derived their information. But in the autumn of 1865 the Brazilian government applied in London for a loan of £4,000,000. Such was the competition for this loan, and such the confidence of English financiers in Brazil, that £30,000,000 were subscribed. The loan, of course, immediately went above par. From that time to this "operators" at Buenos Ayres and Montevideo, one thousand miles from the seat of war, had a motive in sending rumors and partial statements detrimental to the allies by the English steamer to Lisbon, whence their correspondence would be telegraphed to London; and the result would be the depression of the Brazilian loan for a few days, then when the "rise" took place the "operators" and their friends could profit by their former transaction. In regard to the contest, Brazil had no other alternative than war with Lopez, who is as truly a despotic dictator as Francia was. The origin of the war is impartially set forth on page 353. The present position of the allies is very much that of the armies of the United States at the end of 1864, when Sherman made his famous "march to the sea" and Grant was before Richmond. Brazil in 1867 sent an army to the north of Paraguay and retook all the fortified ports seized by Paraguay in 1864; and the allied land and naval forces at the beginning of 1868, after varied experience, were closing upon Humaita, the last stronghold of the Paraguayans, — a fortress far more inapproachable than Sebastopol. 1868 will doubtless see a complete resolution of a struggle whose end is the liberation of Brazilian citizens and the reopening (which Paraguay had guaranteed by solemn treaty obligations) of the great natural highway to the sea for the four nations of Eastern South America.

BRAZILIAN COFFEE. — Brazil has also had her peaceful triumphs. In the great Exposition held at Paris in 1867 Brazil attracted much attention by the display of her material resources. She succeeded in obtaining a number of prizes. To the uninitiated it may seem strange that from all the countries — Arabia, Java, Ceylon, Venezuela, the West Indies, and Central America — contesting for the production of the best coffee, Brazil bore away the palm. But it has long been known to dealers that coffee does *not* depend upon where it grows, but upon the length of time it remains upon the tree and upon the manner of its curing. The Southern and the South-western States became acquainted with coffee imported from Rio de Janeiro fifty years ago, at a time when Brazilians did not know how to cure coffee; but the taste of the South and West has alone kept up the demand for the green, poorly cured coffee known in commerce as "Rio." The Brazilians themselves never use "Rio," and although three fourths of all the coffee imported into the United States come from Brazil, yet much of it is sold as Mocha and Java, or under any other name than "Rio." The English, Americans, and Germans make the poorest drink from coffee in the world, while the Latin nations, who never *boil* their coffee, make the best beverage. For the history and culture of coffee see pages 449, 451.

COTTON can be grown in any portion of the Empire of Brazil. In quality it ranks far above our "uplands," and in the Liverpool market the best Brazilian cottons stand next — though at a distance — to the "Sea-Island." Pernambuco is the chief port for exportation. There are no great cotton plantations, but the most of its culture is carried on by small farmers and by free negroes and half-breeds. An article in the *New York Evening Post*, entitled "Small Farms for Cotton Culture" in our own country, called forth a communication from Mr. Hitch, of the house of Henry, Forster, & Co. of Pernambuco, in which he describes the Brazilian plan of little farms cultivated by from one to six persons. This has an important bearing on cotton culture in the United States. Mr. Hitch shows how the demand caused by the "cotton famine" brought forth the supply to such an extent that

Pernambuco in five years increased her cotton exportation tenfold. See page 525.

GUARANÁ. — At the French Exposition of 1867 a brown chocolate-colored substance figured under the head of the medicines from Brazil. This brown material might at first sight have been taken for chocolate cast in the form of serpents, diminutive turtles, tapirs, &c. It was, however, a remedy which has been used for centuries in Brazil and Bolivia, and which has lately become one of the most fashionable antifebrile remedies in Paris. Guaraná is the indigenous name of this new contribution to civilized Pharmacy. The junior author has often partaken of it on the Amazon; and as many have recently inquired concerning the Guaraná, a short notice of it may be interesting. Dr. Cotting of Roxbury, Mass., gives a brief account of it in the Boston Medical and Surgical Journal for February 7, 1867, pages 20, 21. On the west bank of the river Tapajos (excepting the Madeira, the longest southern affluent of the Amazon) lives a tribe of Indians called the Mauhés or Maués, who prepare from the seeds of a small climbing plant (the *Paullinia sorbilis*) the Guaraná. The plant bears berries somewhat larger than coffee-berries, and two in a capsule, not unlike the coffee. These are roasted, ground, mixed with a little water, made into various shapes, and dried to hardness in an oven. Grated and dissolved in water or lemonade, it is highly esteemed as a refreshing and stimulating drink. It is much used by the inhabitants of Matto Grosso and other interior provinces, and sometimes, it is said, to such an excess as to produce great tremulousness. It is also much used as a remedy in diarrhœa and intermittent fever. Dr. James C. White of Boston, who has analyzed the Guaraná, has given the public his analysis in a very interesting paper.

The authors cannot close this Preface without recording an event which may seem personal to them, but which is also one of sadness to all those who love Brazil in the things that are far beyond her material development. Amongst the devoted men who have gone from the United States to "the land of the Southern Cross," none have been more zealous, wise, and successful in "winning souls" than Rev. A. G. Simonton of Rio de Janeiro. He founded the Presbyterian mission at Rio de Janeiro in 1859. He established the *Imprensa Evangelica*, a religious journal of the most excellent character, and thus, in addition to his constant labors in the pulpit, he did much to furnish Brazil with an Evangelical literature. A few days before his death, when in apparent health, he wrote to the junior author a most cheering letter, stating that during 1867 the fruits of the missions (the American Presbyterian) with which he was connected were no less than 112 conversions, 82 of whom made their profession "before men," and the remainder were soon to follow their example. He died in San Paulo on the 9th of December, 1867. The individual dies, but the Church lives, and his labors will still bring forth fruit to be gathered by the earnest harvesters, his coadjutors, now laboring in the same field.

NEWBURYPORT, MASS., March, 1868.

PREFACE TO THE SIXTH EDITION.

THE favorable reception which five editions of this work have had in the United States, England, and Brazil, indicates a growing interest in the largest and most stable country of South America. It may be that the illustrations accompanying the Preface to the first edition are not so appropriate to-day as they were ten years ago, but there is still too much ignorance of Brazil in Europe and North America. The present edition will give some idea of the material and moral progress of Brazil during the last decade.

While several new volumes on some particular portion of the country have been written since 1857, no other work in our language has given a general view of Brazil and the Brazilians. As much of the political and social life of the Empire centres at Rio de Janeiro, the history and descriptions of the state of affairs at the capital are, to a great extent, those of the whole country. It is for this reason that the reader is detained longer in the city where the Monarch resides and the Parliament holds its sessions.

Since 1857, one of the authors (J. C. F.) has visited Brazil in four different years, passing much time at Rio de Janeiro; sojourning on plantations, and observing the phases of Brazilian slavery; making extensive journeys along the sea-coast, and penetrating the interior. In 1862 he ascended the Amazon to the verge of Peru, — more than two thousand miles up the most marvellous river in the world.

It was the intention of the authors to publish a new edition in 1864; but unexpected duties, both in that and the following years, called the junior colleague to Brazil, and prevented the desired end. The advantages of later information will, it is hoped, more than compensate for the delay.

The experience of the authors in Brazil extends over a long period, and they have endeavored conscientiously and impartially to give their views of the country and its people. They have had no motives to do otherwise. While they have not spared what they deemed faults, whether in religion, slavery, or other matters (see concluding chapter), they have not withheld praise when due, and it has not been from intention if they have failed to bring out the good points of the Brazilians. To foreign merchants in Brazil, unsuccessful in business, or to travellers hastening through the country, ignorant of the Portuguese and French languages, and never associating with the inhabitants, the descriptions of those who have resided long in the Empire, or have travelled extensively through it, seem overwrought. One must always bear in mind the origin of the Brazilians, their newness among the nations of the earth, and the fact that the only true mode of comparing Brazil is not to measure her progress with the United States, England, or France, but with the countries of America whose inhabitants are of the Latin race.

To have detailed with only an ordinary degree of minuteness the changes and progress of Brazil for the last ten years, would require a large volume. As it is, there have been, by emendations and additions, and by notes and appendices, nearly one hundred pages of new matter printed in this edition, while the ordinary text has in many cases been changed and increased. Everything, so far as possible, is brought up to date (1866), by notes at the end of the appropriate chapter. In some cases letters and itinerary are retained, irrespective of date, because they illustrate manners and

customs that remain *in statu quo.* When greater space was needed, the subject is more fully set forth in the Appendix. Thus, in
regard to the present PARAGUAYAN WAR (about whose merits
there has been as much ignorance * in both the United States and
England as there was in Europe concerning the late Rebellion in
North America), the reader will find a brief history of its origin
and progress on page 352.

BRAZILIAN SLAVERY is treated in Chap. VIII., and the most
recent information concerning it is given on page 139.

EMIGRATION TO BRAZIL has of late attracted much attention,
especially in the South of the United States, since 1865. Information under this head can be found on page 333, and in the concluding chapter, in the Speech of Paula Souza, page 592.

Intimately connected with this subject is that of the RELIGIOUS
DISABILITIES OF PROTESTANTS; and no portion of this work is so
indicative of great moral progress as the Speech of Sr. Furquim
d'Almeida (page 595), and the article in Appendix I.

For important METEOROLOGICAL TABLES, see Appendix K.

The MINERAL RICHES of Brazil are known to be considerable.
Diamond and gold mines have been the chief sources of mineral
wealth, but hitherto there has been a deficiency of useful minerals.
The desideratum has at length been supplied. Coal discoveries
have been made in various sea-coast provinces, but the most important in this respect was made by Mr. Nathaniel Plant (from
the British Museum) in Rio Grande do Sul. For a full account
of this rich coal mine see Appendix H. In the same Appendix it
will be seen that the new gold mines of Sr. Tasso, in Northern

---

* In four different issues of the journals of New York and Boston, in August,
1866, it was stated that the Allies were at "the last extremity," and that the Paraguayans were just about to annihilate them. Up to this time (October, 1866), the
Paraguayans, in the four great battles, counting from that of Riachuelo, 1865, have
not won a single victory. The par of gold at Rio is 26d. to the mil reis; the average
in 1865 – 66 was 23½d., which does not look like an "extremity."

Brazil, demonstrate that the precious metal is by no means confined to the region of S. João del Rei.

The leading Brazilian Journals at the capital are noticed on pages 252, 253. Brazilian Literature and literary Brazilians (pages 586, 589) will interest many Anglo-Saxon readers.

While frequent mention.is made of the ability and accomplishments of the Emperor Dom Pedro II., Chapter XIII. is especially devoted to that monarch.

On pages 183–185, some account is given of Statesmen and Political Parties.

In 1865, Professor Agassiz, the well-known *savant*, accompanied by an American scientific corps, visited Brazil at the invitation of the Emperor. The Professor made extensive and most interesting explorations, an account of which is soon to be given to the world. His investigations in the Amazon region have excited a great interest amongst men of science. Major Coutinho, a Brazilian officer of engineers, at the command of the Emperor, accompanied Professor Agassiz in his explorations of the Amazon, and afterward published at Rio, both in Portuguese and English, some account of the wonderful *fauna* discovered by Professor Agassiz in Northern Brazil. The English version of a portion of these letters will be found in Appendix J.

For Population, Commercial Tables, Weights and Measures, and other statistics, see Appendices E, F, and G.

Within a few years several works on the Brazilian Empire have appeared in England, France, Germany, and Brazil. Amongst these may be mentioned Bates's " Naturalist on the Amazons," a charming book, and the best yet published on that wonderful region. The *Deux Années au Brésil* of Biard is, aside from its fine illustrations, the most worthless book ever published on any country. The author seems not to have had one serious reflection. Halfeld's " Survey of the River San Francisco " is a magnificent elephant folio, published at Rio de Janeiro, of which Sir Roderick Murchison said (before the Royal Geographical Society), " Any country might be proud of such a work." It cannot be purchased, but a number of copies have been sent by the Brazilian government to the libraries of the United States and Europe. The articles of M. Elisé Reclus, published in the *Revue des Deux Mondes*, in 1862, show their author to be an earnest friend of liberty, and, also, that they were written after a very brief visit to Brazil. The conclusions are somewhat hasty, especially when based on M. Biard's book, and Dr. Avé Lallemant's interesting but partial *Reise durch sud Brasilien*, Leipsic, 1859. Quite a number of German works have appeared concerning the " colonization " of Germans in Brazil (which was in many cases a shameless piece of jobbery), and the writers are not disposed to look upon anything in Brazil with the least degree of allowance. Mr. Thomas Woodbine Hinchliff's " South American Sketches " (Longmans, London, 1863) is a very pleasing and accurate book. Sr. Pereira da Silva is now writing a complete history of Brazil, in the Portuguese language. Sr. Aguiar (of San Paulo), in a pamphlet entitled *O Brazil e os Brazileiros*, has given some very caustic sketches of his country-men. Sr. Soares, of Rio Grande do Sul, has furnished us with an important statistical work on the resources of Brazil. Sr. M. M. Lisboa has, in his *Romances Historicos*, opened a literary mine, in which it would be well if more Brazilian writers would delve. Sr. A. C. Tavares Bastos, in his *Cartas do Solitario* (Letters of the Hermit), has given the Brazilian public a most important volume on the various political, economical, and moral questions that so deeply concern the well-being of the Empire. A very excel-lent book on the resources of Brazil was published, in 1863, by the Baron of Penedo (Brazilian Minister to England), *apropos* to the contribution of Brazil to the Great Exposition of London, in 1862.

The thanks of the authors are due, for corrections and contributions in preparing this edition, to Mr. Robert William Garrett of Rio de Janeiro; to Mr. Brambier of Parà; to His Excellency Sr. d'Azambuja, Brazilian Minister to the United States; to the Chevalier d'Aguiar, Brazilian Consul-General at New York; and to Professors Gumere and Cope, of Haverford College, near Philadelphia. To the late John S. Gillmer of Bahia the authors were under many obligations.

OCTOBER, 1866.

PREFACE.

THE popular notion of Brazil is, to a certain extent, delineated in the accompanying side-illustrations. Mighty rivers and virgin forests, palm-trees and jaguars, anacondas and alligators, howling monkeys and screaming parrots, diamond-mining, revolutions, and earthquakes, are the component parts of the picture formed in the mind's eye. It is probably hazarding nothing to say that a very large majority of general readers are better acquainted with China and India than with Brazil. How few seem to be aware that in the distant Southern Hemisphere is a stable constitutional monarchy, and a growing nation, occupying a territory of greater area than that of the United States, and that the descendants of the

Portuguese hold the same relative position in South America as the descendants of the English in the northern half of the New World! How few Protestants are cognizant of the fact that in the territory of Brazil the Reformed religion was first proclaimed on the Western Continent!

The following work, by two whose experience in the Brazilian Empire embraces a period of twenty years, endeavors faithfully to portray the history of the country, and, by a narrative of incidents connected with travel and residence in the land of the Southern Cross, to make known the manners, customs, and advancement of the most progressive people south of the Equator.

While "Itineraries" relating to journeys of a few months in various portions of the Empire have been recently published, no general work on Brazil has been issued in Europe or America since the "Sketches" of the senior author, (D. P. K.,) which was most favorably received in England and the United States, but has long been out of print.

Although the present volume is the result of a joint effort, the desire for greater uniformity caused the senior author

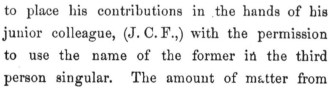

to place his contributions in the hands of his junior colleague, (J. C. F.,) with the permission to use the name of the former in the third person singular. The amount of matter from each pen is, however, more nearly equal than at first sight would appear.

The authors have consulted every important work in French, German, English, and Portuguese, that could throw light on the history of Brazil, and likewise various published memoirs and discourses read before the flourishing "Geographical and Historical Society" at Rio de Janeiro. For statistics they have either personally examined the Imperial and provincial archives, or have quoted directly from Brazilian state papers.

For important services, the authors are happy to acknowledge their indebtedness to Conselheiro J. F. de Cavalcanti de Albuquerque, His Brazilian Majesty's Minister-Plenipotentiary at Washington, and M. le Chevalier d'Aguiar, Brazilian Consul-General at New York; to Hon. Ex-Governor Kent, of Maine, and Ferdinand Coxe, Esq., of Philadelphia, both of whom held high diplomatic positions at Rio de Janeiro; to Hon. Judge J. U. Petit, formerly Consul in one of the most important Northern provinces of Brazil; to Mrs. L. A. Cuddehy, late of Rio de Janeiro; and to Rev. H. A. Boardman, D.D., of Philadelphia. They also express their obligations to Mr. D. Bates, Dr. Thos. Rainey, and to A. R. Egbert, M.D., for valuable contributions to the Appendix.

The numerous illustrations are, with few exceptions, either from sketches, or daguerreotype views taken on the spot. All have been faithfully as well as skilfully executed by Messrs. Van Ingen & Snyder, of Philadelphia. The accompanying map, prepared by Messrs. J. H. Colton & Co., is one of the most perfect ever published of an Empire which has never been surveyed. In 1855 the junior author travelled more than three thousand miles in Brazil, making corrections of this map as he journeyed; and his sincere thanks are heartily given to Senhor John Lisboa, of Bahia, who has devoted himself to the geography of his native land.

In 1866, J. H. Colton & Co. have published a gigantic map of Brazil, which was a work of years by Mr. Rensberg, one of the leading lithographers of Rio de Janeiro. Messrs. Fleuss, Brothers, publishers of the *Semana Illustrada*, have also issued several important maps from their establishment—the *Imperial Instituto Artistico* at Rio.

NOTES FOR THOSE GOING TO BRAZIL.

THE Portuguese language is universally spoken in Brazil. It is not a dialect of the Spanish, but is a distinct tongue: as Vieyra says, it is the eldest daughter of the Latin. Portuguese and French are the Court languages. One-sixth of the population of large cities and towns speak French. Those acquainted with the French, Italian, or Spanish easily acquire the Portuguese. English is taught in all the higher schools; and it is gratifying to the American that at the capital, and in some other important places, the "Class Readers" of George S. Hillard, Esq., (author of "Six Months in Italy,") are the text-books. While Messrs. Trübner & Co., in London, and the Messrs. Appleton, in New York, have published manuals for learning the Portuguese, it may be of advantage to state that if an Englishman or Anglo-American can give to the vowels the Continental sound, learn the contractions, accents, &c., and the peculiarities of two or three consonants, he will find the Portuguese the easiest of all foreign tongues. The termination ão is pronounced almost like *oun* in the English word *noun*. Words ending in ões are pronounced as if an *n* were inserted between the *e* and the *s*. Thus, Camões, English Camoens. Terminations *em* and *im* are very nearly pronounced like eng and ing in English: *e. g.* Jerusalem is pronounced Jerusale*ng*. *X* has the force of *Sh*: thus, one of the great affluents of the Amazon, Xingú, is pronounced *Shingú*.

The word Dom, (*dominus*,) which always precedes the name of the Emperor, is not used indiscriminately like the *Don* of the Spanish, but is a title applied by the Portuguese and their descendants only to monarchs, princes, and bishops.

One *milreis*, (a thousand reis,) about fifty-six cents, or two shillings and sixpence English. The Brazilian unit-coin is always represented by the dollar sign *after* the mil: thus, 5$500 are five mil and five hundred reis,—about three dollars. A conto of reis is little more than £112.

Clothes, of course, should be of a character adapted to the tropics; but always take some woollen garments, for in the interior, and south of Bahia, the thermometer often indicates 60° Fahrenheit. It hardly need be added that a dress-coat is indispensable for those going to the palace. All personal effects, like wearing-apparel, are admitted duty free; but the traveller would do well to remember that he should not be overstocked with cigars. There are many drawbacks at the custom-house in favor of goods belonging to emigrants, as agricultural implements, machinery, &c. &c. (vide page 333 and the concluding chapter of this work.)

As to the PATENT LAWS, mode of obtaining certain privileges for inventions, &c., William V. Lidgerwood, Esq., (United States Chargé d'Affaires in 1865–66,) can give more information than any other person in Brazil. He resides at Rio de Janeiro.

Messrs. Trübner & Co. (60 Paternoster Row, London) have facilities for furnishing Brazilian and Portuguese books to any parts of Europe or North America. We are glad that this house is to publish a new and complete Portuguese and English Dictionary,—a very great *desideratum*, as all such lexicons now extant are exceedingly antiquated.

English and American publications of standard works and light literature are to be obtained of H. M. Lane & Co., 15 Rua Direita, Rio de Janeiro; and of Guelph de Lailhacar & Co., Rua do Crespo, Pernambuco, and in the city of San Paulo.

Carrington & Co.'s United States & Brazilian Express is a very great convenience which has followed the establishment of the United States and Brazil Mail Steamship Company. Messrs. Carrington & Co. (30 Broadway, New York) charge themselves to deliver parcels and money, or to fill orders in Pará, Pernambuco, Bahia, and Rio de Janeiro, and *vice versâ*. Fales and Duncan, Commission Merchants, 57 Rua Direita, are the agents at Rio de Janeiro.

Hotels in Brazil are not equal to those in Europe or the States. At Rio all have high prices, ranging, according to room, from ten shillings English to £1. The Exchange Hotel and Hotel dos Estrangeiros are the best English hotels in the capital. Hotel d'Europa is the best French hotel. Bennett's, an hour from Rio, is the most comfortable place in Brazil. Bahia, Hotel Furtin is a good restaurant, and convenient to those arriving from sea. At Pernambuco, the Hotel Universel has the same recommendations. The hotels of Bahia and Pernambuco are small, compared with those of Rio. The prices of 1855 (pages 295 and 296) are from one-third to one-half higher in 1866,—except at Petropolis, at which place are several good hotels.

6

CONTENTS.

CHAPTER I.

PAGE

The Bay of Rio de Janeiro—Historic Reminiscences—First Sight of the Tropics—
Entrance to the Harbor—Night-Scenes—Impressions of Beauty and Grandeur—
Gardner and Stewart—The Capital of Brazil—Distinction of Rio de Janeiro........ 13

CHAPTER II.

Landing—Hotel Pharoux—Novel Sights and Sounds—The Palace Square—Rua
Direita — Exchange — The "Team" — Musical Coffee-Carriers — Custom-House—
Lessons in Portuguese, and Governor Kent's Opinion of Brazil—Post-Office—Dis-
like of Change—Senhor José Maxwell—Rua do Ouvidor—Shops and Feather-
Flowers—The Brazilian Omnibus can be full—Narrow Streets and Police-Regu-
lations—A Suggestion to relieve Broadway, New York— Passeio Publico—Bra-
zilian Politeness—The "Gondola"—The Brazilian imperturbable—Lack of Hotels
—First Night in Rio de Janeiro.. 24

CHAPTER III.

Discovery of South America—Pinzon's Visit to Brazil—Cabral—Coelho—Americus
Vespucius—The Name "Brazil"—Bay of Rio de Janeiro—Martin Affonso de Souza
—Past Glory of Portugal—Coligny's Huguenot Colony—The Protestant Banner
first unfurled in the New World—Treachery of Villegagnon—Contest between
the Portuguese and the French—Defeat of the Latter—San Sebastian founded—
Cruel Intolerance—Reflections... 46

CHAPTER IV.

Early State of Rio—Attacks of the French—Improvements under the Viceroys—
Arrival of the Royal Family of Portugal—Rapid Political Changes—Departure of
Dom John VI.—The Viceroyalty in the Hands of Dom Pedro—Brazilians dis-
satisfied with the Mother-Country—Declaration of Independence—Acclamation of
Dom Pedro as Emperor.. 61

CHAPTER V.

The Andradas—Instructions of the Emperor to the Constituent Assembly—Dom
Pedro I. dissolves the Assembly by Force—Constitution framed by a Special Com-
mission—Considerations of this Document—The Rule of Dom Pedro I.—Causes of
Dissatisfaction—The Emperor abdicates in favor of Dom Pedro II. 73

CHAPTER VI.

PAGE

The Praia de Flamengo—The Three-Man Beetle—Splendid Views—The Man who cut down a Palm-Tree—Moonlight—Rio "Tigers"—The Bathers—Gloria Hill—Evening Scene—The Church—Marriage of Christianity and Heathenism—A Sermon in Honor of Our Lady—Festa da Gloria—The Larangeiras—Ascent of the Cercovado—The Sugar-Loaf.. 86

CHAPTER VII.

Brotherhoods—Hospital of San Francisco de Paula—The Lazarus and the Rattlesnake—Misericordia—Sailors' Hospital at Jurujuba—Foundling-Hospital—Recolhimento for Orphan-Girls—New Misericordia—Asylum for the Insane—José d'Anchieta, Founder of the Misericordia—Monstrous Legends of the Order—Friar John d'Almeida—Churches—Convents... 107

CHAPTER VIII.

Illumination of the City—Early to Bed—Police—Gambling and Lotteries—Municipal Government—Vaccination—Beggars on Horseback—Prisons—Slavery—Brazilian Laws in favor of Freedom—The Mina Hercules—English Slave-Holders—Slavery in Brazil Doomed 124

CHAPTER IX.

Religion—The Corruption of the Clergy—Monsignor Bedini—Toleration among the Brazilians—The Padre—Festivals—Consumption of Wax—The Intrudo—Processions—Anjinhos—Santa Priscilliana—The Cholera not cured by Processions..... 140

CHAPTER X.

The Home-Feeling—Brazilian Houses—The Girl—The Wife—The Mother—Moorish Jealousy—Domestic Duties—Milk-Cart on Legs—Brazilian Lady's Delight—Her Troubles—The Marketing and Watering—Kill the Rio—Boston Apples and Ice—Family Recreations—The Boy—The College—Common-Schools—Highest Academies of Learning—The Gentleman—Duties of the Citizen—Elections—Political Parties—Brazilian Statesmen—Nobility—Orders of Knighthood.......................... 1..

CHAPTER XI.

Praia Grande—San Domingo—Sabbath-Keeping—Mandioca—Fonte de Arca—View from Ingá—The Armadillo—Commerce of Brazil—The Finest Steamship Voyage in the World—American Seamen's Friend Society—The English Cemetery—English Chapel—Brazilian Funerals—Tijuca—Bennett's—Cascades—Excursions—Botanical Gardens—An Old Friend—Home.. ..

CHAPTER XII.

The Campo Santa Anna—The Opening of the Assemblea Geral—History of Events succeeding the Acclamation of Dom Pedro II.—The Regency—Constitutional Reform—Condition of Political Parties before the Revolution of 1840—Debates in the House of Deputies—Attempt at Prorogation—Movement of Antonio Carlos—Deputation to the Emperor—Permanent Session—Acclamation of Dom Pedro's

PAGE

Majority — The Assembly's Proclamation — Rejoicings — New Ministry — Public Congratulations—Real State of Things—Ministerial Programme—Preparations for the Coronation—Change of Ministry—Opposition come into Power—Coronation postponed—Splendor of the Coronation—Financial Embarrassments—Diplomacy —Dissolution of the Camara—Pretext of Outbreaks—Council of State—Restoration of Order—Sessions of the Assembly—Imperial Marriages—Ministerial Change —Present Condition 211

CHAPTER XIII.

The Emperor of Brazil—His Remarkable Talents and Acquirements—New York Historical Society—The First Sight of D. Pedro II.—An Emperor on Board an American Steamship—Captain Foster and the "City of Pittsburg"—How D. Pedro II. was received by the "Sovereigns"—An Exhibition of American Arts and Manufactures—Difficulties overcome—Visit of the Emperor—His Knowledge of American Authors—Success among the People—Visit to the Palace of S. Christovão—Longfellow, Hawthorne, and Webster. 231

CHAPTER XIV.

Brazilian Literature—The Journals of Rio de Janeiro—Advertisements—The Freedom of the Press—Effort to put down Bible-Distribution—Its Failure—National Library —Museum—Imperial Academies of Fine Arts—Societies—Brazilian Historical and Geographical Institute—Administration of Brazilian Law—Curious Trial............ 251

CHAPTER XV.

The Climate of Brazil—Its Superiority to other Tropical Countries—Cool Resorts— Trip to St. Alexio—Brazilian Jupiter Pluvius—The Mulatto Improvisor—Sydney Smith's "Immortal" Surpassed—A Lady's Impressions of Travel—An American Factory—A Yankee House—The Ride up the Organ Mountains—Forests, Flowers, and Scenery—Speculation in Town-Lots—Boa Vista—Height of the Serra dos Orgões—Constancia—The "Happy Valley"—The Two Swiss Bachelors—Youth renewed — Prosaic Conclusion — Todd's "Student's Manual" — The Tapir—The Toucan—The Fire-Flies—Expenses of Travelling—Nova Fribourgo—Canta Gallo —Petropolis.. 268

CHAPTER XVI.

Preparations for the Voyage to the Southern Provinces—The Passengers—Ubatuba —Eagerness to obtain Bibles—The Routine on Board—Aboriginal Names—San Sebastian and Midshipman Wilberforce—Santos—Brazilians at Dinner—Incorrect Judgment of Foreigners—S. Vincente—Order of Exercises—My Cigar—Paranaguá —H.B.M. "Cormorant" and the Slavers—Mutability of Maps—Russian Vessels in Limbo—The Prima Donna—An English Engineer—Arrive at San Francisco do Sul 303

CHAPTER XVII.

The Province of Paraná—Message of its First President—Maté, or Paraguay Tea— Its Culture and Preparation—Grows in North Carolina—San Francisco do Sul— Expectations not fulfilled—Canoe-Voyage—My Companions not wholly carnivorous—A Travelled Trunk—The Tolling-Bell Bird—Arrival at Joinville—A New Settlement—Circular on Emigration to Brazil... 320

CHAPTER XVIII.

PAGE

Colonia Donna Francisca—The School-Teacher—The Clergyman—A Turk—Bible-Distribution—Suspected—A B C—The Fallen Forest—The House of the Director—A Runaway—The Village Cemetery—Moral Wants—Orchidaceous Plants—Charlatanism—San Francisco Jail—The Burial of the Innocent, and the money-making Padre—The Province of Sta. Catharina—Desterro—Beautiful Scenery—Shells and Butterflies—Coal-Mines—Province of Rio Grande do Sul—Herds and Herdsmen—The Lasso—Indians—Former Provincial Revolts—Present Tranquillity assured by the Overthrow of Rosas and of the Paraguayan Lopez Jr............ 334

CHAPTER XIX.

Journey to San Paulo—Night-Travelling—Serra do Cubatão—The Heaven of the Moon—Frade Vasconcellos—Ant-Hills—Tropeiros—Curious Items of Trade—Ypiranga—City of San Paulo—Law-Students and Convents—Mr. Mawe's Experience contrasted—Description of the City—Respect for S. Paulo—The Visionary Hotel-Keeper.. 354

CHAPTER XX.

History of San Paulo—Terrestrial Paradise—Reverses of the Jesuits—Enslavement of the Indians—Historical Data—The Academy of Laws—Course of Study—Distinguished Men—The Andradas—José Bonifacio—Antonio Carlos—Alvares Machado—Vergueiro—Bishop Moura—A Visit to Feijó—Proposition to abolish Celibacy—An Interesting Book—The Death of Antonio Carlos de Andrada—High Eulogium—Missionary Efforts in San Paulo—Early and Present Condition of the Province—Hospitalities of a Padre—Encouragements—The People—Proposition to the Provincial Assembly—Response—Result—Addenda—Present Encouragements.. 366

CHAPTER XXI.

Agreeable Acquaintance—Old Congo's Spurs—Lodging and Sleeping—Company—Campinas—Illuminations—A Night among the Lowly—Arrival at Limeira—A Pennsylvanian—A Night with a Boa Constrictor—Eventful and Romantic Life of a Naturalist—The Bird-Colony destined to the Philadelphia Academy of Natural Sciences—Ybecaba—Sketch of the Vergueiros—Plan of Colonization—Bridge of Novel Construction—Future Prospects... 396

CHAPTER XXII.

A New Disease—The Culture of Chinese Tea in Brazil—Modus Operandi—The Deceived Custom-House Officials—Probable Extension of Tea-Culture in South America—Homeward Bound—My Companion—Senhor José and a Little Difficulty with him—California and the Musical Innkeeper—Early Start and the Star-Spangled Banner—The Senhores Brotero of S. Paulo—Fourth of July inaugurated in an English Family—"Yankee Doodle" on the Plains of Ypiranga—Lame and Impotent Conclusion—Astronomy under Difficulties—Deliverance—Return to Rio de Janeiro.. 416

CHAPTER XXIII.

PAGE

The Brazilian North—Extent of the Empire—The Falls of Itamarity—Gigantic Fig-Tree—The Keel-Bill—A Plantation in Minas-Geraes—Peter Parley in Brazil—Sweet Lemons—Baronial Style—The Padre—Vesper-Hours—The Plantation-Orchestra—The White Ants obedient to the Church—The Great Ant-Eater—The Paca—The Musical Cart—The Mines and other Resources of Minas-Geraes—Coffee: its History and Culture—The Province of Goyaz—Stingless Bees and Sour Honey—Mato Grosso—Long River-Route to the Atlantic—A New Thoroughfare—Lieutenant Thomas J. Page—The Survey of the La Plata and its Affluents—First American Steamer at Corumba—Steamboat-Navigation on the Paraguay—Officers of the American Navy—Dr. Kane and Lieutenant Strain—Diamond and Gold Mines the Hinderers of Progress—The Difference in the Results from Diamonds and Coffee.. 432

CHAPTER XXIV.

Cape Frio—Wreck of the Frigate Thetis—Campos—Espirito Santo—Aborigines—Origin of Indian Civilization—The Palm-Tree and its Uses—The Tupi-Guarani—The Lingoa Geral—Ferocity of the Aymores—The City of Bahia—Porters—Cadeiras—History of Bahia—Caramuru—Attack on the Hollanders—Measures taken by Spain—The City retaken—The Dutch in Brazil—Slave-Trade—Sociability of Bahia—Mr. Gilmer, American Consul—The Humming-Bird—Whale-Fishery—American Cemetery—Henry Martyn—Visit to Montserrat—View of the City—The Emperor's Birthday—Medical School—Public Library—Image-Factory—The Wonderful Image of St. Anthony—No Miracle—St. Anthony a Colonel—Visit to Valença—Daring Navigation—*In Puris Naturalibus*—The Factory and Colonel Carson—American Machinery—Skilful Negroes—Return Home—Commerce with the United States.. 464

CHAPTER XXV.

Departure from Bahia—The Vampire-Bat—His Manner of Attack—The Bitten Negro—Annoyances magnified—Anacondas—One that swallowed a Horse—The Marmoset—Province of Alagoaz—The Republic of Palmares—Pernambuco—The Amenities of Quarantine-Life—Improvements at the Recife—Peculiarities of Pernambucan Houses—Beautiful Panorama—Various Districts of the City—A Bible-Christian—Extraordinary Fanaticism of the Sebastianists—Commerce of Pernambuco—The Population of the Interior—The Sertanejo and Market-Scene—The Sugar and Cotton Mart—The Jangada—Parahiba do Norte—Natal—Ceará—The Paviola—Temperature and Periodical Rains—The City of Maranham—Judge Petit's Description—The Montaria—Departure.. 503

CHAPTER XXVI.

Magnificence of Nature in the Brazilian North—The City of Pará—The Entrance of the Amazon—The first Protestant Sermon on these Waters—Parallel to the Black-Hole of Calcutta—Effects of Steam-Navigation—Improvements in Pará—The Canoa—Bathing and Market Scenes—Produce of Pará—India-Rubber—Pará Shoes—The Amazon River—Mr. Wallace's Explorations—The Vaca Marina—Cetacea of the Amazon—Turtle-Egg Butter—Indian Archery—Brazilian Birds and Insects—Visit to Rice-Mills near Pará—Journey through the Forest—The Paraense Bishop's Suspicions of Dr. Kidder—State of Religion at Pará... 539

CHAPTER XXVII.

PAGE

Amazonas— Its Discovery—El Dorado—Gonçalo Pizarro—His Expedition—Cruel-
ties—Sufferings—Desertion of Orellana—His Descent of the River—Fable of the
Amazons — Fate of the Adventurer — Name of the River — Settlement of the
Country — Successive Expeditions up and down the Amazon — Sufferings of
Madame Godin—Present State—Victoria Regia — Steam-Navigation—Effects of
Herndon and Gibbon's Report—Peruvian Steamers—The Future Prospects of the
Amazon.. 563

CONCLUSION.. 582

NOTES.. 599

APPENDICES.

APPENDIX A.—Chronological Summary of the Principal Events that have transpired
in the History of Brazil... 601

APPENDIX B.—Abstract of the Brazilian Constitution, sworn to on the 25th of
March, 1824, and revised in 1834.. 603

APPENDIX C.—Lines composed by D. Pedro II..................................... 605

APPENDIX D.—Slavery and the Slave-Trade in Brazil—England and Brazil......... 606

APPENDIX E.—Tables of Brazilian Coins, Weights, and Measures.................. 607

APPENDIX F.—Population—The Yellow Fever of Brazil.............................. 609

APPENDIX G.—Imports, Exports, Revenue, &c. of Brazil............................ 614

APPENDIX H.—Recent Discoveries of Coal in Brazil................................. 617

APPENDIX I.—Religious Disabilities 625

APPENDIX J.—Professor Agassiz's Labors on the Amazon......................... 627

APPENDIX K.—Thermometrical Observations at Rio de Janeiro in 1864............. 634

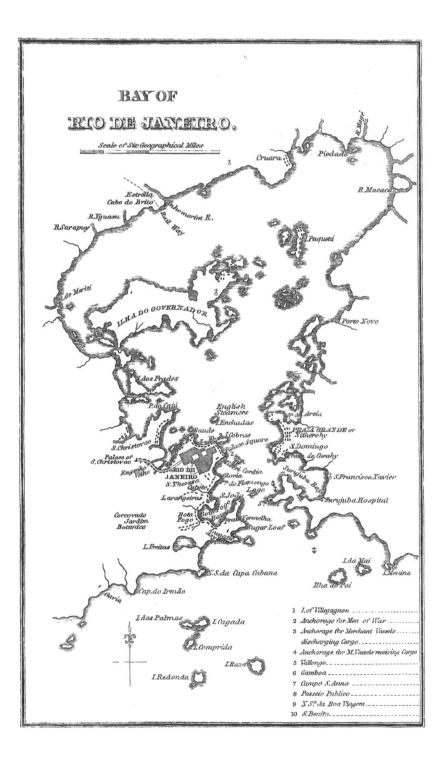

BAY OF
RIO DE JANEIRO.

Scale of Six Geographical Miles

Cruara
Piedade
R.Mage
Estrella
Cabo do Brito
R.Yguasu
Inhomerim R.
R.Sarapuy
Real Way
R.Macacú
I.Paquetá
de Moriti
Porto Novo
ILHA DO GOVERNADOR
dos Frades
P.do Caiú
English
Steamers
Enchadas
Sobras
da Areia
Saude
PRAIA GRANDE or
Nitherohy
S.Christovao
Palace Square
S.Domingo
Praia de Carahy
Palace of
S.Christovao
RIO DE
JANEIRO
Gloria
Cordia
Jurujuba Bay
S.Francisco.Xavier
Engenho
Velho
S.Theresa
Catette
do Flamengo
Lage
Larangeiras
S.João
S.Luiz
Jurujuba Hospital
Corcovado
Jardim
Botanico
Bota
Fogo
Botafogo
Praia Vermelha
Sugar Loaf
L.Freitas
N.S.da Copa Cabana
I.da Mai
Ilha do Pai
I.Menina
Gavia
Cap.do Irmão
I.das Palmas
I.Cagada
I.Comprida
I.Raza
I.Redonda

1   I.of Villegagnon
2   Anchorage for Men of War
3   Anchorage for Merchant Vessels
    discharging Cargo
4   Anchorage for M.Vessels receiving Cargo
5   Vallongo
6   Gamboa
7   Campo S.Anna
8   Passeio Publico
9   N.S.ª da Boa Viagem
10 S.Benito

THE SUGAR-LOAF, (ENTRANCE TO THE BAY OF RIO.)

Brazil and the Brazilians.

CHAPTER I.

THE BAY OF RIO DE JANEIRO — HISTORIC REMINISCENCES — FIRST SIGHT OF THE
TROPICS—ENTRANCE TO THE HARBOR—NIGHT-SCENES—IMPRESSIONS OF BEAUTY
AND GRANDEUR—GARDNER AND STEWART—THE CAPITAL OF BRAZIL—DISTINC-
TION OF RIO DE JANEIRO.

THE Bay of Naples, the Golden Horn of Constantinople, and the
Bay of Rio de Janeiro, are always mentioned by the travelled
tourist as pre-eminently worthy to be classed together for their
extent, and for the beauty and sublimity of their scenery. The first
two, however, must yield the palm to the last-named magnificent
sheet of water, which, in a climate of perpetual summer, is enclosed
within the ranges of singularly-picturesque mountains, and is
dotted with the verdure-covered islands of the tropics. He who,

in Switzerland, has gazed from the Quai of Vevay, or from the windows of the old Castle of Chillon, upon the grand panorama of the upper end of the Lake of Geneva, can have an idea of the general view of the Bay of Rio de Janeiro; and there was much truth and beauty in the remark of the Swiss, who, looking for the first time on the native splendor of the Brazilian bay and its circlet of mountains, exclaimed, *"C'est l'Helvétie Meridionale!"* (It is the Southern Switzerland!)

What a glorious spectacle must have presented itself to those early navigators—De Solis, Majellan, and Martin Affonso de Souza— who were the first Europeans that ever sailed through the narrow portal which constitutes the entrance to Nictherohy, (*Hidden Water*,) as these almost land-locked waters were appropriately and poetically termed by the Tamoyo Indians! Though the mountain-sides and borders of the bay are still richly and luxuriantly clothed, *then* all the primeval forests existed, and gave a wilder and more striking beauty to a scene so enchanting in a natural point of view, even after three centuries of the encroachments of man. De Souza—as the common tradition runs—supposed that he had entered the mouth of a mighty river, rivalling the Orinoco and the Amazon, and named it Rio de Janeiro, (*River of January*,) after the happy month—January, 1531—in which he made his imagined discovery. Whatever may have been the origin of this misnomer, it is not only applied to the large and commodious bay, but to the province in which it is situated, and to the populous metropolis of Brazil, which sits like a queen upon its bright shores.

We all of us know, either by our own experience or by that of others, what is the sight of land to the tempest-tossed voyager. When the broad blue circle of sea and sky, which for days and weeks has encompassed his vision, is at length broken by a shore, —even though that shore be bleak and desolate as the ice-mountains of the Arctic regions,—it is invested with a surpassing interest, it is robed in undreamed-of charms. What, then, must be the emotions of one who, coming from a latitude of stormy winter, beholds around him a land of perpetual summer, with its towering and crested palms, and its giant vegetation arrayed in fadeless green!

In December, 1851, when the Hudson and the Potomac were

bridged by the ice-king, and clouds and snow draped the sky and the land, our good vessel stood out upon a stormy sea. A few weeks of gales and rolling waves, varied by light winds and calms, brought us to Cape Frio. This isolated peak shoots up as steeply as the chalk-cliffs of England, as high as the Rock of Gibraltar, and is covered to its very summit with verdure. No clouds—as I last beheld them in conjunction with *terra firma* — were frowning over this summer-land. The balmiest breezes were blowing, and the palms upon the adjacent hills were gracefully waving above the world of vegetation—so new to me—which gleamed in the warm sunlight. It was in the midst of such a scene that the day, not without evening-glories, faded away. The morning sun shone clearly, and the lofty mountain-range near the entrance to the harbor stood forth in an outline at once bold, abrupt, and beautiful.

The first entrance of any one to the Bay of Rio de Janeiro forms an era in his existence :—

> "an hour
> Whence he may date thenceforward and forever."

Even the dullest observer must afterward cherish sublimer views of the manifold beauty and majesty of the works of the Creator. I have seen the most rude and ignorant Russian sailor, the immoral and unreflecting Australian adventurer, as well as the cultivated and refined European gentleman, stand silent upon the deck, mutually admiring the gigantic avenue of mountains and palm-covered isles, which, like the granite pillars before the Temple of Luxor, form a fitting colonnade to the portal of the finest bay in the world.

On either side of that contracted entrance, as far as the eye can reach, stretch away the mountains, whose pointed and fantastic shapes recall the glories of Alpland. On our left, the Sugar-Loaf stands like a giant sentinel to the metropolis of Brazil. The round and green summits of the Tres Irmãos (*Three Brothers*) are in strong contrast with the peaks of Corcovado and Tijuca; while the Gavia rears its huge sail-like form, and half hides the fading line of mountains which extends to the very borders of Rio Grande do Sul. On the right, another lofty range commences near the principal fortress which commands the entrance of the bay, and, forming curtain-like ramparts, reaches away, in picturesque headlands, to the bold promontory well known to all South Atlantic navigators as Cape Frio. Far through the opening of the bay, and in some places towering even above the lofty coast-barrier, can be discovered the blue outline of the distant Organ Mountains, whose lofty pinnacles will at once suggest the origin of their name.

The general effect is truly sublime; but as the vessel draws nearer to the bold shore, and we see, on the sides of the double mount which rises in the rear of Santa Cruz, the peculiar bright-leaved woods of Brazil, with here and there the purple-blooming quaresma-tree,—and when we observe that the snake-like cacti and rich-flowering parasites shoot forth and hang down even from the jagged and precipitous sides of the Sugar-Loaf,—and as we single out in every nook and crevice new evidences of a genial and pro-lific clime,—emotion, before overwhelmed by vastness of outline, now unburdens itself in every conceivable exclamation of surprise and admiration.

The breeze is wafting us onward, and we pass beneath the white walls of the Santa Cruz fortress. A black soldier, dressed in a light uniform of enviable coolness, leans lazily over a parapet, while higher up on the ramparts a sentinel marches with leisurely tread near the glass cupola which, illuminated at night, serves as a guide to the entering mariner. Immediately an enormous trumpet is protruded from this cupola, and our good ship is saluted by a stentorian voice, demanding, in Portuguese-English, the usual questions put to vessels sailing into a foreign port. We soon glide from under the frowning guns of Santa Cruz, and are just abreast Fort Lage, celebrated as the first spot of the bay ever inhabited by civilized man. The scene which now opens before us is exquisitely beautiful. Far to our left, beneath the Sugar-Loaf, but nearer to the city, is the fortress of St. John, bright amid the surrounding verdure. Passing through a fleet of gracefully shaped canoes and market-boats, manned by half-clad blacks, we cling to the steep right-hand coast, which soon precipitously terminates, and reveals to us the lovely little Bay of Jurujuba,—the "five-fathom" bay of the English. Again looking to the opposite side, beyond St. John, we have a glimpse of the graceful Cove of Botafoga (the Bay of Naples in miniature) and the pretty suburb of the same name, which seems like a jewel set between the smooth white beach and

Gavia.   Corcovado.   Tres Irmaus.   Sugar-Loaf.   Santa Cruz.   I. Pal.   I. Mai.

ENTRANCE TO THE BAY OF RIO DE JANEIRO.

(*After a Sketch by Mrs L. A. Cuddehy.*)

the broad circle of living green. Here too we have another of the many views of the Corcovado and the Gavia, which, as we vary our position, are ever changing and ever beautiful.

Now the vast city looms up before us, extending, with its white suburbs, for miles along the irregular shores of the bay, and running far back almost to the foot of the Tijuca Mountains, diversified by green hills which seem to spring up from the most populous neighborhoods. These combined circumstances prevent a perfect view of Rio de Janeiro from the waters. While gazing upon the domes and steeples, on the white edifices of the city, and the bright verdure-clad Gloria, Santa Theresa, and Castello Hills, we are cut short in our admiration by the cry of a Brazilian official:—"Let go your anchor." The command is obeyed, and we are comfortably lying to under the formidable-looking guns of the Forteleza Villegagnon. Our vessel swings round and reveals to us on the opposite shore the city of Praia Grande, the parti-colored cliff of St. Domingo, and upon a mere rock, which seems a fragment of the adjoining shore, the little church of Nossa Senhora de Boa Viagem, in which Roman Catholic voyagers are supposed to pay their vows, and around which many graceful palm-trees are nodding in the cool ocean-breeze. While awaiting the visit of the custom-house officers we remain upon deck, and tire not of scenes so novel and exciting. Little steamers and graceful falluas* are passing and repassing from Praia Grande and St. Domingo. White sails are dotting the bay as far as the eye can reach, while all around us the serried masts of Brazilian and foreign vessels are evidences that we are in the midst of a vast and busy mart.

The night soon succeeds the short twilight of the tropics, and the city from our ship seems like a land of fairy enchantment. Brilliancy and novelty do not end with the day. Innumerable gaslights line the immense borders of the city down to the very edge of the bay, and are reflected back from the water in a thousand quivering flashes. The very forms of the hills themselves are defined amid the darkness by rows of lamps extending over their verdure-clad summits, and seem like the fabled star-bridges of an Arabian tale. The steam ferry-boats bear various-colored lights,

---

* See engraving on page 60.

2

and each vessel in the harbor has a lamp at its fore; while every turn of the wheel furrows through a diamond sea, and every dash of the oar and every ripple from the gentle evening breeze reveals a thousand brilliant phosphorescent animalculæ illuminating the otherwise darkened waters. When we look above us we behold new constellations spangling the heavens, and *their* queen is the Southern Cross, guarded by her silent and mysterious attendants, the Majellan Clouds. The Great Bear has long since been hidden from us; but just peeping over the natural ramparts of the Organ Mountains, we see an old and a welcome friend in that beaming Orion, who here loses none of his northern splendor, and does not even pale before his rival of the Southern Hemisphere. Amid such scenes who could close their eyelids in sleep? Dr. Kidder on one occasion, returning from the northern provinces, entered the harbor at nightfall during a squall, and thus describes the scene:—

"We passed close under the walls of Fort Santa Cruz; but, just as the vessel was in the most critical part of the passage, the wind lulled, and the current of the ebbing tide swept her back, and by degrees carried her over toward the rocks upon which Fort Lage is constructed. The moment was one of great excitement and danger. Our situation was perceived at the forts, which severally fired guns, and burned white and blue lights, in order to show us their position.

"A more sublime scene can hardly be imagined. The rolling thunders of the cannon were echoed back by the surrounding mountain-peaks, and the brilliant glare of the artificial flames appeared the more intense in the midst of unusual darkness. Happily for the vessel and all on board, the wind freshened in time, and we were borne gallantly up to the man-of-war anchorage, where, at nine o'clock, we were lying moored to not less than seventy fathoms of chain.

"The moon had not yet risen, and the evening remained very dark. This circumstance heightened the beauty of the city and the effect of her thousand lamps, which were seen brightly burning at measured intervals over the hills and praias of her far-stretching suburbs. One young man was so enchanted with the novelty and splendor of the scene, that he remained on deck all night to gaze upon it, notwithstanding rain fell at intervals."

More than one have had to confess that their first twenty-four hours before Rio have been spent in a perpendicular position with the eyes wide open, and could exclaim, with emphasis,—

"Most glorious night!
Thou wert not sent for slumber."

Every thing is so fresh, so novel and awakening, that we are like children on the eve of some great festival or the night before the first journey to some vast city with whose wonders the story-book and the improvisations of the nursery have filled the imagination to the full.

I have again and again entered and quitted the Bay of Rio de Janeiro when the billows were surging and when the calm mantled the deep; and, whether in the purple light of a tropic morning, in the garish noon, or in the too brief twilight of that Southern clime, it has always presented to me new glories and new charms. It has been my privilege to look upon some of the most celebrated scenes of both hemispheres; but I have never found one which combined so much to be admired as the panorama which we have attempted to describe. On the Height of St. Elmo I have drank in as much of beauty from that curvilinear bay of Southern Italy, upon whose bosom float the isles of Capri and Ischia, and upon whose margin nestle the gracefully-shaped Vesuvius, the long arm of Sorrento, and the proverbially-brilliant city of Naples. I have seen very great variety in the blue, isle-dotted Bay of Panama; and I have beheld in the Alps, and in the western entrance to the Straits of Majellan, where the black, jagged Andes are rent asunder, scenes of wildness and sublimity without parallel; but, all things considered, I have yet to gaze upon a scene which surpasses, in combined beauty, variety, and grandeur, the mountain-engirdled Nictherohy.

The above impressions were penned before I had read, with a single exception, one of the many detailed descriptions of the Bay of Rio de Janeiro; and it occurred to me that those who had never seen the natural beauties of this region would not give ready assent to its exaltation above so many other places famous for their scenery. Such might say, "He is an enthusiast, an exaggerator." I have since perused many books, journals, and letters on Brazil; and all—from the ponderous tomes of Spix and Von Martius, down to the ephemeral lines of a contributor to the news-papers—are of one accord in regard to this wonderful bay. Though the works may be devoted to history, science, commerce, or to the epistolary correspondence of friends, in this respect they all bear a resemblance; for all draw the same portrait and from the same original. Indeed, when reading the description given by the late lamented English botanist, Gardner, I half suspected myself a plagiarist, though I had never read his interesting and truly valuable travels until my own account was written.

Describing the entrance of the harbor, this naturalist says,— " Passing through the magnificent portal of the bay, we came to an anchor a few miles below the city, not being allowed to proceed farther until visited by the authorities. It is quite impossible to express the feelings which arise in the mind while the eye surveys the beautifully-varied scenery which is disclosed on entering the harbor,—scenery which is perhaps unequalled on the face of the earth, and on the production of which nature seems to have exerted all her energies. Since then I have visited many places celebrated for their beauty and their grandeur, but none of them have left a like impression on my mind. As far up the bay as the eye could reach, lovely little verdant and palm-clad islands were to be seen rising out of its dark bosom; while the hills and lofty mountains which surround it on all sides, gilded by the rays of the setting sun, formed a befitting frame for such a picture. At night the lights of the city had a fine effect; and when the land-breeze began to blow, the rich odor of the orange and other perfumed flowers was borne seaward along with it, and, by me at least, enjoyed the more from having been so long shut out from the companionship of flowers. Ceylon has been celebrated by voyagers for its spicy odors; but I have twice made its shores, with a land-breeze blowing, without experiencing any thing half so sweet as those which greeted my arrival at Rio."

The description given by the Rev. C. S. Stewart is valuable in showing the impressions of this magnificent bay upon one who had, since his first visit to Brazil, viewed some of the most re-nowned scenes in the world :—

" I was anxious to test the fidelity of the impressions received twenty years ago from the same scenery, and to determine how far the magnificent picture still lingering in my memory was justified by the reality, or how far it was to be attributed to the enthusiasm of younger years and the freshness of less experienced travel. The early light of the morning quickly determined the point. I was hurried to the deck by a message from Lieutenant R——, already there, and do not recollect ever to have been impressed with higher admiration by any picture in still life than by the group of mountains and the coast-scene meeting my eyes on the left. The wildness and sublimity of outline of the Pāo de Assucar, Dous Irmãos, Gavia, and Corcovado, and their fantastic combinations, from the point at which we viewed them, can scarce be rivalled; while the richness and beauty of coloring thrown over and around the whole, in purple and gold, rose-color, and ethereal blue, were all that the varied and glowing tints of the rising day ever impart. No fancy-sketch of fairy-land could surpass this scene, and we stood gazing upon it as if fascinated by the work of a master-hand."

The city of Rio de Janeiro, or San Sebastian, is at once the commercial emporium and the political capital of the nation. While Brazil embraces a greater territorial dominion than any other country of the New World, together with natural advantages second to none on the globe, the position, the scenery, and the increasing magnitude of its capital render it a metropolis worthy of the empire. Rio de Janeiro is the largest city of South America, the third in size on the Western Continent, and boasts an antiquity greater than that of any city in the United States.

Its harbor is situated just within the borders of the Southern Torrid Zone, and communicates, as before described, with the wide-rolling Atlantic, by a deep and narrow passage between two granite mountains. This entrance is so safe as to render the services of a pilot entirely unnecessary. So commanding, however, is the position of the various fortresses at the mouth of the harbor upon its islands and on the surrounding heights, that, if efficiently manned by a body of determined men, they might defy the hostile ingress of the proudest navies in the world.

Once within this magnificent bay of Nitherohy, the wanderer

of the seas may safely moor his bark within hearing of the roar of the ocean-surf.

The aspect which Rio de Janeiro presents to the beholder bears no resemblance to the compact brick walls, the dingy roofs, the tall chimneys, and the generally-even sites of our Northern cities. Its surface is diversified by hills of irregular but picturesque shape, which shoot up in different directions, leaving between them flat intervals of greater or less extent. Along the bases of these hills, and up their sides, stand rows of buildings, whose whitened walls and red-tiled roofs are in happy contrast with the deep-green foliage that always surrounds and often embowers them.

The most prominent eminence, almost in front of us, is the Morro do Castello, which overlooks the mouth of the harbor, and on which is the tall signal-staff that announces, in connection with the telegraph on Babylonia Hill, the nation, class, and position of every vessel that appears in the offing. Upon our right we see the convent-crowned hill of San Bento; and if we could have a bird's-eye view from a point midway between the turrets of the convent and the signal-staff of Morro do Castello, we should see the city spread beneath us, with its streets, steeples, and towers, its public edifices, parks, and vermillion chimneyless roofs, and its aqueducts spanning the spaces between the seven green hills, constituting a gigantic mosaic, bordered upon one side by the mountains, and on the other by the blue waters of the bay.

From the central portion of the city the suburbs extend about four miles in each of the three principal directions, so that the municipality of Rio de Janeiro, containing five hundred thousand inhabitants, covers a greater extent of ground than any European city of the same population.

Here dwell a large part of the nobility of the nation, and, for a considerable portion of the year, the representatives of the different provinces, the ministers of state, the foreign ambassadors and consuls, and a commingled populace of native Brazilians and of foreigners from almost every clime. That which in the popular estimation, however, confers the greatest distinction upon Rio, is not the busy throng of foreign and home merchants, sea-captains, ordinary Government-officials, and the upper classes of society; but it is in the fact that here resides the imperial head of Brazil, the young and gifted Dom Pedro II., who unites the blood of the Braganzas and the Hapsburgs, and under whose constitutional rule civil liberty, religious toleration, and general prosperity are better secured than in any other Government of the New World, save where the Anglo-Saxon bears sway.

Attractive as may be the natural scenery and the beauties of art abounding in any country, it must be confessed that human existence, with its weal or woe, involves a far deeper interest. And the traveller but poorly accomplishes his task of delineating the present, if he leaves unattempted some sketches of the history of the past as an introduction to the scenes and events which have come under his own observation. After glancing rapidly at some of the most striking sights and customs of Rio de Janeiro, I shall introduce a brief sketch of its past history.

HOTEL PHAROUX.

CHAPTER II.

LANDING—HOTEL PHAROUX—NOVEL SIGHTS AND SOUNDS—THE PALACE SQUARE—
RUA DIREITA—EXCHANGE—THE "TEAM"—MUSICAL COFFEE-CARRIERS—CUSTOM-
HOUSE—LESSONS IN PORTUGUESE, AND GOVERNOR KENT'S OPINION OF BRAZIL—
POST-OFFICE—DISLIKE OF CHANGE—SENHOR JOSÉ MAXWELL—RUA DO OUVIDOR—
SHOPS AND FEATHER-FLOWERS — THE BRAZILIAN OMNIBUS CAN BE FULL —
NARROW STREETS AND POLICE-REGULATIONS — A SUGGESTION TO RELIEVE
BROADWAY, NEW YORK—PASSEIO PUBLICO—BRAZILIAN POLITENESS—THE "GON-
DOLA"—THE BRAZILIAN IMPERTURBABLE—LACK OF HOTELS—FIRST NIGHT IN
RIO DE JANEIRO.

THE stranger who, with anxious expectation, has paced the deck of his vessel as it lies at anchor under Villegagnon, knows no more welcome sound than the permission from the Custom-House and health officers to land and roam through the city which for hours before his eyes have visited.   The blacks who have come from the shore now return, pulling their heavy boat lustily along, for they are sure of a treble price from the newly-arrived.   Who that has visited Rio de Janeiro will not at a glance recognise the landing-place depicted in the engraving?   *Hotel Pharoux,* the Palace Stairs, and the Largo do Paço, (Palace Square,) are associated with Rio de Janeiro in the mind of every foreign naval officer who has been on

24

the Brazil station. But changes have taken place, and greater are in contemplation, among this slow-moving people. Hotel Pharoux still lifts its gray walls; but it is modernized, and the old restaurant and stable in the basement have given way to shell-merchants and feather-flower dealers, and the upper stories form a private hospital. We no longer land at the Palace Stairs, where formerly at flood-tide the waters of the bay dashed and foamed against the stone parapet which at this point marked their limit. The square has been extended into the waves, and soon the Government will have fine quays along the whole water-edge in this part of the city.

Instead of the old granite steps, we ascend the wooden stairs at the end of a long jetty. Here our boat has arrived, amid odors that certainly have not been wafted from "Araby the blest," and we learn that the sewerage of Rio is a portable instead of an underground affair. The sense of hearing, too, is wounded by the confused jabbering of blacks in the language of Congo, the shouts of Portuguese boat-owners, and by the oaths of American and English sailors. Once clear of this throng, what novel sights and sounds astonish us! A hackney-coachman, in glazed hat and red vest, invites us to a ride to the Botanical Gardens; a smart-looking mulatto points to his "carriage" hard by the Hotel de France. Before their words are ended, the roll of drums and the blast of bugles attract our attention in another direction. There, in front of the old palace, is drawn up a handful of the National Guard, composed of every imaginable complexion, from white to African; and now, as every day at noon, they remove their helmets, listen for a moment with religious veneration to the strain of music which the black trumpeters puff out from swelling cheeks, and then resume, with the exception of the sentinels, their difficult task of loitering in the corridors of the huge building, or basking in the sunshine, until another sound of the bugle shall call them to change guard or fall into ranks at vespers.

We are not yet ready to try the vehicles of Rio de Janeiro; so we dismiss our would-be coachmen, and look around us in the Largo do Paço.

At the Palace Square the stranger finds himself surrounded by a throng as diverse in habits and appearance, and as variegated in

complexion and costume, as his fancy ever pictured. The majority
of the crowd are Africans, who collect around the fountain to
obtain water, which flows from a score of pipes, and, when caught
in tubs or barrels, is borne off upon the heads of both males and
females.

The slaves go barefooted, but some of them are gayly dressed.
Their sociability when congregated in these resorts is usually
extreme, but sometimes it ends in differences and blows. To prevent disorders of this kind, soldiers are generally stationed near
the fountains, who are pretty sure to maintain their authority
over the unresisting blacks. Formerly there were only a few
principal fountains; now there are large *chafarizes* in all the

THE LARGO DO PAÇO, AND RUA DIREITA, FROM THE PALACE.

squares, and at the corners of every third or fourth street are
smaller streams of the pure element, which flow at the turning of
a stopcock.

The Palace is a large stone building, exhibiting the old Portuguese style of architecture. It was long used as a residence by the vice-roys, and for a time by Dom John VI., but is now appropriated to various public offices, and contains a suite of rooms in which court is held on gala-days. The buildings at the rear of the Palace Square (represented on the left of the engraving) were all erected for ecclesiastical purposes. The oldest was a Franciscan convent, but has long since been connected with the Palace, and used for secular purposes. The old chapel, with its short, thick tower, remains, but has been superseded, in popularity as well as in splendor, by the more recently-erected imperial chapel, which, without belfry, stands at its right. Adjoining the imperial chapel is that of the third order of Our Lady of Mount Carmel, which is daily open, and is used as a cathedral. The steeples of this church during certain festivals are illuminated to the very crosses, and present a splendid appearance from the shipping.

The streets of the city are generally quite narrow; but the *Rua Direita*, which is seen in the above cut beyond the Largo do Paço, is wide, and well paved with small square blocks of stone which are brought from the Isle of Wight. The *Rua Direita* and many of the principal streets of Rio de Janeiro are now as well paved as the finest thoroughfares of London or Vienna, presenting a great contrast to the former irregular and miserable pavement, which was in use up to 1854. The *Rua Direita* and the *Largo do Rocio* are the points whence omnibuses start for every portion of the vast city and its suburbs.

The houses seldom exceed three or four stories; but a four-story house at Rio is equal in height to one of five in New York. Formerly nearly all were occupied as dwellings, and even in the streets devoted to business the first floors only were appropriated to the storage and display of goods, while families resided above. But since 1850 this has greatly changed in the quarter where the wholesale houses are found: proprietors and clerks now reside in the picturesque suburbs of Botafogo, Engenho Velho, and across the bay at Praia Grande or San Domingo. Every evening presents an animated spectacle of crowded steamers, full omnibuses, and galloping horses and mules, all conveying the *negociantes* and *caixeiros* (bookkeepers) to their respective residences.

The distant steeples on our left are those of the Church of

Candelaria, which is situated on a narrow street back from the Rua Direita. It is the largest church in the city, and presents taller spires and a handsomer front than any other.

The *Praça do Commercio*, or Exchange, occupies a prominent position in the Rua Direita. This building, formerly a part of the Custom-House, was ceded by Government for its present purposes in 1834. It contains a reading-room, supplied with Brazilian and foreign newspapers, and is subject to the usual regulations of such an establishment in other cities. Beneath its spacious portico the merchants of eight or nine different nations meet each other in the morning to interchange salutations and to negotiate their general business. The Exchange is not far from the Custom-House, which formerly had its main entrance adjoining the Praça.

THE RIO TEAM (NOW ABOLISHED).

Nothing can be more animated and peculiar than the scenes which are witnessed in this part of the Rua Direita during the business-hours of the day,—viz.: from nine A.M. to three P.M. It is in these hours only that vessels are permitted to discharge and receive their cargoes, and at the same time all goods and baggage must be despatched at the Custom-House and removed therefrom. Consequent upon such arrangements, the utmost activity is required to remove the goods despatched, and to embark those productions of the country that are daily required in the transactions of a vast commercial emporium. There were the black-coated merchants congregated about the Exchange, and there came the negro dray. The *team* consisted of five stalwart Africans pushing, pulling, steering, and shouting as they made their way amid the serried throng, unmindful of the Madeira Islander, who, with an imprecation and a crack of his whip, urged on a thundering mule-cart laden with boxes. Now an omnibus thunders through the crowd, and a large four-wheeled wagon, belonging to Smith's Express for the transportation of "goods," rolls in its wake. Formerly all this labor was performed by human hands, and scarcely a cart or a dray was used in the city, unless, indeed, it was drawn by negroes. Carts and wagons propelled by horse-power are now quite common; but for the moving of light burdens and for the transportation of furniture, pianos, &c. the negro's head has not been superseded by any vehicle until 1862, when Smith's Express, and large wagons called andorinhas, came in vogue, except for pianos.

THE FORMER COFFEE-CARRIERS OF RIO DE JANEIRO.

In 1857, while we were almost stunned by the sounds of the multitude, we had a new source of wonderment. Above all the confusion of the Rua Direita, we heard a stentorian chorus of voices responding in quick measure to the burden of a song. We beheld, over the heads of the throng, a line of white sacks rushing around the corner of the Rua da Alfandega, (*Custom-House Street.*) We hastened to that portion of Rua Direita, and saw that these sacks had each a living ebony Hercules beneath. These were the far-

famed coffee-carriers of Rio.   They usually went in troops, number-
ing ten or twenty individuals, of whom one took the lead and was
called the captain.   These were generally the largest and strongest
men that could be found.   While at work they seldom wore any
other garment than a pair of short pantaloons; their shirt was
thrown aside for the time as an encumbrance.   Each one took a
bag of coffee upon his head, weighing one hundred and sixty
pounds, and, when all were ready, they started off upon a measured
trot, which soon increased to a rapid run.   Since 1860 carts are
used for coffee.

The negro porters of pianos and crockery frequently carry in
their hands musical instruments, resembling children's rattle-
boxes: these they shake to the double-quick time of some wild
Ethiopian ditty, which they all join in singing as they run.
Music has a powerful effect in exhilarating the spirits of the
negro; and certainly no one should deny him the privilege of
softening his hard lot by producing the harmony of sounds which
are sweet to him, though uncouth to other ears.   It is said, however, that an attempt was at one time made to secure greater
quietness in the streets by forbidding them to sing.   As a consequence, they performed little or no work; so the restriction was
in a short time taken off.   Certain it is that they now avail themselves of their vocal privileges at pleasure, whether in singing and
shouting to each other as they run, or in proclaiming to the people
the various articles they carry about for sale.   The impression
made upon the stranger by the mingled sound of their hundred
voices falling upon his ear at once is not soon forgotten.

We now turn from the busy throng of the Rua Direita, and in a
few minutes we ascend the steps of a stately building, over whose
portico we read, in huge green letters, — ALFANDEGA.
We will not stop to trace the origin of this word and many others
in the Portuguese tongue beginning with *Al*, to their Moorish origin,
but will immediately inform the reader that it is the first word he
learns in Brazil, and one which, in various languages, most travellers in foreign countries have occasion to remember.   This is
the Custom-House.   We enter a vast hall of fine architecture,
lighted by a graceful dome.   There are hundreds of despatchers,
merchants, and officers.   But what a contrast to the noisy multi-

tude of the Rua Direita! All are uncovered, and, as each enters
the hall, the hat is removed and not replaced until the portico is
again reached. What a capital discipline for Anglo-American
visitors and for English and North American shipmasters, whose
head-coverings seem to be a portion of their corporeal existence!
I once heard Albert Smith, in one of his delightful conversaziones,
say that in foreign lands an Englishman considers it a part of the
British constitution not to take off his hat except when "God save
the Queen" may accidentally fall upon his ear. The Brazilian is
very strict in the outward observance of politeness; and, as he
would never enter a private residence without removing his hat,
so he considers that he should not enter any of the edifices belonging to the Government of his Emperor without showing the same
respect.

At the end of the hall, on an elevated platform, is the chief-
collector, who is constantly engaged in signing despatches and
various other custom-house papers, which are noiselessly handed
him by sub-officers and clerks. The inspector-in-chief, who presided
over the Alfandega of Rio in 1855, was Senhor S. Paio Vianna,
of Bahia, who, though strict and almost rigorous in the administration of his office, is a gentleman of great intelligence and amenity
of manner. He took a deep interest in the finances of the empire,
and his annual statement was clear and full of important information
to the commercial statistician. His predecessor was Sr. Ferraz, to
whom is greatly due the immense reforms that have taken place
in the custom-house of Rio de Janeiro. Formerly it was most
corruptly administered: bribery was the rule and not the exception. To this day some most wonderful stories are told of the
year 1844, when the treaty between England and Brazil expired,
by limitation, in the month of November. Bales, bags, and boxes
went through the Custom-House with astonishing rapidity; and
there is a tradition that the entire cargo of a schooner entered the
rear of the Alfandega, and in a remarkably short time emerged
from the Portão Grande, (*Great Door.*) But there is no longer
opportunity for such abuses; and the largest custom-house of the
empire is as well conducted as those of Germany or France.

At the left of the chief-collector, in the rear of a row of supporting columns,—is the *guarda mor*,—Sr. Leopoldo Augusto da

Camara Lima, who is known to every ship-captain as Senhor Leopoldo. This gentleman, who speaks the English language most fluently, has been arrayed on the liberal side of Brazilian politics for the last twenty years, and was in the front rank of those who condemn the African slave-trade, which was so completely abolished in 1850. The office of the *guarda mor* in 1865 is nearer the water.

The vast warehouses of the Alfandega extend quite to the seaside.* Here conveniences are constructed for landing goods under cover. Once out of boats or lighters, they are distributed and stored in respective departments, until a requisition is formally made for their examination and despatch. The removal of the various articles within the Custom-House, as well as their transportation.to the great door of exit, is facilitated by means of small iron railways extending to every portion of the many buildings.

That troublesome delays should occasionally occur in the despatch of goods and baggage is not surprising to any one acquainted with the tedious formalities required by the laws; nor would it be strange, if, among the host of *empregados* or sub-officers connected with this establishment upon very limited pay, some are occasionally found who will embarrass your business at every step until their favor is conciliated by a direct or indirect appropriation of money to their benefit; but this is more rare than formerly.

Most of the large commercial houses have a despatching-clerk, whose especial business it is to attend upon the Alfandega; and the stranger who is unaccustomed to the language and customs of the country will always avoid much inconvenience by obtaining the services of one of these persons. From my own experience in passing books and baggage through the different custom-houses of Brazil, I am prepared to say that a person who understands and endeavors to conform to the laws of the country may expect in similar circumstances to meet with kind treatment and all reasonable accommodations. If, however, a glance at your watch tells you, in the midst of your labors and difficulties, that three o'clock

---

* In the "View of Rio de Janeiro from the Island of Cobras," merely the water-front of the Alfandega is seen extending above the entire width of the palm-tree in the foreground.

is near at hand, and you undertake to urge the sub-collector to expedite matters, you are sure to receive in reply, *"Paciencia, senhor."* This is our second lesson in Portuguese; and the third soon follows in response to our demand, "When can these things be despatched?" *"Amanhã"* (to-morrow,) is promptly given. But should you succeed in getting through the *portão grande* about the time that huge door is being closed up for the day, you will witness a lively scene. Boxes, bales, and packages of every species of goods, cases of furniture, pipes of wine, and coils of rope, lie heaped together in a confusion only equalled by the crowd of clerks, feitors, and negroes, who block up the whole Rua Direita in their rush to obtain possession of their several portions, and in their vociferations to hasten the removal of their merchandise.

We are perhaps wishing to expedite the tall Mina blacks whom we have engaged to transport our luggage to its place of destination. By signs manual our meaning is comprehended, but we receive a very cool *"Espera um pouco, senhor,"* (Wait a little, sir,) which completes our studies in Portuguese for the day. And what a lesson we have received!

Paciencia, amanhã, and *espera um pouco!* These words in action stare the nervous, impatient, tearing, fretting Anglo-American, everywhere throughout Brazil. The Hon. Ex-Governor Kent, whose name is associated with the Northeastern boundary and with the politics of New England, was for four years a resident of Rio de Janeiro as U.S. Consul, and for a portion of the time as acting Chargé d'Affaires. It was his deliberate opinion that Brazil was the best place in existence to cool a fervid, speech-making, community-exciting Yankee. I have laughed heartily at his dry humorous manner, as he has unfolded *con amore* this subject:—

"There is to a quietly-disposed, mild man, past the meridian of life, who has seen many of the rough sides of humanity, something agreeable and pleasant in the tranquil, calm, noiseless habits of the Brazilians. To live a whole year and never attend a caucus or an indignation-meeting, to hear nothing about elections, to see no gatherings of the people, to read no placards calling upon the sovereigns to rise and vindicate their rights, to listen to no stump-speeches or dinner-orations, never once to be importuned to walk or ride in a political procession, to see not one torchlight-pageant in honor of a victory which has saved the country and the offices,—in short, to live without politics,—is, to one who is inclined to quiet, or who has been wearied out in the service, soothing and delightful."

Though the nation, by steamships and railroads and general prosperity, is daily becoming more active, yet it may be still predicated that the Brazilian is not accustomed to be startled and shocked by other people's miseries and woes. With a free and well-supported press, his nature demands no thrilling evening editions, filled with long and minute accounts of the last steamboat disaster, fearful accidents, or horrible murders. As a general thing, he thinks the moral, physical, and political worlds will turn on their own axes without his interference. Hence it was, doubtless, that some of the far-seeing and really wide-awake statesmen of Rio proposed a fine of five dollars to be imposed upon each citizen who did not come up to the polls of the municipal election and deposit his vote.

Almost every one who arrives at Rio is expecting letters that have anticipated him by the English steamer, and, as soon as his trunks are relieved from the Custom-House, he makes his way to the *Correio Geral*, or General Post-Office, in the Rua Direita. You pass by a large vestibule, with a stone floor, occupied by several soldiers, either on guard or sleeping on benches at the extremities of the room, and upon inquiry you ascertain that the Postmaster General and the larger portion of his employees are in the rooms above. We enter the front-door of the large apartment adjoining this vestibule. On the right, behind a high counter, are the letters and newspapers of the Post-Office, distributed, not in boxes, according to alphabetical order, but in heaps, according to the places from whence they have come; as, for instance, from the Mines, from St. Paul's, and other important points. Corresponding to this, on the sides of the room, are hung numerical lists of names, arranged under the head of Cartas de Minas, de S. Paulo, &c. The letters, with the exception of those belonging to certain mercantile houses, and to those who pay an annual subscription to have their correspondence sent them, are thrown together promiscuously, and he who comes first has the privilege of looking over the whole mass and selecting such as belong to himself or his friends. This method has been somewhat modified since the establishment of steam-lines to Europe. On the day that the steamer arrives an immense crowd gathers at the Post-Office; but the letters, instead of being investigated by all upon the counter, are carefully kept in the back-part of the hall, where four persons at a time are admitted. There is just cause of complaint in regard to the delivery of the foreign steamers' (except the English) mails. The English have their own mail-agent. The whole system is needlessly clumsy and inconvenient for a city of three hundred thousand inhabitants. I was informed at Rio that some years since Mr. Gordon, of Boston, who was then U.S. Consul, offered to the Brazilian Government to put their chief Post-Office on the same footing of efficiency that existed in the United States. Mr. Gordon was admirably qualified for this, having been for a number of years the postmaster of the largest distributing and seaport office in New England. His offer was not accepted; for the Brazilians, though more progressive than most South American people, still inherit many characteristics from their Portuguese ancestors, and a prominent ·one is dislike of change. The little progress that the mother-country has made during the last few centuries is admirably illustrated in the following well-known story:—Once upon a time Adam requested leave to revisit this world: permission was granted, and an angel commissioned to conduct him. On wings of love the patriarch hastened to his native earth; but so changed, so strange, all seemed to him, that he nowhere felt at home till he came to Portugal. "Ah, now," exclaimed he, "set me down; everything here is just as I left it."

The larger mails, departing coastwise, are very frequent, regular, and swift. This may also be said of the mail to Petropolis by steamboat, railway, and the turnpike of the Union Industry Company. Otherwise, inland transportation of letters is very slow. But when the D. Pedro II. Railway and similar constructions reach far into the interior, there will be of course corresponding improvement in this respect. The inland mails to the distant provinces depart once in five days, and return at corresponding intervals. Their transmission through the country is slow and tedious, being performed on horseback or by foot-carriers, at an average, throughout the empire, of twenty miles in twenty-four hours. Charges for postage are moderate, and a traveller to any

portion of the country is permitted to carry as many epistles as
his friends will intrust to him, provided they have the Government
stamp affixed to them.

There is, however, one exception to the general cheapness of
postage.   It sometimes happens that books or packages which
ought to have passed through the Custom-House find their way to
the Post-Office, and then the expense is extravagant.   There is a
crying evil which ought to be remedied: I refer to the charge by
the post-office clerks on letters which have already been prepaid.
It amounts to downright robbery.   If the officials are not paid
a sufficient salary, let the Parliament reform the thing, so that
official extortion may no longer continue.

In years gone by, we next sought the large commercial *trapiche*
(warehouse) of Messrs. Maxwell, Wright & Co.   This establishment
was long well known as the leading commission-house of Rio de
Janeiro.   It was built up under the supervision of the vigilant and
prompt Mr. Joseph Maxwell, of Gibraltar, and various members
of his family, in connection with the Messrs. Wright of Baltimore.
Few Americans and Englishmen have gone to Rio without receiving
attentions from some one of the principals or employees of this
house.   At the abundantly-spread table in the dining-room of the
trapiche, many have made their first acquaintance with Brazilian
dishes and with the refreshing fruits of the tropics.

In September, 1854, Sr. José Maxwell, the senior partner of this
important firm, died; and probably the funeral of no other private
citizen in the capital or the empire was ever attended by such a
throng as that which followed to the grave the remains of this kind
father and respected citizen.   This firm no longer exists.

We pass, by the Rua do Rosario, again into the Rua Direita, and
continue our promenade up the Rua do Ouvidor, which is the combined Rue Vivienne, Regent Street, and Broadway of Rio.   It is
not, however, either long or broad, but the shops upon it are brilliant and in good taste.   There is no part of the city so attractive
to the recently-landed foreigner as this street, with its print-shops,
feather-flower stores, and jewellery-establishments.   The diamond,
the topaz and emerald can here be purchased in any number, and
are temptingly displayed behind rich plate-glass.   The feather and
insect-flowers manufactured in Brazil are original and most beauti-

ful. The early Portuguese found that the Indians adorned themselves with the rich plumage of the unsurpassingly brilliant birds
of the forest. In the Amazonian regions the aborigines have not
lost either the taste or the skill of their ancestors, and, like the
cultivators of roses, they are not content with the gorgeous colors
which nature has painted, but by artificial means produce new
varieties. Thus, on the Rio Negro, the Uaupé Indians have a head-
dress which is in the highest estimation, and they will only part
with it under the pressure of the greatest necessity. This ornament consists of a coronet of red and yellow feathers disposed in
regular rows and firmly attached to a strong plaited band. The
feathers are entirely from the shoulders of the great red macaw;
but they are not those that the bird naturally possesses, for the
Indians have a curious art by which they change the colors of the
plumage of many birds. They pluck out a certain number of
feathers, and in the various vacancies thus occasioned infuse the
milky secretion made from the skin of a small frog. When the
feathers grow again they are of a brilliant yellow or orange color,
without any mixture of green or blue as in the natural state of the
bird; and it is said that the much-coveted yellow feather will
ever after be reproduced without a new infusion of the milky
secretion.

In the National Museum on the Campo St. Anna, many of the
curious head-dresses and feather-robes of the aboriginal tribes
. attract the attention of the visitor.

There are few curiosities more esteemed in Europe and the
United States than the feather-flowers of Rio de Janeiro and Bahia.
They are made from the natural plumage, though from time to
time the novice has palmed off upon him a bouquet, the leaves of
which, instead of being from the parrot, have been stolen from the
back of the white ibis and then dyed. This deception can, however, be detected by observing the stem of the feather to be colored
green, which never is the case in nature. No one travelling in the
English steamers should postpone his purchases of these beautiful
souvenirs of bright birds and Brazil until he arrives at St. Vincent,
for the numerous pedlars of that island offer an inferior article
made from artificially-colored feathers. Rio de Janeiro is the best
mart for this kind of merchandise. No ornament can surpass the

splendor of the flowers made from the breasts and throats of humming-birds. A lady whose bonnet or hair is adorned with such plumage seems to be surrounded with flashes of the most gorgeous and ever-varying brilliancy. The carnations and other flowers made from a happy combination of the feathers of the scarlet ibis and the rose-colored spoonbill are also very natural, and are highly prized. Bourget, 115 Ouvidor, is the best naturalist.

In these shops we may also find fish-scale flowers, and those manufactured from the wings of insects, and breast-pins which are made by setting a small brilliant beetle in gold.

From the Rua do Ouvidor we turn into the Rua dos Ourives, (Goldsmiths' Street,) where are scores of shops filled with large quantities of silver and gold ornaments, from a spur to a crucifix.

We now wend our way through the Largo de S. Francisco de Paula to the Largo do Rocio, (or Statue Square, as it is termed by the English,) where we take an omnibus for Botafogo. The Brazilian omnibus is very much like its prototype in all parts of the world, with this single and very important exception :—it is not *elastic.* A New York or Philadelphia omnibus is proverbially "never full;" but the same kind of vehicle in Rio *can* be filled, and, when once *complete,* the conductor closes the door, cries " *Vamos embora,*" (Let us be off,) the driver flourishes his long thong and sets his four-mule team into a gallop. Away we go, rattling across gutters as if there were none, and rushing through narrow streets as if negro water-carriers had no existence. It is curious to behold the heavy-laden slaves clearing the street and dodging into open shop-doors as an omnibus appears in sight. Few accidents occur; and, when they do, prompt reparation is made. On one occasion I was in a "gondola" in the narrow Rua S. José. Our four long-eared beasts were plunging on at a fearful rate, and, being much more unmanageable than horses, could not be pulled up until the fore-wheel crunched upon the legs of a poor old mullatress. She was severely but not fatally injured, and was instantly cared for. The gondola-driver, however, I never saw again holding the reins. The House of Correction, or one of the many prisons, was, without doubt, his abode for the next few months.

The streets, with their diminutive sidewalks, are so narrow that in many of them only one vehicle can pass at a time. I was more than once reminded of Pompeii and Herculaneum, not only in some of the commonest utensils and mechanic implements, in the open shop-windows, and in the house of the Brazilian, who demands a fine parlor, (the *atrium*,) and yet will sleep in a windowless alcove like a dungeon's cell; but in nothing was the resemblance more striking than in the narrow *ruas*, which, doubtless, had their origin in the desire to procure shade. Mr. George S. Hillard, in his thought-begetting "Six Months in Italy," says of the narrow thoroughfares of Pompeii, "As each vehicle must have occupied the space between the curbstones, we are left without any means of conjecturing what expedients were resorted to, or what police-regulations were in force, when two carriages, moving in different directions, met each other." If this accomplished author had visited Rio de Janeiro previous to his excursion to the buried cities of Magna Grecia, the mystery would have been solved. In the narrow Ruas Ouvidor, Rosario, Hospicio, Alfandega, S. José, and others, carriages and omnibuses never meet; and so admirable are the police-regulations that no mistakes ever occur. At the corner of each of these streets where it is crossed by another, we see painted, with great distinctness, an *index* immediately under the name of the street. Thus, two of the streets mentioned above are adjacent to and parallel with each other, and are crossed by the Ruas Direita and Quitanda. Upon their Rua Direita corners we behold the following:—

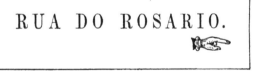

RUA DO ROSARIO.

RUA DO OUVIDOR.

Now, if I am in a carriage at the point where the Ruas Direita and Rosario cross each other, and I wish to visit a shop at the corner of the latter street and the Rua Quitanda, although it is more direct for me to ascend by the Rua do Rosario, yet my Jehu knows that if he should go contrary to the *index* he would be subjected to a heavy fine and forfeiture of certain privileges as a coachman. He therefore whirls through the Direita, up the Rua do Ouvidor, and along the Quitanda, travelling the three sides of the square, and thus avoiding all collision.

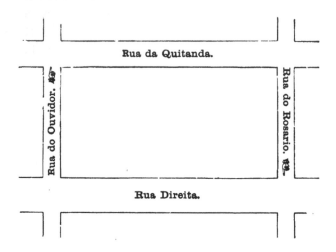

In the city of New York there has been for many years every imaginable proposition for the relief of Broadway, and there is scarcely a citizen or visitor in that vast emporium who has not on more than one occasion been subjected to great inconvenience by the regular "blockade" instituted every day in the lower part of that immense thoroughfare, the whole of which might have been avoided by the simple application of the Brazilian plan, and thus making the innumerable omnibuses, drays, carts, and carriages *descend* Broadway, and those vehicles that are uptownward *ascend* Greenwich Street.

But onward rushes our omnibus at a rapid pace. We whirl by the Carioca Fountain, and, before we can give a second look at the green sides of the Antonio Hill, we are bowling along under the garden-walls of the lofty Ajuda Convent. All seems dismal, with the exception of the foliage that appears above the high enclosure. A turn brings us into the Largo da Ajuda, and at once we have the wonderful view—to Northern eyes at least—of the Passeio Publico, (*Public Promenade*,) and before us the verdant slopes of the Santa Theresa Hill. From beneath the tropic-trees which cover the latter, neat white cottages are peeping, and, for a residence, no elevation within the city is preferable to Santa Theresa. The Passeio Publico, which we are passing, was a favorite resort of mine at Rio; and at all times—whether at night, when it is brilliantly illuminated, or in the brightest hour of the day—it is one of the pleasantest promenades within the precincts of the municipality. Here are overhanging trees, blooming para-

AQUEDUCT, LARGO DA LAPA, AND PASSEIO PUBLICO, FROM THE SANTA THERÈSA.

sites, rare plants, shady walks, and cool fountains. On the side which fronts the bay is a large terrace, from which is a magnificent prospect of the Gloria Hill, the distant Sugar-Loaf, and, far beyond, the rolling ocean.

Having passed this public garden, we are in the square called Largo da Lapa. The palatial building on our right was purchased a few years ago for the National Library, and was formerly one of the most splendid private mansions in Rio.

Over a superbly-paved street our omnibus is hurrying; but from time to time an open gate or a tall Cape of Good Hope pine-tree tells us that gardens are in the rear of forbidding looking walls. We dash along what is called the "Coast of Africa,"—a long row of low houses on our right; while on our left the bay is beneath us, and therefore, the street being unshaded, the appropriateness of the hot cognomen. That large three-story building, formerly the English Embassy, was a foundling hospital. The Chafariz of St. Theresa is built up against a portion of the rock of the jutting hill whence it derives its name. After passing the gardens of the late Barão de Meriti and the Gloria Hill, our passengers begin to descend at the various streets which cross the Catete, which is the widest thoroughfare in this portion of the capital. Each person, as he rises to depart, lifts his hat, and the compliment is returned by every individual in the omnibus, though all may be entire strangers. No one ever enters a large public conveyance in Rio without saluting those within and receiving in return a polite acknowledgment of his presence. Very frequently a pinch of snuff is offered to you by your unknown neighbor. I have seen gentlemen but recently returned from Brazil enter a New York omnibus and deferentially salute the inmates: the polite strangers were received with a smile of derision or looked upon with a stare of contempt.

Each omnibus has painted in large characters upon its sides its capacity: thus, "14 *pessoas*" means that the vehicle is registered at the Bureau de Police to contain that number of persons, and one passenger more than the registered number would subject the company to a heavy fine. I have never seen more passengers within than the figures on the side indicated.

I have more than once mentioned the "gondola,"—that name associated with love-romance and Venice, "moonrise, high midnight, and the voice of song." When I first heard that mellifluous term in Brazil, I fancied that the sharp and graceful little barges of the Queen of the Adriatic had been transported to the bright waters of Rio de Janeiro; but I soon discovered my mistake, and ascertained that this sweet Italian word was used to designate most unpoetic four-wheeled vehicles, drawn by as many kicking, stubborn mules! The gondola in every respect resembles the omnibus, save that no conductor accompanies it. You prepay

Senhor Bernardo or a Senhor somebody else at the Largo do Paço; and if there are any way-fares, these are received by the driver. The gondola does not have the convenience which the New York omnibus possesses, in the shape of the leather strap by which the passenger causes the driver to pull up at the will of the former. In lieu of this, passengers make a very free use of canes, umbrellas, and fists, battering at a terrible rate the end of the gondola nearest the driver; or occasionally the leg of the latter is rather more warmly than affectionately embraced by the individual sitting next to the farther window. Sometimes the gondola cannot be "propelled" by its living oars; and, under such circumstances, when a Scotchman, a Yankee, or a Frenchman will relieve himself of many hard words at the unfortunate Jehu, the Brazilians remain perfectly calm, not once descending to see what is the matter, and conversing with one another as philosophically as if nothing had happened. On one occasion I was witness to a scene which will scarcely be credited. As a gondola full of passengers was turning out of the Rua dos Ourives, it unfortunately "stuck." The driver shouted at his mules, thrashed them with his long raw-hide thong, *tchewed** at them, and stamped his footboard, all to no purpose: the animals could not start the vehicle. Not one passenger got out, but all looked from the windows as if this was a part of the programme for which they had paid their *dous testões*, (five English pence,) and they determined to have their money's worth. The poor driver was in deep distress: quite a crowd collected, but no one offered to aid him, until he, by sundry *vintems*, allured the services of several Africans, whose broad shoulders applied to the wheels, in conjunction with the pulling of the mules, moved gondola, passengers, and all.

Having something of a philological turn, I inquired why these public conveyances were called *gondolas*. I was not long in ascertaining that a monopoly had been granted to certain omnibus companies, which was considered onerous, but the municipal government could not in conscience abolish the contract or confer a new

---

* A sound unrepresentable by letters, similar to that made in the United States in scaring chickens, by which all classes, high and low, in Brazil, call the attention of others. 1866, all the gondolas have conductors.

charter upon another *omnibus* association; however, all scruples were finally overcome by granting privileges to a *gondola company* to carry passengers!

We will end our ride at the Ponta do Catete, and will thence make our way to the Hotel dos Estrangeiros, at the commencement of the Caminho Velho de Botafogo; or we may walk a few steps farther, and enter Johnson's Hotel, on the Caminho Novo. The Hotel dos Estrangeiros is a large house kept on the French plan; the Hotel Johnson is where Englishmen "most do congregate," and where one can find more comfort than at any other establishment for the accommodation of the public in the city. Both are surrounded by verdure, whether we consider the neighboring gardens, or the adjacent hills, whose sides are covered with luxuriantly-foliaged trees and clambering vines. Johnson's Hotel no longer exists.

The stranger at Rio de Janeiro is usually surprised at the scarcity of inns and boarding-houses. There are several French and Italian hotels, with apartments to let; and these are chiefly supported by the numerous foreigners constantly arriving and temporarily residing in the place. But among the native population, and intended for Brazilian patronage, there are only eight or ten inns in a city of three hundred thousand inhabitants, and scarcely any of these exceed the dimensions of a private house. It is almost inconceivable how the numerous visitors to this great emporium find necessary accommodations. It may safely be presumed that they could not, without a heavy draught upon the hospitalities of the inhabitants, with whom, in many instances, a letter of introduction secures a home. In the lack of such a resort, the sojourner rents a room, and, by the aid of his servant and a few articles of furniture, soon manages to live, with more or less frequent resorts to some caza de pasto or restaurant. Most of the members of the National Assembly keep up domestic establishments during their sojourn in the capital. As a consequence of this lack of hotels and boarding-houses, some of the commercial firms maintain a table for the convenience of their clerks and guests. This was once much more common; but, since 1850, probably the greater portion of those formerly thus accommodated club together, rent a house in Botafogo, Praia Grande, or on the Santa Theresa, and keep up an establishment of their own.

Having thus been cicerone of the reader in his rapid whirl through this city of the tropics, I know of no fitter termination to the day than for him to imagine himself in one of the vast rooms of the Hotel dos Estrangeiros.

For many days, in a narrow berth, you have been rudely rocked by the billows, and this is the first night on *terra firma* and a comfortable bed. The windows of your apartment are wide open, and, as you close your eyes, the land-breeze, murmuring softly, bears upon its wings not only the sweet, fresh smell of the earth, but, stealing in its course from the adjacent gardens the fragrance of jessamines, the delicate scent of the flora-pondia, and the odor of the opening orange-blooms, it loads the evening air with the richest aroma. The distant booming of the waves, as they break upon the Praia do Flamengo, is a soothing melody, which lulls you to dreams of scenes not more lovely than those around you, where are

" Larger constellations burning, mellow moons and happy skies,
   Breadths of tropic shade, and palms in cluster, knots of paradise,"—

a land where

" Slides the bird o'er lustrous woodland, swings the trailer from the crag,
   Droops the heavy-blossom'd bower, hangs the heavy-fruited tree,—
   Summer-isles of Eden lying in dark purple spheres of sea."

Note for 1866.—The engravings on pages 28 and 29, though graphically representing the state of things in 1855, are no longer *apropos.* They are kept as a matter of history. The Government has forbidden such exhausting and cruel labor of the slaves. Carts now carry the coffee. But the municipality of Rio should go one step further, and charge three times the amount for license on every cart which does not have springs. To say nothing of the immense weight of the vehicle now in use, whose parallel can only be found in Portugal, it acts like a sledge-hammer to constantly batter the pavement to pieces. A cart with springs could be made just as strong with half the weight, and one mule could propel as much as two with the present brutal street-destroying machine. We are glad to see already a few New York spring-carts at Rio. Have more, and the Fluminenses will pay less for their pavement taxes.

CHAPTER III.

DISCOVERY OF SOUTH AMERICA—PINZON'S VISIT TO BRAZIL—CABRAL—COELHO
—AMERICUS VESPUCIUS—THE NAME "BRAZIL"—BAY OF RIO DE JANEIRO—
MARTIN AFFONSO DE SOUZA—PAST GLORY OF PORTUGAL—COLIGNY'S HUGUENOT
COLONY—THE PROTESTANT BANNER FIRST UNFURLED IN THE NEW WORLD—
TREACHERY OF VILLEGAGNON—CONTEST BETWEEN THE PORTUGUESE AND THE
FRENCH—DEFEAT OF THE LATTER—SAN SEBASTIAN FOUNDED—CRUEL INTOLE-
RANCE—REFLECTIONS.

ALTHOUGH the bay and city of Rio de Janeiro are fraught with
interesting associations to the general student of history, and still
more to the Protestant Christian as that portion of the New World
where the banner of the Reformed religion was first unfurled, yet
I have thought it best to introduce here a brief account of the
early discovery and settlement of Brazil.

Guanihani—that outpost of the New World—was beheld by
European eyes six years before the discovery of South America.
In 1498, Columbus landed near the mouth of the Orinoco. He
recorded, in enthusiastic language, "the beauty of the new land,"
and declared that he felt as if "he could never leave so charming
a spot." The honor, however, of discovering the Western hemisphere south of the equator must be awarded to Vincent Yanez
Pinzon, who was a companion of Columbus, and had commanded
the "Niña" in that first glorious voyage which made known to the
Old World the existence of the New. Pinzon sailed from Palos in
December, 1499, and, crossing the equator, his eyes were glad-
dened, on the 26th of January, 1500, by a green promontory,
which he called Cape Consolation. This is now known as Cape
St. Augustine, the headland just south of the city of Pernambuco.
He sailed thence northward, discovering the vast mouths of the
Amazon, and touched at various points until he reached the
Orinoco.

When Pinzon beheld the palm-groves and densely-foliaged
46

forests, and had scented the spicy breezes which were wafted from the shore, he supposed that he was visiting India-beyond-the-Ganges, and believed that he had already sailed past the renowned Cathay. In the name of Castile he took possession of the goodly land; but, before he reached Spain, Pedro Alvares Cabral, a distinguished Portuguese navigator, had claimed the territory for his own monarch. On the return of Vasco da Gama to Portugal, in 1499, with the certainty of having discovered the route to the Indies by the Cape of Good Hope, the king Dom Emanuel determined to send a large fleet to those famous regions, with instructions to enter into commercial relations with the Eastern sovereigns, or, in case of refusal, to make war upon them and subdue them. The command of this expedition was intrusted to Cabral, and, on the 9th of March, the large fleet, with its fifteen hundred soldiers and mariners, sailed amid grand military and religious ceremonials, the king himself honoring the occasion by his august presence. With this handful of men, intended for the coercion of the Orient to the commercial notions of Portugal, Cabral directed his course to the Cape de Verdes, and thence, in order to avoid the calms which prevail on the African coast, he ran so far to the westward, that, without any intention on his part, he discovered, on the 21st of April, 1500, the same land which, ninety days previously, had been visited by Pinzon. Cabral's discovery was, however, in the present province of Espirito Santo, near Mount Pascal, which is eight degrees south of Cape St. Augustine.

Some Brazilian writers grudgingly mention the voyage of Pinzon; others ignore him altogether, wishing seemingly to ascribe all the glory to one of their own Portuguese ancestors. Doubtless Cabral was led by the trade-winds and by the currents—of which he was not aware—to the coast of Brazil, and thus made his fortunate discovery. To-day, vessels sailing from Europe for the East Indies can (as is well demonstrated by Lieutenant Maury's wind and current charts) make the swiftest voyages by taking advantage of the wonderful trade-winds, steering first toward South America and afterward in the direction of the Cape of Good Hope. Pinzon set forth from Palos with the intention of making Western discoveries; Cabral sailed from Lisbon with instructions to proceed to the Eastern discoveries of Vasco da Gama; but, because a happy accident (some say a fierce storm) forced his fleet to Brazil, and that, too, months after the landing of the Spanish navigator at Cape St. Augustine, there is neither reason nor justice in the national pride which endeavors to take away the priority of discovery from Vincent Yanez Pinzon.

On Easter Sunday mass was celebrated; and on the 1st of May this solemnity was repeated, and, in the presence of thousands of the aborigines, a huge cross was erected, bearing the insignia of Dom Emanuel, and the land, to which they gave the name of *Vera Cruz*, was solemnly taken possession of in the name of the King of Portugal.*

It was the Padre Frei Henrique, of Coimbra, who conducted the religious ceremonies, and in which he was piously joined (so reads the chronicle) by *os indigenos imitando os gestos e movimentos dos Portugezes*, (the savages imitating the gestures and movements of the Portuguese.)

Two convicts were left with the natives, and one of these afterward became of great use as an interpreter. Cabral despatched Gaspar de Lemos to Lisbon, to inform the monarch of the discovery and appropriation of the new land of the True Cross, and then pursued his route to the East Indies. The Pope of Rome laid down a rule regulating the proprietorship of countries discovered by Spain and Portugal, and thus was disposed the question between Pinzon and Cabral.

The king Dom Emanuel was deeply interested in the intelligence brought him by Gaspar de Lemos, and, in May, 1501, sent out to his new dominions three caravellas under the command of Gonçalo Coelho.† In one of these vessels was Americus Vespucius. This expedition partook more of the character of failure than of success, and was replaced, in 1503, by a second, which, consisting of double the number of ships employed in the first, sailed, according to some authorities, under Christopher Jacques;† according to others, under the same Gonçalo Coelho,‡ accompanied

---

* Historia do Brazil, by Gen. J. I. de Abreu Lima. Rio de Janeiro, 1843.

† Ibid. vol. i. chap. ii.

‡ Epitome da Hist. do Brazil, (by Jose Pedro Xavier Pinheiro. Bahia, 1854,) chap. i. p. 27.

again by Americus. Four of these vessels were lost, with the commander-in-chief; but the lucky Florentine escaped, and lived to deprive, indirectly, the new territory of the name conferred upon it by Cabral.[1]

The two remaining ships entered a bay, now supposed to be the spacious *Bahia de Todos os Santos*, and afterward coasted southward two hundred and sixteen leagues, and there remained five months anchored near the land, and maintained amicable relations with the natives. Here they erected a fortress, and left in it twenty-four men.

As the most valuable part of the cargo which Americus Vespucius carried back to Europe was the well-known dyewood, *Cæsalpinia Braziliensis*,—called, in the Portuguese language, *pau brazil*, on account of its resemblance to *brazas*, "coals of fire,"—the land whence it came was termed the "land of the brazil-wood;" and, finally, this appellation was shortened to *Brazil*, and completely usurped the names *Vera Cruz* or *Santa Cruz*.[2] This change was not effected without protestations on the part of some,—not because their taste for euphony was shocked, but on the ground that the cause of religion required a sacred title to the fairest possession of faithful Lusitania in the New World. One of the *reverendissimos* declared that it was through the express interposition of the devil that such a choice and lovely land should be called *Brazil* instead of the pious cognomen given to it by Cabral. Another— a devoted Jesuit—poured forth a jeremiad on the subject, concluding, with emphasis, by stating what a shame it was that "the cupidity of man, by unworthy traffic, should change the wood of the cross, red with the real blood of Christ, for that of another wood which resembled it only in color"!

Other voyages were undertaken at the order of Spain and of Portugal,—thus making known the whole coast of Eastern South America from the Amazon to the Straits of Majellan. Among the navigators at the head of these expeditions were De Solis and Majellan, (Magalhães.) In 1515, De Solis sailed on his Southern voyage, and discovered the Rio de la Plata, which at first bore his own name. On his way thither, he entered the bay now known as Rio de Janeiro. Fernando de Majellan, a Portuguese in the service of Charles I. of Spain, sailed, in 1519, to discover the western passage to the Indies.

4

On the 13th of December he entered the bay previously visited by De Solis, and remained there until the 27th of the same month, and gave to it the name of *Bahia* (bay) *de Santa Luzia,*—the day of his entrance being the anniversary of that saint. He afterward coasted along the continent until he entered those straits which still bear his name, and which were for a century the only known highway to the Pacific. Majellan was the first to circumnavigate the globe.

The usual account of the origin of the term *Rio de Janeiro*, so inappropriately given to a bay, has already been referred to. The facts seem to be adverse to the generally-accepted explanation that Martin Affonso de Souza discovered this sheet of water—which he supposed to be a river—on the 1st of January, 1531. It is incontestable that it was entered twice at least several years previous to his departure from Portugal. Martin Affonso de Souza was a Portuguese gentleman of noble lineage, and of high estimation in the court of Dom John III. The king, having received information of the visits of Spaniards to the coasts which he considered his own, determined to send an expedition, commanded by De Souza, to Brazil. De Souza had plenary powers on land and on sea, and was to fortify and distribute the new territory. He was the first donatory of Portugal in Brazil, and sailed from Lisbon on the 3d of December, 1530. In a few weeks he sighted Cape St. Augustine, near which he encountered three French vessels. He gave them battle, came off victorious, and took them in triumph to the present harbor of Pernambuco. After refitting, he came to *Bahia de Todos os Santos*, where was the little settlement of the shipwrecked Diogo Alvares Corrêa, (Caramuru,) whose romantic history is narrated in another portion of this work. After some delay, he again sailed southward, and, on the 30th of April, 1531, entered the bay which had already been named *Santa Luzia* and *Rio de Janeiro*. By reflecting for a moment upon the time (December 3, 1530) when Martin Affonso de Souza departed from Lisbon, and the various events and delays of the voyage, we can easily perceive that it would be an impossibility to sail more than five thousand miles, (and his were not modern clipper-ships,) fight and capture three vessels, refresh successively at two different ports, and then reach the Bay of Rio de Janeiro on the 1st of January, 1531.[3] Aside from this, we have the direct and simple statement of Pero

Lopes de Souza, brother to the commander, which not only settles the date of their arrival, but the fact that the bay or supposed river was previously known as *Rio de Janeiro,*—viz. : "Saturday, 30th of April, at four o'clock in the morning, we were in the mouth of Rio de Janeiro."[4]

Martin Affonso de Souza formed no settlement on the shores of the magnificent bay which he had entered, but contented himself with remaining there for a few months, where he constructed three brigantines, and then sailed to the coast of the present province of São Paulo.   At a place which possessed no great natural advantages he commenced the first European colony (Vespucius's handful of men and Caramuru's wigwams cannot be called the earliest settlements) in Brazil, and named it *St. Vincent.*   St. Vincent no longer exists, unless its existence may be predicated in the few miserable houses and the broken fountain which mark the spot where was laid the first stone of the proudest colony of Portugal.   On the margin of that spacious and protected harbor which De Souza rejected for an exposed arm of the sea, has sprung up the first commercial city of South America, and the third in the New World.

It will not be uninstructive to glance at the position, at that time, of the kingdom which sent forth Diaz, Vasco da Gama, Cabral, Coelho, Christopher Jacques, Vespucius, and De Souza, upon new and hazardous voyages of discovery.   The territory of European Portugal was then no greater than at present; but her ambitious monarchs and her daring navigators had pushed their conquests and discoveries not only along the whole western and eastern coasts of Africa, but to "the farthest Ind."   Bartholomew Diaz beheld the Cape of Good Hope six years before Columbus discovered America; and Vasco da Gama doubled the same cape ere the great Genoese landed at the mouth of the Orinoco.   Portugal had flourishing colonies in Angola, Loango, and Congo, before Cortez had burned his ships in the Mexican Gulf.   Before the Honorable East India Company was dreamed of, Portuguese viceroys and Portuguese commercial enterprises swayed it over millions in Hindostan and Ceylon.   They trafficked with the distant Peguans and the little-known Burmese, on the banks of the Irrawaddy, three hundred years before Judson proclaimed, near the same river, the gospel of the blessed Saviour. Centuries before the English possessed Hong-Kong or the Americans had opened Japan by commercial treaties, Portugal owned Macao, held intercourse with the curious Chinese, traded with the Japanese, and, through her priests, led more than half a million of those almond-eyed islanders to embrace the doctrines of Rome. Of her immense acquisitions by conquest and discovery, that of Brazil was not to be the least in its importance and future destiny. When we look at what Portugal *was* and what she *is*, we can only exclaim, "How are the mighty fallen!" Portugal has been weighed in the balance and found wanting. Shorn of all her possessions in the East except a territory (comprising Goa and a few unimportant islands) not so large as the State of South Carolina, her commerce is now scarcely known in the Indian Seas. Her dominion west of Asia is limited to her own small European kingdom, to languishing colonies in Africa, and to a few islands in the Atlantic. She owns not an inch of territory in the Western World, where once she had a quarter of the continent. She had not the conservative salt of a pure Christianity to preserve her morality and her greatness. Like Spain, she became at once the patron and the protectress of the Inquisition; and, though the Portuguese are far more tolerant than the Spaniards, yet the Government of Portugal held on to that cursed engine of Roman intolerance until 1821. The contrast between Holland and Portugal forces itself upon the consideration of all. They are both nearly of the same European area and population, both were great maritime nations in the sixteenth century, and both made extensive conquests in the East. But, while neighboring states have created a mercantile marine since the era referred to, Holland, in this respect, still ranks as the third power in Europe and the fourth in the world, and her internal prosperity has not declined. Her credit has always maintained the highest place among the nations of the earth, while Portugal has been more than once on the verge of bankruptcy. Holland to-day governs twenty-two millions of people, who are prosperous and advancing, whether in the Eastern or the Western hemisphere. Portugal, in all her dominions, rules less than one-third of that number. The former is distinguished for tolerance and intelligence; the latter, under the blighting shadow of the Papacy, has, even in the latter half of the nineteenth century, manifested narrowness and bigotry,[5] and her people, as a whole, have been the most ignorant of Europe. The last few years have, however, we trust, been the precursor of a better era for Portugal. Her young and enlightened monarch has come to the throne with enlarged views, and it is fondly hoped that his subjects will be elevated, and that Portugal will assume a position more in accordance with the historical traditions of those days when her kings were energetic, and when her navigators laid at her feet the treasures of the world.

Returning from this digression, let us again watch the progress of events in the new acquisitions of Portugal in the Western World.

Other eyes than those of Spanish navigators were looking toward Brazil, and to that very portion of it which had been slighted by Martin Affonso de Souza. Among the adventurers from France was Nicholas Durand de Villegagnon, a Knight of Malta, a man of considerable abilities, and of some distinction in the French service. He had even been appointed to the gallant post of commander of the vessel which bore Mary, Queen of Scots, from France to her own realms. Villegagnon aspired to the honor of establishing a colony in the New World, and Rio de Janeiro was the chosen spot for his experiment. He had the address, in the outset, to secure the patronage of the great and good Admiral Coligny, whose persevering attempt to plant the Reformed religion in both North and South America was a leading feature in his life up to the time when St. Bartholomew's Eve was written in characters of blood.

Villegagnon proposed to found an asylum for the persecuted Huguenots. Admiral Coligny's influence secured to him a respectable number of colonists. The French court was disposed to view with no small satisfaction the plan of founding a colony, after the example of the Portuguese and Spaniards.

It was in the year 1555 that Henry II., the reigning king, furnished three small vessels, of which Villegagnon took the command and sailed from Havre de Grace. A gale of wind occurred while they were yet on the coast, and obliged them to put into Dieppe, which they accomplished with considerable difficulty. By this time many of the artificers, soldiers, and noble adventurers had become sick of the sea, and abandoned the expedition so soon as they reached the shore.

After a long and perilous voyage, Villegagnon entered the Bay of Nitherohy, and commenced fortifying a small island near the entrance, now denominated Lage, and occupied by a fort. His fortress, however, being of wood, could not resist the action of the water at flood-tide, and he was obliged to remove farther upward, to the island now called Villegagnon, where he built a fort, at first named in honor of his patron, Coligny. This expedition was well planned, and the place for a colony fitly chosen. The native tribes were hostile to the Portuguese, but had long traded amicably with the French. Some hundreds of them assembled on the shore at the arrival of the vessels, kindled bonfires in token of their joy, and offered every thing they possessed to these allies who had come to defend them against the Portuguese. Such a reception inspired the French with the idea that the continent was already their own, and they denominated it La France Antarctique.

It was upon this island that they erected their rude place of worship, and here these French Puritans offered their prayers and sang their hymns of praise nearly threescore years and ten before a Pilgrim placed his foot on Plymouth Rock, and more than half a century before the Book of Common Prayer was borne to the banks of the James River.

On the return of the vessels to Europe for a new supply of colonists, considerable zeal was awakened for the establishment of the Reformed religion in these remote parts. The Church of Geneva became interested in the object, and sent two ministers and fourteen students, who determined to brave all the hardships of an unknown climate and of a new mode of life in the cause. It is interesting to reflect that when the Reformation was yet in its infancy, the subject of propagating the gospel in distant parts of the world was one that engaged the hearts of Christians in the city of Geneva while Calvin, Farel, and Theodore de Beza were still living. It would be difficult to find an earlier instance of Protestant missionary effort.

As the situation of the Huguenots in France was any thing but happy, the combined motive of seeking deliverance from oppression and the advancement of their faith appears to have prevailed

extensively, and induced many to embark. When we look at the
incipient movements of this enterprise, without the knowledge
of its conclusion, there seems as much reason to hope that the
principles of the Reformation would have taken root here, as they
did afterward in North America, where they have produced a
harvest of such wonderful results.

But misfortunes seemed to attend every step of the enterprise.
At Harfleur, the Papist populace rose against the colonists, and
the latter, after losing one of their best officers in the conflict,
were obliged to seek safety in retreat. They had a tedious voyage,
suffering at one time from a violent storm; and, having neared
the Brazilian coast, had a slight encounter with the Portuguese.
However, they were received by Villegagnon with apparent cordiality, and effectual operations began to be undertaken for their
establishment. But it was not long before certain untoward circumstances occurred which developed the real and villanous character
of their leader.

Having gained over to his complete influence a certain number
who cared not for spiritual piety, Villegagnon, under pretence
of changing his religion and returning to the true faith, commenced a series of persecutions. Those who had come to Antarctic
France to enjoy liberty of conscience found their condition worse
than before. They were subjected to abusive treatment and great
hardships. This unnatural defection consummated the premature
ruin of the colony. The newly-arrived colonists demanded leave
to return, which was granted, but in a vessel so badly furnished
that some refused to embark, and the majority, who persisted,
endured the utmost misery of famine. Villegagnon had given
them a box of letters, wrapped in sere-cloth, as was the custom.
Among them was one directed to the chief magistrate of the port
where they might chance to arrive, in which this worthy friend
of the Guises denounced the men whom he had invited out to
Brazil to enjoy the peaceable exercise of the Reformed religion, as
heretics worthy of the stake. The magistrates of Hennebonne,
where they landed, happened to favor the Reformation, and thus
the malignity of Villegagnon was frustrated, and his treachery
exposed. Of those who had feared to trust themselves to a vessel
so badly stored, and so unfit for the voyage, three were put to

death by this persecutor. Others of the Huguenots fled from him to the Portuguese, where they were compelled to apostatize, and to profess a religion which they abhorred.

The homeward-bound colonists were reduced to the greatest extremity, and, from want of food, they not only devoured all the leather,—even to the covering of their trunks,—but in their despair they attempted to chew the hard, dry brazil-wood which happened to be in the vessel. Several died of hunger; and they had begun to form the resolution of devouring each other, when land appeared in view. They arrived just in time to undeceive a body of Flemish adventurers ready to embark for Brazil, and also about ten thousand Frenchmen, who would have emigrated if the object of Coligny in founding his colony had not thus been wickedly betrayed.

Though the Portuguese were so jealous of the Brazilian trade that they treated all interlopers as pirates, yet, by some oversight, they permitted this French colony to remain four years unmolested; and, had it not been for the treachery of Villegagnon to his own party, Rio de Janeiro would probably have been, at this day, the capital of a French colony or of an independent State in which the Huguenot element would have been predominant.

The Jesuits were well aware of this danger, and Nobrega, their chief and provincial, at length succeeded in rousing the court of Lisbon. A messenger was commanded to discover the state of the French fortifications. On the ground of his report, orders were despatched to Mem de Sa Barreto, governor of the colony, and resident at San Salvador, to attack and expel the intruders who remained. Having fitted out two vessels-of-war and several merchantmen, the governor, taking the command in person, embarked, accompanied by Nobrega as his prime counsellor. They appeared off the bar at Rio early in 1560, with the intention of surprising the island at the dead of night. Being espied by the sentinels, their plan was foiled. The French immediately made ready for defence, forsook their ships, and, with eight hundred native archers, retired to their forts.

With reinforcements from St. Vincente, Mem de Sa won the landing-place, and, routing the French from their most important holds, so intimidated them that, under cover of the night, they fled, some to their ships and some to the mainland.

The Portuguese, not being strong enough to keep the position they had taken, demolished the works, and carried off the artillery and stores which they found. A short time after this, new wars, made by the native tribes, broke out against them, and were prosecuted at different points with great ferocity for several years. In the mean time, the French recovered strength and influence. Preparations were again made to extirpate them. A party of Portuguese and friendly Indians, under the command of a Jesuit appointed by Nobrega, landed near the base of the Sugar-Loaf, and, taking a position now known as Praia Vermelha, maintained a series of indecisive skirmishes with their enemies for more than a year. Occasionally, when successful, they would sing in triumphant hope a verse from the Scriptures, saying, " The bows of the mighty are broken," &c. Well might they call the bows of the Tamoyos mighty; for an arrow sent by one of them would fasten a shield to the arm that held it, and sometimes would pass through the body, and continue its way with such force as to pierce a tree and hang quivering in the trunk.

Nobrega at length came to the camp, and at his summons Mem de Sa again appeared with all the succors he could raise at San Salvador. All was made ready, and the attack deferred forty-eight hours, in order to take place on St. Sebastian's Day. The auspicious morning came,—that of January 20, 1567. The stronghold of the French was stormed. Not one of the Tamoyos escaped.

Southey most justly remarks, never was a war in which so little exertion had been made, and so little force employed on both sides, attended by consequences so important. The French court was too busy in burning and massacring Huguenots to think of Brazil, and Coligny, after his generous plans had been ruined by the villanous treachery of Villegagnon, no longer regarded the colony: the day for emigration from his country was over, and they who should have colonized Rio de Janeiro were bearing arms against a bloody and implacable enemy, in defence of every thing dear to man. Portugal was almost as inattentive to Brazil; so that, few and unaided as were the Antarctic French, had Mem de Sa been less earnest in his duty, or Nobrega less able and less indefatigable in his opposition, the former would have retained their place, and perhaps the entire country have this day been French.

Immediately after his victory, the governor, conformable to his instructions, traced out a new city, which he named San Sebastian, in honor of the saint under whose patronage the field was won, and also of the king of the mother-country. The name of San Sebastian has been supplanted by that of Rio de Janeiro.

In connection with the event just narrated, there remains on record a melancholy proof of the cruelty and intolerance of the victors. According to the annals of the Jesuits, Mem de Sa stained the foundations of the city with innocent blood. "Among the Huguenots who had been compelled to fly from Villegagnon's persecution was one John Boles, a man of considerable learning, being well versed both in Greek and Hebrew. Luiz de Gram caused him to be apprehended, with three of his comrades, one of whom feigned to become a Roman Catholic; the others were cast into prison; and there Boles had remained eight years, when he was sent for to be martyred at Rio de Janeiro, for the sake of terrifying his countrymen, if any should be lurking in those parts."

The Jesuits are the only historians of this matter. They pretend that Boles apostatized, having been convinced of his errors by Anchieta, a priest greatly celebrated in the annals of Brazil. But, by their own story, it is not very probable that a man who for eight long years had steadfastly refused to renounce the religion of his conviction would now yield. Boles doubtless proved a stubborn unbending Protestant, and for this suffered a cruel death. And, notwithstanding the statement that he was to be slain as an example to his countrymen, "if any should be found lurking in those parts," it was not the custom of Rome to put to death those who renounced their errors and came into her protecting fold.

When Boles was brought out to the place of execution, and the executioner bungled in his bloody office, "Anchieta hastily interfered, and instructed him how to despatch a heretic as speedily as possible,—fearing, it is said, lest he should become impatient, being an obstinate man, and newly reclaimed, and that thus his soul would be lost. The priest who in any way accelerates the execution of death is thereby suspended from his office; but the biographer of Anchieta enumerates this as one of the virtuous actions of his life."

Though Rio de Janeiro was thus founded in blood, there is no

Roman Catholic country in the world freer from bigotry and intolerence than the Empire of Brazil.

Thus failed the establishment of Coligny's colony, upon which the hopes of Protestant Europe had for a short time been concentrated; and Rio de Janeiro will ever be memorable as the first spot in the Western hemisphere where the banner of the Reformed religion was unfurled. It is true that the attempt was made upon territory which had been appropriated by Portugal; still, a question might arise as to the right of priority in the discovery of this portion of Brazil, for it is certain that the Spaniard, De Solis, and also Majellan, Ruy Faleiro, and Diogo Garcia, Portuguese navigators in the service of Spain, entered the Bay of Nitherohy long before Martin Affonso de Souza. In whatever way this may be settled, the fact of the failure of this Huguenot effort is full of food for reflection; and we can fully sympathize with the remarks of the author of "Brazil and La Plata," in regard to the treachery of Villegagnon, and the consequent defeat of the aims of the first French colonists:—

"With the remembrance of this failure in establishing the Reformed religion here, and of the direct cause which led to it, I often find myself speculating as to the possible and probable results which would have followed the successful establishment of Protestantism during the three hundred years that have since intervened. With the wealth, and power, and increasing prosperity of the United States before us, as the fruits at the end of two hundred years' colonization of a few feeble bands of Protestants on the comparatively bleak and barren shore of the Northern continent, there is no presumption in the belief that had a people of similar faith, similar morals, similar habits of industry and enterprise, gained an abiding footing in so genial a climate and on a soil so exuberant, long ago the still unexplored and impenetrable wilderness of the interior would have bloomed and blossomed in civilization as the rose, and Brazil from the sea-coast to the Andes would have become one of the gardens of the world. But the germ which might have led to this was crushed by the bad faith and malice of Villegagnon; and, as I look on the spot which bears his name, and, in the eyes of a Protestant at least, perpetuates his reproach, the two or three solitary palms which lift their tufted heads above the embattled walls, and furnish the only evidence of vegetation on the island, seem, instead of plumed warriors in the midst of their defences, like sentinels of grief mourning the blighted hopes of the long past."

FORTRESS AND ISLAND OF VILLEGAGNON.

But we should not look too "mournfully into the past;" for though, in the mysterious dealings of Providence no Protestant nation, with its attendant vigor and progress, sways it over that fertile and salubrious land, may we not to a certain extent legitimately consider the tolerant and fit Constitution of the Empire, and its good government, the general material prosperity, and the advancement of the Brazilians in every point of view far beyond all other South American nations, as an answer to the faithful prayers with which those pious Huguenots baptized Brazil more than three centuries ago?

Note for 1866.—The present Emperor has certainly shown himself a friend of toleration. He has aided in the construction of Protestant chapels for colonists; the Government promptly suppressed three riots attempted against Brazilian Protestants, (at Rio de Janeiro, at Bahia, and at Praia Grande;) and other acts might be cited to demonstrate that we have true cause for gratitude at the position of religious toleration in Brazil. But Brazilian legislation should go one step further, and admit to the Parliament all fit men, of whatever religious denomination. Then Brazil will be abreast with the nineteenth century

CHAPTER IV.

EARLY STATE OF RIO — ATTACKS OF THE FRENCH — IMPROVEMENTS UNDER THE
VICEROYS — ARRIVAL OF THE ROYAL FAMILY OF PORTUGAL — RAPID POLITICAL
CHANGES — DEPARTURE OF DOM JOHN VI. — THE VICEROYALTY IN THE HANDS
OF DOM PEDRO — BRAZILIANS DISSATISFIED WITH THE MOTHER-COUNTRY — DE-
CLARATION OF INDEPENDENCE — ACCLAMATION OF DOM PEDRO AS EMPEROR.

FOR one hundred and forty years after its foundation, the city
of San Sebastian enjoyed a state of tranquil prosperity. This
quietness was in happy contrast with the turbulent spirit of the
age, and especially with the condition of the principal towns and
colonies of Brazil; nearly all of which, during the period referred
to, had been attacked by either the English, the Dutch, or the
French. In this interval the population and commerce of the place
greatly increased.

At the commencement of the eighteenth century the principal
gold-mines of the interior were discovered by the Paulistas, the
inhabitants of San Paulo. These gave the name of Minas Geraes
(*General Mines*) to a large inland province, which became then,
as it still remains, tributary to the port of Rio de Janeiro. Gold-
digging was found to produce here effects similar to those which
resulted from it in the Spanish countries. Agriculture was nearly
abandoned, the price of slaves—who had been early introduced—
became enormous, and the general prosperity of the country retro-
graded; while every one who could rushed to the mines, in hope
of speedily enriching himself. We even find that the curious and
abnormal condition of California in 1848 had its counterpart three
centuries ago in Brazil.

Even the Governor of Rio, forgetful of his official character and
obligations, went to Minas Geraes and engaged with avidity in the
search for treasure. The fame of these golden discoveries sounded

abroad, and awakened the cupidity of the French, who, in 1710, sent a squadron, commanded by M. Du Clerc, with the intent of capturing Rio. The whole expedition was ingloriously defeated by the Portuguese, under Francisco de Castro, Governor of Rio de Janeiro. This officer possessed no military ability, but blundered into a victory over the French, and permitted horrid cruelties to be practised upon the prisoners. France was not slow to resent the inhumanity with which her men had been treated.

M. Duguay Trouin, one of the ablest naval officers of the times, sought permission to revenge his countrymen and to plunder Rio de Janeiro. Individuals were found ready to incur the expenses of the outfit, in prospect of the speculation. The project was approved by Government, and an immense naval force was placed at Trouin's disposal.

This expedition was eminently successful. The tactics of the imbecile Castro did not succeed : the city was stormed, taken, and afterward ransomed for a heavy sum. It was during the bombardment that the convent of San Bento was battered by the balls, the marks of which are still visible.

The plunder and the ransom were so great, that, notwithstanding, on the return-voyage of the French, a number of their vessels went down with twelve hundred men and the most valuable part of the booty, there remained to the adventurers a profit of ninety-two per cent. upon the capital they had risked in the outfit.

From the time that Duguay Trouin's squadron weighed anchor on their homeward voyage, no hostile fleet has ever entered the harbor of Rio de Janeiro. Great changes, however, have taken place in the condition of that city.

In 1763 it superseded Bahia as the seat of government, and became the residence of the viceroys of Portugal.

The more substantial improvements of the capital were undertaken at this period. The marshes, which covered a considerable portion of the spot where the town now stands, were drained and diked. The streets were paved and lighted. Cargoes of African slaves, who had hitherto been exposed in the streets for sale, exhibiting scenes of disgust and horror, and also exposing the inhabitants to the worst of diseases, were now ordered to be

removed to the Vallongo, which was designated as a general
market for these unhappy beings.

Fountains of running water were also multiplied. The great
aqueduct which spans the Rua dos Arcos was then constructed;
and in these and various other ways, the health, comfort, and
prosperity of the city were promoted under the successive administrations of the Count da Cunha, the Marquis of Lavradio, and Luiz
de Vasconcellos

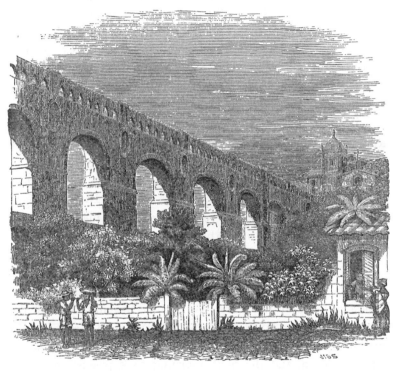

GREAT AQUEDUCT—RUA DOS ARCOS.

The system of government maintained during these periods
throughout Brazil was absolute in the extreme, and by no means
calculated to develop the great resources of the country. Nevertheless, it was anticipated by the more enlightened statesmen of
Portugal that the colony would some day eclipse the glory of
the mother-country. None, however, could foresee the proximity
of those events which were about to drive the royal family (the
house of Braganza) to seek an asylum in the New World, and to

establish their court at Rio de Janeiro.   The close of the eighteenth
century witnessed their development.

The French Revolution and the leading spirit which was raised
up by it involved the slumbering kingdom of Portugal in the
troubles of the Continent.   Napoleon determined that the court
of Lisbon should declare itself against its ancient ally, England,
and assent to the Continental system adopted by the Imperial ruler
of France.   The Prince-Regent, Dom John VI., promised, but hesi-
tated, delayed, and finally, too late, declared war against England.
The vacillation of the Prince-Regent hastened events to a crisis.
The English fleet, under Sir Sidney Smith, established a most
rigorous blockade at the mouth of the Tagus, and the British
ambassador left no other alternative to Dom John VI. than to
surrender to England the Portuguese fleet, or to avail himself
of the British squadron for the protection and transportation of
the royal family to Brazil.   The moment was critical : the army
of Napoleon had penetrated the mountains of Beira ; only an
immediate departure would save the monarchy.   No resource remained to the Prince-Regent but to choose between a tottering
throne in Europe and a vast empire in America.   His indecisions
were at an end.   By a royal decree he announced his intention to
retire to Rio de Janeiro until the conclusion of a general peace.
The archives, the treasures, and the most precious effects of the
crown, were transferred to the Portuguese and English fleets ; and,
on the 29th of November, 1807, accompanied by his family and a
multitude of faithful followers, the Prince-Regent took his departure amid the combined salvos of the cannon of Great Britain
and of Portugal.   That very day Marshal Junot thundered upon the
heights of Lisbon, and the next morning took possession of the
city.   Early in January, 1808, the news of these surprising events
reached Rio de Janeiro, and excited the most lively interest.

What the Brazilians had dreamed of only as a remote possible
event was now suddenly to be realized.   The royal family might
be expected to arrive any day, and preparations for their reception
occupied the attention of all.   The Viceroy's palace was immediately prepared, and all the public offices in the Palace Square
were vacated to accommodate the royal suite.   These not being
deemed sufficient, proprietors of private houses in the neighborhood

VIEW OF RIO DE JANEIRO FROM THE ISLAND OF COBRAS.

were required to leave their residences and send in their keys to the Viceroy.

Such were the sentiments of the people respecting the hospitality due to their distinguished guests, that nothing seems to have been withheld; while many, even of the less opulent families, voluntarily offered sums of money and objects of value to administer to their comfort.

The fleet having been scattered in a storm, the principal vessels had put into Bahia, where Dom John VI. gave that *carta regia* which opened the ports of Brazil to the commerce of the world. At length all made a safe entry into the harbor of Rio, on the 7th of March, 1808. In the manifestations of joy upon this occasion, the houses were deserted and the hills were covered with spectators. Those who could procured boats and sailed out to meet the royal squadron. The prince, immediately after landing, proceeded to the cathedral, and publicly offered thanks for his safe arrival. The city was illuminated for nine successive evenings.

In order to form an idea of the changes that have occurred in Brazil during the last fifty years, it must be remarked, that, up to the period now under consideration, all commerce and intercourse with foreigners had been rigidly prohibited by the narrow policy of Portugal. Vessels of nations allied to the mother-country were occasionally permitted to come to anchor in the ports of this mammoth colony; but neither passengers nor crew were allowed to land excepting under the superintendence of a guard of soldiers. The policy pursued by China and Japan was scarcely more strict and prohibitory.

To prevent all possibility of trade, foreign vessels—whether they had put in to repair damages or to procure provisions and water—immediately on their arrival were invested with a custom-house guard, and the time for their remaining was fixed by the authorities according to the supposed necessities of the case. As a consequence of these oppressive regulations, a people who were rich in gold and diamonds were unable to procure the essential implements of agriculture and of domestic convenience. A wealthy planter, who could display the most rich and massive plate at a festival, might not be able to furnish each of his guests with a knife at table. A single tumbler at the same time might be under the necessity of making repeated circuits through the company. The printing-press had not made its appearance. Books and learning were equally rare. The people were in every way made to feel their dependence; and the spirit of industry and enterprise were alike unknown.

On the arrival of the Prince-Regent the ports were thrown open. A printing-press was introduced, and a Royal Gazette was published. Academies of medicine and the fine arts were established. The Royal Library, containing sixty thousand volumes of books, was opened for the free use of the public. Foreigners were invited, and embassies from England and France took up their residence at Rio de Janeiro.

From this period, decided improvements were made in the condition and aspect of the city. New streets and squares were added, and splendid residences were arranged on the neighboring islands and hills, augmenting, with the growth of the town, the picturesque beauties of the surrounding scenery. The sudden and continued influx of Portuguese and foreigners not only showed itself in the population of Rio, but extended inland, causing new ways of communication to be opened with the interior, new towns to be erected, and old ones to be improved. In fact, the whole face of the country underwent great and rapid changes.

The manners of the people also experienced a corresponding mutation. The fashions of Europe were introduced. From the seclusion and restraints of non-intercourse the people emerged into the festive ceremonies of a court, whose levees and gala-days drew together multitudes from all directions. In the mingled society which the capital now offered, the dust of retirement was brushed off, antiquated customs gave way, new ideas and modes of life were adopted, and these spread from circle to circle and from town to town.

Business assumed an aspect equally changed. Foreign commercial houses were opened, and foreign artisans established themselves in Rio and other cities.

This country could no longer remain a colony. A decree was promulgated in December, 1815, declaring it elevated to the dignity of a kingdom, and hereafter to form an integral part of the United Kingdom of Portugal, Algarves, and Brazil. It is scarcely possible to imagine the enthusiasm awakened by this unlooked-for change throughout the vast extent of Portuguese America. Messengers were despatched to bear the news, which was hailed with spontaneous illuminations from the La Plata to the Amazon. Scarcely was this event consummated when the queen, Donna Maria I., died.

She was mother to the Prince-Regent, and had been for years in a state of mental imbecility, so that her death had no influence upon political affairs. Her funeral obsequies were performed with great splendor; and her son, in respect for her memory, delayed the acclamation of his accession to the throne for a year. He was at length crowned, with the title of Dom John VI. The ceremonies of the coronation were celebrated with suitable magnificence in the Palace Square, on the 5th of February, 1818.

Amid all the advantages attendant upon the new state of things in Brazil, there were many circumstances calculated to provoke political discontent. It was then that bitter feelings toward the natives of Portugal sprang up, which, though modified, still exist throughout the Empire, and made, at a later date, the severance of Brazil from the mother-country more easy of accomplishment than the separation of the thirteen colonies of North America from the crown of Great Britain. There had always been, to a greater or less extent, a certain rivalry between the native Brazilian and the Portuguese; but now it found a new cause of excitement. The Government felt itself bound to find places for the more than twenty thousand needy and unprincipled adventurers who had followed the royal family to the New World. These men cared very little for the welfare of Brazil, either in the administration of justice or in acts for the benefit of the public. Their greatest interest by far was manifested in the eager desire to fleece the country and enrich themselves. Honors were heaped upon those Brazilians who had furnished house and money to the Prince-Regent; and, as he had nothing to give them but decorations, he was soon surrounded by knights who had never displayed either chivalry or learning. The excitement thus aroused in a country where titulary distinctions were hitherto almost unknown was intense. Every one aspired to become a *cavalheiro* or a *commendador*, and the most degrading sycophancy was practised to

obtain the royal favor. Men who had been good traders in imported articles, or successful dealers in mandioca and coffee, once
knighted, could never again return to the drudgery and debasing
associations of commercial life, and must live either on previously-
acquired fortunes or seek Government employment.

On this ground the native Brazilians and the newly-arrived
Portuguese fought their first battles. They were rivals for place,
and, once in office, the Brazilian was as open to every species of
bribery and corruption as the most venal hanger-on of the court
from Lisbon. The Brazilians, however, had one advantage over
their adversaries. The natives sympathized most fully with their
recently-knighted brethren, and listened to their complaints with
a willing ear. These things, together with the wretched state of
morals that prevailed at the court, were calculated to increase the
jealousy of what the Brazilians considered a foreign dominion
over them. The independence of the English North American
colonies and the successful revolutionary struggle of some of the
neighboring Spanish-American provinces still more augmented the
uneasiness of the people; and a consciousness of this increasing
discontent, and a fear that Brazil might be induced to follow the
example of her revolting Spanish neighbors, doubtless had a
powerful influence upon the Government in making the concessions named.

Tranquillity followed the erection of Brazil into a constituent
portion of the kingdom; but it was of short duration. Discontent
was at work. The intended revolt at Pernambuco in 1817 was
betrayed to the Government, and the insurgents were prematurely
compelled to take up arms, and suffered defeat from the troops
sent against them by the Count dos Arcos. From this time there
seems to have been a systematic exclusion of native Brazilians
from commands in the army.

Murmurs were gradually disseminated; but they found no echo—
as in the case of the North American colonies—from the press,
which had, with common schools, followed in the immediate wake
of the English colonists. The first, and at that time the only,
printing-press in the country, was brought from Lisbon in 1808,
and was under the direct control of the royal authorities. Its
columns faithfully recorded for the Brazilian public the health of

all the European princes. It was filled with official edicts, birthday odes, and panegyrics on the royal family; but its pages were
unsullied by the ebullitions of the democracy, or the exposure of
their grievances. As has been well said by Armitage, " to have
judged of the country by the tone of its only journal, it must have
been pronounced a terrestrial paradise, where no word of complaint had ever yet found utterance."

But at length the time arrived when the monotony of the Court
Gazette was interrupted, and the people soon found voices for
their grievances, and in the end substantial redress.

The revolution which occurred in Portugal in 1821, in favor of a
Constitution, was immediately responded to by a similar one in
Brazil.

After much excitement and alarm from the tumultuous movements of the people, the King, D. John VI., conferred upon his son
Dom Pedro, Prince-Royal, the office of Regent and Lieutenant to
His Majesty in the Kingdom of Brazil. He then hastened his departure for Portugal, accompanied by the remainder of his family
and the principal nobility who had followed him. The disheartened
monarch embarked on board a line-of-battle ship on the 24th of
April, 1821, leaving the widest and fairest portion of his dominions
to a destiny not indeed unlooked for by his majesty, but which
was fulfilled much sooner than his melancholy forebodings anticipated.*

Rapid as had been the political changes in Brazil during the last
ten years, greater changes still were about to take place. Dom
Pedro, who now enjoyed the dignity and attributes of Prince-
Regent and Lieutenant of His Majesty the King of Portugal, was
at this period in the twenty-third year of his age. He possessed
many of the essentials of popularity. His personal beauty was
not less marked than his frank and affable manners, and his disposition, though capricious, was enthusiastic. He had decision of
character, and was one who seemed to know when to seize the

* Just as the vessel was ready to sail, the old king pressed his son to his bosom
for the last time, and exclaimed, "Pedro, Brazil will, I fear, ere long separate
herself from Portugal; and if so, place the crown on thine own head rather than
allow it to fall into the hands of any adventurer."

proper moment for calming the populace, as when at Rio, while the King was in the Palace of San Christovão, only three miles away, he, upon his own authority, gave to the people and the troops a decree whereby an unreserved acceptance of the future Constitution of the Portuguese Cortes was guaranteed. He also knew well how to guard his prerogative. The Prince's consort was by lineage and talent worthy of his hand, for Leopoldina was an archduchess of Austria; in her veins coursed the blood of Maria Theresa, and it was her sister Maria Louisa who was the bride of Napoleon. She was not possessed of great personal beauty, yet her kindness of heart and her unpretentious bearing endeared her to every one who knew her.

Dom Pedro had left Portugal when a mere lad, and it was believed that his highest aspirations were associated with the land of his adoption. In the office of Prince-Regent he certainly found scope for his most ardent ambition; but he also discovered himself to be surrounded with numerous difficulties, political and financial. So embarrassing indeed was his situation, that in the course of a few months he begged his father to allow him to resign his office and attributes. The Cortes of Portugal about this time becoming jealous of the position of the Prince in Brazil, passed a decree ordering him to return to Europe, and at the same time abolishing the royal tribunals at Rio. This decree was received with indignation by the Brazilians, who immediately rallied around Dom Pedro, and persuaded him to remain among them. His consent to do so gave rise to the most enthusiastic demonstrations of joy among both patriots and loyalists. The Portuguese military soon evinced symptoms of mutiny.

A conflict seemed inevitable; but the Portuguese commander vacillated in view of the determined opposition manifested by the people, who flew to arms, and offered to capitulate on the condition of his soldiers retaining their arms. This was conceded, on their agreeing to retire to Praia Grande, a city on the opposite side of the bay, until transports could be provided for their embarkation to Lisbon; which was subsequently effected. The measures of the Cortes of Portugal, which continued to be arbitrary in the extreme toward Brazil, finally had the effect to hasten, in the latter country, a declaration of absolute independence. This measure had long been ardently desired by the more enlightened Brazilians, some of whom had already urged Dom Pedro to assume the title of Emperor Hitherto he had refused, and reiterated his allegiance to Portugal But he at length, while on a journey to the province of S. Paulo received despatches from the mother-country, which had the effect of cutting short all delay, and caused him to declare for independence in a manner so decided and explicit that henceforward all retrograde measures would be utterly impracticable.

On the 7th of September, 1822, when he read the despatches, he was surrounded by his courtiers, on those beautiful campinas in sight of San Paulo, a city which had ever been, as it is now, celebrated in Brazil for the liberality and intelligence of its inhabitants. It was then, on the margin of an insignificant stream,—the Ypiranga,—that he made that exclamation, *"Independencia ou morte,"* (Independence or death,) which became the watchword of the Brazilian Revolution; and from the 7th of September, 1822, the independence of the country has since held its official date. It has been truly said that in the eyes of the civilized world it was a memorable circumstance, and must ever form an epoch in the history of the Western continent.

It was indeed a great event, which has led to vast results. It was a grand revolution, begun by one whose very birth and position would have led the contemplative philosopher or statesman to pronounce it impossible that he should become the leader of a popular cause. It was the descendant of a long line of European monarchs who inaugurated that movement which severed the last —the most faithful—of the great divisions of South America from transatlantic rule.

The Prince-Regent hastened to Rio de Janeiro by a rapid journey; and there, so soon as his determination was known, the enthusiasm in his favor knew no bounds.

The municipality of the capital issued a proclamation on the 21st of September, declaring their intention to fulfil the manifest wishes of the people, by proclaiming Dom Pedro the constitutional Emperor and perpetual defender of Brazil. This ceremony was performed on the 12th of October following, in the Campo de Santa Anna, in the presence of the municipal authorities, the functionaries of the court, the troops, and an immense concourse of people. His Highness there publicly declared his acceptance of the title conferred on him, from the conviction that he was thus obeying the will of the people. The troops fired a salute, and the city was illuminated in the evening. José Bonifacio de Andrada, prime minister of the Government, had in the mean time promulgated a decree, requiring all the Portuguese who were disposed to embrace the popular cause to manifest their sentiments by wearing the Emperor's motto— "Independencia ou morte"—upon their arm, ordering also, that all dissentients should leave the country within a given period, and threatening the penalties imposed upon high-treason against any one who should thenceforward attack, by word or deed, the sacred cause of Brazil.

The prime minister was the eldest of three brothers, all of them remarkable for their talents, learning, eloquence, and (though at times factious) for their sterling patriotism. They were uninfluenced by either the adulation of the populace or the favor of the Emperor. José Bonifacio de Andrada combined, to an eminent degree, the various excellencies suited to the emergencies of the incipient stages of the Empire.

The Brazilian Revolution was comparatively a bloodless one. The glory of Portugal was already waning; her resources were exhausted, and her energies crippled by internal dissensions.

That nation made nothing like a systematic and persevering effort to maintain her ascendency over her long-depressed but now rebellious colony. The insulting measures of the Cortes were consummated only in their vaporing decrees. The Portuguese dominion was maintained for some time in Bahia and other ports, which had been occupied by military forces. But these forces were at length compelled to withdraw and leave Brazil to her own control. So little contested, indeed, and so rapid, was this revolution, that in less than three years from the time independence was declared on the plains of the Ypiranga, Brazil was acknowledged to be independent at the court of Lisbon. In the mean time the Emperor had been crowned as Dom Pedro I., and an assembly of delegates from the provinces had been convoked for the formation of a Constitution.

ARMS OF THE BRAZILIAN EMPIRE.

CHAPTER V.

THE ANDRADAS—INSTRUCTIONS OF THE EMPEROR TO THE CONSTITUENT ASSEMBLY— DOM PEDRO I. DISSOLVES THE ASSEMBLY BY FORCE—CONSTITUTION FRAMED BY A SPECIAL COMMISSION—CONSIDERATIONS OF THIS DOCUMENT—THE RULE OF DOM PEDRO I.—CAUSES OF DISSATISFACTION—THE EMPEROR ABDICATES IN FAVOR OF DOM PEDRO II.

THE new state of affairs did not, however, proceed with either smoothness or velocity. Political bitterness, jealousy, and strife were at work. The Andrada ministry* were accused of being arbitrary and tyrannical. Brazil owed her independence, and Dom Pedro I. his crown, chiefly to their exertions; yet their administrations cannot by any means be exempted from censure. Their views were certainly comprehensive, and their intentions patriotic; but their impatient and ambitious spirit rendered them, when in power, intolerant to their political opponents. They were assailed with great energy, and finally compelled to resign; but such were the tumults of the people, and the violent partisan exertions in their favor, that they were reinstated, and José Bonifacio was drawn in his carriage by the populace through the streets of Rio de Janeiro. Eight months afterward a combination of all parties

---

* José Bonifacio was prime minister, and Martin Francisco de Andrada was at the head of the Finance Department.

again effected the ejection of the brothers Andrada from the
ministry, but not from power. They became the most factious
opponents of the Emperor and of the ministry which succeeded
theirs. They were unmitigated in their attacks, both in the
Assembly and through the press.

The Constituent Assembly had done little besides wrangling.
The members were mostly men of narrow views and of little
ability; hence it was that the Andradas, by their eloquence and
knowledge of parliamentary tactics, had such power over their
minds. The Emperor, with great good sense, had, in opening the
sessions, told the Assembly that the recent "Constitutions founded
on the models of those of 1791 and 1792 had been acknowledged
as too abstract and too metaphysical for execution. This has been
proved by the example of France, and more recently by that of
Spain and Portugal." His Imperial Majesty seems to have had a
high standard of constitutional excellence, and one which we would
have deemed it difficult, and perhaps impossible, for the Brazilian
people to have reached. "We have need," he said in his address
from the throne, "of a Constitution where the powers may be so
divided and defined, that no one branch can arrogate to itself the
prerogatives of another; a Constitution which may be an insurmountable barrier against all invasion of the royal authority,
whether aristocratic or popular; which will overthrow anarchy,
and cherish the tree of liberty; beneath whose shade we shall see
the union and the independence of this Empire flourish. In a word,
a Constitution that will excite the admiration of other nations, and
even of our enemies, who will consecrate the triumph of our principles in adopting them." (From the *Falla do Throno*, 3d May,
1823.)

Notwithstanding those instructions, the Constituent Assembly
made no progress in forming a document from which such grand
results were to flow as those depicted by the Emperor. The
Andradas continued their opposition to various measures brought
forward by the Government. His Majesty was irritated by their
continual thrusts at the Portuguese incorporated in the Brazilian
army. An outrage committed by two Portuguese officers upon the
supposed author of an attack upon them was, in the excited state
of public feeling, magnified into an outrage on the nation. The

sufferer demanded justice from the House of Deputies, and the Andradas most loudly demanded vengeance on the Portuguese aggressors. The journal under their control, called the " *Tamoyo*," (from a tribe of Indians who were the bitter foes of the early Portuguese settlers,) was equally violent. It even went so far as to insinuate that if the Government did not turn aside from its antinational course, its power would be of short continuance, and, as a warning to the Emperor, the example of Charles I. of England was alluded to in no unmeaning terms.

But Dom Pedro I. was no weak and vacillating Stuart. He possessed more of the spirit of Oliver Cromwell or of the First Napoleon. The Assembly, through the three brothers, was induced to declare itself in permanent session. The Emperor, finding that they (the Andradas) still maintained their predominance, mounted on horseback, and, at the head of his cavalry, marched to the Chamber, planted his cannon before its walls, and sent up General Moraes to the Assembly to order its instantaneous dissolution.

The Assembly was broken up. The three Andradas were seized, as well as the Deputies Rocha and Montezuma, and were, without trial or examination, transported to France. Thus ended, for a brief period at least, the political career of the eloquent, patriotic, and factious Andradas.

The Emperor issued a proclamation, stating that he had taken the measures recounted above, solely with the view of avoiding anarchy; and the public were reminded that "though the Emperor had, from regard to the tranquillity of the Empire, thought fit to dissolve the said Assembly, he had in the same decree convoked another, in conformity with the acknowledged constitutional rights of his people."

A special commission of ten individuals was convened on the 26th of November, 1823, for the purpose of forming such a Constitution as might meet with the Imperial approval. The members of this commission immediately commenced their labors under the personal superintendence of D. Pedro I., who furnished them the bases of the document which he wished to be framed, and gave them forty days for the accomplishment of the object.

The ten councillors, as a body, were badly qualified for the important task before them; yet several of their number were noted for the excellence of their private characters, and two only for their erudition. One of these two, Carneiro de Campos, was fortunately intrusted with the drawing up of the Constitution, and to him it has been said Brazil is principally indebted for a number of the most liberal provisions of the code,—provisions which he insisted on introducing in opposition to the wishes of many of his colleagues.

It is evident that the drafting-committee of ten could not foresee how liberal were the provisions of this Constitution, for most of them were staunch royalists; yet various providential circumstances conduced to the production of a just and liberal instrument of government. [See Appendix B.]

Its most important features may be stated in a few words. The government of the Empire is monarchical, hereditary, constitutional, and representative. The religion of the State is the Roman Catholic, but all other denominations are tolerated. Judicial proceedings are public, and there is the right of *habeas corpus* and trial by jury. The legislative power is in the General Assembly, which answers to the Imperial Parliament of England or to the Congress of the United States. The senators are elected for life, and the representatives for four years. The presidents of the provinces are appointed by the Emperor. There is a legislative Assembly to each province for local laws, taxation, and government: thus, Brazil is a *decentralized* Empire. The senators and representatives of the General Assembly are chosen through the intervention of electors, as is the President of the United States, and the provincial legislators are elected by universal suffrage. The press is free, and there is no proscription on account of color.

The Constitution thus framed was accepted by the Emperor, and on the 25th of March, 1824, was sworn to by his Imperial Highness, and by the authorities and people throughout the Empire. It is an instrument truly remarkable, considering the source whence it emanated, and we cannot continue the subsequent history of the country without devoting to its merits a few passing reflections.

This Constitution commenced by being the most liberal of all other similar documents placed before a South American people. In its wise and tolerant notions, and in its adaptation to the nation for which it was prepared, it is second only to that which governs the

Anglo-Saxon Confederacy of North America. States and individuals may utter, in their charters of government, fine sentences in regard to equality and right; but if they fail in practicability and in securing those very elements of justice, stability, and progress, the eloquent phrases are but "as sounding brass or a tinkling cymbal." The Brazilian Constitution has, to a great extent, secured equality, justice, and consequently national prosperity. She is to-day governed by the same Constitution with which more than thirty years ago she commenced her full career as a nation. While every Spanish-American Government has been the scene of bloody revolutions,—while the civilized world has looked with horror, wonder, and pity upon the self-constituted bill of the people's rights again and again trampled under foot by turbulent faction and priestly bigotry, or by the tyranny of the most narrow-minded dictators,—the only Portuguese-American Government (though it has had its provincial revolts of a short duration) has beheld but two revolutions, and those were peaceful,—one fully in accordance with the Constitution;* the other, the proclamation of the majority of Dom Pedro II., was by suspending a single article of the Government compact.

Mexico, which, in extent of territory, population, and resources, is more properly comparable to Brazil than any other Hispano-American country, established her first Constitution only one month (February, 1824) earlier than the adoption of the Brazilian charter of government and rights. But poor Mexico has been the prey of every unscrupulous demagogue who could for the moment command the army. Her Constitution has repeatedly been overthrown; the victorious soldiery of a hardier nation placed her at the mercy of a foreign cabinet; her dominion has been despoiled; her commerce crippled and diminished by her own inertness and narrow policy; personal security and national prosperity are unknown, and her people are this day no further advanced than when the Constitution was first set aside in 1835.

Brazil, on the other hand, has been continually progressing. The head of the Empire is in the same family, and governs under

---

* The abdication of Dom Pedro I. in favor of his son, Dom Pedro II., the present Emperor.

the same Constitution that was established in 1824. Her commerce
doubles every ten years; she possesses cities lighted by gas, long
lines of steamships, and the beginnings of railways that are spreading from the sea-coast into the fertile interior; in her borders
education and general intelligence are constantly advancing.

This great contrast cannot be accounted for altogether on the
ground of the difference between the two people and between
their respective forms of government. It is doubtless true that
a Monarchy is better suited to the Latin nations than a Republic;
and it is equally apparent that there is a very great dissimilarity
between the Spaniard and his descendants, and the Portuguese and
his descendants. The Spaniard affects to despise the Portuguese,
and the latter has of late years been underrated in the eyes of the
world.* The child of Castile, take him where you will, is ambitious, chivalric, bigoted, vain, extravagant, and lazy. The son
of Lusitania is not wanting in vanity, but is more tolerant and
less turbulent than his neighbor, and is a being both economical
and industrious.

The reasons, under Providence, of the great divergence in the
results of the Brazilian and Mexican Constitution may be summed
up briefly thus :—Brazil, while providing a hereditary monarchical
head, recognised most fully the democratic element ; while acknowledging the Roman Catholic religion to be that established by the
State, she guaranteed, with the single limitations of steeples and
bells, the unrestricted right of worship to all other denominations;
she established public judicial proceedings, the *habeas corpus*, and
the right of trial by jury.

Mexico, in the formation of her Constitution, copied that of the
United States, but departed from that document, in the two most
important particulars, as widely as the oft-quoted strolling actors
deviated from the original tragedy when they advertised " Hamlet"
to be played *minus* the *rôle* of the Prince of Denmark. The Mexican
Constitution established an exclusive religion with all the rigorous
bigotry of Old Spain ; and public judicial proceedings and the intervention by juries were omitted altogether. The starting-point of

---

* " Strip a Spaniard of all his virtues, and you make a good Portuguese of
him."—SPANISH PROVERB.

Brazil and Mexico were entirely different : the former, happy in a suitable form of government and in liberal principles from the beginning, has outstripped the latter in all that constitutes true national greatness.

Brazil did not, however, attain her present proud position in South America without days of trial and hard experience. Corrupt and unprincipled men were in greater numbers than those who possessed stern and patriotic virtue. The people were ignorant and unaccustomed to self-government, and were often used by unscrupulous leaders to the advancement of their own purposes.

The administration of Dom Pedro I. continued about ten years, and, during its lapse, the country unquestionably made greater advances in intelligence than it had done in three centuries which intervened between its first discovery and the proclamation of the Portuguese Constitution in 1820. Nevertheless, this administration was not without its faults or its difficulties. Dom Pedro, although not tyrannical, was imprudent. He was energetic, but inconstant; an admirer of the representative form of government, but hesitating in its practical enforcement.

Elevated into a hero during the struggle for independence, he appears to have been guided rather by the example of other potentates than by any mature consideration of the existing state and exigencies of Brazil; and hence, perhaps, the eagerness with which he embarked in the war against Montevideo, which certainly had its origin in aggression, and which, after crippling the commerce, checking the prosperity, and exhausting the finances of the Empire, ended only in the full and unrestrained cession of the province in dispute.

It may be remarked, that the defeat of the Brazilians in the Banda Oriental, though a seeming disgrace, was one of the greatest blessings that could have been bestowed upon the Empire. It appears that that war and its disastrous results were the means of preserving Brazil from making such modifications in her Constitution as might, if effected, have terminated in the overthrow of some of her most valuable institutions. The non-success of her arms almost annihilated the thirst for military distinction which was springing up; and the energies of the rising generation were consequently turned more toward civil pursuits, from which resulted social ameliorations that tended to consolidate the well-being of the State.

In addition to the imprudence and inconstancy of the Emperor, it was said—and not without truth—that his habits were extravagant and his morals extremely defective.* And yet, the main cause of his personal unpopularity seems to have consisted in his never having known how to become the man of his people,—in his never having constituted himself entirely and truly a Brazilian.

He was often heard to express the sentiment that the only true strength of a government lay in public opinion; yet, unfortunately, he did not know how to conciliate the public opinion of the people over whom it was his destiny to reign. At the period of the Revolution, he had, under the excitements of enthusiasm, uttered sentiments calculated to flatter the nascent spirit of nationality, and his sincerity had been credited; yet his subsequent employment of a foreign force, his continued interference in the affairs of Portugal, his institution of a secret cabinet, and his appointment of naturalized Portuguese to the highest offices of the State, to the apparent exclusion of natives of the soil, had, among a jealous people, given rise to the universal impression that the monarch himself was still a Portuguese at heart.

The native Brazilians believed that they were beheld with suspicion, and hence became restive under a Government which they regarded as nurturing foreign interests and a foreign party. Opportunities for manifesting their dissatisfaction frequently occurred, and these manifestations were met by more offensive measures. At length, after fruitless efforts to suppress the rising spirit of rebellion in different parts of the Empire, Dom Pedro found himself in circumstances as painful and as humiliating as those which forced his father, Dom John VI., to retire to Portugal. Opposition which had long been covert became undisguised and relentless. The most indifferent acts of the Emperor were distorted to his prejudice, and all the irregularities of his private life were brought

---

* The older citizens of Rio de Janeiro have not yet forgotten the place that the Marchioness of Santos held in the first Emperor's affections; and his slighting treatment of his own spouse—a daughter of the high house of Hapsburg—was notorious. It has been said that, though a bad husband, he was a good father.

before the public. Individuals to whom he had been a benefactor deserted him, and, perceiving that his star was on the wane, had the baseness to contribute to his overthrow. The very army which he had raised at an immense sacrifice, which he had maintained to the great prejudice of his popularity, and on which he had unfortunately placed more reliance than upon the people, betrayed him at last.

After various popular agitations, which had the continual effect of widening the breach between the Imperial party and the patriots, the populace of Rio de Janeiro assembled in the Campo de Santa Anna on the 6th of April, 1831, and began to call out for the dismissal of the new ministry, and for the reinstatement of some individuals who had that very morning been dismissed. Dom Pedro I., on being informed of the assemblage and its objects, issued a proclamation, signed by himself and the existing ministry, assuring them that the administration was perfectly constitutional, and that its members would be governed by constitutional principles. A justice of the peace was despatched to read this to the people; yet scarcely had he concluded, when the document was torn from his hands and trampled under foot. The cry for the reinstatement of the cabinet became louder; the multitude momentarily increased in numbers; and, about six o'clock in the afternoon, three justices of the peace (in Spanish America it would have been a battalion of soldiers) were despatched to the Imperial residence to demand that the "ministry who had the confidence of the people"—as the late cabinet were designated—should be reappointed.

The Emperor listened to their requisition, but refused to accede to the request. He exclaimed, "I will do every thing for the people, but nothing by the people!"

No sooner was this answer made known in the Campo, than the most seditious cries were raised, and the troops began to assemble there for the purpose of making common cause with the multitude. Further representations were made to the Emperor, but were unavailing. He declared he would suffer death rather than consent to the dictation of the mob.

The battalion styled the Emperor's, and quartered at Boa Vista, went to join their comrades in the Campo, where they arrived about eleven o'clock in the evening; and even the Imperial guard

6

of honor, which had been summoned to the palace, followed. The populace, already congregated, began to supply themselves with arms from the adjoining barracks. The Portuguese party, in the mean time, judging themselves proscribed and abandoned, durst not even venture into the streets. The Emperor, in these trying moments, is said to have evinced a dignity and a magnanimity unknown in the days of his prosperity. On the one hand, the Empress was weeping bitterly, and apprehending the most fatal consequences; on the other, an adjutant from the combined assemblage of the troops and populace was urging him to a final answer.

Dom Pedro I. had sent for the Intendant of Police, and desired him to seek for Vergueiro, a noble patriot, who had always been a favorite of the people, and who combined moderation with sterling integrity. Vergueiro could not be found. The envoy from the troops and populace urged his Majesty to give him an immediate decision, or excesses would be committed under the idea that he (the envoy) had been either assassinated or made prisoner. The Emperor replied, with calmness and firmness, "I certainly shall not appoint the ministry which they require: my honor and the Constitution alike forbid it, and I would abdicate, or even suffer death, rather than consent to such a nomination." The adjutant started to give this reply to his general, but he was requested by Dom Pedro (who seemed to be struggling with some grand resolve) to stay for a final answer.

Nothing could be heard from Vergueiro. The populace were growing more impatient, and the Emperor was still firmer in his convictions of that which his position and the Constitution required of him in a moment so critical. But at length, like the noble stag of Landseer, singled out by the hounds, he stood alone. Deserted, harassed, irritated, and fatigued beyond description, with sadness, yet with grace, he yielded to the circumstances, and took the only measure consistent with his convictions and the dignity of his imperial office. It was two o'clock in the morning when he sat down, without asking the advice of any one, or even informing the ministry of his resolution, and wrote out his abdication in the following terms:—

"Availing myself of the right which the Constitution concedes to me, I declare that I have voluntarily abdicated in favor of my dearly-beloved and esteemed son, Dom Pedro de Alcantara.

"BOA VISTA, 7th April, 1831, tenth year ⎱
   of the Independence of the Empire." ⎰

He then rose, and, addressing himself to the messenger from the Campo, said, "Here is my abdication : may you be happy! I shall retire to Europe, and leave the country that I have loved dearly and that I still love." Tears now choked his utterance, and he hastily retired to an adjoining room, where were the Empress and the English and French ambassadors. He afterward dismissed all his ministers save one, and, in a decree which he dated the 6th of April, proceeded to nominate José Bonifacio de Andrada (who, with his brothers, had been permitted to return from exile in 1829) as the guardian to his children.

It was a striking illustration of the ingratitude with which he was treated in the hour of misfortune, that from all those upon whom he had conferred titles and riches he was obliged to turn away to the infirm old man whom, at a former period, he had rejected and cruelly wronged. Finally, after arranging his household affairs, he embarked in one of the boats of the English line-of-battle ship the Warspite, accompanied by the Empress,* and his eldest daughter, the late Queen of Portugal.

It was fortunate for Brazil that she had enjoyed that which no Spanish-American country had ever experienced,—*i.e.* a transition-state. She was not hurried from the colonial condition—an era of childhood—into self-government, which can only be the normal state of nations in their manhood. She had, as we have seen, the monarch of Portugal, with all his prestige, to be her first leader in national existence; afterward the son of the king, who, by peculiar circumstances, was for a time the idol of the people, aided Brazil in coming to a maturity far better fitted for representative-government institutions than any of the neighboring states which had achieved their independence at an earlier date. Had the transition been more violent, the permanence of such institutions would have been endangered. Dom Pedro was certainly, in the hands of God,

---

* The second Empress was the accomplished daughter of Prince Eugene Beauharnais, whom D. Pedro I. had married in 1829.

a prominent agent in giving to Brazil that form of government which this day so wisely rules the Empire.

With all his faults, D. Pedro I. was a great man, and possessed some noble aspirations, coupled with a promptness of action which will be remembered long after his errors have been forgotten. None but a great man could have returned to Europe and have fought the battle of constitutional monarchy against absolutism, as he did in the contest with his brother, Dom Miguel. His brief though chivalric and heroic devotion to the cause of civil and religious freedom in Portugal demands our highest admiration; and the successful placing of the young Queen Donna Maria upon the throne of that country gave quiet to the kingdom, and was one more triumph in Europe of the liberal over the absolute.

As time rolls on, the true merits of D. Pedro I. are more recognised by the Brazilians. Statues and public monuments are erected to his memory; and, though it may not be wholly applicable, yet there is no fulsome adulation, too common in that Southern clime, when they entitle him "*O Washington do Brazil.*"

He loved the country of his adoption; and a few days after the memorable night of his abdication, as he gazed for the last time upon the city of Rio de Janeiro, the magnificent bay, and the lofty Organ Mountains, he poured from a full heart the following touching farewell to his son, Dom Pedro II., in which not only is parental tenderness manifest, but a deep solicitude for the land whose destiny at one time seemed so closely linked with his own:—

"My beloved son and my Emperor, very agreeable are the lines which you wrote me. 1 was scarcely able to read them, because copious tears impeded my sight. Now that I am more composed, 1 write this to thank you for your letter, and to declare that, as long as life shall last, affection for you will never be extinguished in my lacerated heart.

"To leave children, country, and friends is the greatest possible sacrifice; but to bear away honor unsullied,—there can be no greater glory. Ever remember your father; love your country and my country; follow the counsel of those who have the care of your education; and rest assured that the world will admire you, and that I will be filled with gladness at having a son so worthy of the land of his birth. I retire to Europe: it is necessary for the tranquillity of Brazil, and that God may cause her to reach that degree of prosperity for which she is eminently capable.

"Adieu, my very dear son! Receive the blessing of your affectionate father, who departs without the hope of ever seeing you again.   D. PEDRO DE ALCANTARA.

"On board the Warspite frigate. }
      April 12, 1831." }

On the following day D. Pedro I. went on board the English corvette Volage. Before nightfall the Pão de Assucar was cleared, and the ex-Emperor left Brazil forever.

Having thus briefly narrated the history of the Empire to the abdication of the first Emperor, we will again turn our attention to Rio de Janeiro, where most of the preceding events occurred. The establishment of the regency, and the various changes and progress under the new monarch, D. Pedro II., will be found in Chapter XII.

CHAPTER VI.

THE PRAIA DO FLAMENGO—THE THREE-MAN BEETLE—SPLENDID VIEWS—THE MAN
WHO CUT DOWN A PALM-TREE—MOONLIGHT—RIO "TIGERS"—THE BATHERS—
GLORIA HILL—EVENING SCENE—THE CHURCH—MARRIAGE OF CHRISTIANITY AND
HEATHENISM—A SERMON IN HONOR OF OUR LADY—FESTA DA GLORIA—THE
LARANGEIRAS—ASCENT OF THE CORCOVADO—THE SUGAR-LOAF.

My residence at Rio de Janeiro was on the Praia do Flamengo,—a beach so named from its having been in early days
frequented by this beautiful bird. Let the reader imagine the
beaches of Newport, Rhode Island, or of the battle-renowned
Hastings, transferred to the borders of London or New York, so
that, by taking omnibus at Charing Cross or Union Square, in
fifteen minutes he will be on the hard white sands and in the presence of the huge ocean-waves, and he will have an idea of Praia
do Flamengo. Entering one of the *Gondolas Fluminenses* at the
Palace Square, we rattle through various streets until we arrive at
the foot of the Gloria, where, if we wish an up-hill ramble, we
descend from our vehicle and pass over the picturesque eminence,
and are soon cooled by the full blowing sea-breeze; or, if we prefer
a more level promenade, we leave our conveyance at the Rua do
Principe. The noisy wheels, and the equally noisy tongues, have
hitherto prevented any other sounds from occupying our attention;
but now the majestic thunder of the dashing waves breaks upon
our ear. The eye is startled by the foam-crested monsters as they
rear up in their strength and seem ready to devour the whole
mansion-lined shore in their furious rage. The very ground
quakes beneath us, and the air is tremulous with the powerful con-
cussion. But no danger is to be apprehended. The coast, a few
feet from the sands, is rock-bound, and along the whole beach public
and private enterprise have erected strong walls of heavy stone.
Sometimes, however, old Neptune has asserted his rights with

86

such tremendous energy, that masses of rock, weighing tons, have been wrested from their fastenings. In May, 1853, a storm prevailed for several days, and a strong wind blew in the waves of the ocean with great directness against the protecting walls, and the strife was one of the fiercest that I have ever witnessed in contending nature. As they struck the parapet they dashed eighty feet in height, thus showering and flooding the gayly-painted residences, and at the same time, in their retreat, undermining the land-side of the wall, so that for hundreds of feet between the Rua da Princeza and the Rua do Principe the municipality had a heavy job for some favorite contractor. (The paving of the streets was a never-failing source of amusement to me during my first year at Rio. Look at the pavers in the Rua S. José. The paving-ram is the "three-man beetle" of Shakspeare. A trio of slaves are called to their work by a rapid solo executed with a hammer upon an iron bar. The three seize the ram: one—the *maestro*, distinguished by a hat—wails forth a ditty, which the others join in chorus, at the same time lifting the beetle from the ground and bringing it down with a heavy blow. A rest of a few moments occurs, and then the ditty, chorus, and

THE THREE-MAN BEETLE.

thump are resumed: but, as may be imagined, the streets of Rio were by no means rapidly paved.) The damage done to the Praia do Flamengo required more than one year for reparation. A battle between the sea and the land like that of 1853 does not often occur: the rule is peacefulness and amiability, for the huge waves themselves, that seem to foam so angrily, are only joyous in their giant sport, and, once touching the myriad sands, kiss them in their gentlest mood, and hasten silently back to their boisterous companions.

The front of my house looked over the bay to Jurujuba and Praia Grande, and also commanded a view of the long Flamengo Beach, the Babylonia Signal, the lofty Sugar-Loaf, and the entrance to the harbor. Far up the bay were verdant isles, and beyond all towered the lofty Organ Mountains, sometimes gleaming in sunshine, and sometimes half veiled in mist, but always the grandest feature in the landscape. From my back-windows, on my right, I could see the precipitous southern side of the Gloria, and on my left, beyond the red-tiled roofs, upreared the tall Corcovado, whose Rio face is covered with forests. Beneath me was the garden of my neighbor, a plodding Portuguese from Braga. This individual was originally one of those industrious ignorant poor from the mother-country, who in Brazil and elsewhere, by dint of regularity and economy, acquire property, but rarely taste. He had a beautiful stately palm-tree in the centre of his garden. Night after night have I listened to the music of the cool land-breeze as it played through the long, feathery leaves. The sight of it was refreshing when the rays of the noonday sun made the more distant landscape quiver. It was a "thing of beauty," and "a joy," but not "forever." Early one morning I heard the click of an axe; and, rushing to my window, I beheld Sr. M. directing a black, who, with sturdy blows, buried the sharp instrument deep into the trunk of the noble tree, and each succeeding stroke made the graceful summit and the clustering fruit piteously tremble.

> "The ruthless axe that hew'd its silvered trunk
> Cut loose the ties that, tendril-like, had bound
> My love unto the tree; and when it sunk,
> My heart sank with it to the grou.d."
>
> "Woodman, spare that tree,"

sung by the voice of an angel, would not have stayed the work of destruction; and thus the prince of the tropic forest fell by ignominious hands. Sr. M., the regicide, went that morning to his *toucinho* (bacon) and *carne secca* establishment in the Rua do Rosario,

THE GLORIA HILL FROM THE PASSEIO PUBLICO.

congratulating himself, as he stuffed his nostrils with *areia preta*,* that he had gained a few more feet of sunshine for his cabbage-bed, by cutting down a palm-tree that a century would not reproduce.

At evening, the view from the balcony in front of my residence was most charming. On a bright night the heavens were illumined by the Southern Cross, by Orion, and other stellar brilliants; and sometimes, when clouds obscured the lesser celestial lights, the bosom of the bay seemed like a sea of fire. But the most glorious nocturnal sight was to watch the full moon rise above the palm-crowned mountains beyond the Bay of San Francisco Xavier. Mild rays of light would herald the approaching queen, and soon her full round form, emerging, threw upon the distant waters of Jurujuba her silver sheen, while the dashing waves that burst along the whole length of the Praia do Flamengo seemed gorgeous wreaths of retreating moonlight. We are in the height of enjoyment. Perhaps we murmur

" On such a night as this," &c.,

and speak something about chaste Dian "moving in meditation, fancy free," when we are suddenly brought to the sad realization that we are in a sublunary sphere. We rush from the balcony spasmodically, and instantaneously snatch cologne-bottles, *bouquet*, ammonia, or any thing that will relieve our olfactories. The *tigers*† also have opportunities for watching the moon rise. Eight o'clock has arrived, and these odoriferous—not to say savage—beasts come stealthily down the Rua do Principe, and for the next two hours make night hideous, not with yells, but with smells which have certainly been expatriated from Arabia Infelix.

A curious story is generally told the newly-arrived stranger at Rio, of a Fluminensian who on a visit to Paris became exceedingly ill. Every restorative was applied in vain, until a French physician well acquainted with the capital of Brazil was called in, and decided at once that it was impossible to hope for the recovery of the

---

* Literally, *black sand*,—a favorite snuff made in Bahia.

† The sewerage of Rio was formerly very defective, and slaves, nicknamed " tigers," conveyed each night to the water's edge the accumulated offal of the city. and the next tide swept it out to sea.

patient unless he could breathe again his native air; but, as he could not return to Rio, the physician instantly prescribed that there should be concocted in the sick-chamber a compound of the most "villanous smells." To make a long story short, the invalid recovered!

But at the date of writing this nuisance is much more tolerable than formerly, for hermetically-sealed casks have been introduced, and carts during daylight collect them, and their contents are conveyed to some very distant point from the city. Soon Rio will have a good system of sewerage, the plans for which were laid before the Minister of the Empire in 1854. When this is accomplished, no tropic city will surpass it as an abode both healthful and agreeable. 1865, "Rio City Improvements Company" is doing this work.

The Praia do Flamengo, saving this drawback when the wind is in a wrong direction, is one of the most delightful suburbs for the residence of a foreigner. One hour after the tigers have finished their labors, the atmosphere is as free from any thing disagreeable as if naught but the fragrance of orange-flowers had been wafted from the Gloria and the neighboring gardens; and the morning light shines upon a pure white beach.

For five months in the year the Praia do Flamengo is the favorite resort of bathers of both sexes. During the bathing-season, (from November to March,) a lively scene is witnessed every morning. Before the sun is above the mountains a stream of men, women, and children pour down to enjoy a bath in the clear salt water. The ladies who come from a distance are attended by slaves, who bring tents and spread them on the beach for the senhoras, who soon put on their bathing-robes and loose their long black tresses. Men and women, hand in hand, enter the cool, sparkling element, and thus those not skilled in natation resist the force of the huge waves which come toppling in. The senhoras are neatly dressed, in robes made of some dark stuff; but there is not as much coquetry as in a French watering-place, where the ladies study the becoming for the sea as well as for the ball-room. The gentlemen are required by the police-regulations to be decently clad, which still does not impede those who prefer a swimming-bath to the *douche* of the billows.

It is a merry sight to behold Brazilian girls and boys evincing for once some activity,—running on the sand, and screaming with pleasure whenever a heavier wave than before has rolled over a party and sends them reeling to the beach. The prostrate bathers drive their feet convulsively into the sand to prevent being carried back by the receding breakers. Now and then some mischief-makers shout "Shark! shark!" and away dash the senhoras to the shore, to be laughed at by the urchins who raised the cry. There are some traditionary tales about these rough-skinned cannibals, but I never heard a well-authenticated instance of a repast furnished by the bathers of Praia do Flamengo to the dreaded " wolf of the seas."

By seven o'clock the sun is high, and all the busy white throng have departed. Here and there, however, may be seen a curly head popping up and down among the waves, its woolly covering defying the fear of *coup de soleil*. The negresses that accompany the ladies generally enter the water at the same time as their mistresses. On moonlight nights the sea is alive with black specks, which are the *capita* of the slaves in the vicinity, who splash and scream and laugh to their hearts' content. They all swim remarkably well, and it is pleasant to hear their cheerful voices sounding as merrily as if they knew not a sorrow.

The people of Rio are fond of bathing, and on this account are called *cariocas*, which some translate "ducks." Many walk miles to enjoy it. There is a floating bath in the harbor, not far from Hotel Pharoux, for those whose courage is great enough to brave the element which is there called sea-water, but which a truthful narrator, previous to the improved sewerage, would stigmatize by another name.

Nor are the bipeds the only animals that derive benefit from the ablutions on Praia do Flamengo. The horses and mules have allotted to them a certain portion of the beach, where at an early hour they are bathed and brushed. It is a comfort to know that the poor creatures have this chance of cleanliness; otherwise they would suffer greatly from the laziness of their keepers. Gentlemen who care for their horses endeavor to procure English grooms, for a black is proverbially a bad care-taker for any animal. The beautiful horses imported at great expense from the Cape of Good Hope are soon destroyed under the hands of the negroes. It is considered that the climate of Brazil is unfavorable to them, and one can hardly believe that these pampered, delicate animals are of the same race, half English, half Arabian, which at the Cape of Good Hope will endure a journey of sixty or seventy miles a day without other refreshment than a feed of oats and a roll on the sand.* For all useful purposes the horses of the country are better, but they are not so swift or graceful as the imported animals.

It was but a few paces from my front-door to the southern entrance of the Gloria. Here, when the surf was not too high, boats could land, and often were our evenings enlivened by the presence of some of the intelligent officers from the men-of-war whose station was beyond the Fortress Villegagnon.

Once within the gateway at the foot of the hill, we behold a narrow, level strip of ground, occupied by one or two secluded residences and a beautiful private flower-garden. The base of the black rock which rises perpendicularly on the side facing the sea is hidden by large waving banana-trees and overhanging creepers. The diversified summit of the hill is checkered with every evidence of city and country agreeably blended. Narrow paths wind around the hill at different altitudes, leading to the many beautiful residences and gardens by which it is covered to the summit. On either side of the paths are seen dense hedges of flowering mimosas, lofty palms, and the singular cashew-tree, with its bottle-shaped, refreshing fruit, and occasionally other large trees, hung with splendid parasites, while throughout the scene there prevails a quiet and a coolness which could scarcely be anticipated within the precincts of a city situated beneath a tropical sun.

The prettiest residence on the hill was that of the British Consul, Mr. John J. C. Westwood,—a gentleman whom I always found most ready to co-operate in any work of charity or benevolence brought to his notice. In 1864 Mr. Westwood died.

Among the dwellers on the Gloria were two families, (English and Swiss,) who in their tastes and accomplishments were far beyond the mere shopkeeping class so often found in a foreign land. In

---

* When Napoleon was at St. Helena he was supplied with these horses, and their fire exactly suited his style of riding. The old English generals whose duty it was to accompany their "perverse prisoner" had often reason to complain of the pace of the Cape horses.

their pleasant society one was often compensated for the home-circle left far over the billow. The Englishman was an amateur-naturalist of the very first ability, while both families possessed the best periodical and standard literature of England and of France. After the fatigues of the day it was a delightful recrea-tion to spend the even-ing amid such compa-nions and surrounded by such glorious sce-nery. On many moon-light evenings I could enter into the feelings entertained by Dr. Kid-der years before, and, as he expressed it, could realize "the en-chantment of an even-ing-scene so felicitous-ly described by Von Martius."

"A delicate transpa-rent mist hangs over the country; the moon shines brightly amid heavy and singularly-grouped clouds. The outlines of the objects illuminated by it are clear and well defined,

FRUIT AND NUT OF THE CASHEW-TREE.

while a magic twilight seems to remove from the eye those which are in the shade. Scarce a breath of air is stirring, and the neigh-boring mimosas, that have folded up their leaves to sleep, stand motionless beside the dark crowns of the mangueiras, the jaca-tree, and the ethereal jambos. Sometimes a sudden wind arises, and the juiceless leaves of the cashew rustle; the richly-flowered grumijama and pitanga let drop a fragrant shower of snow-white blossoms; the crowns of the majestic palms wave slowly above the silent roof which they overhang like a symbol of peace and tranquillity. Shrill cries of the cicada, the grasshopper, and tree-frog make an incessant hum, and produce by their monotony a pleasing melancholy. At intervals different balsamic odors fill the air, and flowers, alternately unfolding their leaves to the night, delight the senses with their perfume,—now the bowers of paullinias, or the neighboring orange-grove,—then the thick tufts of the eupatoria, or the bunches of the flowering palms, suddenly bursting, disclose their blossoms, and thus maintain a constant succession of fragrance; while the silent vegetable world, illuminated by swarms of fire-flies as by a thousand moving stars, charms the night by its delicious odors. Brilliant lightnings play incessantly in the horizon and elevate the mind in joyful admiration to the stars, which, glowing in solemn silence in the firmament, fill the soul with a presentiment of still sublimer wonders."

Often, while enjoying the scene which the great German naturalist has so eloquently depicted, I was called away from my meditations by the clangor of the bells in the tower of the Gloria Church. Though the worship of Him who made the beautiful nature around me should be ever more elevating than the mere contemplation of the grand and wonderful in the material world, yet the sound of those bells filled me with painful reflections. Whenever I entered that pretty church of Nossa Senhora da Gloria, whenever I gazed upon the kneeling throng and on the evidences of a corrupted Christianity, I could not believe that God was worshipped "in spirit and in truth."

In the interior, the octagonal walls are lined for several feet with large Dutch tiles, representing landscapes and scenes connected with classic heathenism. Actæon and his dogs start the timid deer, or pursue the flying hare; Cupid, too, with arrows in hand, joins the sport. Over the chief altar Nossa Senhora da Gloria, robed like a fashionable lady in silks and laces, looks down upon the scene beneath. She has received many jewels from her devotees, and no gem is esteemed too costly to win her favor. She wears brilliant finger-rings, and diamond buttons fasten the sleeves of her gown. Her bosom and ears are graced with diamond necklaces and rich pendants. An immense diamond brooch sparkles on her breast: this was vowed to the Virgin by Donna Francisca, the consort of Prince de Joinville, in prospective compensation for the restoration of Her Highness's health.   The flowing curls that cluster around Our Lady's brow are also offerings, clipped by some anxious mother from the glossy locks of a favorite child.*

Let us enter the vestry in the rear of the church.   Here we behold a few specimens of what may be seen in every church in Brazil, and which was formerly to be witnessed in almost every heathen temple in old Italia before the days of Constantine the Great.   In the many particulars in which we can trace with certainty the marriage between Christianity and heathenism, none is more curious than the system of *ex votos*.   The ancients who were affected with ophthalmia, rheumatism, boils, defective limbs, &c. &c., prayed to their gods and goddesses for recovery, and at the same time offered on the shrine of the favorite divinity, or suspended near the altar, votive *tablets*, upon which were inscribed a description of the disease and the name of the invalid.   Grateful acknowldgements and miraculous cures were also thus made public for the edification of the faithful worshippers and for the confusion of the incredulous.   Thus, also, in Brazil every church is filled with votive tablets, telling of wonderful cures by Nossa Senhora and innumerable saints with very hard names.

The pious pagans, however, did not limit themselves to mere written thanksgivings and descriptions of the parts affected, but hung up in their temples the handiwork of their mechanicians and artists,—representations in painting and in sculpture of hands, legs, eyes, and other portions of the afflicted body.   In the Gloria Church also may be seen any quantity of wax models of arms, feet, eyes, noses, breasts, &c. &c.   Where the disease is internal, and the seat of pain cannot well be modelled, the subject is gene-

---

* "This wooden *deosa* has a splendid head of hair.   It is the last of a series of rapes of locks committed on her account.   When the brother of Sr. P. L——a, a young gentleman of my acquaintance, was seven years old, his hair reached more than half-way down his back.   His mother, having great devotion to Nossa Senhora, sheared off the silken spoils, and offered them as an act of faith to her, little thinking how literally she was copying the practice of heathen dames.   The locks were sent to a French hairdresser, who wrought them into a wig.   It was then brought to the church and laid in due form before Our Lady, when the priest reverently removed her old wig and covered her with the flowing tresses of the Larangeiras Absalom."—*Ewbank's Sketches of Life in Brazil.*

ralized by representing a bedridden patient: peril by sea is represented by a shipwreck. All proclaim one story,—viz.: the miraculous cure wrought by Nossa Senhora and other saints, through the *ex voto* offering.

We have very early instances of the same mode of procedure among the heathen. The lords of the Philistines, who had seized in battle the ark of the Covenant, were with their people smitten; and, when returning the ark to the children of Israel, the pagan Philistines made golden *ex votos* to accompany their dreadful captive: (1 Sam. vi. 4.)

Mr. Ewbank, who appears to have devoted much attention to comparative archæology and mythology, makes the following quotation from Tavernier, one of the early Roman Catholic travellers in India:—"When a pilgrim goes to a pagod for the cure of disease, he takes with him a figure of the member affected, made of gold, silver, or copper, and offers it to his god." In the second volume of Montfaucon (also a Roman Catholic writer) there is a long account of *ex votos*, "some of which were offered to Neptune for safe voyages, Serapis for health, Juno Lucina for children and happy deliveries: pictures of sick patients in bed, and eyes, heads, limbs, and tablets without number, were offered to Esculapius and other popular medical saints among the heathen."

This sad spectacle of modern heathenism at Rio de Janeiro is somewhat ameliorated by the fact that, whenever the *ex votos* are found in a church consecrated to Nossa Senhora or to some saint, the offerings are mostly brown and dusty with age. Occasionally a fresh pair of eyes or breasts are to be seen, but new wax models are less frequent in the capital than formerly. There must, however, be a demand for them from some portion of the Empire; for one-third of the wax and tallow chandlers (where these objects are obtained) at Rio have an *ex voto* branch in their manufactories. At Tijuca, Mr. M., a planter, informed me that he had just seen one of his neighbors whose arm had been so disabled that its use was lost, until he was advised by some one of the living "saints" to go to a chandlery and purchase a wax model of his unruly member to offer to the Virgin. Suffice to say the arm was completely restored.

On the Sabbath I often passed over the Gloria Hill on my return from the shipping or from the hospitals, where I had been holding service or visiting the sick.   During a festival I mounted the hill as usual, and as I walked beneath the broad platform upon which the church stands, I heard strains of music that were most unlike the solemn chants and the grand anthems of the Romish communion.   They were *polkas* and *dances*, performed by some military band that had been hired for the occasion!   I have recently been informed that this abuse, as well as some others, has been remedied through the direct interposition of the Emperor.

Dr. Kidder thus gives an account of some of the religious exercises at the Gloria, which is applicable to Brazilian church-services in general:—

" Preaching is not known among the weekly services of the church; but I twice listened to sermons delivered here on special occasions.   A small elevated pulpit is seen on the eastern side of the edifice, and is entered from a hall between the outer and inner walls of the building.   In this, at one of the services which occurred during Lent, the preacher made his appearance after mass was over.   The people at once faced round to the left from the principal altar, where their attention had been previously directed.   The harangue was passionately fervid.   In the midst of it the speaker paused, and, elevating in his hand a small wooden crucifix, fell on his knees, and began praying to it as his Lord and Master.   The people, most of whom sat in rows upon the floor, sprinkled with leaves, bowed down their heads, and seemed to join him in his devotions.   He then proceeded, and, when the sermon was ended, all fell to beating their breasts, as if in imitation of the publican of old.

" In the second instance, the discourse was at the annual festa of Our Lady of the Gloria, and was entirely eulogistic of her character.   One of the most popular preachers had been procured, and he seemed quite conscious of having a theme which gave him unlimited scope.   He dealt in nothing less than superlatives:—
'The glories of the Most Holy Virgin were not to be compared with those of creatures, but only with those of the Creator.'
'She did every thing which Christ did but to die with him.'
'Jesus Christ was independent of the Father, but not of his mother.'   Such sentiments, rhapsodically strung together, left no

7

place for the mention of repentance toward God or faith toward the Lord Jesus Christ throughout the whole sermon."

In 1852, on the occasion of a very solemn festival in honor of Our Lady, one of the most eloquent *padres* of Rio was called upon to pronounce the discourse in the Church of Our Lady of Mount Carmel, which adjoins the Imperial Chapel. In the evening of the day referred to, a Roman Catholic gentleman gave me an account of the sermon, one sentence of which I translate for the benefit of the reader :—"The magi of the East and the kings of the Orient came on painful journeys from distant lands, and, prostrating themselves at the feet of Nossa Senhora, offered her their crowns for the bestowment of her hand; but she rejected them all, and gave it to the obscure, the humble but pious St. Joseph !"

During a festival, the faithful (and others, for that matter) can obtain any amount of pious merchandise, in the shape of *medidas* and *bentinhos*,—pictures, images, and medals of saints and of the Pope, &c. &c. These are " exchanged"—never sold—in the church, and fetch round prices. A *medida* is a ribbon cut the exact height of the presiding Lady or saint of the place of worship. These, worn next to the skin, cure all manner of diseases, and gratify the various desires of the happy purchasers. There are certain colors esteemed appropriate to different Nossas Senhoras; and once I ascertained the important fact, that, when some pious *Fluminense* has made a vow to Nossa Senhora, great care must be taken not to permit the wrong color to be used. A lady-member of my family, wishing to make a small present to one of her friends, —a young Roman Catholic mother,—sent a neat pink dress for the little one; but the package was soon returned, with many regrets that the kind offering could not be received, for a vow was upon the mother which had particular reference to her child. She had vowed to a Nossa Senhora (whose favorite colors were like the driven snow and the heavens above) that if her boy recovered from his sickness he should be clothed in nothing but white and blue for the next six months ! At the end of that time, it was added, the present could be accepted.

Bentinhos are two little silken pads with painted figures of Our Lady, &c. upon them. These are worn next to the skin, in pairs, being attached by ribbons, one *bentinho* resting upon the bosom and the othei upon the back. These are most efficacious for protecting the wearer from invisible foes both before and behind.

I visited the Gloria Church during one of these festivals, and the "exchange" of pictures and *medidas* was immense. The price, however, was not always paid in money. I found that wax candles offered to the Virgin were esteemed equal to copper or silver coin. The heat and crowd of the church on this occasion were such that I sought the esplanade in front; and the contrast of the cool night-air and the sweet odors that wafted up from the gardens beneath was as agreeable as refreshing.

The multitude, I soon ascertained, were not confined to the church. Groups were collected around the fountain, and thousands were congregated in the ascent called the *Ladeira da Gloria*, or whiling away their time by eating *doces*, smoking, and conversing in the *Largo*. They were awaiting the fireworks which were to close the festival. The Brazilians are exceedingly fond of pyrotechny, and every festival begins and ends with a display of rockets and wheels. The grand finale surpasses any thing in this line that is ever witnessed in North America; and I doubt if there is a single country in the world, except China, where pyrotechny is so splendid and varied as in Brazil. Not only are there wheels, cones, suns, moons, stars, triangles, polygons, vases, baskets, arches with letters and the usual devices known among us, but, elevated upon high poles, are human figures as large as life, representing wood-sawyers, rope-dancers, knife-grinders, ballet-girls, and whatever vocation of life calls for especial activity. By ingenious mechanism these effigies go through their various parts with remarkable and lifelike celerity. There is nothing *gauche*. The figures are well dressed, even to the gloves of the represented ladies. The wood-sawyer makes the sparks fly, and the knife-grinder whirls a wheel that sends forth a perfect "glory" of scintillations!

There is no *festa* throughout the year that is more enjoyed by the pleasure-loving Fluminenses than that of Nossa Senhora da Gloria. The evening before, the usual number of rockets are sent up,—probably to arouse the attention of the Virgin to the honor that is about to be paid her on the following day, lest, in the multiplicity of her cares, she should forget the approach of this anniversary; for she must have a very wonderful memory if she call to mind each fête-day at which her especial company is requested, seeing that every fourth church in Rio is dedicated to a Nossa Senhora of some kind.

Early on the morning of this festival, the approach to the white temple is crowded with devotees in their gayest attire; for there is nothing in this celebration that requires the usual sombre black. The butterflies themselves, and the golden-breasted humming-birds that flit among the opening jessamines and roses around, are not more brilliant than the senhoras and senhoritas of all ages who flutter about, robed in the brightest colors of the rainbow, and with their long black tresses elaborately dressed and adorned with natural flowers, among which the carnation is pre-eminent. They enter the church to obtain the benefit of the *mass;* and happy they who have strength and lungs and nerve enough to force a way up to the altar through the crowds whom nature has clad in perpetual mourning. Once arrived at this desired spot, they squat upon the floor, and, after saying their prayers and hearing mass, they amuse themselves with chatting to the circle of beaux who, on such occasions, are always in close attendance upon the fair objects of *their* adoration. For be it remarked that most of the praying, as in France, is done by the women; and probably for that reason each man is anxious to secure an interest in the affections of some fair devotee, in order that she may supply his own lack of zeal.

After patiently displaying their charms and their diamonds for some hours, a thrill of excitement passes through the throng, and salvos of artillery announce the approach of the Imperial party, who, when the weather permits, leave their carriages at the foot of the hill, and slowly ascend the steep path that leads to the church. This has been previously strewn with flowers and wild-cinnamon-leaves.

On some occasions, troups of young girls in white, from the different boarding-schools, are in waiting at the top, to kiss the hands of their Majesties. This is the prettiest part of the exhibition,— the Emperor, with his stately form, and the Empress, with her good-humored smile, passing slowly through the lines of bright-eyed girls who are not without a slight idea of their own prominent part in the graceful group.

After the ceremonial in the chapel, the Imperial party descends to the house of Senhor Bahia, a rich Brazilian banker, who has a fine house hard by, where a splendid collation is prepared, and the evening is terminated by fireworks and a ball. The pyrotechnic display is on the road opposite his house; and woe betide any unfortunate wight who would induce a spirited horse to pass that way. There is no other road into the city from Botafogo; so that he may as well take a philosophical resolution, and enjoy, as best he may, the Catherine wheels and the fiery maidens pirouetting in the midst of surrounding sparks.

A distinguishing feature of these gatherings is, that, amid all the thousands present, no scene of rudeness or quarrel is ever witnessed. Perfect good-nature reigns around; and if, in the inevitable pressure, any person is trodden upon or jostled, an instant apology is made, with the hat removed from the head. As water is the only beverage, there is nothing to inflame the bad passions of the multitude. The slaves are not merely respectful in their manners, but evince a joyous sense of liberty for the day; and they ambitiously seek the best places for sight-seeing, which their less active masters in vain wish to attain.

At midnight all is over, and the quiet stars shine down upon the church-crowned and verdure-robed Gloria.

When we descend the Ladeira da Glória and turn to our left, we are in a finely-paved—and in some places macadamized—thoroughfare called the Catete, a wide and important street, leading from the city to Botafogo. About half-way between the town and the last-mentioned suburb, we enter the Largo Machado, which is the commencement of the Larangeiras, or the valley of orange-groves. There were formerly many trees of the *Laranga da terra*,* or native orange, in this lovely spot; and, although the most of them have disappeared, their places have been filled with their sweeter relatives, the *Laranga selecta*, and the night-air is laden with the rich perfume of their flowers. Some of the prettiest gardens—which,

---

* Gardner is of the opinion that the *Laranga da terra*, or bitter orange, is not indigenous.

instead of thick stone walls, are surrounded by open iron railing —and the most beautiful residences in Rio nestle in this quiet valley.

A shallow but limpid brook gurgles along a wide and deep ravine, lying between two precipitous spurs of the Corcovado Mountain. Passing up its banks, you see scores of *lavandeiras*, or washerwomen, standing in the stream and beating their clothes upon the boulders of rock which lie scattered along the bottom. Many of these washerwomen go from the city early in the morning, carrying their huge bundles of soiled linen on their heads, and at evening return with them, purified in the stream and bleached in the sun. Fires are smoking in various places, where they cook their meals; and groups of infant children are seen playing around, some of whom are large enough to have toddled after their mothers; but most of them have been carried there on the backs of the heavily-burdened servants.

LAVANDEIRAS.

Female slaves, of every occupation, may be seen carrying about their children as on page 167; but the *lavandeiras* no longer work in a semi-nude state.

One is reminded by their appearance of the North American Indian pappoose riding on the mother's back; but the different methods of fastening the respective infants in permanent positions produce very different effects. The straight board on which the young Indian is lashed gives him his proverbially-erect form; but the curved posture in which the young negro's legs are bound around the sides of the mother often entails upon him crooked limbs for life.

Up the valley of the Larangeiras is a mineral spring, which at certain seasons of the year is much frequented. It is denominated *Agoa Ferrea,*—a name indicating the chalybeate properties of the water. Near this locality you may enter the road which leads up the Corcovado.

An excursion to the summit of this mountain is one of the first that should be made by every visitor to Rio. You may ascend on horseback within a short distance of the summit; and the jaunt should be commenced early in the morning, while the air is cool and balmy, and while the dew yet sparkles on the foliage. The inclination is not very steep, although the path is narrow and uneven, having been worn by descending rains. The greater part of the mountain is covered with a dense forest, which varies in character with the altitude, but everywhere abounds in the most rare and luxurious plants. Toward the summit large trees become rare, while bamboos and ferns are more numerous. Flowering shrubs and parasites extend the whole way.

I once made the excursion in company with a few friends. Our horses were left at a rancho not far from the summit, and a few minutes' walk brought us through the thicket. Above this the rocks are covered with only a thin soil, and here and there a shrub nestling in the crevices. What appears like a point from below is in reality a bare rock, of sufficient dimensions to admit of fifty persons standing on it to enjoy the view at once, although its sides, save that from which it is reached, are extremely precipitous. In order to protect persons against accidents, iron posts have been inserted, and railings of the same material extend around the edge of the rock. This has been done at the expense of the Government. If we except this slight indication of art, all around exhibits the wildness and sublimity of nature.

The elevation of the mountain—twenty-three hundred and six feet—is just sufficient to give a clear bird's-eye view of one of the richest and most extensive prospects the human eye ever beheld. The harbor and its islands; the forts, and the shipping of the bay; the whole city, from S. Christovão to Botafogo; the botanical garden, the Lagoa das Freitas, the Tijuca, the Gavia, and the

Sugar-Loaf Mountains, the islands outside the harbor, the wide-rolling ocean on the one hand and the measureless circle of mountains and shores on the other, were all expanded around and beneath us. The atmosphere was beautifully transparent, and I gazed and gazed with increasing interest upon the lovely, the magnificent panorama.

From the sides of this mountain various small streamlets flow toward the Larangeiras. By means of artificial channels, these are thrown together to supply the great aqueduct. In descending, we followed this remarkable watercourse until we entered the city, at the grand archway leading from the Hill of Santa Theresa to that of San Antonio, as depicted on page 63. Nor is this section of the route less interesting to those fond of nature. From time to time negroes are met, waving their nets in chase of the gorgeous butter-flies and other insects which may be seen fluttering across the path and nestling in the surrounding flowers and foliage.

Many slaves were formerly trained from early life to collect and preserve specimens in entomology and botany, and, by following this as a constant business, gathered immense collections. These are favorite haunts for amateur naturalists, who, if imbued with the characteristic enthusiasm of their calling, may still find them as interesting as did Von Spix and Von Martius, whose learned works upon the natural history of Brazil may be compared with those of Humboldt and Bonpland in Mexico and Colombia.

The aqueduct is a vaulted channel of mason-work, passing some-times above and sometimes beneath the surface of the ground, with a gentle declivity, and air-holes at given distances. The views to be enjoyed along the line of this aqueduct are, beyond measure, interesting and varied. Now you look down at your right upon the valley of the Larangeiras, the Largo do Machado, the Catete, the mouth of the harbor, and the ocean; anon, verging toward the other declivity of the hill, you may survey the Campo St. Anna, the Cidade Nova, the splendid suburb of Engenho Velho, and, in the distance, the upper extremity of the bay, surrounded by moun-tains and dotted by islands. At length, just above the Convent of Santa Theresa, you will pause to contemplate a fine view of the town. But for the Hill of S. Antonio and the Morro do Castello the greater portion of the city would here be seen at once. The glimpse, however, that is perceptible between these eminences is perhaps sufficient, and the eye rests with peculiar pleasure upon this unusually-happy combination of the objects of nature and of art.

Probably no city in the world can compare with Rio de Janeiro in the variety of sublime and interesting scenery in its immediate vicinity. The semicircular Bay of Botafogo and the group of mountains surrounding it form one of the most picturesque views ever beheld. We are on the Corcovado; before us stands the far-famed Sugar-Loaf; and far behind us appears an immense truncated cone of granite. When seen at a distance, this mountain is thought to resemble the foretopsail of a vessel, and hence its name, the *Gavia*. Between this and the Sugar-Loaf remains a group of three, so much resembling each other as to justify the name of *Dous Irmãos*, or Two Brothers. The head of one of the brothers stretches above his juniors, and also looks proudly down upon the ocean which laves his feet. At the base of the Sugar-Loaf is Praia Vermelha, a fertile beach, named from the reddish color of the soil. It extends to the fortress of S. João on the right, and to that of Praia Vermelha on the left, of the Sugar-Loaf. The latter is a prominent station for new recruits to the army; and many are the poor Indians from the Upper Amazon who have here been drilled to the use of arms. This also was the scene of a bloody revolt of the German soldiery in the time of the First Emperor.

The beach of the ocean outside the Sugar-Loaf is called Copa Cabana. A few scattered huts of fishermen and a few ancient dwellings belonging to proprietors of the land accommodate all the present inhabitants of this locality. Once it used to be far more populous, according to the recollections of Senhor Domingos Lopez,—a garrulous sexagenarian with whom Dr. Kidder became acquainted on one of his visits there, and who detailed to him the monstrous changes that had transpired since his boyhood, when the site of S. Francisco de Paula was a frog-pond, and all the city beyond it not much better, although built up to some extent with low, mean houses. The sand of this beach is white, like the surf which dashes upon it. Whoever wishes to be entertained by the low but heavy thunder of the waves, as they roll in from the green Atlantic, cannot find a more fitting spot; and he that has once enjoyed the sublime companionship of the waves, that here rush to pay their homage at his feet, will long to revisit the scene.

In beholding the Sugar-Loaf for the first time, I was seized with an almost irresistible desire to ascend its summit. This wish was never carried into action. As my countrymen, however, have shared largely in this species of ambition, I shall be more excusable.

It is said by some, that a Yankee midshipman first conceived and executed the hazardous project of climbing its rocky sides. Nevertheless, this honor is disputed by others in behalf of an Austrian midshipman. Belonging to whom that may, it was reserved for Donna America Vespucci, in 1838, to be the first lady who should attempt the exploit; but the Donna failed to accomplish what her ambitious mind determined. Several persons of both sexes have, since this failure, made the attempt, and, at the peril of life and limb, some have succeeded in scrambling to the very top. On the 4th of July, 1851, Burdell, an American dentist, accompanied by his wife, a French coiffeur *et sa dame,* and a young Scotchwoman, made the ascent. From the latter I received an account of that adventurous night, when at times they seemed ready to dash into the foaming ocean beneath. Their toil and danger were of no small magnitude, and, when success finally crowned their foolhardiness, they sent up rockets and built a bonfire, to the astonishment of the gazing Fluminenses. The last ascent of this singular mountain, which is almost as steep as Bunker Hill Monument, was performed by a young American, who, without a companion or the usual appliances and skill of a seafaring man, worked his way up to the very summit, under the full blaze of a burning sun. He was, however, so disgusted with his adventure, that he begged his friends never to mention the subject.

Note for 1866.—Great amelioration in the condition of the streets has taken place since 1855. This is owing to a better system of paving, the stone for the pavement being quarried from the hills that abound in the city. The Rio City Improvements Company, against great natural obstacles, are now prosecuting the sewerage of the capital, which, when completed, will render Rio the finest city of the tropics. Nor should we overlook the Government works and docks, under the direction of Charles Neat, Esq., in front of the Custom-House, on the island of Cobras, and elsewhere. One of Mr. Neat's efficient aids is Mr. Bullman, of Newcastle, England, to whom I am indebted for meteorological notes.

CHAPTER VII.

BROTHERHOODS—HOSPITAL OF SAN FRANCISCO DE PAULA—THE LAZARUS AND THE
RATTLESNAKE—MISERICORDIA—SAILORS' HOSPITAL AT JURUJUBA—FOUNDLING-
HOSPITAL—RECOLHIMENTO FOR ORPHAN-GIRLS—NEW MISERICORDIA—ASYLUM
FOR THE INSANE—JOSÉ D'ANCHIETA, FOUNDER OF THE MISERICORDIA—
MONSTROUS LEGENDS OF THE ORDER—FRIAR JOHN D'ALMEIDA—CHURCHES—
CONVENTS.

To turn from the contemplation of nature to the works of man
is not always the most pleasing transition; and Bishop Heber's
well-known and oft-cited lines—

> "Though every prospect pleases,
> And only man is vile"—

seem doubly true in South America, where the grand and the
beautiful are so wonderfully profuse and in such strong contrast
with the shortcomings of earth's last and highest creature. But
the philanthropy and practical Christianity embodied in the hospitals of Rio de Janeiro are in happy dissimilitude with the
mummeries and puerilities which the Roman Catholic Church has
fostered in Brazil. These institutions, in their extent and efficiency, command our highest respect and admiration.

Among the hospitals of the capital there are a number which
belong to different *Irmandades* or Brotherhoods. These fraternities
are not unlike the beneficial societies of England and the United
States, though on a more extended scale. They are generally
composed of laymen, and are denominated Third Orders,—as, for
example, Ordem Terceiro do Carmo, Da Boa Morte, Do Bom Jesus
do Calvario, &c. They have a style of dress approaching the clerical in appearance, which is worn on holidays, with some distinguishing mark by which each association is known. A liberal
entrance-fee and an annual subscription is required of all the members, each of whom is entitled to support from the general fund in

sickness and in poverty, and also to a funeral of ceremony when dead. The brotherhoods contribute to the erection and support of churches, provide for the sick, bury the dead, and support masses for souls. In short, next after the State, they are the most efficient auxiliaries for the support of the religious establishment of the country. Many of them, in the lapse of years, have become rich by the receipt of donations and legacies, and membership in such is highly prized.

The extensive private hospital of S. Francisco de Paula belongs to a brotherhood of that name. It is located in an airy position, and built in the most substantial manner. Each patient has an alcove allotted to him, in which he receives the calls of the physician and the necessary care of attendants. When able to walk, he has long corridors leading round the whole building, in which he may promenade, or from the windows enjoy the air and a sight of surrounding scenes. There are also sitting-rooms in which the convalescent members of the fraternity meet to converse.

The Hospital dos Lazaros is located at St. Christovão, several miles from the city, and is entirely devoted to persons afflicted with the elephantiasis and other cutaneous diseases of the leprous type. Such diseases are unhappily very common at Rio, where it is no rare thing to see a man dragging about a leg swollen to twice its proper dimensions, or sitting with the gangrened member exposed as a plea for charity. The term "elephantiasis" is derived from the enormous tumors which the affection causes to arise on the lower limbs, and to hang down in folds or circular bands, making the parts resemble the legs of an elephant. The deformity is frightful in itself; but the prevailing belief that the disease is contagious imparts to the beholder an additional disgust.

It was an act of true benevolence by which the Conde da Cunha appropriated an ancient convent of the Jesuits to the use of a hospital for the treatment of these cases. It was placed, and has since remained, under the supervision of the Irmandade do Santissimo Sacramento. The average number of its inmates is about eighty. Few in whom the disease is so far advanced as to require their removal to the hospital ever recover from it. Not long since a person pretended to have made the discovery that the elephantiasis of Brazil was the identical disease which was cured among the ancient Greeks by the bite of a rattlesnake. He published several disquisitions on the subject, and thus awakened public attention to his singular theory. An opportunity soon offered for testing it. An inmate of the hospital, who had been a subject of the disease for six years, resolved to submit himself to the hazardous experiment.

A day was fixed, and several physicians and friends of the parties were present to witness the result. The afflicted man was fifty years old, and, either from a confident anticipation of a cure, or from despair of a happier issue, was impatient for the trial. The serpent was brought in a cage, and into this the patient introduced his hand with the most perfect presence of mind. The reptile seemed to shrink from the contact, as though there was something in the part which neutralized its venom. When touched, the serpent would even lick the hand without biting. It became necessary at length for the patient to grasp and squeeze the reptile tightly, in order to receive a thrust from his fangs. The desired infliction was at length given, near the base of the little finger.

So little sensation pervaded the member that the patient was not aware he was bitten until informed of it by those who saw the act. A little blood oozed from the wound, and a slight swelling appeared when the hand was withdrawn from the cage; but no pain was felt. Moments of intense anxiety now followed, while it remained to be seen whether the strange application would issue for the better or for the worse. The effect became gradually manifest, although it was evidently retarded by the disease which had preoccupied the system. In less than twenty-four hours the Lazarus was a corpse!

The most extensive hospital in the city, and indeed in the Empire, is that called the Santa Casa da Misericordia, or the Holy House of Mercy. This establishment is located upon the sea-shore, under the brow of the Castello Hill, and is open day and night for the reception of the sick and distressed. The best assistance in the power of the administrators to give is here rendered to all, male and female, black or white, Moor or Christian,—none of whom, even the most wretched, are under the necessity of seeking influence or recommendations in order to be received.

From the statistics of this establishment it appears that more than seven thousand patients are annually received, of whom more than one thousand die.

In this hospital are treated vast numbers of English and American seamen, the subjects of sickness or accident on their arrival, or during their stay in the port.   There are few nations of the world which are not represented among the inmates of the Misericordia of Rio de Janeiro.   Free access being always granted to its halls, they furnish an ample and interesting field for benevolent exertions in behalf of the sick and dying.

THE JURUJUBA HOSPITAL.

The years 1850, '51, '52, and '53 were those of great mortality among foreigners on account of the first and only known visit of the yellow fever to Rio de Janeiro and the coast of Brazil.   The number of deaths among the natives was much exaggerated, and in no portion of the Empire was the mortality ever so great as in those parts of the United States which have so often been visited by the same disease.   In 1854, '55, and '56, no cases of the yellow fever occurred, and its appearance and disappearance have been equally mysterious. The reader curious in such matters will find this subject treated in the appendix. (Dr. P. Candido died in 1864.)

New hospitals were arranged for the reception of foreign mariners stricken down with this fell malady; but none have been so well appointed, so well regulated, and so eminently successful, as the hospital at Jurujuba, under the supervision of an able medical committee, of which Dr. Paulo Candido was the chief. The principal visiting and attending physician was Dr. Correo de Azevedo, a gentleman of great affability and experience, speaking ten different languages with fluency, and who was a universal favorite among his patients from all parts of the world. Every day during the year the little steamer "Constancia," bearing the physician and his assistants, passes through the entire shipping, receiving the sick, and then transports them to the southern shores of the St. Xavier's or Jurujuba Bay. The hospital is situated in the midst of perpetual verdure, and where the ocean and land breezes are uncontaminated by the many impurities of a vast city. Here are excellent and kind nurses, who co-operate with the physicians in promoting the recovery of the invalids. Dr. Azevedo now resides at Theresopolis.

Jurujuba Hospital was for me a place of frequent visitation during the prevalence of the dreaded yellow fever. How many a poor wayfarer of the deep have I seen here and on shipboard, far away from country, home, and relatives, go down to the grave! How often, too, have I witnessed the power of that " hope which maketh not ashamed," as I have caught from dying lips the last loving messages sent to a distant father, mother, or sister, or as I have listened to the triumphant hymn which proclaimed the victory over the last foe to man!

Although there was free transit to all who wished to go to the hospital, I never met a single Brazilian or Portuguese priest in my many visits to Jurujuba. It could not be pleaded in extenuation that it was an institution for English and American mariners, for a very large proportion were Portuguese, Spanish, French, and Italian sailors. The only Roman Catholic ecclesiastic of any grade that I ever saw at Jurujuba was one of the devoted Italian Capuchins who seem at Rio to be ever on errands of mercy, through tropic heats and rains, while the lazy, lounging, greasy, acclimated *frades* of San Antonio, San Bento, and of Carmo, live at ease in their huge conventual buildings, situated in the loveliest and healthiest portions of the city.

Before the erection of Jurujuba Hospital nearly all the necessitous foreign invalids were accommodated in the Misericordia.

The benevolence of this latter hospital is not confined to those within its infirmaries, but extends to the different prisons of the city, most of whose inmates receive food and medicines from the provisions of the Misericordia.

Besides the public hospital, the institution has another for foundlings, and a Recolhimento, or Asylum for Female Orphans. The Foundling-Hospital* is sometimes called *Casa da Roda,* in allusion to the wheel in which infants are deposited from the streets and by a semi-revolution conveyed within the walls of the building. This wheel occupies the place of a window, facing the thoroughfare, and revolves on a perpendicular axis. It is divided by partition into four triangular apartments, one of which always opens without, thus inviting the approach of any who may be so heartless as to wish to part with their infant children. They have only to deposit the foundling in the box, and by a turn of the wheel it passes within the walls, they themselves going away unobserved.

That such institutions are the offspring of a mistaken philanthropy is as evident in Brazil as it can be in any country. Not only do they encourage licentiousness, but they foster the most palpable inhumanity. Out of three thousand six hundred and thirty infants exposed in Rio during ten years anterior to 1840, only one thousand and twenty-four were living at the end of that period. In the year 1838–39, four hundred and forty-nine were deposited in the wheel, of whom six were found dead when taken out; many expired the first day after their arrival, and two hundred and thirty-nine died in a short period.

The report of the Minister of the Empire for the official year 1854–55 gives the following alarming statistics and the comments of the minister :—

---

* The Foundling-Hospital was formerly the large three-story building seen on the right-hand side of the "View of the Gloria Hill from the Terrace of the Passeio Publico." The hospital is now in the Rua dos Barbonas.

"In 1854, 588 infants were received, in addition to 68 already in the establishment. Total, 656: died, 435; remaining, 221.

"In 1853, the number of foundlings received was 630, and of deaths 515.(!)

"There was, therefore, less mortality in the past than in the former year. Still, the number of deaths is frightful.

"Up to the present time it has not been possible to ascertain the exact causes of this lamentable mortality, which with more or less intensity always takes place among such infants, notwithstanding the utmost effort and care that has been used to combat the evil."

Well might one of the physicians of the establishment, in whose company a gentleman of my acquaintance visited several departments of the institution, remark, "Monsieur, c'est une boucherie!"

What must be the moral condition or the humane feelings of those numerous persons who deliberately contribute to such an exposure of infant life? One peculiar circumstance connected with this state of things consists in the alleged fact that many of the foundlings are the offspring of female slaves, whose masters, not wishing the trouble and expense of endeavoring to raise the children, or wishing the services of the mothers as wet-nurses, require the infants to be sent to the *engeitaria*, where, should they survive, they of course are free. A large edifice for the accommodation of foundlings is being erected on the *Largo da Lapa*.

The Asylum for Female Orphans is a very popular establishment. It is chiefly supplied from the Foundling-Hospital. The institution not only contemplates the protection of the girls in its care during their more tender years, but provides also for their marriage, and confers on them dowries of from two to four hundred milreis each. On the 2d of July, every year, when the Romish Church celebrates the anniversary of the Visitation of St. Elizabeth, by processions, masses, and the like, this establishment is thrown open to the public, and is thronged with visitors, (among whom are their Imperial Majesties,) some of whom bring presents to the recolhidas, and some ask for them in marriage.

The new buildings of the Misericordia are upon a grand scale, and the view of it to those entering the harbor is, architecturally considered, truly magnificent. It is constructed of stone, and is six hundred feet in length. There is only the half of the immense

8

structure presented to the eye as we look at the sketch below, engraved from a daguerreotype; and the reader will be astonished at the size of this noble beneficiary edifice when he is informed that it is a double building, and that its twin-brother is in the rear of it; but it is so connected as to form several airy quadrangular courts. With its modern improvements, insuring superior ventilation, light, and cleanliness,—with its flower-gardens and shrubberies for the recreation and exercise of the convalescent,—with its cool

St. Luzia's Chapel.   Morro do Castello.   Arsenal of War.

MISERICORDIA.

fountains, its spacious apartments, kind attendants, and beautiful situation,—this hospital is, as has been well said, "a credit to the civilization of the age, and a splendid monument of the munificence and benevolence of the Brotherhood of Mercy."

The Lunatic Asylum, or, as it is officially called, the Hospicio de Pedro II., situated on the graceful Bay of Botafogo, is a splendid, palace-like structure, inaugurated in 1852. The accommodation for the insane is here upon a scale of comfort and splendor only equalled by the Misericordia, whose noble dome lifts itself above the Praia da Santa Luzia. The French Sisters of Charity are the nurses here as well as in the house of the Brothers of Mercy. The Emperor, after whom the hospital at Botafogo is named, is one of its most liberal supporters.

The annual expenses of the Misericordia are about one hundred and fifty thousand dollars. A small portion of its receipts are provided for by certain tributes at the Custom-House, another portion by lotteries, and the balance by donations and the rent of properties which belong to the institution through purchase and legacies The Foundling-Hospital and Recolhimento have been in existence about a hundred years. The original establishment of the Misericordia dates back as far as 1582, and took place under the auspices of that distinguished Jesuit, José de Anchieta. About that time there arrived in the port a Spanish armada, consisting of sixteen vessels-of-war, and having on board three thousand Spaniards, bound to the Straits of Majellan. During the voyage very severe storms had been experienced, in which the vessels had suffered greatly, and sickness had extensively broken out on board. Anchieta was at the time on a visit to the college of his order, which had been founded some years previously, and whose towers still surmount the Castello Hill. Moved by compassion for the suffering Spaniards, he made arrangements for their succor, and in so doing laid the foundation of an institution which has continued to the present day enlarging its charities and increasing its means of alleviating human suffering.

It is impossible to contemplate the results of such an act of philanthropy without a feeling of respect toward its author. How many tens of thousands, during the lapse of more than two hundred and fifty years, have found an asylum within the walls of the Misericordia of Rio de Janeiro,—how many thousands a grave! Anchieta was among the first Jesuits sent out to the New World, and his name fills a large space in the history of that order. His earlier labors were devoted to the Indians of S. Paulo, and along that coast, where he endured great privations and exerted a powerful influence; but he finally returned to Rio de Janeiro, and there ended his days.

His self-denial as a missionary, his labor in acquiring and methodizing a barbarous language, and his services to the State, were sufficient to secure to him an honest fame and a precious memory; but in the latter part of the ensuing century he was made a candidate for saintship, and his real virtues were made to pass for little in comparison with the power by which it was pretended that he had wrought miracles.   Simon de Vasconcellos, Provincial of Brazil, and historian of the province, composed a narrative of his life, which is one of the greatest examples of extravagance extant.

It may be interesting to pass from the Santa Casa da Misericordia, so happily associated with his name, up the steep paved walk which leads to the old Jesuits' College on the Morro do Castello, where Anchieta died. Here we may contemplate the huge antiquated structure, which, although long since perverted from its original use, remains, and is destined to remain perhaps for ages to come, a monument of the wealth and power of the order founded by Ignatius de Loyola, whose name the college bore.

It is sickening to turn our attention from the good which Anchieta did, to the absurd inventions in regard to the founder of the Misericordia after he had been for a hundred years slumbering in the tomb. It is only one of those monstrous legends invented by the priests, approved by the Inquisition, and ratified by the church, which were for centuries palmed off upon the credulity of the people, as a means of advancing the interests and the renown of rival monastic orders.

Mr. Southey remarks:—"It would be impossible to say which order has exceeded the others in Europe in this rivalry, each having carried the audacity of falsehood to its utmost bounds; but in Brazil the Jesuits bore the palm."

Of this few will doubt who read the following. "Some," says Vasconcellos, "have called him [Anchieta] the second Thaumatourgos; others, the second Adam,—and this is the fitter title; because it was expedient that, as there had been an Adam in the Old World, there should be one in the New, to be the head of all its inhabitants and have authority over the elements and animals of America, such as the first Adam possessed in Paradise.

"There were, therefore, in Anchieta, all the powers and graces with which the first Adam had been endowed, and he enjoyed them not merely for a time, but during his whole life; and for this reason, like our common father, he was born with innocence, impassibility, an enlightened mind, and a right will.

"Dominion was given him over the elements and all that dwell therein. The earth brought forth fruit at his command and even gave up the dead, that they might be restored to life and receive baptism from his hand. The birds of the air formed a canopy over his head to shade him from the sun. The fish came into his net when he required them. The wild beasts of the forest attended him in his journeys and served him as an escort. The winds and waves obeyed his voice. The fire, at his pleasure, undid the mischief which it had done, so that bread which had been burnt to a cinder in the oven was drawn out white and soft by his interference.

"He could read the secrets of the heart. The knowledge of hidden things and sciences was imparted to him; and he enjoyed daily and hourly ecstasies, visions, and revelations. He was a saint, a prophet, a worker of miracles, and a vice-Christ; yet such was his humility, that he called himself a vile mortal and an ignorant sinner.

"His barret-cap was a cure for all diseases of the head. Any one of his cilices, [wire shirts,] or any part of his dress, was an efficacious remedy against impure thoughts. Water poured over one of his bones worked more than two hundred miracles in Pernambuco, more than a thousand in the South of Brazil; and a few drops of it turned water into wine, as at the marriage in Galilee. Some of his miracles are commended as being more fanciful and in a more elegant taste [*sic*] than those which are recorded in the Scriptures."

The book in which these assertions are made, and which is stuffed with examples of every kind of miracles, was licensed by the various censors of the press at Lisbon,—one of whom declares, that, as long as the publication should be delayed, so long would the faithful be deprived of great benefit, and God himself of glory!

The same author, who has collected and attested all the fables which credulity and ignorance had propagated concerning Anchieta, has produced a far more extraordinary history of Friar Joam d'Almeida, his successor in sanctity. It was written immediately after Almeida's death, when the circumstances of his life were fresh

in remembrance, and too soon for the embellishment of machinery
to be interwoven.

This remarkable person, whose name appears originally to have
been John Martin, was an Englishman, born in London during the
reign of Elizabeth.   In the tenth year of his age he was kidnapped
by a Portuguese merchant, apparently for the purpose of preserving
him in the Catholic faith; and this merchant, seven years afterward, took him to Brazil, where, being placed under the care of
the Jesuits, he entered the company.

Anchieta was his superior, then an old man, broken down with
exertion and austerities and subject to frequent faintings.   Almeida
used to rub his feet at such times, in reference to which he was
accustomed to say that, whatever virtue there might be in his
hands, he had taken it from the feet of his master.   No volup-
tuary ever invented so many devices for pampering the senses as
Joam d'Almeida did for mortifying them.   He looked upon his
body as a rebellious slave, who, dwelling within-doors, eating at
his table, and sleeping in his bed, was continually laying snares
for his destruction; therefore he regarded it with the deepest
hatred, and, as a matter of justice and self-defence, persecuted,
flogged, and punished it in every imaginable way.   For this purpose he had a choice assortment of scourges,—some of whipcord,
some of catgut, some of leathern thongs, and some of wire.   He
had cilices of wire for his arms, thighs, and legs, one of which was
fastened around the body with seven chains; and another he called
his good sack, which was an under-waistcoat of the roughest horse-
hair, having on the inside seven crosses made of iron, the surface
of which was covered with sharp points, like a coarse rasp or a nut-
meg-grater.   Such was the whole armor of righteousness in which
this soldier of Christ clad himself for his battles with the infernal
enemy.   It is recorded among his other virtues that he never disturbed the mosquitos and fleas when they covered him; that, whatever exercise he might take in that hot climate, he never changed
his shirt more than once a week; and that on his journeys he put
pebbles or grains of maize in his shoes.

His daily course of life was regulated in conformity to a paper
drawn up by himself, wherein he promised "to eat nothing on
Mondays, in honor of the Trinity,—to wear one of his cilices,

according to the disposition and strength of the poor beast, as he called his body, and to accompany it with the customary fly-flapping of his four scourges, in love, reverence, and remembrance of the stripes which our Saviour had suffered for his sake. Tuesdays, his food was to be bread and water, with the same dessert, to the praise and glory of the archangel Michael, his guardian angel, and all other angels. Wednesdays, he relaxed so far as only to follow the rule of the company. On Thursdays, in honor of the Holy Ghost, the most holy sacrament, St. Ignatius Loyola, the apostles, and all saints, male and female, he ate nothing. Fridays, he was to bear in mind that the rules of his order recommended fasting, and that he had forsworn wine except in cases of necessity. Saturday, he abstained again from all food, in honor of the Virgin, and this abstinence was to be accompanied with whatever might be acceptable to her; whereby exercises of rigor as well as prayer were implied. On Sundays, as on Wednesdays, he observed the rules of the community."

The great object of his most thankful meditations was to think that, having been born in England,* and in London, in the very seat and heart of heresy, he had been led to this happy way of life. In this extraordinary course of self-torment, Friar Joam d'Almeida attained the great age of fourscore and two. When he was far advanced in years, his cilices and scourges were taken from him lest they should accelerate his death; but from that time he was observed to lose strength, as if his constitution was injured by the change: such practices were become necessary to him, like a perpetual blister, without which the bodily system, having been long accustomed to it, could not continue its functions. He used to entreat others, for the love of God, to lend him a whip or a cilice, exclaiming, "What means have I now wherewith to appease the Lord? What shall I do to be saved?" Such are the works which a corrupt church has substituted for faith in Christ and for the duties of genuine Christianity.

Nor must this be considered as a mere case of individual madness. While Almeida lived, he was an object of reverence and

---

* On one side of his portrait is the figure of England, on the other that of Brazil, and under them these words :—"Hinc Anglus, hinc Angelus."

admiration, not only to the common people of Rio de Janeiro, but to persons of all ranks. His excesses were in the spirit of his religion, and they were recorded after his death for edification and example, under the sanction of the Superiors of an order which at that time held the first rank in the estimation of the Roman Catholic world.

During his last illness the convent was crowded with persons who were desirous to behold the death of a saint. Nothing else was talked of in the city, and the Fluminenses accosted each other with condolences as for some public calamity. Solicitations were made thus early for scraps of his writing, rags of his garments or cilices, and, indeed, any thing which had belonged to him; and the porter was fully employed in receiving and delivering beads, cloths, and other things which devout persons sent, that they might be applied to the body of the dying saint and imbibe from it a healing virtue. He was bled during his illness, and every drop of the blood was carefully received upon cloths, which were divided as relics among those who had most interest in the college.

When the bell of the college announced his death, the whole city was as greatly agitated as if the alarm of an invasion had been given. The governor, the bishop-administrator, the magistrates, nobles, clergy, and religious of every order, and the whole people, hastened to his funeral. Every shop was shut. Even the cripples and the sick were carried to the ceremony. Another person died at the same time, and it was with great difficulty that men could be found to bear the body to the grave.

An official statement of the proceedings of the day was drawn up, to be a perpetual memorial; and the admiration of the people for Friar Joam d'Almeida was so great, especially in Rio de Janeiro, that they used his relics in diseases with as much faith as if he had been canonized, and with as much success. For a while they invoked no other saint, as if they had forgotten their former objects of devotion!

The practical rules of our Saviour, in the Sermon on the Mount, in regard to cheerfulness and absence of ostentation in religion, are very far from coinciding with the above practices; and one would judge that there was no need of a Mediator for the man who thus worked out his own salvation.

There are within the city of Rio and its suburbs about fifty churches and chapels. They are generally among the most costly and imposing edifices of the country, although many of them have but little to boast as regards either plan or finish. They may be found of various form and style. Some are octagonal, some are in the form of the Roman and some of the Grecian cross, while others are merely oblong. The Church of the Candelaria* was originally designed to be a cathedral for the diocese of Rio de Janeiro. It was commenced about seventy years ago, but is not yet entirely finished. Like nearly every other building for ecclesiastical purposes in the country, it stands as a memento of past generations. The erection of a new church in Brazil is not an event of frequent occurrence.

The chapels of the convents are in several instances larger, and probably more expensive, than any of the churches. That of the Convent of San Bento† is one of the most ancient, having been repaired, according to an inscription it bears, in 1671. The exterior of the edifice is rude but massive; its windows are heavily barred with iron gratings, more resembling a prison than a place of worship. The sides of the chapel are crowded with images and altars. The roof and ceiled walls exhibit paintings designed to illustrate the history of the patron saint, the relics of whose miracles are here carefully preserved. Unnumbered figures of angels and cherubs, carved in wood and heavily gilded, look down upon you from every corner in which they can be fastened: in fact, nearly the whole interior is gilt. The order of the Benedictines is by far the richest in the Empire, possessing houses and lands of vast extent, though the number of monks is at present quite small. In the convent proper, a large square area is surrounded by corridors open on one side, and exhibiting the doors of the several dormitories of the monks on the other. An accessible apartment is devoted to the library, composed of about six thousand volumes. The sombre and melancholy air which pervades

---

* The tall spires of this church may be seen in the general "View of Rio de Janeiro from the Island of Cobras," rising above the right of the central palm-tree.

† The turrets of this convent are those seen farthest to the right, in the "View" referred to in the note above.

this monastic pile is in perfect contrast with the splendid scene to be enjoyed in front of it, and with the neat and modern appearance of the Naval Arsenal, located at the foot of the eminence on which it stands.*

A striking peculiarity in the aspect of Rio de Janeiro is derived from the circumstance that all the most elevated and commanding

FRADES OF ST. ANTHONY.

sites of the city and its vicinity are occupied by churches and convents. Of these may be next mentioned the Convent of St. Anthony, a mendicant order, whose shovel-hat monks, although sworn to eternal poverty, have contrived to obtain a very valuable site and to erect a most costly edifice. The building, since they can possess nothing themselves, belongs, very conveniently, to the Pope of Rome. In it are two immense chapels and a vast cloister, with scarcely enough friars to keep them in order.

· On a hill opposite that of S. Antonio is the nunnery of Santa Theresa, occupying a situation more picturesque, perhaps, than that of either of the monasteries mentioned; and yet, as if to render the appearance of the building as offensive as possible in the midst of scenery ever breathing the fragrance of opening flowers and smiling in beauty, its contracted windows are

---

* On the island of Cobras, nearly opposite the Convent of S. Bento, is an immense copper ring near the water's edge, put down by the celebrated Captain Cook in his last voyage.

not only barred with iron gratings, but even these gratings are set
with bristling spikes.

The Convent of Nossa Senhora da Ajuda, which is overlooked
from the Hill of Santa Theresa, completes the list of monastic institutions in the capital of Brazil.   In this last-mentioned were formerly many inmates who had not taken the veil.   The jealousy
of the Portuguese and their descendants was such, that in other
years it was not uncommon for a gentleman, when making a visit
to the mother-country, to incarcerate—or, more politely, "procure
lodgings" for—his wife in the convent, where she remained during
his entire absence.   I have understood that this shameful practice
has been forbidden by the present Emperor.   The monasteries may
all be considered unpopular, and could never again be erected at
any thing like their present expense.

The churches of all descriptions are generally open every morning.   At this time masses are said in most of them.   Ordinarily
but few persons are in attendance, and these are principally women.
Upon the great holidays, several of which occur during Lent, the
churches are thronged, and sermons are occasionally delivered;
but nothing like regular preaching on the Sabbath or any other day
is known in any part of the country.

Note for 1866.—As the subject of health is mentioned in connection with hospitals in this chapter, I add that I have been deeply interested in the report for
1864 of the sanitary condition of the Empire, published by the President (Dr.
J. P. Regos) of the *Junta Central de Hygiene Publica.*   It shows that, under the
wise and skilful treatment of the Brazilian faculty, cholera and the yellow fever
have almost entirely disappeared from Brazil.   I am indebted to Dr. Manoel
Pacheco da Silva, of Rio de Janeiro, for this valuable pamphlet.   Brazil has suffered the greatest medical loss by the death of Dr. Paulo Candido, who did more
than any other man to make and publish close observations in regard to the epidemics of his country.   He died at Paris in 1864; and his loss was felt in the
eminent medical circles of Europe as much as in Brazil.

CHAPTER VIII.

ILLUMÍNATION OF THE CITY—EARLY TO BED—POLICE—GAMBLING AND LOTTERIES
—MUNICIPAL GOVERNMENT—VACCINATION—BEGGARS ON HORSEBACK—PRISONS—
SLAVERY — BRAZILIAN LAWS IN FAVOR OF FREEDOM — THE MINA HERCULES —
ENGLISH SLAVE-HOLDERS—SLAVERY IN BRAZIL DOOMED.

THE streets of few cities are better lighted than those of Rio de Janeiro. The gas-works on the Atterrado sends its illuminating streams to remote suburbs as well as through the many and intricate thoroughfares of the Cidade Velha and the Cidade Nova. They have not the convenient fiction which city governments so often palm off upon themselves in the United States,—viz.: that the moon shines half the year; for in Rio, whether Cynthia is in the full, or whether shorn of her beams by unforeseen storms, the lamps continue to shed their brilliant light. The coal for the gas comes from England.

After ten o'clock at night few people are seen in the streets. The Brazilians are eminently an "early to bed, early to rise" people. When the great bells ring out the hour of ten, every slave "heels it;" and woe be to him that is caught out after the tocsin tolls the time when the law prescribes that he should be in his master's house; for, if dilatory, the police seize José and commit him to durance vile until his owner ransom him by a smart fine.

The same rule does not hold good in regard to freemen; yet one would think that it was equally in force without regard to class, for the Fluminensians, as a general thing, retire at ten P.M. Nothing is more surprising to a stranger from the North, to whom the night is so attractive, with its coolness, its fragrance, and its brilliancy, than to find the streets and the beautiful suburbs of the city almost as tenantless and silent as the ruins of Thebes or Palmyra.

The police of Rio de Janeiro is military, and is well disciplined by officers of the regular army. They are fortified with plenty

124

ot authority, and take care to use it.   Great difficulties have sometimes occurred between the constabulary and foreigners, where, on
some occasions, the former have been to blame; but it was good
for "Young America," when going "round the Horn" on his way
to California, to be held in wholesome restraint by these "yellow
Brazilians," whom he affected to despise.   The police is armed.
During the day you may see them singly or in pairs, having their
positions in convenient localities for watching the slaves and all
others suspected of liability to disorder.   Now the policeman, with
three or four of his com-
panions, strolls along by
Hotel Pharoux to have an
eye upon the foreign sailors;
or again, with a single *con-
frère*, he takes his stand by
the Carioca fountain; or,
again, his undress-cap,
his blue uniform, his
sword, and his brace of
pistols, are wholesomely
displayed at a corner *venda*,
where the *tamanca**-shod
Sr. Antonio from Fayal
sells *cachaça*, (rum,) pig-
tail tobacco, *carne secca*,
mandioc-flour, red Lisbon
wine, and black beans
The above-mentioned sta-
ples are the articles of
stock and consumption for
the low grocer and the low

POLICEMAN AND VENDA.

class that patronize him.   Sometimes he will get a little higher in
the provision-line, and add butter, brought from Ireland, lard
from the United States, onions from Portugal, sardines, a few hams,
and sausages.   Then, too, he is somewhat of a lumber-merchant;

---

* A sort of wooden-soled slipper much worn by the lower class of whites and the
free blacks.

for he purchases a few bundles of finely-split wood, which, together with charcoal, is the small accompaniment of the kitchen-battery in Brazil. At these vendas is the only hard drinking (except that done by English and Americans) in Rio, and that imbibing is by the slaves. Often Congo or Mozambique becomes eloquent under the effects of cachaça, and then the policeman is an effectual arbiter.

I have found few cities more orderly than Rio de Janeiro; and the police are so generally on the alert, that, in comparison with New York and Philadelphia, burglaries rarely occur. I felt greater personal security at a late hour of the night in Rio than I would in New York. Yet there are occasions when the police receive a strong hint through the public press for their remissness. The following, taken from a late *Correio Mercantil*, is an illustration:— "Night before last, after eight o'clock, an individual named Mauricio was attacked by a band of *capoeiras*,* who fell upon him with clubs, striking him upon the forehead, and gashing his thigh in such a manner as to injure the artery. The victim, bathed in blood, was taken to the drug-store of Sr. Pires Ferão, and there received the necessary succors, which were afforded him by Dr. Thomas Antunes de Abreu, who rushed to the aid of the poor man as soon as he was called. No police-authority appeared to take cognizance of this criminal deed!" Such outrages are exceptions, and a few articles based on facts like the above soon arouse the police to their duty.

There are some offences against the good of society which the police occasionally winked at during my residence in Rio,—*i.e.* gambling. The *jogo* seems an inveterate habit of some Brazilians; and when I have been cooped up with them in quarantine I have had opportunities for watching how every class represented in the Lazareto, from the *padre* down, gave itself up to the gambling-passion. At Rio the laws are very stringent against gambling-houses; and there are times when their owners are earnestly ferreted out by the police. But in the Rua Princeza, during 1852 and '53, a certain lawyer each Saturday night constituted his house a rendezvous where gamblers met,—the regular professional

---

* Africans, who with daggers *run a muck* in the streets, but not often at the present day in Rio. See page 137.

blackleg, (including the lawyer,) and the young pigeon who came to be plucked. When I went to my religious services at nine o'clock on Sabbath morning, their carriages would be still standing before the door, and their sleepy servants yawning and swearing on every side. Policemen regularly marched down the Catete at all hours of the night and in the daytime; yet month after month passed, and the den was not broken up until their operations were for a time suspended by the suicide of one of the parties concerned.

There is another species of gambling most deleterious in its effects, which is countenanced and supported by the Government. I refer to lotteries. They are not "sham" concerns, but prizes are put up, and, if drawn, *paid.* If it is a church, a theatre, or some other public building, to be erected, the Government grants a lottery. There are always six thousand tickets at 20$000 (twenty milreis) each; the highest prize is 20,000$000, (or about ten thousand dollars,) and the second prize is half that sum: there are then two thousand more tickets, which draw prizes of 20$000 (ten dollars) and upward. Everywhere in the city are offices for selling the tickets, and in the country there are equestrian ticket-venders who go from house to house with the risking billets. There is no fraud in awarding the prizes, and there is such a rage for this kind of gambling that the tickets are sold in a few days. The effects are bad; for the poorest whites and the shabbiest blacks will rake, scrape, and steal, until they have sufficient to purchase the quarter part of a billet, and then run with it to the shop where the flaming wheel-sign with *Anda a roda hoje* (The wheel turns to-day) tells them that this is the road to fortune. When such a spirit is engendered by the State, it becomes rather difficult for the municipal authorities to put down private gambling.

The head-quarters of the police are in an ancient public building in the Rua do Conde.

The city government, consisting of nine aldermen, who compose the *Camara municipal,* are elected by the people of Rio (*i.e.* those possessing 100$,—about fifty dollars income) once in four years.

The City Hall, which is called the *Camara* Municipal, is situated on the Campo Santa Anna. The General Government enforces vaccination, and it is on the lower floor of this building where all who present themselves on Thursdays and Saturdays are vaccinated free of charge : the patients, however, are obligated to return after eight days. A portion of the report of the Minister of the Empire is devoted to this subject, and in the report of 1854–55 the minister says that in the cities and large towns it is easy to enforce the law, but in the villages and the country it is difficult to overcome the obstacles which superstition throws in the way.

There is a class, confined to no portion of the world, which comes under the especial surveillance of the police. Every Saturday the beggars have their harvest. Mr. Walsh remarked, in 1828, that beggars were seldom seen in the streets of Rio. This was far from being the case in 1838, when Dr. Kidder resided there. Through the lenity or carelessness of the police, great numbers of vagrants were continually perambulating the streets and importuning for alms ; and mendicants of every description had their chosen places in the thoroughfares of the town, where they regularly waited and saluted the passers-by with the mournful drawl of *Favorece o seu pobre pelo amor de Deos*. If any, instead of bestowing a gift, saw fit to respond to this formula with its counterpart, *Deos lhe favorece,* (God help you,) they were not always sure to escape without an insult. When this state of things was at its height, and it was known that numerous rogues were at large under the disguise of beggars, the chief of the police suddenly sprung a mine upon them. He offered the constables a reward of ten milreis for every mendicant they could apprehend and deliver at the House of Correction. In a few days not less than one hundred and seventy-one *vagabundos* were delivered, over forty of whom were furnished with employment at the marine arsenal. The remainder were made to labor at the penitentiary till they had liquidated the expense of their apprehension. This measure had a most happy effect, and the streets were thenceforward comparatively free from mendicity, although persons really deserving charity were permitted to ask for aid at their pleasure.

But in 1855 the evil had again become a crying one. All shades of beggars seemed to abound everywhere. At length it was discovered that poor, old, worn-out slaves—those afflicted with blindness and elephantiasis—were sent out by their masters to ask alms. A new *chef de police,* however, made an onslaught upon such mendicants. He had them arrested and examined. No slave was

thenceforth allowed to beg, as he rightly deemed that the owner
who had enjoyed the fruit of his labor during his days of health
could well afford to take care of him when overtaken by old age
and sickness.* Twelve mendicants were considered real objects
of charity, and had licenses given them. These beggars, being
either blind or lame, have now the monopoly of the eleemosynary
sympathies of the good people of Rio; and I believe it is found to
be a most profitable business. Some of them are carried in a rede
by two slaves or drawn by one; one worthy rejoices in a little
carriage pulled by a fat sheep, and another—a footless man—rides

THE BEGGAR.

on a white horse. Sometimes, in the country-parts of Brazil, beggars whose pedal extremities are free from all derangement play
the cavalier, altogether disdaining to foot it, and seem to receive
none the less charity than if they trudged from door to door
Upon one occasion, a female beggar, adorned with a feather in her
bonnet and mounted on horseback, rode up to a friend of mine at
St. Alexio, and, demanding alms, was exceedingly indignant at any
inquiries as to the consistency of her costume. The English proverb is not remarkably complimentary to such mendicants; but

---

* The proverb in Portuguese is very forcible:—"He who has enjoyed the meat
may gnaw the bones."

a like application is never heard in the land of the Southern Cross.

The House of Correction, referred to on a previous page, is located under the brow of a high hill, between the suburbs of Catumby and Mata Porcos. The grounds pertaining to it are surrounded by high granite walls, constructed by the prisoners, who have long been chiefly employed on various improvements of the premises. On the hill-side is a quarry, and numbers are employed in cutting stone for more extended walls and buildings. Others are made to carry earth in wooden trays upon their heads, sometimes from one part of the ground to another, or to fill the cars of a tram-railway, which runs from within the walls to the borders of a marsh nearly a mile distant, which is by this process being reclaimed from the tide-water and converted into valuable ground. The more refractory criminals are chained together, generally two and two, but sometimes four or five go along in file, clanking a common chain, which is attached to the leg of each individual.

The House of Correction is as fine a building, in an architectural point of view, as any similar edifice in the United States. The Director, (Sr. Falcão,) however, finds fault with its plan. It is not yet completed; and it is gratifying to see that the Brazilian Government is taking every measure to bring about an entire reform in prison-buildings and prison-discipline. It is one of those evidences of progress in a nation which is unmistakable. In 1852, Sr. Antonio J. de M. Falcão—who, by his intelligence and enlarged views, was admirably fitted for his office—was sent to the United States to inspect our various prison-systems. The report of Sr. Falcão to the Minister of Justice (Sr. J. Thomas Nabuco de Araujo) is incorporated in one of the Relatorios of the nation for 1854–55, and is full of interest. It seems strange to read, in the official message of a Brazilian Minister, familiar and sensible discussions in regard to the systems of Auburn and Pennsylvania; and it is a deserved compliment to Sr. Falcão that his able report has been fully reprinted in our own country, in the "Journal of Prison Discipline," so ably conducted by F. A. Packard, Esq., of Philadelphia. Sr. Falcão gives his preference to the system of Pennsylvania. The Relatorio of the Minister of Justice for the year mentioned is overflowing with instructive and interesting details in regard to penitentiaries and prisons. It is not, however, a mere dry narration of facts, but wise suggestions and feasible improvements are laid before the nation in a manner at once clear, attractive, and forcible.

The city prisons known as the Aljube and the Xadres da Policia all have been in a sad state: bad ventilation, bad food, and miserable damp cells, have called forth the denunciations of Sr. Falcão and other enlightened philanthropists in Rio, and these evils will soon be remedied.

Besides the prisons now enumerated, there are places of confinement in the different forts; those of Santa Cruz and the Ilha das Cobras being the principal.

Many of the prisoners are slaves, though the Brazilian law is not at all dainty as to color or condition. In the Relatorio of the Minister of Justice for the year 1854–55 I find that from the 7th of September, 1853, to the 16th of March, 1855, forty slaves and twenty-one free persons (which includes whites and blacks) were, for murder, condemned to death. The punishment of fourteen of the slaves was commuted, and that of but four of the freemen.

One department of the *Casa da Correcção* is appropriated to the flogging of slaves, who are sent thither to be chastised for disobedience or for common misdemeanors. They are received at any hour of the

THE LOG, IRON COLLAR, AND TIN MASK.

day or night, and retained free of expense as long as their masters choose to leave them. It would be remarkable if scenes of extreme cruelty did not sometimes occur here.

The punishments of the Casa da Correcão are not, however, the only chastisements which the refractory slave receives. There are private floggings; and some of the most common expiations are the tin mask, the iron collar, and the log and chain. The last two denote runaways; but the tin mask is often placed upon the visage to prevent the city-slave from drinking cachaça and the country-slave from eating clay, to which many of the field-negroes are addicted. This *mania*,—for it can be called nothing else,—if not checked, causes languor, sickness, and death.

The subject of slavery in Brazil is one of great interest and hopefulness. The Brazilian Constitution recognises, neither directly nor indirectly, color as a basis of civil rights; hence, once free, the black man or the mulatto, if he possess energy and talent, can rise to a social position from which his race in North America is debarred. Until 1850, when the slave-trade was effectually put down, it was considered cheaper, on the country-plantations, to use up a slave in five or seven years and purchase another, than to take care of him. This I had, in the interior, from intelligent native Brazilians, and my own observation has confirmed it. But, since the inhuman traffic with Africa has ceased, the price of slaves has been enhanced, and the selfish motives for taking greater care of them have been increased. Those in the city are treated better than those on the plantations: they seem more cheerful, more full of fun, and have greater opportunities for freeing themselves. But still there must be great cruelty in some cases, for suicides among slaves—which are almost unknown in our Southern States—are of very frequent occurrence in the cities of Brazil. Can this, however, be attributed to cruelty? The negro of the United States is the descendant of those who have, in various ways, acquired a knowledge of the hopes and fears, the rewards and punishments, which the Scriptures hold out to the good and threaten to the evil: to avoid the crime of suicide is as strongly inculcated as to avoid that of murder. The North American negro has, by this very circumstance, a higher moral intelligence than his brother fresh from the wild freedom and heathenism of Africa; hence the latter, goaded by cruelty, or his high spirit refusing to bow to the white man, takes that fearful leap which lands him in the invisible world.

In Brazil every thing is in favor of freedom;* and such are the facilities for the slave to emancipate himself, and, when emancipated, if he possess the proper qualifications, to ascend to higher eminences than those of a mere free black, that *fuit* will be written against slavery in this Empire before another half-century rolls around. Some of the most intelligent men that I met with in Brazil—men educated at Paris and Coimbra—were of African descent, whose ancestors were slaves. Thus, if a man have freedom, money, and merit, no matter how black may be his skin, no place in society is refused him. It is surprising also to observe the ambition and the advancement of some of these men with negro blood in their veins. The National Library furnishes not only quiet rooms, large tables, and plenty of books to the seekers after knowledge, but pens and paper are supplied to such as desire these aids to their studies. Some of the closest students thus occupied are mulattoes. Formerly a large and successful printing-establishment in Rio— that of Sr. F. Paulo Brito—was owned and directed by a mulatto. In the colleges, the medical, law, and theological schools, there is no distinction of color. It must, however, be admitted that there is a certain—though by no means strong—prejudice existing all over the land in favor of men of pure white descent.

In some intestate cases, a slave may go before a magistrate, have his price fixed, and purchase himself; and I was informed that a man of mental endowments, even if he had been a slave, would be debarred from no official station, however high, unless it might be that of Imperial Senator.

The appearance of Brazilian slaves is very different from that of their class in our own country. Of course, the house-servants in the large cities are decently clad, as a general rule; but even these are almost always barefooted. This is a sort of badge of slavery. On the tables of fares for ferry-boats, you find one price for persons wearing shoes, (*calçadas*,) and a lower one for those *descalças*, or

---

* A Southern lady (the wife of the very popular United States Consul at Rio during the administration of President Pierce) used to say that "the very paradise of the negroes was Brazil;" for there they possess a warm climate, and, if they choose, may make their way up in the world.

without shoes.   In the houses of many of the wealthy Fluminenses
you make your way through a crowd of little woolly-heads, mostly
guiltless of clothing, who are allowed the run of the house and the
amusement of seeing visitors.   In families that have some tincture
of European manners, these unsightly little bipeds are kept in the
background.   A friend of mine used frequently to dine in the
house of a good old general of high rank, around whose table
gambolled two little jetty blacks, who hung about their *"pai"* (as
they called him) until they received their portions from his hands,
and that, too, before he commenced his own dinner. Whenever the
lady of the house drove out, these pets were put into the carriage,
and were as much offended
at being neglected as any
spoiled only son.   They
were the children of the
lady's nurse, to whom she
had given freedom. Indeed,
a faithful nurse is generally
rewarded by manumission.

PRETO DE GANHO AND QUITANDEIRA.

The appearance of the
black male population who
live in the open air is any
thing but appetizing. Their
apology for dress is of the
coarsest and dirtiest de-
scription.   Hundreds of
them   loiter   about   the
streets with large round
wicker-baskets ready to
carry any parcel that you
desire conveyed. So cheaply
and readily is this help obtained, that a white servant
seldom thinks of carrying
home a package, however small, and would feel quite insulted if
you refused him a *preto de ganho* to relieve him of a roll of calico
or a watermelon.   These blacks are sent out by their masters, and
are required to bring home a certain sum daily.   They are allowed

a portion of their gains to buy their food, and at night sleep on a mat or board in the lower purlieus of the house. You frequently see horrible cases of elephantiasis and other diseases, which are doubtless engendered or increased by the little care bestowed upon them.

Formerly the coffee-carriers were the finest race of blacks. They were almost all of the Mina tribe, from the coast of Benin, and were athletic and intelligent. They worked half clad, and their sinewy forms and jetty skins showed to advantage as they hastened at a quick trot, seemingly unmindful of their heavy loads. This work paid well, but soon broke them down. They had a system among themselves of buying the freedom of any one of their number who was the most respected. After having paid their master the sum required by him daily, they clubbed together their surplus to liberate the chosen favorite. There was a Mina black in Rio remarkable for his height, who was called "The Prince," being, in fact, of the *blood-royal* of his native country. He was a prisoner of war, and sold to Brazil. It is said that his *subjects* in Rio once freed him by their toil: he returned, engaged in war, and was a second time made prisoner and brought back. Whether he ever regained his throne I know not; but the loss of it did not seem to weigh heavily on his mind. He was an excellent carrier; and, when a friend of mine embarked, the "Prince" and his troop were engaged to transport the baggage to the ship. He carried the largest case on his head the distance of two miles and a half. This same case was pronounced unmanageable in Philadelphia by the united efforts of four American negroes, and it had to be relieved of half its contents before they would venture to lift it up-stairs.

From time to time the traveller will meet with negroes from those portions of Africa of which we know very little except by the reports of explorers like Livingstone, Barth, and Burton. I have often thought that the slaves of the United States are descended not from the noblest African stock, or that more than a century of bondage has had upon them a most degenerating effect. We find in Brazil very inferior spiritless Africans, and others of an almost untamable disposition. The Mina negro seldom makes a good house-servant, for he is not contented except in breathing the fresh air. The men become coffee-carriers, and the women *quitandeiras*, or street pedlars.

These Minas abound at Bahia, and in 1838 plunged that city into a bloody revolt,—the last which that flourishing municipality has experienced. It was rendered the more dreadful on account of the secret combinations of these Minas, who are Mohammedans, and use a language not understood by other Africans or by the Portuguese.

When the delegation from the English Society of Friends visited Rio de Janeiro in 1852, they were waited upon by a deputation of eight or ten Mina negroes. They had earned money by hard labor and had purchased their freedom, and were now desirous of returning to their native land. They had funds for paying their passage back again to Africa, but wished to know if the coast were really free from the slavers. Sixty of their companions had left Rio de Janeiro for Badagry (coast of Benin) the year before, and had landed in safety. The good Quakers could scarcely credit this last information, thinking it almost impossible that any who had once been in servitude "should have been able and bold enough to make so perilous an experiment;" but the statement of the Minas was confirmed by a Rio ship-broker, who put into the hands of the Friends a copy of the charter under which the sixty Minas sailed, and which showed that they had paid four thousand dollars passage money. (See Appendix.) A few days after this interview, Messrs Candler & Burgess received from these fine-looking specimens ot humanity "a paper beautifully written in Arabic by one of their chiefs, who is a Mohammedan."

In Rio the blacks belong to many tribes, some being hostile to each other, having different usages and languages. The Mina negroes still remain Mohammedans, but the others are nominal Roman Catholics.

Many of them, however, continue their heathen practices. In 1839, Dr. Kidder witnessed in Engenho Velho a funeral, which was of the same kind as those curious burial-customs which the African traveller beholds on the Gaboon River. You can scarcely look into a basket in which the *quitandeiras* carry fruit without seeing a *fetisch*. The most common is a piece of charcoal, with which, the abashed darkey will inform you, the "evil eye" is driven away.

There is a singular secret society among the negroes, in which the highest rank is assigned to the man who has taken the most lives. They are not so numerous as formerly, but from time to time harm the unoffending. These blacks style themselves *capoeiros*, and during a festa they will rush out at night and rip up any other black they chance to meet. They rarely attack the whites, know-ing, perhaps, that it would cost them too dearly.

The Brazilians are not the only proprietors of slaves in the Empire. There are many Englishmen who have long held Africans in bondage,—some for a series of years, and others have purchased slaves since 1843, when what is called the Lord Brougham Act was passed. By this act it is made unlawful for Englishmen to buy or sell a slave in any land, and by holding property in man they are made liable, were they in England, to prosecution in criminal courts. The English mining-company, whose stockholders are in Great Britain, but whose field of operations is S. João del Rey in Brazil, own about eight hundred slaves, and hire one thou-sand more. 1865, the English government has remedied this.

Frenchmen and Germans also purchase slaves, although they have not given up allegiance to their respective countries.

If it be asked, " Who will be the laborers in Brazil when slavery is no more?" the reply is, that, though the slave's bonds are broken, the *man*, and a better man, still exists; and emigrants will come from Germany, Portugal, the Azores and Madeira. 1865, many are emigrating from the South of the United States.

It is a striking fact that emigrants did not begin to arrive from Europe by thousands until 1852. In 1850 and '51 the African slave-trade was annihilated, and in the succeeding year commenced the present comparatively vigorous colonization. Each year the number of colonists is increasing, and the statesmen of the Empire are now devoting much attention to discover the best means for thus pro-moting the advancement of the country.

Almost every step in Brazilian progress has been prepared by a previous gradual advance: she did not leap at once into self-government. She was raised from a colonial state by the residence of the Court from Lisbon, and enjoyed for years the position of a constituent portion of the Kingdom of Portugal. The present peaceful state of the Empire under D. Pedro II. was preceded by

the decade in which the capabilities of the people for self-government were developed under the Regency. The effectual breaking
up of the African slave-trade is but the precursor of a more important step.

Slavery is doomed in Brazil. As has already been exhibited, when
freedom is once obtained, it may be said in general that no social
hinderances, as in the United States, can keep down a man of
merit. Such hinderances do exist in our country. From the warm
regions of Texas to the coldest corner of New England the free
black man, no matter how gifted, experiences obstacles to his elevation which are insurmountable. Across that imaginary line which
separates the Union from the possessions of Great Britain, the
condition of the African, socially considered, is not much superior.
The Anglo-Saxon race, on this point, differs essentially from the
Latin nations. The former may be moved to generous pity for
the negro, but will not yield socially. The latter, both in Europe
and the two Americas, have always placed merit before color.
Dumas, the mulatto novel-writer, is as much esteemed in France
as Dickens or Thackeray are in England. An instance came under
my own observation which confirms most strongly the remark
made above. In 1849, it was my privilege to attend with a large
number of foreigners a soirée in Paris, given by M. de Tocqueville,
then French Minister of Foreign Affairs. I was introduced to a
visitor from the United States, who for the first time looked upon
the scenes of the gay capital, and as we proceeded to the refresh-
ment-room his arm rested on mine. I found that this clergyman,
by his intelligence, common sense, and modesty, commanded the
admiration of all with whom he came in contact. A few weeks
afterward a European university of high repute honored him with
the degree of Doctor of Divinity. In England he was looked upon
with interest and curiosity; but, had he proposed a social alliance
equal to his own station, I doubt if success would have attended
his offer. In 1856, the same clergyman was ejected from a New
York railway-omnibus, by a conductor who daily permitted, without molestation, filthy foreigners of the lowest European class
to occupy seats in the identical car. When the matter was
submitted to the courts of justice, the decision sustained the
conductor. There was no attempt to place the case on any

other ground than that the plaintiff was a man of African descent.

Note for 1866.—The laws and the treatment of slaves have greatly changed for the better since 1850. It is estimated that, by the emancipation by will, by the purchasing of their own freedom, and by the liberation of what were termed *Africanos livros*, (those taken from captured slave-vessels and apprenticed out for fourteen years,) the number of slaves has decreased one million, so that to-day there are not 2,000,000 at the highest calculation. Slavery is now mostly confined to the central sea-coast provinces. But the emancipated were not lost to labor, as some of the advocates of slavery would have us believe. From 1850 to 1860, inclusive, the great tropical staples of coffee, sugar, cotton, and tobacco actually increased more than 90 per cent. One of the latest notable cases of emancipation was by the Emperor, who, on the occasion of the marriage (October 16, 1864) of the Imperial Princess to the Count d'Eu, liberated the slaves that were hers by dower. Sr. Silveira da Motta, Senator from Goyaz, has been a far-sighted statesman in this respect. He has repeatedly brought in bills to limit slavery; and in the session of 1865, after the collapse of the so-called "Confederate States," his efforts, with those of the venerable Senator Visconde de Jequitinhonha, have brought this subject most prominently before the Brazilian people; and slavery, which (if the "institution" had longer survived in North America) would have died in twenty years without special legislation, will doubtless soon be so limited by law that it will be extinguished at an early date. A. C. Tavares Bastos, in the Chamber of Deputies, has been a persevering advocate of emancipation. The case is a difficult one in many respects; and the prayer of every philanthropist is that the Brazilians may have the wisdom to remove this great ulcer from their body politic.

CHAPTER IX.

RELIGION—THE CORRUPTION OF THE CLERGY—MONSIGNOR BEDINI—TOLERATION
AMONG THE BRAZILIANS—THE PADRE—FESTIVALS—CONSUMPTION OF WAX—
THE INTRUDO—PROCESSIONS—ANJINHOS—SANTA PRISCILLIANA—THE CHOLERA
NOT CURED BY PROCESSIONS.

THE "Roman Catholic Apostolic" is the religion of the State in Brazil; yet, by the liberal Constitution, and by the equally-liberal sentiments of the Brazilians, all other denominations have the right to worship God as they choose, whether in public or in private, with the single limitation that the church-edifice must not be *exterior de templo,*—in the form of a temple,—which has been defined by the supreme judges to be a building "without steeples or bells." Roman Catholicism in Brazil has never been subject to the influences with which it has had to contend in Europe since the Reformation. It was introduced contemporaneously with the first settlement of the country as a colony, and for three hundred years has been left to a perfectly free and untrammelled course. It has had the opportunity of exerting its very best influences on the minds of the people, and of arriving at its highest degree of perfection. In pomp and display it is unsurpassed even in Italy. The greatest defender of the Church of Rome must admit that South America has been a fair field for his ecclesiastical polity; and if his religion could have made a people great, enlightened, and good, it has had the power to have made Spanish and Portuguese America a moral, as it is a natural, Paradise. Spain and Portugal, at the time of the appropriation of their possessions in the New World, were equal, if not superior, to the English in all the great enterprises of the fifteenth and sixteenth centuries: but how widely different have been the results which have flowed from the colonies founded by both! Brazil is in every respect the superior State of South America just so far as she has abandoned the exclusiveness of Romanism. Since the Independ-

140

ence, the priest-power has been broken, and the potent hierarchy of Rome does not rule over the consciences and acts of men as in Chili or Mexico. On numerous occasions, measures have been taken in the Assemblea Geral to curtail the assumptions of the triple-crowned priest of the Eternal City; and once,* at least, it was proposed to render the Brazilian Church independent of the Holy See.

It may be said that the advancement in liberality which the Empire has displayed has been owing to political considerations. Granted: but every reader of history knows that the commencement of the English Reformation was largely implicated with politics, and England's independence of the Papal power was the beginning of her greatness as a state, and paved the way for the rapid moral advancement which characterizes England to-day.

In Brazil, however, other than political views must be taken of the present freedom from bigotry. The priests, to some extent, owe the loss of their power to their shameful immorality. There is no class of men in the whole Empire whose lives and practices are so corrupt as those of the priesthood. It is notorious. The *Relatorios* (messages) of the Minister of Justice and the Provincial Presidents annually allude to this state of things. Every newspaper from time to time contains articles to this effect; every man, whether high or low, speaks his sentiments most unreservedly on this point; no traveller, whether Romanist or Protestant, can shut his eye to the glaring facts. In every part of Brazil that I have visited I have heard, from the mouths of the ignorant as well as from the lips of the educated, the same sad tale; and, what is worse, in many places the priests openly avow their shame. Dr. Gardner, the naturalist, lived in Brazil from 1836 to '41, and the greater part of that time in the interior, where foreigners are very rarely found. In speaking of the banishment of the laborious and indefatigable Jesuits, whose lives in this portion of America were without reproach, this distinguished botanist says, "What different men they must have been from the degraded race who now undertake the spiritual welfare of this nation! It is a hard thing to say,

---

* This was during the Regency, when *Padre* Antonio Maria de Moura was nominated to the vacant bishopric of Rio de Janeiro.

but I do it not without well considering the nature of the asser-
tion, that *the present clergy of Brazil are more debased and immoral
than any other class of men.*"*

Though we should lament immorality in any man or class of
men, yet the combination of circumstances mentioned has had its
effect in rendering the people, as well as the Government, tolerant.

A few years ago, Monsignor Bedini (Archbishop of Thebes, and
late Pope's Legate in the United States and in other *partibus infidelium*) was the Nuncio of Pius IX. at the Court of Brazil. In
July, 1846, the nuncio went to the mountain-city of Petropolis,
(about forty miles from Rio,) where are many German Protestants,
who have a chapel of their own, which, as well as the chapels in
other colonies, is protected under the broad shield of the Constitution, and receives a portion of its support directly from the Government. There had been certain mixed marriages; and Monsignor
preached a furious sermon, in which he declared that all Romanists
so allied were living in concubinage,—their marriages were void, and
their children illegitimate. A storm of indignation, both at Petropolis and Rio, fell upon the head of the nuncio, whose arrival in
Brazil had been preceded by the rumor of an assurance to the Pope
that he would bind this Empire "faster than ever to the chair of
St. Peter." The *Diario do Rio de Janeiro*, a conservative journal
always considered the *quasi* organ of the Government, denounced
M. Bedini in firm but respectful language, and insisted that it was

---

* I was once dining with a Roman Catholic gentleman in the province of Rio de
Janeiro, and, of his own accord, he said to me, "How can I obey the injunctions
of my priest? he reads us the Decalogue, and yet he is the greatest breaker of the
seventh commandment." In the province of Bahia I made the acquaintance of a
Roman Catholic who had a number of female operatives under his charge, and a
chapel connected with his establishment. The priest (who was one of the few
moral ecclesiastics in Brazil) died. The proprietor then made known his wish for
a new chaplain. Five candidates presented themselves. Four were men whose
lives were of such a grossly-immoral character that I dare not insult my readers
by the particulars which I received from a member of the Romish Church. The
fifth was an old man of good repute, but not very active. As a *dernier ressort* he
was engaged to fill the chaplaincy; but only a few months elapsed before he was
discovered to be living in open concubinage with an abandoned character, and on
remonstrance would not give up this sinful union.

the highest imprudence thus to kindle the fires of religious intolerance. Its columns contained sentiments in regard to this subject of which the following is a specimen:—"Propositions like those emitted from the Chair of Truth by a priest of the character of M. Bedini are eminently censurable."

The nuncio was put down, but not until one of his friends published what were probably the sentiments of Monsignor, in which he complains of the Emperor for "not taking sides in the controversy and using his influence to prevent the spread of Protestant heresies."

There is no country in South America where the philanthropist and the Christian have a freer scope for doing good than Brazil. So far from its being true that a Protestant clergyman is always *tabooed*, and that the people "entertain a feeling toward him bordering on contempt,"—as one writer on Brazil has expressed it,—I can testify to the strongest friendship formed with Brazilians in various portions of the Empire,—a friendship which did not become weakened by the contact of years or by the plain manifestations and defence of my belief; and I can subscribe to the remark put forth by my colleague in 1845, when he says,—

"It is my firm conviction that there is not a Roman Catholic country on the globe where there prevails a greater degree of toleration or a greater liberality of feeling toward Protestants.

"I will here state, that in all my residence and travels in Brazil in the character of a Protestant missionary, I never received the slightest opposition or indignity from the people. As might have been expected, a few of the priests made all the opposition they could; but the circumstance that these were unable to excite the people showed how little influence they possessed. On the other hand, perhaps quite as many of the clergy, and those of the most respectable in the Empire, manifested toward us and our work both favor and friendship.

"From them, as well as from the intelligent laity, did we often hear the severest reprehension of abuses that were tolerated in the religious system and practices of the country, and sincere regrets that no more spirituality pervaded the public mind."

To one who looks alone at the empty and showy rites of the

Roman Catholic Church in Brazil, there is no future for the
country.   But when we consider the liberal and tolerant sentiments that prevail,—when we reflect upon the freedom of debate,
the entire liberty of the press, the diffusion of instruction, and the
workings of their admirable Constitution,—we cannot believe that
future generations of Brazilians will retrograde.   Intellectuality
without morality is, we are aware, an engine of tremendous power
wanting a balance-wheel; but we have faith that God, who has
blessed Brazil so highly in other respects, will not withhold from
her the greatest boon, however untoward at present may be the
prospect of such a bestowment.

A faithful narrator cannot pass over this subject without giving
a brief notice of some of the peculiarities connected with worship
at the capital, which, to a
certain extent, are those
witnessed in every pro-
vince of the Empire.

There is no mistaking
a priest or any species
of ecclesiastics in Brazil.
The *frades*, (monks,) the
Sisters of Charity, as well
as the priests, have their
peculiar costumes,—most
of them exceedingly inconvenient in a warm climate.
You cannot be an hour in
the streets of Rio de Ja-
neiro without beholding
the *padre*, with his large
hat and his closely-but-
toned and long gown,
moving along with per-
fect composure under a
hot sun that makes every

THE PADRE.

one else swelter.   In the churches, where there generally pervades
a cool atmosphere, the padre, with his uncovered, tonsured head,
with his thin gowns and airy laces, seems prepared for a tropic

clime; but, when the mass is said and his duties are finished, he doffs his garment of common-sense thickness and dons that which would be comfortable in a Northern winter.

The padre's office is not onerous in Brazil, unless he choose to make it such; and very few are thus inclined. There are no poor families to visit through rude snow-storms; there is no particular cure of souls, beyond repeating masses in the cool of the morning, the carrying of the Host to the hopeless sick, and attendance at a funeral, for which the carriage and fee are always provided. The confessional does not trouble him greatly, for the people are not much given to confession, knowing too well the character of the confessor. If he is of an ambitious turn of mind, he becomes a candidate for the Chamber of Deputies,—perchance he succeeds in securing a seat in the Senate,—and there he will pour out more eloquence, in *ore rotundo* Lusitanian, than he has ever delivered from the pulpit. Perhaps formerly his heaviest duties were in getting up festivals. They have been wonderfully abridged as to number, but still there is a very respectable share of them, which gives work to the padres and the alms-collectors, and holidays to clerks, school-children, and slaves.

Bishop Manuel do Monte Roderigues d'Araujo, when professor at Olinda, published a compendium of moral theology, and he states that the number of holidays observed in the Empire of Brazil is the same as that decreed by Pope Urban VIII. in 1642, with the addition of one in honor of the patron saint of each province, city, town, and parish, for which Urban's decree also provides. These holidays are divided into two general classes :—*Dias santos de guarda*, or whole holidays, in which it is not lawful to work; and *Dias santos dispensados*, or half-holidays, in which the ecclesiastical laws require attendance upon mass, but allow the people to labor. The number of the former varies from twenty to twenty-five, according as certain anniversaries fall on a Sabbath or on a weekday; while the number of the latter is from ten to fifteen. The celebration of these holidays by festivals and processions engages universal attention throughout the country; and the North American is constantly reminded of the 4th of July *minus* the patriotic enthusiasm. The number of festivals were curtailed within a few years; yet some five or six during the

10

year arrest the course of commerce and material duties generally.

It is particularly observable that all the religious celebrations are deemed interesting and important in proportion to the pomp and splendor which they display. The desirableness of having all possible show and parade is generally the crowning argument urged in all applications for Government patronage, and in all appeals designed to secure the attendance and liberality of the people.

The daily press of Rio de Janeiro must annually reap enormous sums for religious advertisements, of which I give one or two specimens.

The announcement of a festival in the Church of Santa Rita is thus concluded :—

" This *festa* is to be celebrated with high mass and a sermon, at the expense of the devotees of the said Virgin, the Most Holy Mother of Grief, who are all invited by the Board to add to the *splendor* of the occasion by their presence, since they will receive from the above-named Lady due reward."

The following is the advertisement of a *festa* up the bay, at Estrella, and is as clumsily put together in Portuguese as it appears in the literal English translation which I have given :—

" The Judge and some devout persons of the Church of Our Lady of Estrella, erected in the village of the same name, intend to hold a festival there, with a chanted mass, sermon, procession in the afternoon, and a *Te Deum*,—all with the greatest pomp possible,—on the 23d instant; and at night there will be a beautiful display of fireworks. The managers of the feast have asked the Director of the Inhomerim Steamboat Company to put on an extra steamer that will leave the Praia dos Mineiros at eight o'clock in the morning and return after the fireworks.

" It is requested that all the devotees will deign to attend this solemn act, to render it of the most brilliant description.

                               "Francisco Pereira Ramos, *Secretary.*
" Estrella, Sept. 17, 1855."

The following will be to Northern Christians as novel as it is irreverent :—

" The Brotherhood of the Divine Holy Ghost of San Gonçalo (a small village across the bay) will hold the feast of the Holy Ghost, on the 31st instant, with all possible splendor. Devout persons are invited to attend, to give greater pomp to this act of religion. On the 1st proximo there will be the feast of the Most Holy Sacrament, with a procession in the evening, a *Te Deum*, and a sermon. On the 2d,—the feast of the patron of San Gonçalo,—at three P.M. there will be *brilliant horse-racing* [!]; after which, a *Te Deum* and magnificent fireworks."

But it is not the Church alone which advertises the *festas.* The tradesmen, having an eye to business, freely make known their ecclesiastic wares through the agency of public journals. The following is a specimen :—

"Notice to the Illustrious Preparers of the Festival of the Holy Spirit.—In the *Rua dos Ourives,* No. 78, may be found a beautiful assortment of Holy Ghosts, in gold, with glories, at eighty cents each; smaller sizes, without glories, at forty cents; silver Holy Ghosts, with glories, at six dollars and a half per hundred; ditto, without glories, three dollars and a half; Holy Ghosts of tin, resembling silver, seventy-five cents per hundred."

The language of the last two advertisements seems to us like blasphemy; but, with the Brazilian public, there is a levity and a want of veneration in holy things shocking to all whose religious impressions are derived from the word of God.

In some particulars the festivals of all the saints are alike. They are universally announced, on the day previous, by a discharge of skyrockets at noon and by the ringing of bells at evening. During the *festa*, also,—whether it continue one day or nine,—the frequent discharge of rockets is kept up. These missiles are so constructed as to explode high up in the air, with a crackling sound, after which they descend in beautiful curves of white smoke if in the daytime, or like meteoric showers if at night. Dr. Walsh, who had resided a number of years in Turkey, thought that the Brazilians quite equalled the Turks of Constantinople in their fondness for exploding gunpowder on festival occasions. He, moreover, gives an estimate, by which it would appear that "about seventy-five thousand dollars are annually expended in Rio for gunpowder and wax,—the two articles which enter so largely into all these exhibitions of pomp and splendor." The wax is consumed in vast quantities of candles that are kept burning before the different shrines, interspersed with artificial flowers and other decorations.

Great care is bestowed upon this manner of adorning churches, by day as well as by night. Sometimes regular rows of blazing tapers are so arranged in front of the principal altars as to present the appearance of semicones and pyramids of light streaming from the floor to the roof of the edifice. These tapers are all made of wax, imported from the coast of Africa for this express use. No animal-oils are used in the churches of Brazil: that which supplies the lamps is made from the olive or from the palm-nut. The tapers are manufactured from vegetable and bees' wax.

Nothing is more imposing than the chief altar of the Candellaria Church, when illuminated by a thousand perfumed tapers, which shed their light amid vases of the most gorgeous flowers. Dr. Walsh states that on a certain occasion he counted in the chapel of S. Antonio eight hundred and thirty large wax flambeaux burning at once, and the same night, in that of the Terceira do Carmo, seven hundred and sixty; so that, in consideration of the number of chapels from time to time illuminated in a similar way, his estimate hardly appears extravagant.

Sometimes, on the occasion of these festivals, a stage is erected in the church, or in the open air near by, and a species of dramatic representation is enacted for the amusement of the spectators. At other times an auction is held, at which a great variety of objects, that have been provided for the occasion by purchase or gift, are sold to the highest bidder. The auctioneer generally manages to keep the crowd around him in a roar of laughter, and, it is presumed, gets paid in proportion to the interest of his entertainment.

Epiphany is celebrated in January, and is styled the day of kings. The occurrence of this holiday is not likely to escape the mind of the most indifferent, for in the morning your butcher kindly sends your beef *gratis*. The festa on that day is in the Imperial Chapel, the Emperor and Court being in attendance to give it a truly royal character. The 20th of January is St. Sebastian's day, on which it is customary to honor the "glorious patriarch" under whose protection the Indians and the French were routed, and the foundations of the city laid. The members of the municipal chamber, or city fathers, take especial interest in this celebration, and by virtue of their office have the privilege of carrying the image of the saint in procession from the Imperial Chapel to the old Cathedral.

The Intrudo, answering to the Carnival in Italy, extends through the three days preceding Lent, and is generally entered upon by the people with an apparent determination to redeem time for amusement in advance of the long restraint anticipated.

The Intrudo, however, is no more celebrated as it was when I first went to Rio. It was then a saturnalia of the most liquid character, and every one,—men, women, and children,—gave themselves up to it with an *abandon* most strongly in contrast with their usual apparent stiffness and inactivity. Before it was suppressed by the police it was a marked event. It was not with showers of sugar-plums that persons were saluted on the days of the Intrudo, but with showers of oranges and eggs, or rather of waxen balls made in the shape of oranges and eggs, but filled with water. These articles were prepared in immense quantities beforehand, and exposed for sale in the shops and streets. The shell was of sufficient strength to admit of being hurled a considerable distance, but at the moment of collision it broke to pieces, bespattering whatever it hit. Unlike the somewhat similar sport of snowballing in cold countries, this *jogo* was not confined to boys or to the streets, but was played in high life as well as in low, in-doors and out. Common consent seemed to have given the license of pelting any one and every one at pleasure, whether entering a house to visit or walking in the streets.

In fact, whoever went out at all on these days expected a ducking, and found it well to carry an umbrella; for in the enthusiasm of the game the waxen balls were frequently soon consumed: then came into play syringes, basins, bowls, and sometimes pails of water, which were plied without mercy until the parties were thoroughly drenched.

Men and women perched themselves along the balconies and windows, from which they not only threw at each other, but also at the passers-by. So great indeed were the excesses which grew out of this sport that it was prohibited by law. The magistrates of the different districts formally declared against the Intrudo from year to year, with but little effect until 1854, when a new *chef de police* with great energy put a stop to the violent Intrudo and its peltings and duckings. It is now conducted in a dry but humorous manner, more in the style of Paris and Rome. The origin of the Intrudo was for a long time considered to have some remote connection with baptism; but Mr. Ewbank has been the first to trace clearly its beginning, and in a very interesting archæological article follows it up to India, that storehouse of many of the practices of the Latin Church.

The procession on Ash-Wednesday is conducted by the third order of Franciscans from the Chapel of the Misericordia, through tne principal streets of the city, to the Convent of S. Antonio. Not less than from twenty to thirty stands of images are borne along on the shoulders of men. Some of these images are single; others are in groups, intended to illustrate various events of scriptural history or Roman Catholic mythology. The dress and ornaments of these effigies are of the most gaudy kind. The platforms upon which they are placed are quite heavy, requiring four, six, and eight men to carry them; nor can all these endure the burden for a long time. They require to be alternated by as many others, who walk by their side like extra pall-bearers at a funeral. The streets are thronged with thousands of people, among whom are numbers of slaves, who seem highly amused to see their masters for once engaged in hard labor. The senhors indeed toil under their loads. The images pass into the middle of the street, with single files of men on either side, each one bearing a lighted torch or wax candle several feet in length. Before each group of images marches an angel (anjinho) led by a priest, scattering rose-leaves and flowers upon the path.

As the reader may be anxious to know what kind of angels take part in these spectacles, I must explain that they are a class created for the occasion, to act as tutelary to the saints exhibited. Little girls, from eight to ten years old, are generally chosen to serve in this capacity, for which they are fitted out by a most fantastic dress. Its leading design seems to be to exhibit a body and wings; wherefore the skirt and sleeves are expanded to enormous dimensions, by means of hoops and cane framework, over which flaunt silks, gauzes, ribbons, laces, tinsels, and plumes of diverse colors. On their head is placed a species of tiara. Their hair hangs in ringlets down their faces and necks, and the triumphal air with which they march along shows that they fully comprehend the honor they enjoy of being the principal objects of admiration.

Military companies and bands of martial music lead and close up the procession. Its march is measured and slow, with frequent pauses, as well to give the burdened brethren time to breathe, as to give the people in the streets and windows opportunity to gaze and wonder. Few seem to look on with any very elevated emotions. All could see the same or kindred images in the churches when they please; and, if the design is to edify the people, a less troublesome and at the same time more effectual mode might easily be adopted. There appears but little solemnity connected with the scene, and most of that is shared by the poor brethren who tug and sweat under the platforms: even they occasionally endeavor to enliven each other's spirits by entering into conversation and pleasantry when relieved by their alternates.

THE ANJINHO.

When the Host is carried out on these and other occasions, but a small proportion of the people are seen to kneel as it passes, and no compulsion is used when any are disinclined to manifest that degree of reverence.*

---

* In 1852 John Candler and Wilson Burgess, two philanthropic Englishmen belonging to the Society of Friends, went to Brazil for the purpose of presenting to the Emperor "an address on slavery and the slave-trade." Their singular costume attracted much notice in the streets; "and on one occasion," they say in their narrative, "as we were walking in the Rua Direita, a Brazilian gentleman accosted us in imperfect English, informing us that he had been in England, and knew the Quakers. 'They [the Brazilians] ask me,' he continued, 'who you are; I tell them Friends,—very good people.' Finding him disposed to be familiar, we told

No class enter into the spirit of these holiday parades with more zeal than the people of color. They are, moreover, specially complimented from time to time by the appearance of a colored saint, or of Nossa Senhora under an ebony skin. *"Lá vem o meu parente,"* (There comes my kindred,) was the exclamation heard by Dr. Kidder from an old negro, as a colored effigy, with woolly hair and thick lips, came in sight; and in the overflow of his joy the old man had expressed the precise sentiment that is addressed by such appeals to the senses and feelings of the Africans.

Palm Sunday in Brazil is celebrated with a taste and effect that cannot be surpassed by any artificial ornaments. The Brazilians are never indifferent to the vegetable beauties by which they are surrounded, since they make use of leaves, flowers, and branches of trees on almost every public occasion; but on this anniversary the display of the real palm-branches is not only beautiful, but often grand.

Holy Week, by which Lent is terminated, is chiefly devoted to religious services designed to commemorate the history of our Lord; but so modified by traditions, and mystified by the excess of ceremonies, that few, by means of these, can form any proper idea of what really took place before the crucifixion of Christ. The days are designed in the calendar as Wednesday of darkness, Thursday of anguish, Friday of passion, and Hallelujah Saturday.

Maunday Thursday, as the English render it, is kept from the noon of that day till the following noon. The ringing of bells and the explosion of rockets are now suspended. The light of day is excluded from all the churches; the temples are illuminated within

---

him we were seeking the National Library. 'I will go with you,' he said. Taking us by the arm, he took us by a narrow paved court-way which we had just avoided. A Roman Catholic church, in which high mass was performing, opened by its principal entrance into the court, and a number of persons stood bareheaded before the doors. We requested him not to take us that way, as we could not take off our hats in honor of the service, and we desired not to give offence. ' Never mind,' was his rejoinder; 'leave that to me.' On coming to the people he took off his own hat, and as we passed through them he said, ' These are my friends; you must give dispensation ;' and we were suffered to go on without molestation. Such dispensation is not permitted in Portugal."—*Narrative of a recent visit to Brazil by John Candler and Wilson Burgess.* London, 1853: Edward Marsh

by wax tapers, in the midst of which, on the chief altars, the Host is exposed. Two men stand in robes of red or purple silk to watch it. In some churches the effigy of the body of Christ is laid under a small cloister, with one hand exposed, which the crowd kiss, depositing money on a silver dish beside it at the same time. At night the people promenade the streets and visit the churches. This is also an occasion for a general interchange of presents, and is turned greatly to the benefit of the female slaves, who are allowed to prepare and sell confectionery for their own emolument.

Friday continues silent, and a funeral-procession, bearing a representation of the body of Christ, is borne through the streets. At night occurs a sermon, and another procession, in which *anjinhos*, decked out as has already been described, bear emblematic devices alluding to the crucifixion. One carries the nails, another the hammer, a third the sponge, a fourth the spear, a fifth the ladder, and a sixth the cock that gave the warning to Peter. Never are the balconies more crowded than on this occasion. There is an interest to behold one's own children performing a part, which draws out hundreds of families who otherwise might remain at home. There is no procession more beautiful and imposing than this. As I gazed at the long line of the gown-clad men, bearing in one hand an immense torch, and leading by the other a brightly-decked *anjinho*,— as from time to time I saw the images of those who were active or silent spectators of that sad scene which was presented on Calvary eighteen hundred years ago,—as I beheld the soldiers, helmet in hand and their arms reversed, marching with slow and measured tread,—as I heard the solemn chant issuing from the voice of childhood, or as the majestic minor strains of the *marche funèbre* wailed upon the night-air,—the æsthetic feelings were powerfully moved. But when a halt occurred, and I witnessed the levity and the utter indifference of the actors, the effect on myself vanished, and I could at once see that the intended effect upon the multitudes in the street and in the neighboring balconies was entirely lost.*

---

* In Brazil, all veneration is taken away by the familiarity of the most sacred things of our holy religion. At Bahia I learned, through a number of Roman Catholic gentlemen, of an occurrence which took place in 1855, in the province of Sergipe del Rey. It was at a festival, and there was to be a powerful sermon

Hallelujah Saturday is better known as "Judas's day," on account of the numerous forms in which that "inglorious patriarch" is made to suffer the vengeance of the people. Preparations having been made beforehand, rockets are fired in front of the churches at a particular stage of the morning service. This explosion indicates that the hallelujah is being chanted. The sport now begins forthwith in every part of the town. The effigies of poor Judas become the objects of all species of torment. They are hung, strangled, and drowned. In short, the traitor is shown up in fireworks and fantastic figures of every description, in company with dragons, serpents, and the devil and his imps, which pounce upon him.

KILLING JUDAS.

Besides the more formal and expensive preparations that are made for this celebration by public subscription, the boys and the negroes have their Judases, whom they do feloniously and mali-

---

preached on the crucifixion. A civilized Indian, by the promise of *muita cachaça*, (plenty of rum,) consented to personify our Saviour on the cross. His position was a trying one, and at the foot of the crucifix stood a bucket filled with rum, in which was a sponge attached to a long reed. The individual whose duty it was to refresh the *caboclo* forgot his office while carried away by the florid eloquence of the *Padre*. The Indian, however, did not forget his contract, and, to the astonishment as well as amusement of the audience, shouted out, "*O Senhor Judêo*, SENHOR *Judeio, mais fel!*" (O Mr. Jew, Mister Jew, a little more gall!)

ciously drag about with ropes, hang, beat, punch, stone, burn, and drown, to their hearts' content.

Lent being over, Easter Sunday is ushered in by the quick and joyous strains of music from fine bands or large orchestras; by illuminating the churches with unwonted splendor; and by the triumphal discharge of rockets in the air, and of artillery from the forts and batteries.

On Whitsunday the great feast of the Holy Spirit is celebrated. In preparation for this, begging-processions go through the streets, a long while in advance, in order to secure funds. In these expeditions the collectors wear a red scarf (*capa*) over their shoulders: they make quite a display of flags, on which forms of a dove are embroidered, surrounded by a halo or gloria. These are handed in at windows and doors, and waved to individuals to kiss: they are followed by the silver plate or silk bag, which receives the donation that is expected from all those, at least, who kiss the emblem. The public are duly notified of the approach of these august personages by the music of a band of tatterdemalion negroes, or by the songs and tambourine accompaniments of sprightly boys who sometimes carry the banner.

Collections of this stamp are very frequent in the cities of Brazil, inasmuch as some festa is always in anticipation. Generally a miniature image of the saint whose honor is contemplated is handed around

COLLECTORS FOR CHURCH FESTIVALS.

with much formality, as the great argument in favor of a donation. The devotees hasten to kiss the image, and sometimes call up their children and pass it round to the lips of each. These collectors, and a class of females called *beatas*, at times become as troublesome as were the common beggars before they were accommodated at the House of Correction. Occasionally but one or two of these individuals go around, crying out, with a most nasal twang, in the street and at every corner, "*Esmolas* [alms] *para nossa Senhora*" of this or that church. (1866, this begging is greatly curtailed.)

On the preceding page we behold a pair of these semi-ecclesiastic gentlemen-beggars who may be seen returning along the Praia da Santa Luzia after one of their collecting-excursions.

The expeditions for *Espirito Santo* assume a very peculiar and grotesque character in remote sections of the Empire. The late Senator Cunha Mattos describes them, in the interior, under the name of *fuliões cavalgadas*. He mentions in his Itinerario having met one between the rivers of S. Francisco and Paranahiba, composed of fifty persons, playing on violins, drums, and other instruments of music, to arouse the liberality if not the devotion of the people; and also prepared with leathern sacks and mules, to receive and carry off pigs, hens, and whatever else might be given them.

Among the Indians in the distant interior, the live animals are frequently promised beforehand to some particular saint; and often, when a traveller wishes to buy some provisions, he is assured, "That is St. John's pig;" or, "Those fowls belong to the Holy Ghost."

The procession of *Corpus Christi* is different from most of the others. The only image exposed is that of St. George, who is set down in the calendar as the "defender of the Empire." How this "godly gentleman of Cappadocia" became the defender of Brazil I have not been able to ascertain; but his festival—falling as it does on *Corpus Christi* day—is celebrated with great pomp. It is a daylight affair, and occurs in the pleasantest season of the year. St. George is always carried around the city on horseback. He is ruddy and of a fair countenance, with a flowing wig of flaxen curls floating on his shoulders. He flourishes in armour and a red velvet mantle. For the day some devout person of his name lends the saint his jewels; but when the festival is over he is stripped of his glories and put away for the moths till the following year. He is not remarkable for his horsemanship: his stiff legs stick out on each side, and two men hold him to the saddle. If his prototype had been no better equestrian, the dragon would have been un-killed to the present day.

The Emperor walks bareheaded, and carrying a candle, in this procession, in imitation of the piety of his ancestors, and is attended by the Court, the cavalheiros, or knights of the military orders, and the municipal chamber in full dress, with their insignia and badges of office. Whenever the Emperor goes out on these occasions, the inhabitants of the streets through which he is to pass rival each other in the display of rich silk and damask hangings from the windows and balustrades of their houses.

In 1846, a certain Brazilian had the distinguished honor of trans-porting from Rome to Rio the holy remains of the martyr-virgin St. Priscil-liana. This was deemed a most auspicious acquisition for the city by some, but by others it was highly condemned as an egregious humbug. Nevertheless, she was inaugurated. In order that the bones might not appear as repulsive as those of the renowned "eleven thousand virgins" in the Church of St. Ursula at Cologne, the frail remains of St. Priscilliana were en-cased in wax by some clever artist at Rome at the time her saintship was said to have been removed from the catacombs where she had been buried more than a thousand years!

SANTA PRISCILLIANA.

St. Priscilliana's likeness was engraved, and the picture was "*exchanged;*" and the above engraving is a fac-simile of the one "exchanged" while I resided in Rio de Janeiro. She is represented with a sword stuck unpleasantly through her delicate neck, which means, as the Bishop of Rio de Janeiro* hath it, that the Emperor Julian the Apostate had her put to death in this manner! The erudite bishop does not give us any of his authorities; but the faithful are expected never to entertain the least doubt when a high prelate speaks. I know not what miracles she has performed at Rio, for very little is heard concerning her at present, and it is certain that she did not prevent the yellow fever and cholera from visiting the capital of the Empire. It may, however, be asserted, on the other hand, that this was not the department of St. Priscilliana; as St. Sebastian is supposed to have the city under his especial charge.

When the cholera visited the coast of Brazil, though not so fatal as in Europe and the United States, yet its ravages were somewhat extensive among the slaves, who had escaped the yellow fever which in former years had attacked the whites. When the cholera made its appearance at Rio, the city was in a universal wail of terror: charms and amulets were eagerly sought after, and superstitious preventives were invented every hour. Prayers of saints were worn next to the skin, as they are among the Mohammedans of Arabia or the heathen of India. Badly-executed pictures of St. Sebastian were *"exchanged"* for a few vintems, and a star, with a prayer to the Virgin Mary, called "The miraculous Star of Heaven," was considered a certain safeguard to any person who possessed it. Advertisements like the following appeared in the daily papers:—

ORAÇÃO PARA BENZER AS CASAS

contra a epidemia reinante, ornada de emblemas religiosos, troca-se por 80 rs., na Rua dos Latoeiros n. 59.

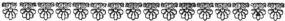

"*A Prayer for blessing residences* against the reigning epidemic, adorned with religious emblems, is exchanged for four cents at No. 59 Rua dos Latoeiros."

---

* Pastoral letter published March, 1846, at Rio de Janeiro. Also *Noticia Historica da Santa Priscilliana* in the *Annuario do Brazil* for 1846.

The succeeding announcement, however, must have been from some money-making fellow without church-policy in his head, or he would have advertised his holy ware as *troca-se* instead of *vende-se*.

PALAVRAS SANTISSIMAS

E

ARMAS DA IGREJA

contra o terrivel flagello da peste, com a qual se tem applacado a Divina Justiça, como se vio no caso que succedeu no real mosteiro de Santa Clara de Coimbra em 1480. Vende-se na Rua da Quitanda n. 174. Preço, 320 rs.

[Translation.] *"Most holy words and arms of the Church* against the terrible scourge of the pest, with which Divine Justice chastises, as seen in the case which succeeded in the royal monastery of St. Claire of Coimbra in 1480. To be sold at No. 174 Rua da Quitanda. Price, 16 cents."

What Dr. Paulo Candido, Dr. Meirelles, Dr. Sigaud, Dr. Pacheco da Silva, and other eminent physicians, thought of such remedies we know not; but we believe that both they and many of the people of Rio de Janeiro looked upon this religious quackery in the right light. Nevertheless, there was, in the general alarm, a great summoning of the church militant, and the newspapers of September, 1855, are full of long-sentenced notices of penitential processions.

Such appeals to the faithful were not in vain. The images were removed and carried through the streets; and torchlight-processions of immense length—in which marched delicate ladies bare-foot—were of frequent occurrence. With all these precautions, the pestilence did not cease, though business went on as usual. Common sense, however, had not left Rio, notwithstanding the panic which prevailed. The secular authorities, urged on by the able editor of the principal newspaper of the city, at last forbade all processions, as the exposure consequent thereon tended to promote the spread of disease; so the saints had no more promenades by lamplight, and the young ladies kept their bare feet at home.

It is pleasing to contemplate at this crisis the conduct of the monarch. The Emperor and his family remained at their palace near the city, in order to inspirit the people, although it was the usual time of removal to their mountain-residence of Petropolis. His Majesty visited the hospitals, and superintended the sanatory regulations, besides contributing largely to the fund for the sick poor.

We cannot devote more space to religion in Brazil,—this interesting but painful subject,—painful to every true Christian and well-wisher to his race. If we look at Brazil in the *point de vue religieuse*, we are overwhelmed at the amount of ignorance and superstition that prevails. Let any one read Mr. Ewbank's Sketches, and they will see, archæologically considered, how close is the relation between heathen Rome and Christian Rome. If we grant that this corrupt church at one time had the only light and knowledge, there is no necessity that we should remain in modified darkness or use the glimmer of lamplight when we may have the clear effulgence of the noonday sun. May that light beam upon Brazil!

Note for 1866.—There are several native Protestant churches now in Brazil. The regularly ordained pastors of these churches are legally authorized to perform the marriage ceremony. The clause of the constitution in regard to religious toleration has been fully tested on three different occasions, and is shown to be a "living letter."—The *Imprensa Evangelica* is the bi-monthly religious journal of the Protestants at Rio, and is reasonably successful. Several faithful preachers of the gospel from Europe and North America are now laboring with encouraging success. In Appendix I will be found a remarkable article from the Roman Catholic editor of the "Anglo-Brazilian Times," on the necessity of removing from Brazilian law disabilities in regard to civil and mixed marriages, and to Protestant representation in Parliament.

CHAPTER X.

THE HOME-FEELING—BRAZILIAN HOUSES—THE GIRL—THE WIFE—THE MOTHER—MOORISH JEALOUSY—DOMESTIC DUTIES—MILK-CART ON LEGS—BRAZILIAN LADY'S DELIGHT—HER TROUBLES—THE MARKETING AND WATERING—KILL THE *BIXO*—BOSTON APPLES AND ICE—FAMILY RECREATIONS—THE BOY—THE COLLEGIO—COMMON-SCHOOLS—HIGHEST ACADEMIES OF LEARNING—THE GENTLEMAN—DUTIES OF THE CITIZEN—ELECTIONS—POLITICAL PARTIES—BRAZILIAN STATESMEN—NOBILITY—ORDERS OF KNIGHTHOOD.

THE German, the Englishman, and their descendants, have no characteristic more marked than the home-feeling. The fireside-circle, with its joys and cares, does not belong to the Gaul or to the Italian. The Southern European has much in his delicious climate to make him an out-of-door being. The old Roman was one who lived in public. His existence seemed to be a portion of the forum, the public bath, the circus, and the theatre. "Without books, magazines, and newspapers, without letters to write, and with a fine climate always attracting him into the open air, there was nothing to call him home but the requisitions of eating and sleeping." The city of Pompeii probably contained not more than twenty-five thousand inhabitants, and only one-sixth of its space has been exhumed. In that small district there have been found public edifices merely for theatrical entertainment, which will seat seventeen thousand spectators. Most of the nations descended from the Romans are, like them, without the endearing associations connected with the word *home*. There is, however, an important exception to this rule in the case of the Portuguese nation, which in every other respect is more Roman than any living people. The home and the family exist; and doubtless the Lusitanians owe this to the Moors, who engrafted upon the Latin stock something of Oriental exclusiveness. The Portuguese and their American descendants to this day watch with a jealous eye their private abodes, and, spending many of their hours within those precincts which are

11   161

their castles, the home-attachments and family associations have been cherished and perpetuated.

I propose in this chapter to consider the residence and the family,—to trace the education of the children to that age when they go forth to occupy the position of adult years.

The city-home is not an attractive place; for the carriage-house and stable are upon the first floor, while the parlor, the alcoves, and the kitchen are in the second story. Not unfrequently a small area, or court-yard occupies the space between the coach-house and the stable, and this space separates, on the second floor, the kitchen from the dining-room.

OLDER BRAZILIAN DWELLING-HOUSE.

The engraving represents one of the older city-residences at Rio. The access to the staircase is through the great door whence the carriage thunders out on festas and holidays. At night it is shut by iron bars of prison-like dimensions. Every lock, bolt, or mechanical contrivance seem as if they might have come from the Pompeiian department of the Museo Borbonico at Naples. The walls, composed of broken bits of stone cemented by common mortar, are as thick as those of a fortress.

In the daytime you enter the great door and stand at the bottom of the staircase; but neither knocker nor bell announce your presence. You clap your hands rapidly together; and, unless the family is of the highest class, you are sure to be saluted by a slave from the top of the stairs with *"Quem é?"* (Who is there?) If you should behold your friends in the balcony, you not only, if intimate, salute by removing the hat, but move quickly the fingers of your hand, as if you were beckoning to some one.

The furniture of the parlor varies in costliness according to the degree of style maintained; but what you may always expect to find is a cane-bottomed sofa at one extremity and three or four chairs arranged in precise parallel rows, extending from each end of it toward the middle of the room.   In company the ladies are expected to occupy the sofa and the gentlemen the chairs.

The town-residences in the old city always seemed to me gloomy beyond description.   But the same cannot be said of the new houses, and of the lovely suburban villas, with their surroundings of embowering foliage, profusion of flowers, and overhanging fruits.   Some portions of the Santa Theresa, Larangeiras, Bota-fogo, Catumby, Engenho Velho, Praia Grande, and San Domingo, cannot be surpassed for beautiful and picturesque houses.   I cite particularly the homes of M. Maximo de Souza, and Mr. Ginty.

There are various classes of society in Brazil as well as else-where, and the description of one would not hold good for another; but, having sketched the house, I shall next endeavor to trace the inmates from infancy to adult life.

The Brazilian mother almost invariably gives her infant to a black to be nursed.   As soon as the children become too trouble-some for the comfort of the senhora, they are despatched to school; and woe betide the poor teachers who have to break in those viva-cious specimens of humanity!   Accustomed to control their black nurses, and to unlimited indulgence from their. parents, they set their minds to work to contrive every method of baffling the efforts made to reduce them to order.   This does not arise from malice, but from want of parental discipline.   They are affectionate and placable, though impatient and passionate,—full of intelligence, though extremely idle and incapable of prolonged attention.   They readily catch a smattering of knowledge : French and Italian are easy to them, as cognate tongues with their own.   Music, sing-ing, and dancing suit their volatile temperaments; and I have rarely heard better amateur Italian singing than in Rio de Janeiro and Bahia.   Pianos abound in every street, and both sexes become adept performers.   The opera is maintained by the Government, as it is in Europe, and the first musicians go to Brazil.   Thalberg triumphed at Rio de Janeiro before he came to New York.   The manners and address of Brazilian ladies are good, and their carriage is graceful.   It is true that they have no fund of varied knowledge to make a conversation agreeable and instructive; but they chatter nothings in a pleasant way, always excepting a rather high tone of voice, which I suppose comes from frequent commands given to Congo or Mozambique. Their literary stores consist mostly of the novels of Balzac, Eugene Sue, Dumas *père et fils*, George Sand, the gossipping *pacotilhas* and the *folhetins* of the newspapers. Thus they fit themselves to become wives and mothers.

Dr. P. da S——, a gentleman who takes a deep interest in all matters of education, and whose ideas are practically and successfully applied to his own children, who possess solid acquirements as well as graceful accomplishments, once said to me, "I desire with all my heart to see the day when our schools for girls will be of such a character that a Brazilian daughter can be prepared, by her moral and intellectual training, to become a worthy mother, capable of teaching her own children the elements of education and the duties which they owe to God and man: to this end, sir, I am toiling." Such schools are increasing, and some are very excellent; but, in eight cases out of ten, the Brazilian father thinks that he has done his duty when he has sent his daughter for a few years to a fashionable school kept by some foreigner: at thirteen or fourteen he withdraws her, believing that her education is finished. If wealthy, she is already arranged for life, and in a little time the father presents to his daughter some friend of his own, with the soothing remark, "*Minha filha*, this is your future husband." A view of diamonds, laces, and carriages dazzles her mental vision, she stifles the small portion of heart that may be left her, and quietly acquiesces in her father's arrangement, probably consoling herself with the reflection that it will not be requisite to give her undivided affections to the affianced companion,—that near resemblance of her grandfather. Now the parents are at ease. The care of watching that ambitious young lady devolves on her husband, and thenceforth he alone is responsible. He, poor man, having a just sense of his own unfitness for such a task, places some antique relative as a duenna to the young bride, and then goes to his counting-house in happy security. At night he returns and takes her to the opera, there to exhibit the prize that his *contos**

---

* A *conto* of *reis* is one thousand milreis,—equal to five hundred dollars. The Brazilian never reckons a man's wealth by saying, "He is worth so many thousand *milreis*;" but, "He has so many *contos*."

have gained, and to receive the congratulations of his friends on the lovely young wife that he has bought. "'Tis an old tale;" and Brazil has not a monopoly of such marriages.

Then the same round of errors recommences: her children feel the effects of the very system that has rendered the mother a frivolous and *outward* being. She sallies forth on Sundays and festas, arm-in-arm with her husband or brother, the children preceding, according to their age, all dressed in black silk, with neck and arms generally bare, or at most a light scarf or cape thrown over them, their luxuriant hair beautifully arranged and orna-

GOING TO MASS.

mented, and sometimes covered with a black lace veil: prayer-book in hand, they thus proceed to church. Mass being duly gone through and a contribution dropped into the poor-box, they return home in the same order as before.

It is often matter of surprise to Northerners how the Brazilian ladies can support the rays of that unclouded sun. Europeans glide along under the shade of bonnets and umbrellas; but these church-going groups pass on without appearing to suffer. The bonnet is, however, becoming the prevailing mode.

You remark, in these black-robed, small-waisted young ladies, a contrast to the ample dame who follows them. A Brazilian matron generally waxes wondrously broad in a few years,—probably owing to the absence of out-door exercise, of which the national habits deprive her. It cannot be attributed to any want of temperance; for we must always remember that Brazilian ladies rarely take wine or any stimulant. On "state occasions," when healths are drunk, they only touch it for form's sake. During many years of residence, I cannot recall a single instance of a lady being even suspected of such a vice, which, in their eyes, is the most horrible reproach that can be cast upon the character. *Está bebado,* (He is drunk;)—pronounced in the high and almost scolding pitch of a Brazilian woman,—is one of the severest and most withering reproaches. In some parts of the country the expression for a dram is *um baeta Inglez,* (an English overcoat;) and the term for an intoxicated fellow, in the northern provinces, is *Elle está bem Inglez,* (He is very English.) The contrast between the general sobriety of all classes of Brazilians and the steady drinking of some foreigners and the regular "blow-out" of others is painful in the extreme.

Wives in Brazil do not suffer from drunken husbands; but many of the old Moorish prejudices make them the objects of much jealousy. There is, however, an advance in this respect; and, far more frequently than formerly, women are seen out of the church, the ballroom, and the theatre.

Nevertheless,—owing to the prevailing opinion that ladies ought not to appear in the streets unless under the protection of a male relative,—the lives of the Brazilian women are dull and monotonous to a degree that would render melancholy a European or an American lady.

At early dawn all the household is astir, and the principal work is performed before nine o'clock. Then the ladies betake themselves to the balconies for a few hours, to "loll about generally," to gossip with their neighbors, and to look out for the milkman and for the *quitandeiras.* The former brings the milk in a cart of novel construction to the foreigner,—or at least he has never seen such a vehicle used for this purpose before going to Brazil. The cow is the milk-cart! Before the sun has looked over the mountains, the *vacca,* accompanied by her calf, is led from door to door by a Portuguese peasant. A little tinkling bell announces her presence. A slave descends with a bottle and receives an allotted portion of the refreshing fluid, for which he pays about sixpence English.   One would suppose that all adulteration is thus

avoided.   The inimitable Punch says, if in the human world the "child is father to the man," in the London world the pump is father to the cow,—judging from the results, (*i.e.* the milk sold in that vast metropolis.)   Alas! mankind is the same in Brazil that it is in London.   Milk may be obtained pure from the cow if you stand in the balcony and watch the operation; otherwise your bottle is filled from the tin can carried by the Oportoense, and which can has oftentimes a due proportion of the water that started from the top of Corcovado and has gurgled down the aqueduct and

THE QUITANDEIRA.

through the fountain at the corner of the street.

The *quitandeiras* are the venders of vegetables, oranges, guavas, maracujas, (fruits of the "passion-flower,") mangoes, *doces*, sugar-cane, toys, &c.   They shout out their stock in a lusty voice, and the different cries that attract attention remind one of those of Dublin or Edinburgh.   The same nasal tone and high key may be noticed in all.   Children are charmed when their favorite old black tramps down the street with toys or doces.   Here she comes, with her little African tied to her back and her tray on her head. She sings,—

> "Cry meninas, cry meninos,
> Papa has money in plenty,
> Come buy, ninha, ninha, come buy!"—

and, complying with the invitation, down run the little meninos and meninas to buy doces doubly sugared, to the evident destruction of their gastric juices and teeth.   Be it remarked, *en passant*, that no profession has more patronage in Rio than that of dentistry.

At length there appears at the head of the street that charm of a Brazilian lady's day,—the pedlar of silks and muslins.   He announces his approach by the click of his *covado*, (measuring-stick,) and is followed by one or more blacks bearing tin cases on their heads.   He walks up-stairs sure of a welcome; for, if they need nothing of his wares, the ladies have need of the amusement of looking them over.   The negroes deposit the boxes on the floor and retire. Then the skilful Italian or Portuguese displays one thing after another; and he manages very badly if he cannot prevail on the economical lady to become the possessor of at least one cheap bargain.   As to payment, there is no need of haste: he will call again next week, or take it by instalments,— just as the senhora finds best; only he should like senhora to have *that* dress,

THE BRAZILIAN LADY'S DELIGHT.

—it suits her complexion so well; he thought of the senhora as soon as he saw it; and the price,—a mere *nada*.   Then, too, he has a box of lace, some just made,—a new pattern for the ends of towels,—insertion for pillow-cases, and trimmings for under-garments.

Some families have negresses who are taught to manufacture this lace,—the thread for which is brought from Portugal,—and

their fair owners make considerable profit by exchanging the products of their lace-cushions for articles of clothing. One kind of
needlework in which they excel is called *crivo*. It is made by
drawing out the threads of fine linen and darning in a pattern.
The towels that are presented to guests after dinner are of the
most elaborate workmanship, consisting of a broad band of *crivo*
finished by a trimming of wide Brazilian thread-lace.

These Italian and Portuguése pedlars sell the most expensive
and beautiful articles. A Brazilian lady's wardrobe is almost
wholly purchased at home. Even if she do not buy from the
mascate, she despatches a black to the Rua do Ouvidor or Rua
da Quitanda, and orders an assortment to be sent up, from which
she selects what is needed. The more modern ladies begin to wear
bonnets, but these are always removed in church. Almost every
lady makes her own dresses, or, at least, cuts them out and
arranges them for the slaves to sew, with the last patterns from
Paris near her. She sits in the midst of a circle of negresses, for
she well knows that "as the eye of the master maketh the horse
fat," so the eye of the mistress maketh the needle to move. She
answers to the description of the good woman in the last chapter
of Proverbs:—"She riseth up while it is yet night, and giveth a
portion to her maidens; she maketh fine linen [crivo and lace] and
selleth it;" and, though her hands do not exactly lay hold on the
spindle and distaff, yet "she looketh well to the ways of her household, and eateth not the bread of idleness," always excepting that
taken on the balcony.

We may infer that the habits of servants were the same in Solo-
mon's time as in Brazil at the present day, judging by the amount
of trouble they have always given their mistresses. A lady of
high rank in Brazil declared that she had entirely lost her health
in the interesting occupation of scolding negresses, of whom she
possessed some scores, and knew not what occupation to give them
in order to keep them out of mischief. A lady of noble family
one day asked a friend of mine if she knew any one who desired
to give out washing, as she (the senhora) had nine lazy servants at
home for whom there was no employment. She piteously told her
story, saying, "We make it a principle not to sell our slaves, and
they are the torment of my life, for I cannot find enough work to

keep them out of idleness and mischief." Another, a marchioness, said that her blacks "would be the death of her."

Slavery in Brazil, setting aside any moral consideration of the question, is the same which we find the "world over,"—viz.: It is an expensive institution, and is, in every way, very poor economy. When I have looked upon the careless, listless work of the bond-man, and have watched the weariness of flesh to the owner, I have sometimes thought the latter was most to be pitied. Any cruelty that may be inflicted upon the blacks by the whites is amply avenged by the vices introduced in families, and the troublesome anxiety given to masters.

One of the trials of a Brazilian lady's life is the surveillance of the slaves who are sent into the streets for the purpose of market-ing and carrying water.

The markets in Rio are abundantly supplied with all kinds of fish and vegetables. Of the former there are many delicate species unknown in the North. Large prices are given for the finer kinds. One called the garopa is much sought for as a *pièce de résistance* for the supper-table on a ball-night. Fifty milreis (about twenty-five dollars) are given on such occasions. A fish is always the sign of a *casa de pasto*, or common restaurant, at Rio.

The market near the Palace Square is a pleasant sight in the cool of the morning. Fresh bouquets shed a fragrance around, and the green vegetables and bright fruits contrast well with the dark faces of the stately Mina negresses who sell them. "What is the price of this?" "What will the senhor give?" is the common reply; and woe betide the first efforts of a poor innocent ship's-steward in his early attempts at negotiation with these queenly damsels, whose air seems to indicate that with them to sell or not to sell is equally indifferent and beneath their notice.

The indigenous fruits of the country are exceedingly rich and various. Besides oranges, limes, cocoanuts, and pineapples, which are well known among us, there are mangoes, bananas, fruitas de conde, maracuja, pomegranates, mammoons, goyabas, jambos, araças, cambocas, cajus, cajas, mangabas, and many other species whose names are Hebrew to Northern ears, but which quickly convey to a Brazilian the idea of rich, refreshing, and delicate fruits, each of which has a peculiar and a delicious flavor.

With such a variety to supply whatever is to be desired, in view of either the necessaries or luxuries of life, none need complain. These articles are found in profusion in the markets, and also hawked about through the town and suburbs by slaves and free negroes, who generally carry them in baskets upon the head. Persons who wish to purchase have only to call them by a suppressed whistle, (something like pronouncing imperfectly the word *tissue,*) which they universally understand as an invitation to walk in and display their stock.

THE EDIBLE PALM, (EUTERPE EDULIS.)

In an outer circle of the market mentioned you find small shops filled with birds and animals. Here gay macaws and screaming parrots keep up a perpetual concert with chattering apes and diminutive monkeys. At a little distance outside are huge piles of oranges, panniers of other fruits ready to be sold to the retailer and the *quitandeiras,* wicker-baskets filled with chickens and bundles of palmito for cooking. It makes one sad to think that the procuring of these palmito-sticks has destroyed a graceful palm, (*Euterpe edulis;*) but what is there that we are not ready to sacrifice to that Maelstrom, the stomach? One of those beautiful trees I sketched at Constancia, fifty miles from Rio. It was not straight, as we usually find it, but gracefully curved; and, as it lifted its slender form and tufted summit above the tropic forest, it presented a picture of such uncommon loveliness, that day after day I visited the spot to drink my fill of beauty.

Here comes the black cook, José, or Cæsar, basket on arm, counting with his fingers, and bent on beating down to the lowest

A BARGAIN.

price the white-teethed Ethiopian who presides, in order that he may have a few vintems, filched from his master, to spend, as he returns home, in the purchase of a little cachaça, "*para matar o bixo,*" ("to kill the beast.") What this much-feared animal is has never been ascertained; but certainly, judging from the protracted effort that is required to kill him, he must be possessed of remarkable tenacity of life,—a sort of phœnix among animals! The fish, vegetables, fruit, and indispensable chickens, being purchased to his satisfaction, he next goes to the street appropriated to the butchers. Here he buys some beef, lean but not ill-flavored, an apology for mutton easily mistaken for patriarchal goat, or a soft, pulpy substance, considered a great delicacy, (appropriately termed, by the Emerald Islanders, "staggering Bob,")—the flesh of an unfortunate calf that had scarcely time to look at the blue sky ere it was consigned to the butcher's knife. Then he proceeds to the venda to purchase the little dose for his *bixo*, and wends home, in high good-humor, to prepare breakfast.

In many families a cup of strong coffee is taken at sunrise, and then a substantial meal later in the morning. Dinner is usually served about one or two o'clock,—at least where the hours of foreigners have not been adopted. Soup is generally presented, and afterward meat, fish, and pastry at the same time. Except at dinners of ceremony, an excellent dish, much relished by foreigners, always finds a place on a Brazilian table. It is compounded of the feijão, or black beans of the country, mingled with some *carne secca* (jerked beef) and fat pork. *Farinha*, or mandioca-flour, is sprinkled over it, and it is worked into a stiff paste. This farinha is the bread for the million, and is the principal food of the blacks throughout the country, who would consider it much deteriorated by being eaten in any other manner than with the fingers. It is an excellent and nutritious diet, and with it they can endure the hardest labor. Coffee or maté are often taken after dinner, and the use of tea is becoming more common. The "cha nacional" bids fair to rival that of China; but the maté, though not generally used in the Middle and Northern provinces, is considered more wholesome than tea, being less exciting to the nerves. Some families have supper frequently of fish; but in others nothing substantial is taken after dinner, and they retire very early to rest. Rio is as quiet at ten o'clock P.M. as European cities at two in the morning. Even the theatre-goers make but little noise, as they are generally on foot,—at least if they reside in the city. So much do the places of public amusement depend on the pedestrians, that if the evening is decidedly rainy it is usual to postpone the performance until another night. It must be remembered that half an hour's rain transforms the streets of Rio into rushing canals, all the drainage being on the surface. On a drenching day, the *pretos de ganho*, or porters, who lounge at the corner of every street, make a good harvest by carrying people on their backs across these impromptu streams. Sales are often announced with this condition :—"The weather permitting."

One of the greatest delights for the black population of Rio is the necessity of carrying water from the *chafariz* or public fountain, or from the water-pipe which is at the corner of almost every street. Blackey lazily lounges out with his *barril* under his arm, and happy is Congo if he espies a long *queue* of his compatriots awaiting their turn at the stopcock. Here the news of their little

world is told amid bursts of Ethiopian laughter; or a small flirtation is carried on with Rosa or Joaquinha from the next street; or perhaps there is an upbraiding lecture administered by some jetty damsel from Angola, whose voice, to his consternation, is by no means *pianissimo*. There is another out-door affair much more congenial: *i.e.* many a sly attempt to kill the bixo is made at the adjoining venda while the water pours into the *barrils* of the earlier comers.

Some mistresses, however, who find that their cooks have *always* to wait

THE ANGOLIAN'S REPROACH.

for the water, make arrangements with the water-carriers, who perambulate the streets with an immense hogshead mounted on wheels and drawn by a mule. This vehicle, during a fire, (not a frequent occurrence,) is required to supply the fire-engines. These men are generally natives of Portugal or the Azores, and seem eminently qualified by nature to be hewers of wood and drawers of water. They carry the water up-stairs and pour it into large earthen jars, which bring to mind the waterpots at the marriage of Cana in Galilee. The huge earthen vases are arranged on stands in places where there is a current of air, and the liquid element in them thus acquires a coolness which, though not equal to the iced water of the United States, possesses a delightful frigidity. Ice is in Brazil an expensive luxury, brought solely from North America, and not in general use even in Rio, and, of course, unknown in the country. Boston apples and ice are both in the highest esteem; but the latter was rejected, as altogether unwholesome, upon its introduction in 1833, and the first cargo was a total loss to the adventurers. At the present time both command a good price; and in the month of January the quitandeiras may be heard crying out lustily, " Maçãs Americanas," (American apples,) which they sell for five or six vintems each

THE ILHEO WATER-VENDER.

The Fluminensian lady has occasionally some respite from slave-watching and household cares, when the senhor takes her to Petropolis or Tijuca, or perhaps gives her a few weeks of fresh air at Constancia or Nova Fribourgo. Such visits are not, however, so frequent as one would wish, and the senhora must content herself with festas, the opera, and a ball, as a relief from her usual round of duties. An evening-party in Rio generally means a ball. Familiar intercourse with the higher families is difficult of attainment by foreigners; but when the stranger is admitted he is received en famille, and all ceremony is laid aside. In such home-circles the evenings are often spent in music, dancing, and games of romps.

Here men of highest position are sometimes seen unbending their stiff exteriors, and joining heartily in innocent mirth. A game called "*pilha tres*" is a favorite, and is quite as wild and noisy as "pussy wants a corner." An American gentleman informed me that on one occasion he joined in this play with a Minister of the Empire, the Viscountess, (his wife,) two Senators, an ex-Minister-plenipotentiary, three foreign Chargés d'Affaires, and the ladies and children of the family. No one feared any loss of dignity by thus laying aside, for the moment, his ordinary gravity, and all seemed to enjoy themselves in the highest degree.

The Brazilians have large families, and it is not an uncommon thing to find ten, twelve, or fifteen children to a single mother. I saw a gentleman—a planter—in the province of Minas-Geraes, who was one of twenty-four children by the same mother. I afterward was presented to this worthy matron at Rio de Janeiro.

I am persuaded that there is much of the home-element among the Brazilians. Family fête-days and birthdays are celebrated with enthusiasm. Though the standard of general morality is very much lower than that of the United States and England, I believe it to be above that of France, and there is a *home-feeling* diffused among all classes, which tends to render the Brazilian a more order-loving man than the Gaul. With a pure religion his excellencies would make him infinitely superior to the latter.

The education of the Brazilian boy is better than that of his sister. There is, however, a great deal of superficiality: he is made a "little old man" before he is twelve years of age,—having his stiff black silk hat, standing collar, and cane; and in the city he walks along as if everybody were looking at him, and as if he were encased in corsets. He does not run, or jump, or trundle hoop, or throw stones, as boys in Europe and North America. At an early age he is sent to a *collegio*, where he soon acquires the French language and the ordinary rudiments of education in the Portuguese. Though his parents reside in the city, he boards in the *collegio*, and only on certain occasions does he see his father or mother. He learns to write a "good hand," which is a universal accomplishment among the Brazilians; and most of the boys of the higher classes are good musicians, become adepts in the Latin, and many of them are taught to speak English with creditable fluency.

CASCATE GRANDE—TIJUCA.

(By Hamilton, after a Sketch by Mrs. L. A. Cuddehy.)

The examination was formerly a great anniversary, when the little fellows were starched up in their stiffest clothes and their minds were "crammed" for the occasion.   The boys acted their parts, and the various *professores,* in exaltation of their office, read or delivered *memoriter* speeches to the admiring parents; and the whole was wound up by some patron of the school crowning with immense wreaths the "good boys" who stood highest during the session. The *collegio* then took a vacation of a few weeks, and commenced again with its boarders, the "very young gentlemen" students But these things have greatly changed for the better, and many collegios are ably conducted.

The principals of these establishments, when gifted with good administrative capacities, reap large sums.   One with whom I was acquainted had, after a few years' teaching, 20,000$000 (ten thousand dollars) placed out at interest.   The *professores* do not always reside in the *collegio,* but teach by the hour for a stipulated sum, and are thus enabled to instruct in a number of schools during the day.   The English language has become such a *desideratum* at Rio, that every collegio has its *professor Inglez.*

There has recently been a great improvement in the collegios as well as in the public schools.   The *professores* were summoned, by a commission under the Superintendent of Public Instruction, to appear at the Military Academy, and there to be examined as to their qualifications for giving instruction.   If they passed their examination, which was most rigid, they received a license to teach, for which they had to pay a certain fee.   The principals also were required to undergo an examination, if the commission should think it proper; and they were not permitted to carry on their *collegios* without a certificate.   The educational authorities also asserted their right to visit these private academies at any hour of the day or night, to examine the proficiency of the scholars at any time during the term, to investigate their sleeping-apartments, their food, and whatever appertained to their mental or physical well-being.   This was not a mere threat, but schools were actually visited, and some were reformed more rapidly than agreeably.   The system of "cramming" was in a measure broken up, and the Empire thus took under its control the instruction given in the private as well as in

the public *aulas*. This educational innovation at the capital is
owing to the energetic measures taken by the Visconde de Itaborahy, and Dr. Manoel Pacheco da Silva, who is at present the
President of the first classical institution of Rio de Janeiro, the
Imperial College of D. Pedro II. The note of reform was sounded;
every duty connected with teachers or scholars was fully investigated, and the revolution was made, notwithstanding the
complaints of *prófessores* who were degraded as incompetent,
and parents who found their children rigidly examined and only
promoted in the public schools after convincing proofs of real
progress.

There is a common-school system throughout the Empire, more
or less modified by provincial legislation. The General Government
during the years 1854–55 educated 65,413 children: there were
probably as many more of whom we have no Government report,
who were educated by private tuition and under provincial
authority. When, therefore, we consider the number of slaves and
Indians in Brazil, and also when we reflect that the common-school
system is in its infancy, it is an encouraging proportion. There
are great defects in these elementary schools, but each year they
are improving. There seems to be an inquiry among the educated
men and the statesmen as to the plan best adapted to the country.
This inquiry is not always confined to the highest class of citizens.
Once in the interior I was aroused from my slumbers by a loud
knocking at the door. I hastily opened it, and saw a respectably-
dressed Brazilian, who informed me that he was a school-teacher,
and, learning that an American was in the village and would leave
that morning, he had made bold to come at this early hour (the
sun was just peeping over the palm-trees) to ask me if I could
either give him an account of the American system of teaching, or
could send him documents on that subject. In the same place
another teacher spoke to me of Horace Mann's reports on the common schools of Massachusetts!

Great ignorance prevails in a large portion of the population, and,
though many years may elapse before a tolerable degree of knowledge will be properly diffused, yet the *beginning* has been made,
and the French proverb is true in this as in other things, "*Ce n'est
que le premier pas qui coûte.*" (It is only the first step that costs.)

In the city of Rio, instruction can be divided into the following classes:—the primary, the secondary, (*instrução secundaria*,) and the private schools, (*collegios.*) The College of Pedro II., the Military and Naval Academies, the Medical College, and the Theological Seminary of St. Joseph, are also under the direction of the State. In the private schools are nearly five thousand scholars.

Through some one of these establishments the juvenile Brazilian ascends the hill of knowledge. An institution already referred to, which of late has awakened more interest than any other in the capital of Brazil, was organized in the latter part of 1837, under the name of Collegio de Dom Pedro II. It is designed to give a complete scholastic education, and corresponds, in its general plan, to the lyceums established in most of the provinces, although in endowment and patronage it is probably in advance of any of those. There was at the opening an active competition for the professorships, eight or nine in number. All of them are said to have been creditably filled. The concourse of students was very considerable from the first organization of the classes. A point of great interest connected with this institution is the circumstance that its statutes provide expressly for the reading and study of the Holy Scriptures in the vernacular tongue. For some time previous to its establishment, copies of the Scriptures had been used in the other schools and seminaries of the city, where they were not likely to be less prized after so worthy an example on the part of the Emperor's College. The Rev. Mr. Spaulding (who was the clerical colleague of Dr. Kidder at Rio de Janeiro) had an application to supply a professor and an entire class of students with Bibles; to which he cheerfully acceded, by means of a grant from the Missionary and Bible Societies

The Military and Naval Academies are for the systematic instruction of the young men destined to either branch of the public service. At fifteen years of age, any Brazilian lad who understands the elementary branches of a common education, and the French language so as to render it with facility into the national idiom or Portuguese, may, on personal application, be admitted to either of these institutions. I have never witnessed a more interesting scene than the assembling of these young men for their morning recitations. It carried me back to the Northern uni versities, so much vigor and spirit did the *Brazileiro* students manifest in their sports and repartees, or in their explanations to each other of difficult points of geometry and engineering which were soon to be brought before their professors.

The regular army of Brazil is about twenty-two thousand men. The national guard consists nominally of more than four hundred thousand men.

The Naval Academy was formerly on board a man-of-war at anchor in the harbor, and introduced its pupils at once to life upon the water. 1865, the academy is removed to the city.

The Imperial Academy of Medicine occupies the large build- ings near the Morro do Castello, and is attended by students in the different departments, to the number of more than three hun- dred. A full corps of professors, several of whom have been edu- cated in Europe, occupy the different chairs, and, by their reputa- tion, guarantee to the Brazilian student an extensive course of lectures and study. The institution is in close connection with the Hospital da Misericordia, which at all times offers a vast field for medical observation.

The Theological Seminary of St. Joseph has less attraction for the Brazilian youth than any other educational establishment at Rio.

The young *Brazileiro*, (of course we speak of the gentleman's son,) after leaving his *collegio*, enters the Medical Academy, or, having a warlike inclination, becomes a middy or a cadet, or he possibly may enter the Seminary of St. Joseph. If he has a legal turn, he is sent to the Law Schools at S. Paulo or Pernambuco. The young Brazilian likes nothing ignoble : he prefers to have a gold lace around his cap and a starving salary to the cares and toils of the counting-room. The Englishman and German are the wholesale importers, the Portuguese is the jobber, the Frenchman is the coiffeur and fancy dealer, the Italian is the pedlar, the Portu- guese islander is the grocer, the Brazilian is the gentleman. Every place in the gift of the Government is full of young *attachés*, from the diplomatic corps down to some petty office in the custom- house. The Brazilian, feeling himself above all the drudgery of life, is a man of leisure, and looks down in perfect contempt upon the foreigner, who is always grumbling, fretting, and busy. The Brazilian of twenty-five is an exquisite. He is dressed in the last Paris fashion, sports a fine cane, his hair is as smooth as brush can make it, his moustache is irreproachable, his shoes of the smallest and glossiest pattern, his diamonds sparkle, his rings are unexceptionable: in short, he has a high estimation of himself and his clothes. His theme of conversation may be the opera, the next ball, or some young lady whose father has so many *contos*.

In spite of all drawbacks, many of these men, in after-life,— whether in the diplomatic circle, in the court-room, in the House of Deputies, or in the Senate,—show that they are not deficient in talent or in acquirements. They can almost all turn a sentence well, rhyme when they choose, or make a fine *ore rotundo* speech, echoed by the *apoiados* of their companions. Some few become fine scholars, and more of them are readers than are generally supposed. Many of them travel for a year or two, and are educated in Europe or in the United States. The interest which the Brazilians, with D. Pedro II. at their head, are now manifesting in learned societies, —whose ranks are recruited from the very class mentioned,—demonstrates that the "little old men" of twelve have not all turned out "froth;" though too much of the vain, the light, and the superficial must be predicated of the Brazilian, who looks upon cards, balls, and the opera as essential portions of his existence. From such men you would not expect much of the "sterner stuff" which enters into the structure of great statesmen. Nevertheless, the country has made wonderful progress; and it must be added, that from time to time there have arisen from the lower ranks of society men of power, who have become leaders. There is nothing in the origin or the color of a man that can keep him down in Brazil.

It must be borne in mind that the Brazilian thus described is not the portrait of the large majority of the citizens of the Empire, but of one from the higher classes as generally found in the cities. There are exceptions; but the same religion and the same mode of thinking have, to a greater or less degree, given a similarity to all who comprise the upper ranks of society, and from whom come the magistrates, officers, diplomatists, and legislators. Their greatest defect is not the want of a polished education, but of a sound morality, a pure religion. Without these, a man may be

amiable, refined, ceremonious; but their absence makes him irresponsible, insincere, and selfish. As nations are made up of individuals, it should be the ardent desire of every Christian and
philanthropist that this Southern people, which have so favorably
set out in their national career, may have that which is far higher
than mere refinement or education.

The duties of the Brazilian citizen are clearly defined in the
Constitution and by-laws of the Empire. Each male citizen who
has attained his majority is entitled to a vote if he possess an
income of one hundred milreis. Monks, domestics, individuals not
in the receipt of 100$000 rent, and, of course, minors, are excluded
from voting. Deputies to the *Assemblea Geral* are chosen, through
electors, for four years. The Senator, who holds his position for
life, is elected in a manner somewhat different from the *Deputado*.
Electors, chosen by popular suffrage, cast their ballots for candidates aspiring to the senatorial office. The names of the three
who stand highest on the list are handed to the Emperor, who
selects one; and thus he who has been chosen through the people,
electors, and the Emperor, takes his chair for lifetime in the Brazilian Chamber of Peers. There seems to have been great wisdom
in all these conservative measures, and their excellencies are the
more enhanced when we examine the various laws and qualifications that pertain to elections and candidates in the States of
Spanish America. The Chamber of Deputies consists of one
hundred and eleven members, and the Senate, according to the
Constitution, must contain half that number. The provincial
legislators are chosen directly by the people.

An election in Brazil is not very dissimilar to an election in the
United States. Rio de Janeiro is divided into ten or twelve parishes
(*freguezias*) or wards. A list of voters in each parish is posted up
for some weeks before an election, and the Government designates
clerks and inspectors for the various *freguezias*. The elections are
held in churches. Upon an American expressing to a Brazilian his
surprise in regard to this seeming inconsistency in a Roman Catholic country,—where the importance put upon the visible temple
is as great as if it were the very gate of heaven,—no satisfactory
reply was obtained. The only theory by which the Fluminensian
attempted to account for it was on the supposition that when the

Constitutional Government was adopted it was deemed advisable to give a solemnity to the act of voting,—that men in the sacred edifice and before the altar would be restrained from acts of violence, and would be otherwise more guarded than in a secular building. Experience, however, has shown that political rancor will ride over all religious veneration; for it is said that on certain occasions, in some of the provinces, the exasperated electors have seized the tall candlesticks and the slender images from the altar to beat conviction into the heads of their opponents.

A ballot-box, in the shape of a hair trunk, is surrounded by the clerks and inspectors; the vote is handed to the presiding officer; the name of the voter is checked, and the ballot is then deposited. Groups of people, active electioneerers and vote-distributers, may be seen in and around the church, like the crowds of the "unterrified" near the polls in the United States. The Government has great power in the elections through the numerous office-holders in its employ; but ofttimes it suffers a defeat. The supreme authorities have the right to set aside an election in cases of violence or fraudulent procedure.

The parties are the *ins* and the *outs*, or Government and Opposition. The party-lines were formerly more closely drawn, under the names of *Saquarèmas*, (the Conservatives,) and *Luzias*, (the Progressives.) These names are derived from two unimportant freguezias in the provinces of Rio de Janeiro and Minas-Geraes. 1866, the Liberals are now the *ins*.

These parties for some years contended for power and principle, and so warm were their struggles that at times they seemed to battle more for rule than for the success of principles. The Luzias endeavored to promote the welfare of Brazil by adopting laws and regulations for which the Saquarèmas did not think the country yet prepared. Both struggled for many years, and alternately held the reins of government: at last the Saquarèma party triumphed, and from 1848 to 1864 was at the head of affairs. (1866, parties are now called Liberal and Conservative.)

In 1854 the two parties were nearly reconciled, there being few dissidents. This was owing to the wise policy of the Saquarèmas. They made very good use of their great influence; they adopted some of the ideas of their opponents; and they promoted to

Government employment a number of the Luzias who were men of acknowledged ability and probity.

This reconciliation was mostly owing to the political tacticn of the late Marquis of Paraná, who was a most skilful politician and a fluent speaker. He was an instance of a man of talent reaching by his industry and energy the highest position in the gift of the monarch and people. He knew well how to employ intrigue, and his moral character was by no means spotless; yet at his death, in September, 1856, party-spirit was laid aside, the faults of the man were covered, and the energy and talent of the states-man only were remembered.

Among the distinguished politicians and orators of Brazil may be counted the Marquis of Olinda, (Pedro de Araujo Lima,) who was educated at the Portuguese University of Coimbra, and has dedicated more than thirty years of his life to the service of his country. He was Regent during the minority of the Emperor, and has been at various times a member of the Cabinet.

The Marquis d'Abrantes, (Miguel Calmon du Pin,) a skilful diplo-matist, consummate financier, and a distinguished orator, was at different periods a member of the Cabinet, and made himself still better known by a volume giving an account of his diplomatic mission in Europe. The Marquis d'Abrantes was President of one of the most useful and important societies in Brazil,—A Sociedade Auxilia-dora da Industria Nacional,—a voluntary company of gentlemen whose object is to advance the agricultural and mechanical and mineral interests of the country, by importing model implements, by correspondence with agriculturalists and manufacturers in all parts of the world, by combating indifference and indolence and every unprofitable routine of cultivation, and by developing the resources of the country. He died in 1865.

Among the veteran statesmen may be mentioned Senator Ver-gueiro, (once Regent during the minority of D. Pedro II.,) who has materially advanced the prosperity of his country by promoting, at his own expense, European immigration. A fuller sketch of this noble octogenarian is found in another chapter. (Died 1860.)

The Visconde de Uruguay (Paulino José Soares de Souza) was formerly a leader in Brazilian politics, and was Minister of Foreign Affairs when the cruel Dictator Rosas was overthrown by the combined Brazilian and Argentine armies and was expelled from Buenos Ayres. He died in 1866.

The Visconde de Itaborahy (Joaquim José Rodriges Torres) is a skilful financier, who has been frequently a member of the Cabinet; and it is to him that are due the reforms in the public treasury and the creation of a national bank. He has recently been engaged in promoting the interests of education, and in reforming public institutions. His views about steamers and railways are narrow.

The Visconde de Abactè (Antonio Paulino Limpo de Abreo) has been many times Minister of Foreign Affairs, and is a brilliant and persuasive orator.

The Visconde de Sepetiba, (Aureliano de Souza Oliveira,) who has also been frequently a member of the Cabinet, was one of the first who promoted the organization of companies to execute different enterprises of internal improvement. (Died 1855.)

The present (1857) Minister of Marine (João Mauricio Wanderly) was President for three years of the province of Bahia, and directed its affairs with so much energy and prudence that he fully earned the honor of being called by the Emperor to take part in the Cabinet. (He is now Baron de Cotegipe.)

Zacarias de Goes e Vasconcellos, former President of the new province of Paranà, is a brilliant orator, and was called to a place in the Cabinet which went out in 1853. (Premier in 1864. and 66.)

Luis Pedreira do Coutto Ferraz, though comparatively a young man, has been called to places of high honor and trust, and in 1854–57 filled the important post of Minister of the Empire.

The Marques de Caxias—the Minister of War in the Cabinet which had so long been at the head of affairs—was, at the death of the Marques of Paraná, placed by the Emperor over the Department of Finance. He is a gentleman of ability, affable in his manners, and distinguished as the commander-in-chief of the Brazilians in the war against Rosas, and, in 1865–66, of the forces against Paraguay.

The Visconde de Jequitinhonha, (Montezuma,) as a politician, diplomatist, and lawyer, ranks among the first men of the Empire. (1865, he made strong Emancipation speeches.)

Brazil has always been well represented in foreign lands, and her diplomatic corps is not, like that of the United States, recruited from mere political partisans, but its members are fitted for their posts by education, discipline, and graduation, in the same manner as the diplomatic ranks of England and France.

Among them no one stands higher than the Baron of Penedo, who represented Brazil in the United States from 1852 to 1855. This gentleman distinguished himself as an advocate at Rio de Janeiro, and is now Brazilian minister at the court of St. James. He is a man of varied culture and enlarged, statesmanlike views.

These are only a few of the leading men of the Empire, and want of space alone prevents the mention of many more.

Titles of nobility have been often used in the foregoing pages, and demand a further explanation.

Nobility in Brazil is not hereditary, but *bene merito*, and has no landed interest or political influence. If a Brazilian has distinguished himself by his statesmanship, his valor, or his philanthropy, and he receives patent of nobility from the Emperor, his son does not thereby become noble. The title is lost to the family at the death of its possessor. The titles of nobility are six,—viz.: Marques, Count, Viscount *com grandeza*, Baron *com grandeza*, Viscount, and Baron. There are six orders of knighthood.

Note for 1866.—We have elsewhere mentioned the names of other Brazilian statesmen, but we cannot forbear to refer to Visconde de Camaragibe, a conservative leader of Pernambuco; to Sr. Sinimbú, a judge, senator, and minister; Sr. Paranhos, distinguished as senator, diplomatist, minister, and orator; and Sr. Souza Franco, senator from Pará. Sr. Octaviano ranks high as a writer, speaker, and diplomatist. The brothers Ottoni are well known, one as a political leader, the other as a promoter of internal improvements. Silveira da Motta has great honor for steps taken to abolish slavery. Nabuco, as minister and senator, is a gentleman of great prominence. Saraiva, late Minister of Foreign Affairs, is a man of ability, frankness, and energy. He was master of the situation; and the affairs of Brazil in the war with Paraguay received a new impulse by the change of cabinet which brought Saraiva into power. Sr. Sá e Albuquerque is a promising young statesman. The Baron of Prados, a man of rare scientific attainments, presided with dignity over the Deputies in 1864–65. José Bonifacio, Pedro Luiz, Junqueira, and many others, are rising men. Martinho Campos is a deputy of great independence, and of enlarged views. A. C. Tavares Bastos, the youngest Brazilian statesman, has attracted much attention at home and abroad by his liberal policy in regard to commerce, education, and administrative reform. Among the leading diplomats not already mentioned are the Lisboas, at Paris and Brussels, and Sr. d'Azambuja, so long Brazilian Under-Secretary of State, now at Washington. Sr. C. F. Ribeiro, deputy from Maranham, is an alumnus of Yale

CHAPTER XI.

PRAIA GRANDE—SAN DOMINGO—SABBATH-KEEPING—MANDIOCA—PONTE DE AREA—VIEW FROM INGÁ—THE ARMADILLO—COMMERCE OF BRAZIL—THE FINEST STEAM-SHIP VOYAGE IN THE WORLD—AMERICAN SEAMEN'S FRIEND SOCIETY—THE ENGLISH CEMETERY—ENGLISH CHAPEL—BRAZILIAN FUNERALS—TIJUCA—BENNETT'S—CASCADES—EXCURSIONS—BOTANICAL GARDENS—AN OLD FRIEND—HOME.

RIO DE JANEIRO, sometimes called *A Corte* (the Court) by the Brazilians, while situated within the province of the same name, is only the capital of the Empire. Praia Grande, on the opposite side of the bay, is the capital of the province of Rio de Janeiro. The latter city is in a neutral district, like the District of Columbia in the United States, and all the laws of this metropolis, as those of Washington, emanate from the General Government.

Ferry-boats, resembling those in the United States, run half-hourly between the Court and Praia Grande, touching at the white little village of San Domingo. The passage is made in thirty minutes, and gives a fine view of the entrance to the harbor, the whole water-line of Rio, and the various anchorages for the shipping. These American boats were introduced by Dr. Rainey.

Praia Grande and San Domingo stretch around a semicircular bay, and probably contain about sixteen thousand inhabitants. On account of the quietness and cheaper rents, many prefer this side of the water to the *urbs fluminis* as a place of residence. I here frequently held religious services, and the Sabbath seemed more like a day of rest than in Rio, where so many shops are open and the people generally given to amusement. In regard to the holy keeping of the day of rest the Brazilians are no more scrupulous than their co-religionists in France or Italy. Military parades are as frequent upon that day as any other; and operas, theatres, and balls are probably more crowded than during the evenings of secular time. The foreign wholesale establishments are closed;

187

but many of the native shopkeepers, and nearly all of the small French dealers, make as great a display, in the morning at least, as on Monday or Saturday. It must, however, be admitted to the credit of the Brazilians that they have made great improvements in this respect. Formerly there was no closing of the smaller places of business on Sunday, and that day, until within a few years, was the favorite of the week for holding auction-sales. This the authorities suppressed by edict; and in 1852, a number of the Brazilian jobbers, by an agreement, (*convenio*,) for a while abstained from Sunday dealings; but this move was by no means so apparent as the suppression of the auctions. In the discussion which arose in regard to Sabbath-keeping, the Bishop of Rio de Janeiro, and the leading journals, took an active part. Notwithstanding all these ameliorations, the Lord's day is one of amusement and business, so far as Brazilians are concerned; and its profanation is such as to shock even those who are not accustomed to the decent observance of that portion of time in England, Scotland, or the United States.

In Praia Grande and S. Domingo there are beautiful *chacaras*, (country-seats,) and quiet, shady nooks, whose delicious fragrance and coolness contrast refreshingly with the hot landing-place of the steam ferry-boat.

Twenty minutes' walk from the *praia* (beach) will bring us into the sparsely-inhabited environs, where we may see the coffee-tree, with its cherry-like berries, the noble dome-shaped mangueira, whose fruit is esteemed so highly by the English in the East Indies, and orange-trees, whose rich, yellow burdens never become wearisome to the eye or cloying to the palate. There, too, we may see fields of the mandioca, which plant has been and is as much associated with the sustentation of life in Brazil as wheat in more northern climes. This vegetable, (*Jatropha manihot L.*,) being the principal farinaceous production of Brazil, is deserving of particular notice. Its peculiarity is the union of a deadly poison with highly-nutritious qualities. It is indigenous to Brazil, and was known to the Indians long before the discovery of the country. Southey remarks:—"If Ceres deserved a place in the mythology of Greece, far more might the deification of that person have been expected who instructed his fellows in the use of mandioc." It is difficult to imagine how savages should have ever discovered that a wholesome food might be prepared from this root.

Their mode of preparation was by scraping it to a fine pulp with oyster-shells, or with an instrument made of small sharp stones set in a piece of bark, so as to form a rude rasp. The pulp was then rubbed or ground with a stone, the juice carefully expressed, and the last remaining moisture evaporated by the fire. The operation of preparing it was thought unwholesome, and the slaves, whose business it was, took the flowers of the *nhambi* and the root of the *urucu* in their food, "to strengthen the heart and stomach."

The Portuguese soon invented mills and presses for this purpose. They usually pressed it in cellars, and places where it

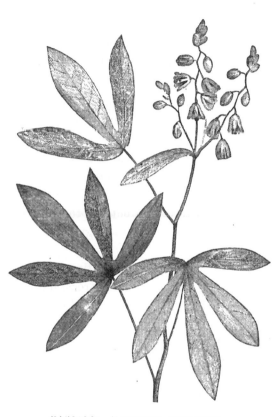

MANDIOCA, (JATROPHA MANIHOT.)

was least likely to occasion accidental harm. In these places it is said that a white insect was found generated by this deadly juice, itself not less deadly, with which the native women sometimes poisoned their husbands, and slaves their masters, by putting it in their food. A poultice of mandioc, with its own juice, was considered excellent for imposthumes. It was administered for worms, and was applied to old wounds to eat away the diseased flesh. For some poisons, also, and for the bite of certain snakes, it was esteemed a sovereign antidote. The simple juice was used for cleaning iron. The poisonous quality is confined to the root; for the leaves of the plant are eaten, and even the juice might be. made innocent by boiling, and be fermented into vinegar, or inspissated till it became sweet enough to serve for honey.

The crude root cannot be preserved three days by any possible care, and the slightest moisture spoils the flour. Piso observes, that he had seen great ravages occasioned among the troops by eating it in this state. There were two modes of preparation, by which it could more easily be kept. The roots were sliced under water, and then hardened before a fire. When wanted for use, they were grated into a fine powder, which, being beaten up with water, became like a cream of almonds. The other method was to macerate the root in water till it became putrid, then hang it up to be smoke-dried; and this, when pounded in a mortar, produced a flour as white as meal. It was frequently prepared in this manner by savages. The most delicate preparation was by pressing it through a sieve and putting the pulp immediately in an earthen vessel on the fire. It then granulated, and was excellent when either hot or cold.

The native mode of cultivating it was rude and summary. The Indians cut down the forest-trees, let them lie till they were dry enough to burn, and then planted the mandioc between the stumps. They ate the dry flour in a manner that baffled all attempts at imitation. Taking it between their fingers, they tossed it into their mouths so neatly that not a grain was lost. No European ever tried to perform this feat without powdering his face or his clothes, to the amusement of the savages.

The mandioc supplied them also with their banqueting-drink. They prepared it by an ingenious process, which savage man has often been cunning enough to invent, but never cleanly enough to reject. The roots were sliced, boiled till they became soft, and set aside to cool. The young women then chewed them, after which they were returned into the vessel, which was filled with water, and once more boiled, being stirred the whole time. When this process had been continued sufficiently long, the unstrained contents were poured into earthen jars of great size, and buried up to the middle in the floor of the house. The jars were closely stopped, and, in the course of two or three days, fermentation took place. They had an old superstition that if it were made by men it would be good for nothing. When the drinking-day arrived, the women kindled fires around these jars, and served out the warm potion in half-gourds, which the men came dancing and singing to receive, and always emptied at one draught. They never ate at these parties, but continued drinking as long as one drop of the liquor remained, and, having exhausted all in one house, removed to the next, till they had drank out all in the town. These meetings were commonly held about once a month. De Lery witnessed one which lasted three days and three nights. Thus, man, in every age and country, gives proof of his depravity, by converting the gifts of a bountiful Providence into the means of his own destruction.

Mandioca is difficult of cultivation,—the more common species requiring from twelve to eighteen months to ripen. Its roots have a great tendency to spread. Cut slips of the plant are inserted in large hills, which at the same time counteract this tendency, and furnish it with a dry soil, which the mandioca prefers. The roots, when dug, are of a fibrous texture, corresponding in appearance to those of the long parsnip. The process of preparation is first to wash them, then remove the rind, after which the pieces are held by the hand in contact with a circular grater turned by water-power. The pulverized material is then placed in sacks, several of which, thus filled, are subjected to the action of a screw-press for the expulsion of the poisonous liquid. The masses thus solidified by pressure are beaten fine in mortars. The substance is next transferred to open ovens, or concave plates, heated beneath, where it is constantly and rapidly stirred until quite dry. The appearance of the farinha, when well prepared, is very white and beautiful, although its particles are rather coarse. It is found upon every Brazilian table, and forms a great variety of healthy and palatable dishes. The fine substance deposited by the juice of the mandioca, when preserved, standing a short time, constitutes the *tapioca* of commerce, so well known in the culinary departments of North America and Europe, and is now a valuable export from Brazil.

Another species, called the *Aipim*, (manihot Aipim,) is common. It is destitute of all poisonous qualities, and is boiled or roasted, and is but little inferior to the potato or the large Italian chestnut. It has further the advantage of requiring but eight months to ripen, although it cannot be converted into farinha.

Not far from Praia Grande is the foundry, engine-manufactory, and ship-yard of Ponte da Arêa, where four or five hundred mechanics and laborers, under European and Brazilian supervision, are turning out works of importance and magnitude. In the year 1854, besides kettles, stills, and boilers, this establishment constructed four steamers with their engines, and two more steamers and a bark were upon the stocks.

But the most attractive part of this side of the water is the peaceful and beautiful Rua da Ingá and the Praia de Carahy. We wind through a thoroughfare—if it can be so called—overhung by graceful shade-trees; and on either side, almost hidden by hedges of mimosa, creeping and flowering vines, huge plants and cacti in gorgeous bloom, are the vermilion roofs and the blue arabesques of Brazilian cottages. In a few minutes we reach the Praia de Carahy, where the fanning sea-breeze dashes the waves in foaming brightness against the shell-paved beach. The scene beyond is indescribable in its beauty and its grandeur; and the view of the surrounding mountains and Rio de Janeiro nestling at their base has often reminded me of the observations of Mr. Hillard in regard to Naples and Edinburgh, when he says, "The works of man's hands are subordinate to the grand and commanding features of nature around and above them: . . . . the magnificent lines and sweeps of the landscape eat up the city itself."

When I gazed from the craggy cliff of Ingá upon the rolling surf beneath,—the graceful lake-like Bay of Jurujuba on our left, the islet of Boa Viagem before us, crowned with its picturesque chapel, dear to mariners and kissed by the breeze-swayed palm-tree, and as with silent wonder I beheld far across the water the giant groupings of the Pão de Assucar, the Tres Irmãos, the wide-topped Gavia, the columnar Corcovado, and the distant Tijuca,—I could realize the emotions of the same polished and forcible writer when acknowledging the utter impossibility of describing the Italian scene to which the Brazilian landscape is equal in beauty and superior in sublimity. What Mr. Hillard has said of

Pão d'Assucar.   Ft. St. John.   Botafogo.   Gavia.   Coreovado.   N. S. da Boa Viagem.   VILLAGEN CONDER

VIEW FROM INGÁ, ST. DOMINGO, BAY OF RIO DE JANEIRO.

the glorious environs of Naples is doubly true of the view from Ingá:—"What words can analyze and take to pieces the parts and details of this matchless panorama, or unravel that magic web of beauty into which palaces, villas, forests, gardens, the mountains and the sea, are woven? What pen can paint the soft curves, the gentle undulations, the flowing outlines, the craggy steeps, and the far-seen heights, which, in their combination, are so full of grace, and, at the same time, expression? Words here are imperfect in struments, and must yield their place to the pencil and the graver. But no canvas can reproduce the light and color which play around this enchanting region. No skill can catch the changing hues of the distant mountains, the star-points of the playing waves, the films of purple and green which spread themselves over the calm waters, the sunsets of gold and orange, and the aerial veils of rose and amethyst which drop over the hills from the skies of morning and evening."

Such scenes can be *felt*, not described.

If we now turn from the white beach and the magnificent *Vista de Ingá*, and seek the reddish-colored hills which are beyond the Bay of Jurujuba, we shall in our rambles frequently meet portions of the earth freshly thrown up. This has been done by the armadillo; for the pointed snout and the strong claws of this little buckler-clad animal admirably adapt him for burrowing, which operation he performs with such astonishing rapidity that it is almost impossible to get at him by digging. The hunters, in such a case, resort to fire, and smoke the armadillo out of his den. Not being able to stand the fumes of burning wood, the little fellow rushes through the

THE ARMADILLO.

new-made aperture, rolls himself up, is easily captured, and his delicate flesh is soon consigned to the kitchen. This power of enveloping himself so completely in his shell that he appears like a round stone or a cocoanut, is a provision of a kind Providence. The armadillo cannot run with any degree of rapidity, and, when attacked by birds of prey, he rolls himself up like a hedgehog, and offers only a solid uniform surface impervious to beaks and talons.

Or again, if set upon by a dog or some small quadruped, he "swallows himself" and rolls down a hill. I have before me a specimen

of the armadillo that was seized in his doubled-up state and thrust immediately into boiling water, which has preserved him in that position. So little does it resemble the live animal or his natural elongated appearance, that no friend to whom I have shown him could divine what it was, nearly every one taking him to be some strange Brazilian nut. The engravings afford a perfect likeness of him from two different points of view: neither head nor tail can be made of him, unless the triangular piece is his os frontis.

In returning to Rio de Janeiro, it is often an agreeable variety to make the passage in a *falua*.* This is a species of boat with lateen sails, and may be of twen'.y or forty tons' burden. They are manned by a captain, who steeri, takes the three-cent fare, and scolds the poor blacks. When it is calm, the more than half-naked negroes slowly pull at the long oars, which are so heavy, that, in order to obtain a " purchase," they are obliged to step up on a sort of bench before them, and thus, rising and falling to a monotonous African ditty, they form one of the peculiar sights of Rio. Many of the poorer classes go as passengers on these *faluas;* but they are mostly used for the transportation of light cargoes to various towns on the bay. If we take a *falua* to the Saude, we pass through vast quantities of shipping.

The great interests of Brazilian commerce draw an immense number of vessels from all portions of the globe. Brazil itself possesses the second navy of the Western World, and her steam-frigates and her sloops-of-war rendered essential service in the overthrow of the tyrant Rosas at Buenos Ayres.

Since 1839, Brazil has had steamship-lines running along the

---

* The sail-boats in the engravings on pages 60 and 201 are *faluas.*

whole of her four thousand miles of sea-coast, but it was not until 1850 that steam-communication was established to Europe. It was then that the Royal British Mail Steamship Company, whose vessels start from Southampton, began their monthly voyages. In 1857 Brazil had for a short time six different lines of steamers, connecting her with England, France, Hamburg, Portugal, Belgium, and Sardinia. The United States, which hitherto had been the great commercial rival of Great Britain in Brazil, had not a single line of steamers to any portion of South America; and, while England was reaping golden harvests, the balance of trade was each year accumulating against us. With all this so evident, it did seem strange that the General Government of the Union, which had aided in extending our mercantile interests by subsidies to steamships running to other lands, had been so tardy in regard to South America, and especially unmindful of Brazil. England's commerce with Brazil since the establishment of her first steam-line in 1850 has increased her exports more than one hundred per cent., while the United States required *thirteen years* to make the same advance. Her entire commerce with Brazil, imports and exports, advanced two hundred and twenty-five per cent. since her first steam-line was established. Each year the balance of trade was increasing rapidly against us. In 1860–61 the United States exported to Brazil $6,018,394, while in return the United States imported from Brazil $22,547,091; or, in other words, only a year's trading with Brazil left against us the cash balance of $16,528,697, which we had to pay at heavy rates of exchange. England, in 1864, sold Brazil $40,612,985, and bought of her in return only $33,079,755, thus leaving the latter her debtor. Why was there such a disastrous account against us? British steamers, energy, and capital, and our neglect, had thus advanced the commerce of England. Our Government and our merchants, notwithstanding their boasted enterprise, did next to nothing to foster the trade with Brazil. Purchasing as we do half her coffee crop and the greater portion of her India-rubber, there should have been an effort on our part to introduce effectually the many productions of our country which we can furnish as well as Great Britain Our common cottons are better than the imitations of the same manufactured at Manchester, England, and yet labelled "Lowell

drillings" and "York Mills, Saco, Me." We can furnish many
kinds of hardware and other items cheaper and better than
England. The few efforts made by single individuals (as in the
case of several American merchants) to introduce the labor-saving
machines of our country have already resulted in the establishment of a number of Brazilian houses in Rio de Janeiro, where
one can purchase various articles under the comprehensive name
of *Generos Norte Americanos.* In 1856, the United States purchased
one-third of all the exports of Brazil, but the imports from the
United States into the Empire were not one-*tenth* of the Brazilian
imports. This subject demands investigation from individuals and
from our Government. It does not fall within my province to
extend this to greater length in this portion of the work, but the
statistician and the political economist, as well as those who are
engaged in commerce, will find in statistical works much information in regard to our business-relations with Brazil: however, in
this connection I will mention the efforts of several persons who
were among the earliest to foresee the benefits arising from steam
communication between Brazil and the United States. To William
Wheelwright, Esq., (the energetic founder and *entrepreneur* of the
Pacific Mail steam-line on the west coast of South America, the
builder of the Copiapo Railway, and now the principal contractor
for constructing the great Argentine Central Railway,) belongs the
honor of first suggesting the steam-line from New York to Rio de
Janeiro. John Gardiner, Esq., for many years a merchant at Rio,
actually made propositions to the United States Congress of 1851–52
for effecting this desired object. In 1854, Dr. Thomas Rainey,
now Director-in-Chief of the Ferry Company at Rio, devoted particular attention to this subject. At a pecuniary loss to himself,
he travelled twice from Washington to Rio de Janeiro,—visiting
the Amazon and the West Indies,—going before the Executive
heads and the statesmen of each Government, and calling attention
to the important facts which he had elucidated after patient investigation. These facts, printed in former editions of this work,
were very striking and convincing, and afforded to friends of both
lands some of the strongest arguments for uniting by steam the
two greatest American countries. Dr. Rainey, in connection with
R. M. Stratton, S. L. Mitchell, and others, urged before the United

States Congress of 1857–58 a proposition to establish a line from New York and Savannah to Pará or Maranham, so as to unite at either of those ports with the Brazil Packet Company's steamers. This measure was defeated by only eight votes.

In 1852, the junior author was so impressed with the evidence before him at Rio that the commerce of Brazil was gliding away from the United States, that he wrote a letter on the subject of steam communication to the New York *Journal of Commerce*, and from that time forward he continued to agitate in the press, before Chambers of Commerce and popular audiences in the United States, and by visits to Brazil and by correspondence with Brazilian statesmen, until there was no further necessity for agitation.

Hon. A. C. Tavares Bastos (the young Brazilian statesman referred to on page 186), by his essays entitled *Cartas do Solitario*, (Letters of a Hermit,) by his communications to the daily press of Rio, and by his persistent advocacy in the Parliament, did much among his countrymen to bring about a correct public opinion on this subject.

It was a favorite idea with the friends of this measure that the interests of the Western continent should be united; that the policy of the North and South American States should be as far as possible American, and not European, and that to this end they should be locked in the closest embraces by steam; that by this alone they could cultivate those intimate relations of friendship and that mutual confidence which would result in the material advancement of the New World. The communication with Brazil, and, consequently, with all South America, was exceedingly difficult. We had no means of sending letters and passengers except by sailing-vessels, which are slow, unreliable, and but little disposed to accommodate the interests of rivals. Nearly all passengers and letters went to Liverpool, thence to Southampton or the Continent, and thence to Brazil, La Plata, and the Windward Islands,—a distance of nearly nine thousand miles. Our commercial men not only had to send by this most unnatural transit, but were compelled to submit to the most harassing disadvantages, and were almost at the mercy of European rivals. It is, therefore, to be regretted that the Congress of 1857–58 did not have time to act upon the report laid before that body. It was, however, only a work of time. In June, 1865, the Senate of Brazil passed the bill, (brought into the Chamber of Deputies in 1864,) based on the law of the United States Congress, signed by President Lincoln, May 28, 1864, to the following effect: that Brazil unite with the Government of the United States in granting a joint subsidy to a line of steamers making twelve round voyages per annum, from New York to Rio de Janeiro, touching at St. Thomas, Pará, Pernambuco, and Bahia. "The United States and Brazil Mail Steamship Company" obtained the contract.

Behind the island of *Enxadas* are the Royal Mail, the French, and the Liverpool steamers, which have come over the pleasantest route, save one, known in ocean-navigation. I have sailed on many seas, but only one other voyage which, all things considered, is comparable to that from Rio de Janeiro to England. We are out of sight of land but six days at the longest stretch, (from Pernambuco to the Cape de Verds;) while the average number of days at sea without stopping are two and a half. From Rio to Bahia there are but three days' steaming over summer waters; and the ten or twelve hours at the second city of the Empire gives plenty of time for refreshing promenades or rides into the country. In less than two days we land at Pernambuco, where we spend from twelve to twenty hours, lay in a stock of fine oranges and pine-apples, (capital anti-nauseatics,) and perhaps purchase a few scream-ing parrots or chattering monkeys to present to our European friends. We then steam for St. Vincent, (Cape de Verds,) where we remain a few hours, and, next steering northward, in forty-eight hours we behold, one hundred and fifty miles at sea, the tall Peak of Teneriffe lifting itself more than thirteen thousand feet from the bosom of the ocean. Here we revel in peaches, pears, figs, and luscious clusters of grapes,—in short, all the fruits of the temperate zone. We pass through the Canaries, and in thirty hours are at Funchal, where the fruit-dose is repeated; a walk upon the shore (if health-bill clean) is permitted, and, after being bored a few hours by the pedlars and grape-venders, we bid farewell to picturesque Madeira, and, at the end of three days, sail up the mouth of the Tagus and anchor before Lisbon. When we leave Portugal, we steam along its coast and that of Spain, and in three days we land at Southampton. No such steamer-voyage exists in the world; and those who are in quest of the new, the strange, and the beautiful, can nowhere so easily and so cheaply gratify their wishes in those respects as by the trip from Southampton to Rio, or *vice versâ*. 1866, Teneriffe and Madeira are no longer ports of call.

The steam-voyage from New York to Rio (*via* the tropic Isle of St. Thomas, Pará on the Amazon, and the bright cities of Pernambuco and Bahia) equals in pleasantness the route from Europe to Brazil. The United States steamers anchor at Enxadas.

From the island of Enxadas, on either hand, over vessels from the coasting-smack to the largest freighting-ships, may be seen the flags of Spain, Portugal, Italy, Russia, Hamburg, France, Belgium, Bremen, Austria, Denmark, Sweden, England, the United States, the South American Republics, and Brazil. These vessels are required to anchor at sufficient distance apart to swing clear of each other in all the different positions in which the ebbing and flowing tide may place them: thus, boats may pass among them at pleasure.

Situated accessibly as the port of Rio de Janeiro is, upon the great highway of nations, with a harbor unrivalled, not only for beauty, but also for the security it affords to the mariner, it becomes a touching-point for many vessels not engaged in Brazilian commerce. Those that suffer injury in the perils of the sea between the equator and the Cape of Good Hope generally put in here for repairs. Many sons of the ocean, with dismasted or waterlogged vessels, have steered for this harbor as their last hope. At the same time, nearly all men-of-war and many merchantmen, bound round Cape Horn or the Cape of Good Hope, put in here to replenish their water and fresh provisions. Thus, in the course of business and of Providence, missionaries, either outward or homeward bound, were in various instances thrown among us for a brief period; and we scarcely knew which to value most,—the privilege of enjoying their society and counsel, or that of extending to them those Christian hospitalities not always expected on a foreign shore. We enjoyed many such visits that will long be remembered, and we seemed to be brought directly in contact with Russia, India, the Sandwich Islands, and Central and South Africa, —the countries where the individuals met with had severally labored.

Such circumstances beautifully illustrate the central position and the important character of the harbor of Rio de Janeiro, which forms a converging-point for vessels from any port of the United States and Europe, and for returning voyages from Australia, California, and the islands of the Pacific.

Annually more than twelve thousand mariners, sailing under the flags of England and the United States, are gathered at Rio de Janeiro. This class of men demands the earnest attention of the philanthropic Christian. If pestilence visits Rio, they are sure to fall before it sooner than any other men who resort thither. The improvidence of sailors is proverbial, and their general dissipation and recklessness are well known. A greater proportion of these men die annually than of those who follow any other calling. They therefore really call for most earnest effort in their behalf, both morally and physically.

The exertions that have been made among sailors at Rio from time to time have not been entirely in vain. The American Seamen's Friend Society—a noble institution, which has carried the church over the world for Americans and Englishmen—established a chaplaincy at this port more than twenty years ago. No chapel was ever erected, because the peculiar regulations of the port are such that vessels lie at anchor away from the shore; hence it has been usual to hold services on board various vessels that might be in the harbor. The Bethel flag, with its white dove, would be hoisted to the main, and, when unfurled to the breeze, like a church-bell, though mute, would call the hardy mariners from the various anchorages to come up to the floating tabernacle, there to join in the hymn of praise, or to listen, in this distant clime, to the lessons of sacred truth. During a number of years it was my privilege, in connection with duties on shore, to fill the post of American Chaplain. It was my custom, when the port was healthy, to visit the English and American vessels each Friday, conversing with the officers, dropping a word of advice to the sailors, and placing in the hand of each a tract to announce the ship over which the Bethel flag would float on the following Sunday. When the yellow fever prevailed, I daily attended the hospitals and boarded the ships to administer the comforts of the gospel to the sick and dying sailors. Poor fellows! Many passed

from time into eternity without being able to send a parting message to their distant friends; but, whenever I could ascertain the
address of their relatives, I forwarded their dying words, which
were frequently the outpourings of their faith and hope in Christ.

In this round of duties I was materially aided by Senhor Leopoldo, the *guarda-mor*, who, with great kindness, made an exception
in favor of the chaplain, allowing me to visit all the vessels in port
without the special daily permit.*

From the loading-ground to the British Cemetery at Gamboa
the distance by water is little more than a mile; and often have I

ENGLISH CEMETERY AT GAMBOA.

had to lead the mournful procession from the landing-place up the
green walks of this quiet and retired resting-place for the dead.
In this beautiful and secluded spot sleep more than one minister-
plenipotentiary and admiral. Men of eminent station, as well as the
unknown English and American citizen, the German, the Frenchman, the Swede, and the representatives of the commercial marine
of almost every nation, here slumber in death. No portion of Rio

---

* This courtesy can be better appreciated when the reader is informed that, by
the narrow and restricted port-laws of Brazil, no one except a custom-house officer
can visit, without permit, a vessel that is discharging. The penalty for each
offence is a fine of fifty dollars.

was ever more impressive to me, whether it was in reading the
solemn funeral-service in the hearing of many, or when, with none
but the sexton, I stood by the new-made grave, or when alone
I wandered through the shady walks. This cemetery belongs to
the English; but the application of any consul for the burial of a
deceased person of another nation is never rejected.

While Englishmen either at home or at Rio have done so much
toward preparing and beautifying a suitable resting-place for the
dead, they have sadly neglected the living who come to this mart.
There is regular service for those who reside in the city; but for
the six thousand mariners who sail hither under the English flag,
no provision has been made. The duties of the English chaplain
confine him to the shore; and, though occasionally English officers
and masters go to the chapel, the sailor is neglected. It may be
said, "There stands the chapel; let him go thither." Men who are
not accustomed to the sound of the church-going bell, and whose
proclivities are not particularly God-ward, have some hesitation to
row one mile upon the water, and then, in a tropic clime, to walk
another, in a strange city, to a house of worship with which they
do not feel associated by ordinary local ties. For such men, either
the English Bethel Union, or some benevolent association connected
with the Established Church or with Dissenters, should make provision for regular worship. If men will not come to the gospel, we
must take it to them; and the most earnest workman in the vine-
yard of our Master will find enough to do among the English sailors
in the harbor of Rio de Janeiro. The lower class of English laborers,
either in the mines or engaged in the construction of railways, is
annually increasing, and it is hoped that the effort for ameliorating
the moral condition of the resident workmen, so auspiciously begun
at the Saúde, may be followed up on the vast water-parish which
is ever to be found floating on the commodious bay. I am aware
that there are those who look upon it as a more hopeful task to
labor for the good of souls among the heathen than for seamen.
While I would not have a single soldier called in from the distant
outposts, I do believe that, under the circumstances, no distant
field is more encouraging than caring for the spiritual welfare of
those who "go down to the sea in ships." They may be termed a
"hard set;" but they have noble and generous qualities and great

temptations. It therefore becomes the English Christian not to rest until in every important foreign port he establishes worship for the sailor. (1866, the Gamboa Cemetery is exclusively English.)

The English Chapel is situated in Rua dos Barbonos, near the Largo da Mai do Bispo. This neat little edifice was erected in 1823, almost immediately after the achievement of Brazilian Independence. Service is held here each Sunday morning at eleven o'clock, and the English resident experiences a homelike feeling when he finds himself surrounded by his countrymen, and listens to the sacred and beautiful service to which he was accustomed in the

THE ENGLISH CHAPEL.

land of his birth. It is, however, painful to reflect that so few avail themselves of the opportunity which this chapel affords for hearing the truth. The attendance is better since Rev. Mr. Preston's arrival. Compared with all other English chapels which I have visited in many foreign lands, that of Rio de Janeiro is the least frequented.

There are a number of Roman Catholic cemeteries in the vicinity of the city, which belong to the different brotherhoods. The Brazilian funerals are conducted with much pomp. Formerly interments took place in the churches; but, since 1850, there have been no intermural burials. Carriages and outriders, and a long train of friends in vehicles, make up the procession. There are not, to a great extent, those peculiar customs and ceremonies which were

formerly consequent upon a death in a Brazilian family. There is
more parade than upon the Continent, and probably more, since the
burial-reform, than in England. The deceased child, often decked
with flowers, is borne to the grave in an open hearse with gilded
pillars. The driver of the hearse, the footman, and the four outriders, upon white horses, are in red livery. Custom forbids the
presence of women at a funeral, and also the attendance of very
near relatives. If the deceased be above ten years of age, the immediate relatives remain at home for eight days, during the first
of which a profound silence is maintained. When friends come to
offer their sympathy, the customary salutation of those who enter
is, "Will you permit me to offer my condolence for the loss you
have sustained?" Silence is then preserved by both parties, and,
after some minutes, the visitor withdraws.

From the cemetery of Gamboa is a vista of the Serra de Tijuca;
and among the many jaunts near the city, none surpasses in interest the ride up these mountains. Passing through the long street
of Engenho Velho, which is lined with the residences of wealthy
families, each surrounded with its chacara or grounds, that glow
with the fadeless verdure of mangueiras, orange-groves, and palms,
interspersed with flowers of the brightest hues, we reach the foot
of the mountain. Here are many picturesque villas, each having
piazzas in front, and often approached by a large stone gateway,
where, in the evening, the family sit to amuse their listless hours
by watching the passers-by. These country-residences are built in
a style that accords well with the glowing climate. The pediments
and cornices of the houses are ornamented with arabesques on a
ground of vivid blue. No ugly clusters of smoking chimneys
deform the roofs. The white walls glitter amid the dark foliage,
or stand in strong relief against the steep mountain-sides. The
native families generally live on the plain, and near the ever-
attractive road; but the Englishman, true to his national character,
climbs the mountain and builds an eyrie among the clouds.

On arriving at a mineral spring, called *Agoa Ferrea*, you quit
the railway for the more agreeable mode of travel afforded by
horse or mule. It is true that invalids and hard-hearted people
may cause four mules to drag them up the steep ascent. But no
one possessing eyes, taste, and health, should miss the opportunity

VIEW OF RIO DE JANEIRO FROM TIJUCA.

of a horse-back ride. It is difficult to speak calmly of the scenery about Rio. No pen can do justice to the view that meets the eye half-way up the mountain. A good cicerone will keep your attention fixed on the flowers that adorn the left bank of the road until he reaches a low part of the brushwood and pulls in his horse, exclaiming, "Look!" A wondrous view it is that bursts upon you. There, unfurled before you, like a fairy panorama, are the bay with its islands, the distant mountains blending with the clear blue sky,—a dark precipitous cliff on the right, pouring down its tiny cascades in silvery lines, that relieve its barren sternness, and on the left a high hill, covered with glossy-leaved coffee-plants: on the plain below rises a single mound, and beyond is the gleaming city,—its white edifices peacefully encircling the green hills of Conception, San Bento, and Antonio. Nothing but a large oil-painting can convey any just idea of this view; and it was here that an English painter took his stand for his tropic landscape. Leutsinger has the best photographs of Rio scenery.

After a long gaze you turn away only half satisfied, and immediately lose sight of all on that side of the mountain, but soon discover the open sea beyond the opposite descent. A few minutes more brings you to the residence of Mr. Bennett, an intelligent Englishman, who has erected in this beautiful spot a boarding-house, where many of the foreign residents pass the hot months. Here, while only eight miles from the Praca do Commercio, far from the heat and noise of the busy city, we could spend our days and nights in ease and comfort. No mosquitoes fright away sleep with their fierce war-whoops; no cockroaches—or baratas, as they are called—crawl over your feet as you sit in the piazza. But do not imagine that there is total stillness. On the contrary, the air is vocal with the sounds of that portion of animated nature which loves to disturb nocturnal hours. Pre-eminent above all is the staccato music of the blacksmith-frog, whose substantial body a man's hands could not enclose, and every sound that he produces rings upon the ear like the clang of a hammer upon an anvil, while the tones uttered by his congeners strikingly resemble the lowing of distant cattle.

Not far from Bennett's are the coffee-plantations of Mr. Lescene and of Mr. Moke, which are among the very first that were cultivated in Brazil; and, as they are the only fazendas near to the city, no stranger should omit an early walk to the lovely valley where they are found.

The excursions from the boarding-house are most varied and interesting. To climb the Pedra Bonita and gaze upon the moun. tain-landscape and the far-off meeting of sky and ocean 's the delightful work of a few hours. The charm of Tijuca is that, while its climate is unchanging June, and its verdure tropical, it

BENNETT'S, TIJUCA.

possesses the sparkling cascades and thundering waterfalls of Switzerland. If we wander from Bennett's toward Rio, and turn to our left, a few moments will bring us to a limpid stream which hangs like a ribbon down the mountain-side, and sends up

> " Brave notes to all the woods around,
> When morning beams are gathering fast,
> And hush'd is every human sound."

This beautiful fall is said to come from a height of three hundred feet, and reminded me of the leaping brooks of the Valley of the Rhone, or the graceful cascade of Arpenaz, that swings from an Alpine cliff into the sweet vale of Maglan. Or again, if we ride for a half-hour in the opposite direction from the mountain boarding-house, we reach a wild and verdant spot, where, dismissing our horses, we climb up through banana-fields and forest, and reach the foaming waters of the *Cascata Grande*. Here the Tijuca River leaps for sixty feet or more over a rocky inclined plain, presenting, when the volume is increased, an imposing appearance; but, when the stream is only supplied by the clear springs of the Serra, it glides down in a transparent sheet, revealing the shining rock beneath. The river pursues its way over a rock-bed down the mountain, and loses itself in the lake which mirrors the giant Gavia.

Mr. Ewbank, who is usually very correct in his facts, has curiously departed from his accustomed precision in the statement that it was "in this secluded retreat that the Bishop of Rio lay concealed during the troubles with the French Protestants of Coligny's time." No "Bishop of Rio" was in existence "during the troubles of Coligny's time." The only bishopric in Brazil for many years was that of Bahia. The French were finally expelled from the Bay of Rio de Janeiro in 1567, and it was not until this was effected that the city of San Sebastian or Rio de Janeiro was founded. Mr. Ewbank was doubtless misled by some one informing him that the remains near the *Cascata Grande* were those of walls erected for the bishop when the French took possession of Rio. This is perfectly correct; for in 1711, after the disastrous defeat of the French commander Du Clerc, (in 1710,) Du Guay Trouin came with an avenging squadron to Rio de Janeiro, and on such a scale were his preparations that the inhabitants fled to the mountains of Tijuca, and there remained until the city was taken and sacked, and did not return before Trouin had sailed away with his heavy ransom.

But if Mr. Ewbank has been led into error so far as a date is concerned, he has more than made up for it by his beautiful and graphic painting of the bright Falls of Tijuca, as it appeared to him when taking a picnic-dinner upon the glistening stones:—"Our table extended into the channel; and there we banqueted and reclined amid scenery far excelling that which Pliny's Laurentinum dining-chamber opened on. Shielded from the sun by nature's parasols, far from the busy scenes of artificial life, not a carking care to trouble us, and our spirits airy as our dresses, we laughed and talked and dipped our cups in the crystal stream as people did in the golden age. Flora adorned the hanging shrubbery; Pomona, from the distance, looked on; zephyrs played round us; and naiads—if naiads there be—frisked in the falls and threw spray at us as they glided by."

From Tijuca there is a very fine excursion around the base of the Gavia, high up whose steep sides are certain curious hieroglyphics, which have long occupied the attention of the learned. These characters seem like Roman letters; but the best explanation of their existence upon this precipitous wall is that nature has chiselled them by rains and sun, and, perhaps in times remote, by little shrubs, whose seeds, deposited by wandering birds, have grown in the crevices until their swelling roots have aided the rain in prying off friable portions of the rock.

This excursion can be extended upon the wave-washed beach around to the Botanical Gardens, above which, from one of the lesser hills, is a prospect not excelled by the views of Como and Maggiore. The abrupt Corcovado presents a new face as it looks down upon the calm Lagoa das Freitas. The stately palms of the Jardim Botanico seem from our elevation like the trees of a child's toy garden. The Serra, across the Bay of Rio, takes every shade of purple and blue during the daytime, and, as the sun at eventide darts his rays athwart the Pão de Assucar and the Irmãos, the distant white fortress of Santa Cruz stands out from waters and mountains of rose. A lady friend, who sketched for me the opposite engraved scene, accompanied the gift with this remark in regard to the exquisite tints of that tropic region:—"Years of familiarity never destroyed for me the loveliness and marvellousness of these hues, which a painter would hesitate to put upon canvas for exhibition to the inhabitants of a less genial zone." There is less difficulty, however, in transferring to the sketch-book the bold outlines of those peculiar-shaped mountains which abound throughout almost every league of the capital province of the Empire; and the many scenes presented in this portion of " Brazil and the Brazilians," which were taken to support no argument of mine, will expose the absurdity as well as the inaccuracy of the descriptions given, even in the latest American edition of McCulloch, of " the neighborhood of Rio de Janeiro," which " consists in a great measure of plains" !

The Botanical Gardens, to which we can now easily descend, is situated in this romantic spot, and is reached from the city by a fine turnpike which leads through Botafogo and under the shadow of Corcovado. It is not a flower-garden, but rather a *Jardin des Plantes,* where rare exotics, from the tiniest parasite up to the loftiest palm, come under our inspection. Here you may behold groves of cinnamon and clove trees, acres of Chinese tea, the *Nogaras da India,* the bread-fruit, cacáo and camphor trees, besides many others that are objects of great curiosity. There was one tree, half hidden

LAGOA DE FREITAS.

by the dome-shaped *manqueiras,* that I often visited with peculiar emotions of pleasure. It was a small North American maple. As I looked upon that little tree,—an exotic in this distant land, where no wintry blasts would strip it of its foliage, where not even an autumnal frost would robe it in those gorgeous hues which the flowers of this summer clime hardly surpass,—I could sympathize with the Bedouin of the desert who, upon beholding the palm-tree in the *Jardin des Plantes* of Paris, was transported far over mountain and sea to the country of his nativity. The most surprising sight to the Northern stranger in the Botanical Gardens is the long avenue of the Palma Real, (*Oreodoxa regia,*) which we enter from the great gate, and which, in its regularity, extent, and beauty, is

14

unrivalled. It is a colonnade of natural Corinthian columns, whose
graceful, bright-green capitals seem to support a portion of the
blue dome that arches above.

But the sun's last rays are empurpling the granite peaks around
us, and, after a gallop through the villa-lined San Clemente, we
reach Botafogo. The lamps are already twinkling, and throw
their light upon the edge of that graceful little bay where the gay
regatta holds its annual festivity. Five minutes more, we dismount
at the Hotel dos Estrangeiros; and thus we have accomplished tho
entire circuit of the city San Sebastian de Rio de Janeiro.

Note for 1866.—No one who has not visited Tijuca since 1855, can have a just
idea of the many improvements that have taken place in that charming spot. The
Tijuca railway conveys passengers from the Praca da Constitucão to the foot of
the mountain, a distance of seven miles. Near the terminus are several fine Brazilian chacaras and villas, amongst them, the most conspicuous by its size and
good taste, is that of Militão Maximo de Souza. From the terminus horses can
be obtained, and after a fine up-hill ride, over a new road, you arrive at Boa
Vista, where a number of elegant English and Brazilian residences have been
recently erected. By a winding road, at the right of the main highway, you
reach the finest house for picturesque situation, comfort and solidity in Brazil.
This is the residence of William Ginty, Esq. Where else in the world will you
find all the adjuncts of gas, (from the city mains,) running water, fine gardens
on a verdure-clad peak 1300 feet above the level of the sea! Mr. Bennett has
more than doubled the accommodations of his most excellent hotel, and, by his
tasteful horticultural adornments and other important additions, he has rendered
his mountain and valley home more attractive than ever. To the geologist a new
attraction is found, in the fact that it was in front of Mr. Bennett's that Prof,
Agassiz, in May, 1865, first discovered erratic boulders and drift—the evidence
of glaciers in the tropics at some remote geologic period of time.

CHAPTER XII.

THE CAMPO SANTA ANNA—THE OPENING OF THE ASSEMBLEA GERAL—HISTORY OF
EVENTS SUCCEEDING THE ACCLAMATION OF DOM PEDRO II.—THE REGENCY—
CONSTITUTIONAL REFORM — CONDITION OF POLITICAL PARTIES BEFORE THE
REVOLUTION OF 1840—DEBATES IN THE HOUSE OF DEPUTIES—ATTEMPT AT
PROROGATION—MOVEMENT OF ANTONIO CARLOS—DEPUTATION TO THE EMPEROR
—PERMANENT SESSION—ACCLAMATION OF DOM PEDRO'S MAJORITY—THE ASSEM-
BLY'S PROCLAMATION—REJOICINGS—NEW MINISTRY—PUBLIC CONGRATULATIONS
—REAL STATE OF THINGS—MINISTERIAL PROGRAMME—PREPARATIONS FOR THE
CORONATION—CHANGE OF MINISTRY—OPPOSITION COME INTO POWER—CORONA-
TION POSTPONED — SPLENDOR OF THE CORONATION — FINANCIAL EMBARRASS-
MENTS—DIPLOMACY—DISSOLUTION OF THE CAMARA—PRETEXT OF OUTBREAKS—
COUNCIL OF STATE—RESTORATION OF ORDER—SESSIONS OF THE ASSEMBLY—
IMPERIAL MARRIAGES—MINISTERIAL CHANGE—PRESENT CONDITION.

THE usual carriage-route to and from Gamboa is through the Campo de Santa Anna. Many important public buildings are upon the side of this large square. The railway station, an extensive garrison, the Camara Municipal, the National Museum, the Palace of the Senate, the Foreign Office, and one of the large opera-houses, are to be found on different portions of the park. It presents an animated scene on the 3d of May, when the session of the As- semblea Geral is opened by the Emperor in person. The procession from St. Christovão to the Palace of the Senate is not surpassed in scenic effect by any similar pageant in Europe. The foot-guards, (halberdiers,) with their battle-axes,—the dragoons and the hussars in picturesque and bright uniforms,—the mounted military bands,— the large state-carriages, with their six caparisoned horses and liveried coachmen and postillions,—the chariot of the Empress, drawn by eight iron-grays,—the magnificent Imperial carriage, drawn by the same number of milk-white horses decked with Prince-of-Wales plumes,—and the long cavalcade of troops,—form a pageant worthy of the Empire. The six coaches-and-six are for the officers of the Imperial household. Her Majesty Dona Tl eresa is surrounded by her maids of honor in their robes and trains of green and gold. Believing that some fair readers will be gratified with the details of Dona Theresa's toilette, one who is better acquainted than I am with ladies' costume says that the *habillement* of the Empress, on state-occasions, is an under-dress of white satin, heavily embroidered with gold, with a profusion of rich lace falling deeply over the corsage and forming its sleeves. These are looped up with diamonds magnificent in size and lustre. The train is of green velvet, with embroideries in gold corresponding with those of the skirt. Her head-dress, with the hair worn in long ringlets in front, is a wreath of diamonds and emeralds in the shape of flowers rising into the form of a coronet over the forehead, and from which a white ostrich-feather falls gracefully to the shoulder. A broad sash, the combined ribbons of different orders,—scarlet, purple, and green,—crosses the bust from the right shoulder to the waist, above which a mass of emeralds and diamonds of the first water sparkles on her bosom. Her smile is one of engaging sweetness, which is not assumed on mere state-occasions, but is seen habitually, whether this Neapolitan princess is accompanying her august spouse in an afternoon ride, or whether with a single attendant she grants a private audience to those who desire to pay their homage to her majesty.

The Emperor is indeed a Saul,—head and shoulders above his people; and in his court-dress, with his crown upon his fine, fair brow, and his sceptre in his hand, whether receiving the salutes of his subjects or opening the Imperial Chambers, he is a splendid specimen of manhood. His height, when uncovered, is six feet four inches, and his head and body are beautifully proportioned: at a glance one can see, in that full brain and in that fine blue eye, that he is not a mere puppet upon the throne, but a man who *thinks*.

The opening of the Chambers is always performed by His Majesty in person. He reads a brief address from the throne, setting forth the condition and necessities of the Empire, and then, pronouncing the session *aberta*, descends from the dais, followed in procession to his Imperial carriage by all the dignitaries of court and members of the Assembly. The cortège returns to San Christovão through streets that are decorated with hangings of crimson silk and satin brocade. There is not the enthusiasm attending this ceremony which is manifested at the inauguration of a new President of the United States, but the circumstances are different : the opening address of the Emperor corresponds to the annual message of the President, and there is no occasion for the jubilatic proceedings which are the concomitant parts of an inauguration. The monarchial principle is deeply imbedded in the heart of the Brazilian, and, in its adaptation to them and their country, it is infinitely superior to republicanism.

It is appropriate, in connection with the opening of the Assemblea Geral, to give a sketch of the events succeeding those which brought the present Emperor to the throne of Brazil.

It will be remembered that it was in the Campo de Santa Anna that the citizens assembled in April, 1831, and demanded D. Pedro I. to restore the ministry which was the favorite of the people. Upon the refusal of the monarch to this request, repeatedly and respectfully urged through proper magistrates, several divisions of the army and the national guard joined the populace. An adjutant was sent to the Palace of San Christovão for a final answer, which was given in the abdication of the monarch under circumstances which command our highest admiration.

The Adjutant (Miguel de Frias Vasconcellos) returned at full gallop from San Christovão with the decree of abdication in his hand. It was received with the liveliest demonstrations of joy, and the morning air rang with "vivas" to Dom Pedro the Second.

At an early hour all the Deputies and Senators in the metropolis, together with the ex-Ministers of State, assembled in the Senate-House and appointed a provisional Regency, consisting of Vergueiro, Francisco de Lima, and the Marquis de Caravellas, who were to administer the government until the appointment of the permanent Regency provided for by the Constitution. The son in favor of whom this abdication was made was not six years old: nevertheless, he was borne in triumph to the city, and the ceremony of his acclamation as Emperor was performed with all imaginable enthusiasm. During the progress of these events, the *corps diplomatique* had assembled at the house of the Pope's nuncio, to determine on what course they should take in the progressing revolution. Mr. Brown, the American *chargé d'affaires*, declined being present at this meeting, apprehending that its special design was to protect the common interests of royalty. Those who met, however, agreed to present an address to the existing authorities, in which, after stating that the safety of their several countrymen was perilled in the midst of the popular movements then taking place, they de· manded for them the most explicit enjoyment of the rights ai.d immunities conceded by the laws and treaties of civilized nations They furthermore resolved to wait upon the ex-Emperor in a body, to learn from his own lips whether he had really abdicated !

These measures were highly offensive to the new Government, being considered in the light of an uncalled-for interference. That Government was at the same time highly pleased with the course pursued by Mr. Brown, and also by Mr. Gomez, the chargé from Colombia, who dissented from the policy of the monarchial diplomatic agents. The Minister of State remarked that their conduct was that of "true Americans."

The 9th of April was appointed as the first court-day of Dom Pedro II., while the ex-Emperor still remained in the harbor. A Te Deum was chanted in the Imperial Chapel. The troops appeared in review; and an immense concourse of people, wearing leaves of the "arvore nacional" as a badge of loyalty, filled the streets. They detached the horses from the Imperial carriage, so that they might draw their infant sovereign with their own hands. When he had been conveyed to the palace he was placed in a window, and the unnumbered multitude passed before him. After this he received the personal compliments of the *corps diplomatique,* none of whom were absent, notwithstanding the recent excursion on board the Warspite.

The new Government courteously offered Dom Pedro I. the use of a public ship. He declined it, on account of the delay and expense that would be necessary to its outfit; remarking, at the same time, that his good friends, the Kings of Great Britain and France, could well afford him the conveyance for himself and family which had been offered by their respective naval commanders on that station.

On the 17th of June the Assemblea Geral proceeded to the election of the permanent Regency. The individuals elected were Lima, Costa Carvalho, and João Braulio Muniz. The General Assembly was occupied during this session by exciting debates on the subject of constitutional reform.

Senhor Antonio Carlos de Andrada presided in the Chamber of Deputies. José Bonifacio, who had been appointed by the ex-Emperor as tutor to his children, was recommissioned by the Assemblea, that body having decided that the former appointment was invalid. On accepting his charge, that distinguished Brazilian declared that he would receive no compensation for the services he might render in that important capacity,—which declaration he maintained in the spirit of a true patriot.

Notwithstanding the magnitude of the revolution that had so suddenly transpired, the public tranquillity was scarcely at all disturbed.

On the 7th of October official despatches arrived, bringing the congratulations of the Government of the United States upon the new order of things. This was the first demonstration of the sentiments of other nations that was communicated at the Brazilian court, and as such was received with peculiar satisfaction.

In the month of April, 1832, two military riots occurred in Rio de Janeiro, and in July following the Minister of Justice, in his public report, seized the occasion to denounce the venerable José Bonifacio, on suspicion of his having connived at the preceding disturbances. The report of a committee in the Camara dos Deputados demanded his dismission without a hearing. The Camara agreed to this by a bare majority, but the Senate dissented, and that plot for degrading Andrada failed. The Regents sent in their resignation to the General Assembly. A deputation from the Chamber of Deputies besought them to remain in office. They consented, but immediately organized a new ministry.

The next year, however, the opposition triumphed, not in verifying these unjust accusations, but in deposing the old patriot as tutor to the young Emperor.

The year 1834 was celebrated on account of the important changes that were made in the Constitution of the Empire. One of these created annual assemblies in the provinces, instead of the general councils before held. The members of the provincial assemblies were to be elected once in two years. Another abolished the triple Regency, and again conferred that office upon a single individual, to be elected once in four years.

After the election for Sole Regent took place, the Senate delayed for a long time the announcement of the successful candidate; but at length it was made known that Diogo Antonio Feijo, of San Paulo, had received a large majority of the electoral votes. Feijo, although a priest, had been for many years engaged in political life, and only two years before had been elected a Senator. One of the last acts of the preceding administration had been to appoint him Bishop of Mariana, a diocese including the rich province of the Minas. Feijo was installed Sole Regent on the 12th of October, 1835. On the 24th he issued a judicious proclamation to the Brazilian people, setting forth the principles that he intended to observe in his administration.

The agitated question of the Regency being settled, affairs assumed a more permanent aspect. Several foreign nations, at this juncture, advanced their diplomatic agents to the highest grade. The United States were desired to do the same, but did not consent.

In 1836 the Government, among other suggestions for the public good, proposed to employ Moravian missionaries to catechize the Indians of the interior. This measure, together with every other originated by this administration, was opposed with the utmost rancor and bitterness by Vasconcellos, a veteran politician of great abilities and uncommon eloquence, but of doubtful principles and bad morals. Notwithstanding the arts and power of Vasconcellos, the leading measure of the administration prevailed. This was a loan of two thousand contos of reis (£200,000) for the temporary relief of the treasury. Open and active rebellions were at this time in progress in Rio Grande do Sul, and also in Pará. Their influence, however, was scarcely apparent at the capital, where every thing seemed quiet and prosperous. The General Assembly was slow in making provision to suppress these outbreaks, and when they were about to adjourn Feijo prolonged the session a month, "that the members might do their duty." Movements for the abolition of the Regency, and the installation of the young Emperor, had already commenced, even at this early day. At times, and in favorable circumstances, they became more apparent.

Feijo's administration was not calculated to be popular. His character partook of the old Roman sternness. When he had once marked out a course for himself, he followed it against all opposition. Disinclined to ostentation himself, he did not countenance it in others. He neither practised nor abetted the usual arts of flattering the popular will. He sometimes changed his ministers, but his advisers seldom or never. At length, so embarrassed did he find himself between the rebellion of Rio Grande and the factious opposition that checked his measures for repressing it, that he determined to retire from his office.

On the 17th of September, 1837, Feijo abdicated the Regency, and the opposition party came into power. Pedro Araujo Lima, then minister of the Empire, assumed the Regency by virtue of a provision of the Constitution, although Vasconcellos was the prime mover in the new order of affairs. No commotion took place, and it was evident that the strength of the new Government consisted in union. A different policy was adopted toward the boy Emperor. Feijo had been distant and unceremonious; the new administration became over-attentive. More display was made on public occasions, and the inclinations of a people passionately fond of the pomp and circumstance of royalty began to be fully gratified. In October, 1838, the votes of the new election were canvassed, and Lima was installed Regent. His term of office was to cover the minority of the Emperor.

Whether the Regent himself expected such a result or not, it soon became apparent that the dignity of his office was quite eclipsed by the new honors with which the young sovereign was complimented. The frequent changes of ministry hitherto had embarrassed the diplomacy of the Brazilian Government, and had caused much dissatisfaction to foreign powers, who were unwilling to see their claims neglected from any cause. By degrees, however, the foreign as well as the internal affairs of the Government became more permanently adjusted.

The year 1840 was signalized in Brazil by a new and startling political revolution, which resulted in the abolition of the Regency. The Emperor, Dom Pedro II., was now in his fifteenth year; and the political party opposed to the Regent and the existing ministry espoused the project of declaring his minority expired, and of elevating him at once to the full possession of his throne. This project had been occasionally discussed during the last five years. But it had always been characterized as premature and absurd. It was argued that the Constitution limited the minority of the sovereign to the age of eighteen years, and that was early enough for any young man to have the task of governing so vast an Empire. On the other hand, it was urged that, as to responsibility, the Constitution expressly provided that none should attach itself to the Emperor under any circumstances. Hence an abolition of the Regency would, as matter of course, devolve the powers of the regent upon some other officer. There would be one difference, however. The Regent, as such, enjoyed the privileges of royalty itself, being also perfectly irresponsible. This circumstance was urged as a great and growing evil. However desirable it was for a sovereign to possess the attribute of irresponsibility, it was a dangerous thing for a citizen, accidentally elevated to office, to have the power of dispensing good or evil without expecting to answer for his conduct. As these subjects were discussed, much feeling was aroused; but the best-informed persons supposed that the Regent would be able to defeat the plan laid for his overthrow.

The debate upon the motion in the House of Deputies to declare the Emperor of age began early in July, and at first turned principally upon constitutional objections. The legislature had, in fact, no power to amend or overstep the Constitution. But the plan was arranged, minds were heated, and the passions of the people began to be enlisted. Violence of language prevailed, and personal violence began to be threatened. Antonio Carlos de Andrada, already described as a man of great learning and eloquence, but at the same time fiery and uncontrollable, stood forth as the champion of the assailing party, accusing the Regent and his ministry of usurpation, especially since the 11th of March, when the Imperial Princess, Donna Januaria, became of age. His efforts were powerfully resisted, but his cause rapidly gained favor both in the Assembly and among the people.

Galvão, until recently attached to the other party, made an impressive speech on the side of immediate acclamation as inevitable.

Alvares Machado demanded that party trammels should now be abandoned. "The cause of the Emperor was the cause of the nation, and ought to receive the approbation of every lover of the country."

Navarro, a young but powerful member from Matto Grosso, followed in a violent and denunciatory speech, in which he stigmatized the Regent, and all his acts, in the most opprobrious language. While in the heat of his harangue, he suddenly exclaimed, "*Viva a maioridade de sua Majestade Imperial!*" The crowded galleries had hitherto observed the most religious silence; but this exclamation drew forth a burst of enthusiastic and prolonged applause. Navarro, no longer able to make himself heard, drew his handkerchief from his bosom to respond to the vivas from the gallery. Members of the other party sitting near him imagined they saw a dagger gleaming in his hand, and, not knowing whose turn might come first, began to flee for their lives. One seized Navarro to keep him quiet; but he, not perceiving the reason of the assault, furiously repelled it. For a few moments the most intense and uncontrollable excitement prevailed; but order was soon restored.

Crowds of people now assembled out of doors, demanding the elevation of the young Emperor. Some went so far as to proclaim his majority in the public squares of the city. The ministerial party desperately resisted these strange movements in the House, but they were unable to stave off the debate.

Limpo de Abrêo, (afterward Visconde de Abacté,) an ex-minister, was in favor of the Revolution, but he wished it to be a deliberate and consistent one,—at least preceded by the report of a committee justifying the step. After much opposition to the measure, the committee was appointed, and a momentary calm ensued. During the night both parties reviewed their positions. The clubs and lodges held their sessions, and the opposition met in caucus. The Regent and his ministry were also in conclave. Vasconcellos, the Senator from Minas-Geraes, the veteran politician, but a man who had long been obnoxious on account of great moral delinquencies, was called in as their counsellor.

The session of the Chamber of Deputies next day was opened in the midst of the deepest anxiety. The galleries were crowded with people. The report of the committee was anxiously looked for, and indeed imperiously demanded, but did not appear.

Navarro accused the majority of the committee of treacherously intending delay. He urged the immediate and unceremonious declaration of the Emperor's majority. He appealed to the galleries, and received a deafening response of vivas to Dom Pedro II. Indescribable confusion ensued. The President of the Chamber attempted to call up the order of the day; but it was impossible. The absorbing question must be discussed. The more moderate of the Opposition wished the young Emperor's elevation deferred till his birthday,—the 2d of December. The more violent exclaimed vehemently against any delay whatever. The debate was protracted to an unusual length. In the midst of it a messenger entered, bearing documents from the Regent. They were read by the Secretary. The first was a nomination of Bernardo Pereira de Vasconcellos as Minister of the Empire! At the mention of the name of Vasconcellos, irrepressible sensations of indignation were apparent throughout the House. The Secretary proceeded to read the second document, which proved to be an act of prorogation, adjourning the General Assembly over from that moment to the 20th of November following.

Confusion and indignation were now at their height. The people in the galleries could not be restrained. They poured down a torrent of imprecations upon the administration, mingled with vivas to the majority of Dom Pedro II. Antonio Carlos, Martin Francisco, (the two Andradas,) Limpo de Abrêo, sprang to their feet, and one after the other entered their vehement protests against this act of madness on the part of the Government. They charged the Regent with treason, and declared that every Brazilian should resist his high-handed measures. They represented Lima as clutching, with a death-grasp, the power that was about to escape from his hands. They denounced him as a usurper, willing to sacrifice the monarch and the throne, at the hazard of lighting up the flames of civil war in every corner of the Empire. Vasconcellos was portrayed as a monster whose name was significant of every vice and crime, and withal the worst enemy the Emperor had; but it was into his hands that the young monarch was now betrayed!

The President of the House attempted to enforce the Act of Prorogation, but was prevented. Antonio Carlos de Andrada now started forth, and called upon every Brazilian patriot to follow him to the halls of the Senate,—situated upon the Campo de Santa Anna, and nearly a mile distant. His friends in the House, and the people *en masse*, accompanied him. The multitude increased at every step. On the arrival of the Deputies at the Senate, the two Houses instantly resolved themselves into joint session, and appointed a deputation, with Antonio Carlos at its Lead, to wait upon the Emperor and obtain his consent to the acclamation. During the absence of the deputation, several of the Senators endeavored to calm the passions of the people. The multitude without had increased to the number of several thousand. No soldiers appeared; but the cadets of the Military Academy, in the heat of their juvenile enthusiasm, rushed to arms and prepared to defend their sovereign.

Presently the deputation returned, and announced that, after its members had represented to the Emperor the state of affairs which existed at the present crisis, His Majesty had consented to assume the reins of government, and had ordered the Regent to revoke his obnoxious decrees and to pronounce the Chambers again in session. Thunders of applause followed this announcement. The enthusiasm of the people knew no bounds. The country was saved, and no blood was shed! The citizens proceeded to congratulate one another upon this peaceful triumph of public opinion.

The discussions of the Assembly turned upon the manner of consummating the revolution which had thus singularly commenced. Lima was now stigmatized as the *ex*-Regent, and was pronounced incompetent to reassemble the body which he had tried to prorogue. The Marquis of Paranaguá, President of the Senate, declared that neither House was now in session, but that the members of both composed an august popular assemblage, personifying the nation, demanding that their Emperor be considered no longer a minor. It was finally resolved to remain in permanent session until His Majesty should appear and receive in their presence the oath prescribed by the Constitution. The Assembly consequently remained in the Senate-House all night. A body of the National Guards, the alumni of the Military Academy, and numerous citizens, also remained to guard them.

At daylight the people generally began to reassemble. By ten o'clock not less than eight or ten thousand of the most respectable citizens surrounded the palace of the Senate. At that hour the President of the Assembly made a formal declaration of the objects of the present convocation. The rolls of both Houses were then called, and the legal number, both of Senators and of Deputies, being found present, the President arose and said:—

"I, as the organ of the Representatives of this nation in General Assembly convened, declare that His Majesty Dom Pedro II. is from this moment in his majority, and in the full exercise of his constitutional prerogatives. The majority of His Majesty Senhor Dom Pedro II.! Viva Senhor Dom Pedro II., constitutional Emperor and perpetual defender of Brazil!! Viva Senhor Dom Pedro II.!!!"

Millions of vivas from the members of the Assembly, from the spectators in the gallery, and from the multitude in the Campo, now rent the air in response, and were prolonged with indescribable enthusiasm and delight. Deputations were appointed to wait upon His Majesty when he should arrive, and to prepare a proclamation for the Empire. At half-past three o'clock the Imperial escort appeared. His Majesty was preceded by the dignitaries of the palace, and followed by his Imperial sisters. On beholding the young Emperor, the enthusiasm of the crowd exceeded any former limit. Nothing but a reiteration of vivas could be heard in the Campo during the whole ceremony. His Majesty was received with all possible formality, and conducted to the throne, near which the members of the diplomatic corps were already seated in their court-uniform. The Emperor now knelt down and received the oath prescribed by the Constitution; and, after the *auto de juramento* was read aloud and solemnly signed, the following proclamation, already drafted by Antonio Carlos de Andrada, and approved by the Assembly, was now uttered:—

"Brazilians!—The General Legislative Assembly of Brazil, recognising that happy intellectual development with which it has pleased Divine Providence to endow his Imperial Majesty Dom Pedro II., recognising also the inherent evils which attach themselves to an unsettled government,—witnessing, moreover, the unanimous desire of the people of this capital, which it believes to be in perfect accordance with the desire of the whole Empire,—viz.: to confer upon our august monarch the powers which the Constitution secures to him; therefore, in view of such important considerations, this body has, for the well-being of the country, seen fit to declare the majority of Dom Pedro II., so that he may enter at once upon the full exercise of his powers as constitutional Emperor and perpetual defender of Brazil. Our august monarch has just taken in our presence the solemn oath required by the Constitution.

"Brazilians! The hopes of the nation are converted into reality. A new era has dawned upon us. May it be one of uninterrupted union and prosperity! May we prove worthy of so great a blessing!"

After the ceremonies of the occasion had been completed, His Majesty proceeded to the city palace, accompanied by the National Guards and the people. In the evening a numerous and splendid reception took place, and the joy of the whole city was manifested by a spontaneous and most brilliant illumination.

To the astonishment of every one, the revolution was now complete. The Regency was abolished; perfect tranquillity prevailed; and Dom Pedro II.—the boy who, when six years old, had been acclaimed sovereign of a vast Empire—was now at fifteen invested with all the prerogatives of his Imperial throne. The youthful Emperor was very tall for his age, but not of the handsome proportions for which he is now so distinguished. His mind was of an exceedingly mature cast. As a student he was, it may be said without any exaggeration, most remarkable in his tastes, application, and rapid advancement. The study of the natural sciences —not a mere smattering of them, but the most thorough and abstruse investigation—was his delight; and his facility for acquiring language was such, that this day he can converse in the principal tongues of Europe. It was therefore no empty phrase which Antonio Carlos de Andrada used when he spoke of the "happy intellectual development" of His young Imperial Majesty. He was not a mere "boy Emperor."

The preceding year had witnessed the inauguration of steam-navigation along the whole Brazilian sea-coast, so that the news of the recent events at Rio de Janeiro was soon made known in every town of the extensive Atlantic board, and by special couriers in a few weeks the most remote parts of the wide Empire were sending up their vivas for Dom Pedro II.

Congratulations were the order of the day. Every society, every public institution, every province, and nearly every town, from the capital to the remotest parts of the Empire, hastened, on the reception of the news, not only to celebrate the event with extravagant rejoicing, but also to send a deputation to utter, in the presence of the Emperor, their most profound sentiments of joy at his elevation to the sovereignty, and their cherished hopes of his prosperity and happiness.

Thus was accomplished, without bloodshed, the third popular revolution of Brazil. The Constitution, with the exception of the article relating to the majority of the Emperor, remained intact.

In regard to the peculiar form of rule of the preceding nine years, it may be said that there can be no doubt that the government of the Regency was a benefit to Brazil. During the entire period of its existence it had to struggle with serious financial difficulties, and also with the formidable rebellion of Rio Grande do Sul, besides temporal outbreaks in other provinces. Nevertheless, improvement became the order of the day, and, in various ways, was really secured.

The personal rule of the Emperor commenced under auspicious circumstances. He was the object of an enthusiasm which has never waned. The two leaders of his first Cabinet were Antonio Carlos and Martin Francisco Andrada. Their elder brother, José Bonifacio, was no more. In 1833, upon his deposition as tutor to the Emperor, he withdrew from public life, and retired to the beautiful island of Paquetá in the Bay of Rio, where he remained until a short time before his death at Nictherohy in 1838.

Antonio Carlos at the very outset frankly and lucidly set forth the principles upon which the ministerial action would be based under the new order of things. Those principles were safe and consistent; and from the known energy of the Andradas, together with their associates, it may be presumed that no efforts were spared to put them in practice.

The nation at large was exhilarated with the idea of the glorious revolution that had transpired; but the legislature, tired by its recent paroxysms, soon fell back into its old method of doing business. The first leading measure of the opposition was the appointment of a Council of State, the members of which were to hold the office of special advisers to the Emperor. It became an immediate and protracted subject of discussion, but did not succeed till late in the following year. Things throughout the Empire moved on in their ordinary course, save that, when the subject of the Emperor's elevation lost its novelty, that of his approaching coronation became the theme of universal interest and of unbounded anticipation.

The early part of the year 1841 was fixed upon for the coronation. Preparations for that event were set on foot long in advance of the time. Expectants of honors and emoluments attempted to rival each other in parade and display. Extraordinary embassies were sent out from the different courts of Europe, in compliment to the Brazilian throne.

While diplomatists and politicians were intent upon sharing the honors of this occasion, the artisans and shopkeepers of the metropolis displayed quite as much tact in securing the profits of it. Exorbitant prices were demanded for every article of ornament and luxury; but those articles had now become necessary, and aspiring poverty, not less than grudging avarice, was compelled to submit to extortion.

Before the next session of the General Assembly difficulties had occurred which seriously embarrassed the administration. Several of the provinces had resisted the new appointments of presidents, and in so doing had manifested tendencies to revolution. But the most serious evil grew out of the long-standing rebellion in Rio Grande do Sul. In the anxiety of the Cabinet to bring this internal war to a close, Alvares Machado had been appointed an agent of the Government to treat with the rebels. Much confidence had been reposed in his personal influence with those in revolt, and he had been invested with extraordinary and unconstitutional powers. But, with all the facilities offered them, the insurgents refused to compromise. Machado was then appointed President of the province.

In this office, instead of wielding a rod of iron, as his predecessors had done, or had attempted to do, he adopted conciliatory measures, and rather entreated a negotiation. This attitude was stigmatized as dishonorable to the Empire, and such an outcry was made in regard to it as to excite general alarm lest the interests of the throne should be betrayed. This outcry was aimed at the ministry. A change was demanded, and was at length obtained. On the 23d of March the Andradas and their friends, with a single

15

exception, were dismissed; and thus those who had brought about the new order of things were supplanted, just in time for their opponents to secure the decorations and the emoluments that were soon to be distributed.

Mortifying as this circumstance may have been in some of its bearings, it caused no grief to the Andradas in view of their personal wishes. They could point to the early days of their political prosperity, in proof of their disinterested devotion to their country. They could now, as then, retire in honorable poverty, preserving the claim of pure patriotism as a more precious treasure than wealth or titles. Theirs was the distinction that would cause posterity to inquire why they did not receive the honors they had deserved. Other men were welcome to the ignominy of wearing titles they had never merited.

When the General Assembly convened in May, it was found expedient to postpone the coronation. Thus, for two months longer this anticipated fête continued to be the all-engrossing topic of conversation and of preparation in every circle, from the Emperor and Princesses down to the lowest classes. That anxiously-looked-for event transpired at length, on the 18th of July, 1841. It was magnificent beyond the expectations of the most sanguine. The splendor of the day itself,—the unnumbered thousands of citizens and strangers that thronged the streets,—the tasteful and costly decorations displayed in the public squares and in front of private houses,—the triumphal arches,—the pealing salutes of music and of cannon,—the perfect order and tranquillity that prevailed in the public processions and ceremonies of the day, together with nearly every thing else that could be imagined or wished,—seemed to combine and make the occasion one of the most imposing that ever transpired in the New World. The act of consecration was performed in the Imperial Chapel, and was followed by a levée in the palace of the city. The illuminations at night were upon a splendid scale, and the festivities of the occasion were prolonged nine successive days.

So far as pomp and parade could promote the stability of a Government and secure a lasting respect for a crown, every thing was done for Brazil on that day that possibly could be done without greater means at command. There were circumstances, however, connected with the monarchial pomp and the lavish expenditures of this coronation, which could not fail to be very embarrassing to those who had to struggle with them. The finances of the Empire were at the very lowest ebb, and constantly deteriorating. The money used in support of this grand fête, including an expense of one hundred thousand dollars for an Imperial crown, was borrowed, and added to an immense public debt. In addition to this, the Government was far from being stable and settled. Its councils were divided, and its policy vacillating. The existence of this state of things formed a principal pretext for the splendid demonstration alluded to. It was thought to be an object of the first importance to surround the throne with such a degree of splendor as would forever hallow it in the eyes of the people.

After the coronation, the sessions of the General Assembly were resumed. On the 23d of November a law was passed establishing the *Conselho de Estado*. This body was modelled upon the double basis of the ordinary and extraordinary Privy Councils of Great Britain. Among the gentlemen composing this Council were Lima, Calmon, Carneiro Leão, and Vasconcellos. The very individuals who opposed the Andradas at the period of the young Emperor's elevation, and who were then put down by acclamation, had, in the short space of a year, not only managed to get back into public favor, but also to secure life-appointments of the most influential kind.

Vasconcellos, it is true, sought no titles. They were playthings which he could easily dispense with for the gratification of his fellow-partisans. But he loved power, and neither mortifications nor defeat diverted him an instant from its pursuit. He finally gained a position which probably suited his inclinations better than any other, and in which, as the master-spirit of the body, his influence must be widely felt.

On the 1st of January, 1842, the Honorable Mr. Hunter,* United States Chargé d'Affaires at Rio de Janeiro, presented to His Majesty the Emperor his credentials as envoy-extraordinary and minister-

---

* No foreign diplomatist in Brazil left warmer friends than the late Honorable Mr. Hunter, of Rhode Island. His accomplishments as a scholar and his affability as a gentleman won the hearts of all.

plenipotentiary, to which rank he had been advanced. This compliment was speedily reciprocated by the appointment of the
Honorable Mr. Lisboa as the minister of Brazil at Washington.

In continuance of the present historical sketch of Brazilian
affairs, it is painful to add that the year 1842 was marked by
repeated and serious disturbances in different parts of the Empire.
They commenced with the elections for deputies. Various frauds
had been enacted, by suddenly changing the day, hour, and places
of elections. What was worse, bodies of troops and armed men
were introduced to influence votes, while crowds of voters were
brought in from other districts. In short, bribery, corruption, and
force triumphed over the free exercise of public opinion. It is
not to be presumed that one party was guilty of these measures
alone; but it appeared, in the issue, that the opposition had succeeded and that the ministerial party was in the minority. The
conduct of the ministry was such—though they acted with some
degree of plausibility in regard to preventing the regular meeting
of the Assembly and in issuing a decree for an extraordinary
session—that the sounds of rebellion were heard in parts of the
Empire which hitherto had been the most faithful and the most
tranquil. San Paulo and Minas-Geraes were in commotion and
disorder. The utmost consternation prevailed, and even at the
capital an incendiary proclamation was·posted up at the corners
of the streets, calling upon the people to free the Emperor from
the domination which had been imposed upon him, and to rescue
both the throne and the Constitution from threatened annihilation.

It is interesting to note that the Brazilians, in their internal
commotions, put the blame in the right place, and have ever
rallied around D. Pedro. He, on the other hand, has always
proved, by his character and by his measures, worthy of their
devotion. The power of the Emperor of Brazil is not like that
of the monarch of Russia, but is as limited as that of the sovereign of the British realm.

The Government was now driven to extreme measures. The ·
militia was called out, and martial law was proclaimed in the
three disturbed provinces. The supremacy of the law was maintained. The prospects of the Empire were for a short time very
gloomy and unpromising, but by degrees the storm blew over.

Order was gradually restored without actual hostilities or the loss of many lives. The worst consequences of the rebellion were experienced in the districts where it occurred, although public confidence and the national revenue suffered severely.

The elections at the close of the year occurred with more quietness, and on the 1st of January, 1843, the Emperor opened the General Assembly in person, and a new ministry was appointed. From that time to this there has been a softening down of parties and factions; and, though there has always been a certain amount of corruption and unscrupulousness in the political affairs of the nation, no great disturbances have affected its welfare, and there has been a constant tendency to obedience to law. In connection with this, financial difficulties were diminished and national prosperity increased.

The most remarkable public events that transpired at Rio during the year 1843 were the Imperial marriages. They were celebrated with great rejoicings and all possible splendor.

As early as July, 1842, the Emperor Dom Pedro II. had ratified a contract of marriage with Her Royal Highness the Most Serene Princess Senhora Donna Theresa Christina Maria, the august sister of His Majesty the King of the Two Sicilies. The marriage was duly solemnized at Naples, and, on the 5th of March, a Brazilian squadron, composed of a frigate and two corvettes, sailed from Rio de Janeiro to the Mediterranean, to conduct the Empress to her future home.

In the mean time, on the 27th of March, a French squadron arrived, under the command of His Royal Highness Prince de Joinville, son of Louis Philippe. This was Joinville's second visit to Brazil. Soon after his arrival he made matrimonial propositions to Her Imperial Highness Donna Francisca, the third sister of the Emperor. The customary negotiations were closed with despatch. On the 1st of May the marriage was solemnized at Boa Vista. On the 13th of May the Prince and his Imperial bride sailed for Europe.

The Empress Donna Theresa arrived at Rio on the 3d of September, and was received not only with magnificent ceremonies, but also with sincere cordiality on the part of the Brazilians.

It may be mentioned here that the eldest sister of D. Pedro II.,

Donna Maria, Queen of Portugal, had previously taken, as her royal consort, Prince Fernando Augusto, of Saxe-Coburg Gotha; and on the 28th of April, 1844, Her Imperial Highness Donna Januaria was also married to a Neapolitan prince,—the Count of Aquilla, brother to the Empress of Brazil and the King of the Two Sicilies. Thus, in the course of a single year, the Imperial family of Brazil contracted honorable and flattering alliances with the courts of Europe.

In 1844, Brazil was rejoiced by the birth of the Imperial Prince Dom Affonso; but his untimely death the following year brought mourning upon the nation. In 1846, the Princess Isabella (the present heir-presumptive) was born, and, in 1847, her sister, the Donna Leopoldina. In case of the death of these princesses, and the demise of the Emperor without other issue, the Constitution provides that the eldest child (Donna Januaria) shall be heir to the Imperial throne.

In 1850, the slave-trade (which had continued despite solemn treaties) was effectually put down; and, soon after, a number of the leading dealers in the inhuman traffic—men who had hitherto held high position in society—were banished.

The same year witnessed the first steamship-line to Europe; and now the Empire is united to the Old World by no less than three lines; and one line links together the two Americas.

For the last ten years the progress of Brazil has been onward. Her public credit abroad is of the highest character. Internal improvements have been projected and are being executed on a large scale; tranquillity has prevailed, undisturbed by the slightest provincial revolt; party spirit has lost its early virulence; the attention of all is more than ever directed to the peaceful triumphs of agriculture and legitimate commerce; public instruction is being more widely diffused; and, though much is yet required to elevate the masses, still, if Brazil shall continue to carry out the principles of her noble Constitution, and if education and morality shall abound in her borders, she will in due time take position in the first rank of nations.

Note for 1866.—In October, 1864, the Princess Imperial, Donna Isabella, was married to Prince Louis Philippe M. F. Gaston d'Orléans, Count d'Eu, eldest son of the Duc de Nemours. They spent the greater portion of 1865 in Europe. In December, 1864, the second Princess, Donna Leopoldina, was married to Prince Auguste of Saxe-Coburg Gotha, whose mother was Clementine d'Orléans, so that both the Princesses married grandsons of Louis Philippe.

CHAPTER XIII.

THE EMPEROR OF BRAZIL — HIS REMARKABLE TALENTS AND ACQUIREMENTS — NEW YORK HISTORICAL SOCIETY — THE FIRST SIGHT OF D. PEDRO II. — AN EMPEROR ON BOARD AN AMERICAN STEAMSHIP — CAPTAIN FOSTER AND THE "CITY OF PITTSBURG" — HOW D. PEDRO II. WAS RECEIVED BY THE "SOVEREIGNS" — AN EXHIBITION OF AMERICAN ARTS AND MANUFACTURES — DIFFICULTIES OVERCOME — VISIT OF THE EMPEROR — HIS KNOWLEDGE OF AMERICAN AUTHORS — SUCCESS AMONG THE PEOPLE — VISIT TO THE PALACE OF S. CHRISTOVÃO — LONGFELLOW, HAWTHORNE, AND WEBSTER.

WE naturally turn with interest and a laudable curiosity to look at the character and abilities of the monarch who has been called by Providence to the head of a growing nation. The Emperor of Brazil, by the various limits of the Constitution, has not the scope for kingcraft that is the heritage of Alexander II. or the achievement of Napoleon III. The life of some crowned heads is only an official one; very few of the *Dei gratia* rulers possess intrinsic merit: they are educated, refined, and may or may not be affable. In the eye of the legitimist their chief distinction is the blood which has coursed through the veins of generations of kings. He who is situated half-way between the legitimist and the red republican regards with a greater or less degree of veneration the representative of executive power which he beholds in the ruler, and is possibly excited to a certain admiration by the amiable and benevolent character which he who sits upon the throne may possess. But it is very rare, in the history of nations, to find a monarch who combines all that the most scrupulous legitimist would exact, who is limited by all the checks that a constitutionalist would require, and yet has the greatest claim for the respect of his subjects and the admiration of the world, in his native talent and in his acquisitions in science and literature. These rare combinations meet in Dom Pedro II. In his veins courses the united blood of

231

the Braganzas, the Bourbons, and the Hapsburgs. By marriage
he is related to the Royal and Imperial families of England,
France, Russia, Spain, and Naples. His father (Dom Pedro I.)
was an energetic Braganza; his mother (Donna Leopoldina) a
Hapsburg, and sister-in-law to Napoleon I. His relatives, it will
be seen, are of every grade,—from the constitutional monarch to
the most absolute ruler.

His powers, modified by the Brazilian Constitution, have already
been considered; and it remains to point out his chief and commanding title to the regard of his nation and the world.

He has devoted much time to the science of chemistry, and his
laboratory at San Christovão is always the scene of new experiments. Lieutenant Strain, the noble hero of the Darien Expedition,—whose science is as well known as his kindness and bravery,
—informed me that, on a visit to Rio de Janeiro more than ten
years ago, he found the Emperor a thorough devotee to the studies
of natural phenomena. Dr. Reinhardt—who has spent many years
in Brazil as a naturalist—visited the capital of the Empire when
D. Pedro II. was not yet out of his teens : the latter heard that an
American *savant* was about to enter upon a scientific exploration
of the Empire, and sent for him to aid him in performing certain
new chemical experiments, accounts of which had been perused by
his Majesty in the European journals of science. Dr. Reinhardt
further added, that the young monarch, in his enthusiasm, paid no
attention to the time that flew by as they, in a tropic clime and a
close room, were cooped up for hours over fumigating chemicals.

It is well known at Rio de Janeiro that he is a good topographical engineer, and his theoretical knowledge of perspective is
sometimes put in practice; for the German Prince Adalbert, in
the published account of his visit to Brazil, states that the Emperor
presented him with a very creditable painting from the Imperial
palette. He has a great *penchant* for philological studies. I have
heard him speak three different languages, and know, by report,
that he converses in three more; and, so far as translating is concerned, he is acquainted with every principal European tongue.
His library abounds in the best histories, biographies, and encyclo-
pedias. Some one has remarked that a stranger can scarcely start
a subject in regard to his own country that would be foreign to

CAMERA DO SENADO.

The Annual Opening of the Assemblea Geral by D. Pedro II.

Dom Pedro II. There is not a session of the Brazilian Historical Society from which he is absent; and he is familiar with the modern literature of England, Germany, and the United States, to a degree of minuteness absolutely surprising. When Lamartine's appeal for assistance was wafted over the waters, it was the Emperor of Brazil who rendered him greater material aid than any other, by subscribing for five thousand copies of his work, for which he remitted to the sensitive *littérateur* one hundred thousand francs. His favorite modern poet is Mr. Longfellow, for whom he has an unbounded admiration.

In literature and science he is not, however, confined to large tomes, but a portion of each morning is allotted to the perusal of foreign periodicals and journals, as well as the publications of Brazil. That which emanates from his own pen is rarely seen; but I have before me some original lines by the monarch, which a member of the diplomatic corps at Rio copied from the album of one of the Imperial household. They were doubtless never intended for the public eye; but the justness of their sentiment in English, if not the mellifluousness of their Portuguese, is appreciable by every reader of this work. (See Appendix.)

In 1856, the Honorable Luther Bradish, the accomplished and dignified presiding officer of the New York Historical Society, at the June meeting of that association, proposed Dom Pedro II. as an honorary member of that learned body. The proposition was seconded by Marshal S. Bidwell, Esq., and I need hardly add that the vote was carried by acclamation. The same society, on a subsequent evening, was briefly addressed by the Rev. Dr. Osgood, whose remark in regard to the Emperor of Brazil is as true as it is forcible:—"Dom Pedro II., by his character, and by his taste, application, and acquisitions in literature and science, ascends from his mere fortuitous position as Emperor, and takes his place in the world as a *man*."

The Brazilian ruler receives his talents in a direct line: Dom Pedro I. was a man of great energy and ability, and Donna Leopoldina was not without some of that power which characterized Maria Theresa. The early studies of Dom Pedro II. were conducted by the Franklin of Brazil,—José Bonifacio de Andrada; and we know not how much his tastes for science may have been influenced by that ardent admirer of the study of nature. His mind early became imbued with such pursuits, and, when growing up to manhood, as we have already seen, he omitted no opportunity for making additions to his store of knowledge.

The first time that I saw the Emperor he was in citizen's dress, accompanied by the Empress. They were in a coach-and-six, preceded and followed by horse-guards. He likes a rapid movement, and, whether on horseback or in a carriage, his chamberlains and guards are kept at a pace contrary to the usual manifestations of activity among the Brazilians. Two of the dragoons precede the coach at full gallop, and, at the blast of their bugles, the street is cleared of every encumbrance in the shape of promenaders and vehicles. It has, however, occurred to me that the neck-muscles of their Majesties must be exceedingly fatigued after their frequent city and suburban rides, for the humblest subject who salutes them is reciprocated in his attention. Their usual afternoon-drive is through the Catete and Botafogo to the Botanical Garden.

A combination of circumstances brought me afterward into a much closer relation with his Majesty than as a mere spectator of his fine form when he passed rapidly by. In 1852, during the temporary absence of Mr. Ferdinand Coxe, the Secretary of the United States Legation at Rio de Janeiro, I was chosen to fill his place, and finally, after his resignation, I was appointed Acting Secretary. In September, 1852, it became my duty to go to the Palace of San Christovão in company with Governor Kent, who, in the absence of the Minister-Plenipotentiary, held the post of Chargé d'Affaires in addition to that of American Consul. The occasion that demanded this official visit of Governor Kent was, in accordance with court-etiquette, to thank his Majesty for having accepted the invitation of the American Captain Foster to visit the "City of Pittsburg." This large merchant-steamer was on its way to California viâ the Straits of Majellan, and, while stopping for coals in the harbor of Rio de Janeiro, the captain invited the Emperor and his court to an excursion on board the splendid specimen of American naval architecture under his command. The Emperor having signified his acceptance, all was made ready, and, at eleven o'clock, the guns of the forts and of the men-of-war told that the Imperial party were embarking in the state-barges for the steamer

The day was most beautiful, and Captain Foster spared no pains in adorning his fine steamer in a manner worthy of his guests. Flags and streamers were suspended from every mast, the standards of the North American Republic and the South American Empire floated in unison, while a full orchestra from the flower-strewn deck sent forth the national anthems of Brazil and the Union. When the barges reached the "City of Pittsburg," Captain Foster, with the American Chargé d'Affaires by his side, received the Emperor, and, when welcoming him on board, placed the steamer at his Majesty's order.

Dom Pedro II. was accompanied by the Empress, and also by the Cabinet Ministers, the Imperial household, and the chief officers of the army and navy. All were in full court-dress, with the exception of their Majesties.

The excursion was of unusual interest. The fine steamer of twenty-two hundred tons ploughed her way through the various anchorages until she reached the men-of-war; the cannon of the forts saluted her as she passed, and the vessels-of-war not only sent forth their booming salvos, but the yards were manned, and the sailors shouted their loud vivas to D. Pedro II. In the mean while, the Emperor examined the "City of Pittsburg" from the coal-bunkers to her engine; and, as it fell to my duty to make many of the explanations, it afforded an opportunity for observing the *man* and forgetting the unbending features of the Emperor. He was not content with beholding the mere upper-works of the machinery, but descended into the hot and oily quarters of the lower part of the ship, where the most intricate portion of the engine was situated: a half-hour was afterward devoted to studying the engraved plan of the machinery, which was further explained by the chief engineer of the steamer, and by Mr. Grundy, an English engineer, who has long been connected with the Brazilian navy.

When the investigation of the engine was concluded, the Emperor wished to visit the forward-deck. Now, Americans are the vainest people in the world, and we were all afraid that on this part of the vessel Dom Pedro would not only be shocked with the appearance of some very rough specimens of humanity on their way to the gold-regions of the Pacific, but that the said specimens would not give His Majesty the reception which was due to his station as

the Executive head of the most powerful South American Government. The Emperor's attention, however, could not be diverted to
a different point; and the captain, fearing and trembling, was led
to the forward-deck. There, upon the taffrail, sat representatives
of the New York "Mose," the Philadelphia "Killer," and the Baltimore "Plug-ugly." The captain's heart sank within him: he was
proud of his ship, proud of his illustrious guest, but he had very
little to be proud of in some of his passengers,—especially the
unkempt and unterrified, who were even more picturesque after
their voyage than upon election-day. The Emperor now approached the sovereigns,—ay, near enough to have them "betwixt
the wind and his nobility." Then occurred a scene, rich beyond
description, which could never have taken place with others than
Americans for actors. One of the unshaven, whose tobacco had,
up to this time, occupied the greater portion of his mouth and
thoughts, suddenly tumbled from the taffrail, discharged his quid
into the ocean, and, hat in hand, yelled forth, in a well-meaning
but terrific voice, "Boys, three cheers for the Emperor of the
Brazils!" In a twinkle of an eye every Californian was upon his
feet; and never, in their oft-fought battles for the "glorious Democracy," did they send forth such round and hearty huzzas as they
did that day to D. Pedro II. The suddenness, the earnestness, the
good intention, and the enthusiasm of the whole procedure were
most mirth-provoking. The captain's fears subsided: his *pons asi-
norum* was crossed, and he took breath and laughed freely. The
Emperor returned the impromptu salute with great respect, and,
for the occasion, with becoming gravity.

The Empress and her suite were not less pleased with the commodious saloons and richly-decorated cabins of the steamer than
her Imperial spouse had been with all its mechanical appointments.

The "City of Pittsburg" was at the command of the Emperor;
but on we steamed, notwithstanding a portion of the court became
exceedingly sea-sick. His Majesty was too well pleased with his
new floating-dominion to resign it so soon; and thus we passed ten
miles beyond the Sugar-Loaf before the order was given to return.
The panorama of coast-mountains never appeared to me more
magnificent than on that bright September day.

The captain had prepared a sumptuous collation, but there was an obstacle which seemed more difficult to surmount than the forward-deck. The Imperial pair were not even in the habit of dining with their suite, and, except on rare state-occasions, eminent Ministers-Plenipotentiary had never been invited to partake of a repast in the same room with their Majesties. There was no precedent of a collation having been given on the deck of an American vessel, and, above all, on board of a mere commercial ship. No one liked the idea of consulting the Emperor about an affair apparently so trifling as to the manner in which he desired to eat, and therefore Captain Foster, who is as modest as he is hospitable, took the whole matter into his own hands and made a precedent. The "City of Pittsburg" was constructively a part of the United States, and the captain was determined to do the honors of his country as he would have done them on the banks of the Hudson. Their Majesties were accommodated with an entire table to themselves, which, like six others in the ship, was separated from its fellows by the space of two feet. The American party occupied the adjoining table; the ministers and noblemen were seated at another in a different part of the saloon, while the chamberlains stood near the Emperor. Perhaps D. Pedro had no objection to the proximity of the Americans, considering that they were· all "sovereigns." Captain Foster, who spoke French, proposed, with a dignity becoming the occasion, the health of their Majesties; and all passed off as easily and as happily as if there had been a thousand and one ceremonies and precedents to have been supported and followed.

We entered the harbor amid the booming of cannon, and at sunset the Imperial party again embarked in the state-barges, having spent what they afterward declared to have been one of the most agreeable days of their lives. Again and again have I heard their Majesties express their remembrance of that excursion; and none of Captain Foster's personal friends felt a deeper sympathy for him than did D. Pedro II. and Donna Theresa when they learned, through the public journals, the sad fate of the "City of Pittsburg" in the harbor of Valparaiso.

In 1854, I returned for a few months to the United States. Having often had occasion while in Brazil to remark the ignorance which prevailed in regard to my own country, and the reciprocal ignorance of the people of the United States in regard to Brazil, I desired to do all that was in the power of a single individual to remove erroneous impressions and to bring about a better understanding between the two countries.   There were higher objects in view than the mere diffusion of knowledge and the promotion of commerce; and, now that two years have elapsed since this little effort was undertaken, I have the satisfaction of knowing that new avenues of reciprocity have been opened, that school-books have been prepared for Brazil in the American style, and that thousands of dollars' worth of some of the articles displayed have been ordered since 1855.

I shall here introduce, even at the hazard of some repetition, the greater part of a letter addressed to the "New York Journal of Commerce" and the "Philadelphia Ledger," which gives an account of the effort to which I have referred.   It is on my part due to others to premise that many did not fully understand the proposed enterprise, and, after hearing of its success, regretted that they had not had an opportunity of being represented in the "Exposition" at the capital of Brazil.

"RIO DE JANEIRO, May 23, 1855.

"MESSRS. EDITORS:—[After a few preliminary remarks, I wrote as follows:]   The motives which prompted me to undertake this affair were simply the good of the United States and Brazil.   When laboring for several years as a missionary-chaplain at Rio de Janeiro, I found great ignorance in regard to our country, its progress, and its producing-resources.   I also discovered a reciprocal ignorance in the United States concerning Brazil.   In the latter country *we* were known as a bold, hardy race, which consumed two-thirds of the Brazilian coffee-crop, for which we sent, in return, flour and a few articles of no great note.   In the United States, Brazil was often classed among the Spanish countries of America: few people were aware that the Portuguese language was spoken, and that here was the only monarchy in America, and the only other constitutional Government on the Western continent which has marched forward in tranquillity and material prosperity.   I here found English, German, and French goods and publications, with some few exceptions, the *mode*,—and this, too, when many of the

same articles were to be bought cheaper in the United States; and
I also ascertained that our ships often came in ballast for coffee,
paying for it cash at most exorbitant rates of exchange, when
European vessels brought cargoes at a profit in payment for the
chief staples of Brazil.

"In Brazil I found a very great want of school-books. In Chili
and New Grenada I saw Spanish books published by Messrs. Appleton, and I desired to see the same for the youth of Brazil, where
very great attention is awakening to the subject of education. I
observed here scientific societies which rank, in dignity and devotion to *belles-lettres*, with the New York Historical Society, and like
associations of our own land.

"It was my ardent wish, first, to see this seven millions of
tolerant people possessing sound morality and true religion. It
was my next desire to see men of science and learning in Brazil
linked with the kindred spirits of our vigorous land; to behold
good school-books in the hands of Brazilian children; and to see
our manufactures taking their stand in this country, which is so
great a consumer.

"In 1854, on account of the ill health of a member of my family,
I was compelled to leave suddenly my field of labor for the United
States. There, after several months, it became evident that I
should have to abandon the land of my adoption. It was, however, necessary for me to return to Brazil, in order to settle up my
affairs. It was then that, through the public journals, I offered
my services to convey to Rio de Janeiro, free of charge to the
donors, any articles that might be sent to my address. These
objects I solicited for the Emperor, for scientific and literary associations, and for exposition to the public. I was a clergyman, and
I thought that no one could accuse me of speculation. For two
months was I, more or less, engaged at my own expense in making
solicitations in person, as well as by the press and by letters. I
regret to say that many persons who should have been interested
in such an enterprise did not choose to respond to the solicitations
of an unknown name, and thus the Exposition was not so rich
in some departments as it otherwise would have been, although
I with pleasure record that there were some influential men who
lent the weight of their names to the project.

"At length a number of artists, publishers, merchants, and manufacturers were induced to send specimens of books, engravings,
sculpture, and manufactures; but these were few in comparison to
those who might have contributed to their own future benefit.

"Messrs. Corner & Sons, of Baltimore, generously placed their
bark at my disposal for a free passage. In the month of March,
the good bark 'Huntingdon' left Baltimore with my packages on
board. Robert C. Wright, Esq., of that city, and his first clerk,
Mr. W. R. Jackson, did every thing in their power to facilitate the
enterprise, and to them more than to others I am indebted for
the successful consummation of my desired object. In April we
arrived at Rio de Janeiro, and for three weeks I had such vexation
and delay that I almost despaired of a prosperous termination.
Through the kindness of the Baron of Penedo, then Brazilian
Minister at Washington, and by a letter from Hon. William
Trousdale, the American Minister, my boxes and packages were
admitted free of duty. The custom-house regulations of this
country are exceedingly strict, and I had to give an account of
every thing that I had brought for the statistical purpose of the
Minister of Finance. As I had no list of the articles nor of their
values, as many of the boxes contained one hundred different
tightly-made packages, and as there were many objects of a fragile
nature, and as *every thing* had to be opened by officers who might
not be the most careful, I suffered mentally and physically both
before and after the examination. It was no easy matter to undo
so many parcels, and it was hard to restore again some fine specimens after a clumsy underling had put a nail through them.

"The chief collector of the custom-house believed, from the day
that I arrived until the day of the examination, that I was meditating some plot against the finances of the country, and openly
told some of the merchants that I intended to sell these things.
[That gentleman afterward became a very warm and an attentive
friend.] But when I had patiently assisted in opening for examination box after box, and we came to one containing the splendid
photographs of Fredericks & Gurney, the chief examiner said to
one of the others, 'Go call the second collector.' He came, and,
after expressing his astonishment at such perfection in photography,
he sent for the collector-in-chief. This latter gentleman left his

platform in the large public hall of the custom-house, and found his
way to the store-room. His admiration knew no bounds when he
saw the large life-sized photograph of Webster,—the last likeness
of the great statesman. From this time onward, his suspicions in
regard to my project ceased. He looked with great pleasure into
Colton's fine maps, and delighted in a critical examination of the
exquisite bank-note engraving of Danforth & Wright and that of
Toppan & Carpenter, who had contributed some most beautiful
specimens of this mingling of the beautiful with the useful in art.
The examination and noting down the contents of the boxes went
on very swiftly from the time of this visit of the chief collector.

THE NATIONAL MUSEUM.

"One week after the custom-house was cleared, I received an order
from the Minister of the Empire, granting me a large hall in the
National Museum, for the purposes of an Exposition. The same
day I went to the palace, and communicated to the Emperor that
I should be ready to receive him at eleven A.M. next day, (May 16,)

at the Museum. His Majesty received me, it seemed to me, with more amiability than his usual serious countenance indicated, and I soon discovered, from a remark which he made, that I was indebted to His Excellency Senhor Carvalho de Moreiro for a full explanation to His Majesty of my project, which was on my part far more philanthropic than commercial.

"That night sleep did not visit me, so busily was I engaged in the arrangement of the whole affair. The next day, at five minutes before eleven, (His Majesty is noted for his punctuality,) I heard the well-known bugle-blast of the Imperial horse-guards; and, before my assistants had time to withdraw, the coaches containing Dom Pedro II. and the chamberlains drew up at the Museum.

"By the aid of some kind friends, I had so disposed the six hundred different objects that the exhibition was not wanting in an imposing appearance. The American and Brazilian flags fell in graceful folds over the portrait of Washington and the likenesses of the Emperor and his father. The maps of Colton and others, and engravings from New York, Philadelphia, and Boston, covered the walls. Books and small manufactured articles occupied tables; beautifully-designed wall-papers and sample-books of mousseline de laines were suspended; and large agricultural implements were arranged on platforms provided for the occasion.

"His Majesty commenced at one end, and with great earnestness and interest examined every thing in detail. He made many inquiries, and manifested a most intimate knowledge with the progress of our country. He was filled with admiration at the specimens of books, steel engravings, chromo-lithography, (of Philadelphia,) and agricultural implements. Every now and then you might have heard him calling to some of his noblemen or chamberlains to come and admire with him this or that work of the useful or beautiful arts. He was not, however, indiscriminate in his praise, but was perfectly frank in his criticism.

"Being himself a thorough student of physical science, and a good engineer, he examined with minuteness the splendid edition of the United States Coast Survey, from the bureau of the United States Coast Survey, Washington; and he appreciated at their just value the various scientific works that occupied a conspicuous table.

"For half an hour he pored over Youman's Atlas of Chemistry, and praised its thorough excellence and simplicity. While examining a work on physiology, I heard him remarking upon the superiority of the Craniology by the late Dr. Morton; and he informe ̇ me that he possessed the writings of that eminent student of the human frame. He was also well read in the immense tomes of the pains-taking, erudite, and conscientious Schoolcraft, whose works on the aborigines of North America were sent out by the Chief of the Bureau of Indian Affairs at Washington.

"His Majesty was deeply interested in the various maps, geographies, and school-books sent out by Colton, Appletons, Woodford & Brace, T. Cowperthwait, and Barnes. The finely-illustrated publications of the various benevolent societies of our land were sent out for the Imperial family, and attracted deserved attention. The Emperor was much pleased with the only specimens of wood-engraving, which were forwarded by Mr. Van Ingen, of the firm of Van Ingen & Snyder, whose skill has illustrated this work.

"The earnest examination which he gave the machinery, manufactures, and agricultural implements justified the reputation which Dom Pedro II. enjoys in this respect. Howell's wall-papers, after drawings by the students of the Philadelphia Academy of Design, and the beautiful silk manufactures of Horstman, and Evans,—which ought to be classed among works of art,—called forth much praise.

"He next approached the table where were the books presented by the Appletons and Parry & McMillan. Taking up the 'Republican Court,' he said, 'I am astonished at such perfection in binding.' I replied, 'And none of those volumes were bound expressly for your Majesty." The binding of Appletons' books was superb. He opened the 'Homes of the American Authors,' and surprised me by his knowledge of our literature. He made remarks on Irving, Cooper, and Prescott,—showing an intimate acquaintance with each. His eye falling on the name of Longfellow, he asked me, with great haste and eagerness, 'Avez-vouz les poëmes de Monsieur Longfellow?' It was the first time that I ever saw Dom Pedro II. manifest an enthusiasm which, in its earnestness and simplicity, resembled the warmth of childhood when about to possess itself of some long-cherished object. I replied, 'I believe not, your Majesty.' 'Oh,' said he, 'I am exceedingly sorry, for I have sought in every bookstore of Rio de Janeiro for Longfellow, and I cannot find him. I have a number of beautiful *morceaux*, but I wish the whole work; I admire him so very much.' That evening I found, among the books sent by Parry & McMillan, the 'Poets and Poetry of America.' In this volume is a biographical sketch of Longfellow, as well as some of the choicest selections from his pen. This, with T. Buchanan Read's 'New Pastoral,' were afterward commented on and received with the most visible pleasure by His Majesty.

"I was absent from the part of the hall where Dom Pedro II. was looking at the engravings of the American Bank-Note Company,) and when I returned I found him engaged in a discussion with his first chamberlain as to John Quincy Adams,—the chamberlain (as the majority of even well-educated foreigners) supposing that John Quincy Adams was the elder Adams. The Emperor insisted that John Quincy Adams was not the early advocate of liberty and the 'comrade,' as he termed him, of Washington,—but that he was the son of John Adams, and, like his father, was a President of the United States. And soon after he gave a very thorough re-examination of the 'Republican Court,' and pointed out to the chamberlain the distinguished mother of John Quincy Adams. He was very anxious to see a portrait of Jefferson. One of my assistants found a very neatly-engraved portrait of the sage of Monticello from the burin of Toppan & Carpenter. When he received it, you should have heard him, without pedantry or affectation, expatiate with great minuteness, correctness, and judgment on the character of Jefferson as compared with that of Washington.

"Approaching some very fine lithographs published by Williams & Stephens, of New York, I introduced His Majesty to 'Young America,' that handsome but independent-looking lad, and to 'Uncle Sam's Youngest Son, Citizen Know-Nothing.' I thought that I had now a subject of which His Majesty really knew nothing; but I found that I was mistaken, as he recounted to some one the pranks that this young fellow had been playing, and added that he was a citizen of some power and knowledge, judging from the recent (1855) elections in the United States.

"Thus the whole day was occupied in the examination and explanation of the American collection.

"A few days after the Exposition was closed, I had the many things destined for the Imperial family taken to the large *palacete* of the Marquis d'Abrantes, situated in one of the most charming environs of Rio,—viz.: the shore of the Neapolitan-shaped Bay of Botafogo. His Majesty was spending some weeks here for the benefit of sea-bathing. I passed the guards at the gate, and as I ascended the steps the Emperor saw me, and, meeting me at the door, thanked me heartily for what I had done, I desired him to allow me to remain a few moments until the boxes arrived, as I must give him some explanations as to the secret lock of the most excellent trunk sent him by Peddie & Morrison, of Newark, N. J. With his permission I went into the beautiful garden, where were the richest and rarest of flowers in a land of perpetual bloom. The air was truly loaded with sweet fragrance. There were fountains and statuary, many brilliant-plumaged birds, and, indeed, every thing in nature and in art to please and to gratify those alive to the beautiful. When looking upon a scene so enchanting I could only desire that this land, for which God has done so much in a natural point of view, might possess the solid mental and moral advantages which belong to our more rugged North through the instrumentality of education and religion.

"The blacks soon arrived with the heavy boxes and the nicely-finished plough, (sent by B. Myers, of Newark, N. J.,) all of which, by the order of the chamberlain, were placed in the ante-room, where His Majesty again examined and admired them. The first thing that he inquired for was '*My Longfellow*,' (in the 'Poets and Poetry of America;') the next, 'Youman's Atlas of Chemistry:' he then asked for the beautiful specimens of chromo-lithography, (by Sinclair, and Duval, of Philadelphia,) and finally inquired after the steam fire-engine which made its travels from Cincinnati to Boston last spring. I furnished him with a plan of it which had been given me by a clerk in the *Baltimore Sun* office. He instantly took it, and began to explain its operations to a French *savant* who was visiting the palace. For one hour he was engaged in a review of the products of our country. He called the Empress, who also expressed her gratification in the highest terms as I displayed the beautiful books sent for herself and the princesses. Her Majesty was not only pleased with what had drawn forth the praises of her Imperial spouse, but she, as well as her maids of honor, displayed the woman in the delight manifested at the fancy soaps and other articles of toilette sent out by H. P. & W. C. Taylor, of Philadelphia, and Colgate & Co., of New York. Many thanks were given to me for those who had been so kind in remembering the Imperial family of Brazil, and I left the palace, feeling that, so far as the head of the Brazilian Government was concerned, all was most successful.

"With His Majesty's subjects the enterprise was not less fortunate. On the 17th and 18th the Museum was visited by some thousands, and astonishment and admiration were constantly upon the lips of the Brazilians. Each evening I was completely worn out by answering the many questions that were propounded from every side. I have no doubt that a proper exhibition of American arts and manufactures, arranged by business-men and those who have means to carry it out, would redound a thousandfold to the benefit of American commerce. For, during my walks among those who were examining the various articles, I heard remarks which convinced me that it only required to have our country's productions known to cause a large importation. During and since the Exposition, I have had many orders for books, engravings, wall-papers, and Manchester prints; and this morning I had an application for a *sugar-crushing machine*, and a *large lithographic printing-press*. My reply in all cases has been, 'I am not a commercial man; I am not here for that purpose; I have no pecuniary interest whatever in this matter: but there are houses here which have correspondents in America.'

"Upon the evening of the 16th, the Statistical Society of Brazil held its meeting in the same hall where were the products of the United States. The Viscount Itaborahy presided, and invited me to address the Society. I was very glad to have the opportunity of explaining my plans to such a body of gentlemen, and found them most sympathetic: they freely expressed their desire to see the United States and Brazil more closely united. These remarks were reported for the press, and my motives were thus more widely made known to the people.

"The contributions from Washington, from the Bureau of the Coast Survey, and from the Patent-Office, and the splendid work on the North American Indians, to which Schoolcraft has devoted his life, were looked upon by the Historical and other Societies as a very great acquisition to their libraries. In this connection I must not omit to mention some important medical works sent out by Lippincott, Grambo & Co., which were presented to the Imperial Academy of Medicine. From these associations I received letters of thanks, showing that the contributions of the various donors are justly appreciated. The Brazilian Historical and Geographical Society published in the daily press the list of historical and other works and library-catalogues that had been thus added to their own increasing literary stores.

"I have already occupied too much of your space, and I must still beg leave to add a few remarks.

"I do not claim the 'Exposition' to have been a perfect collection of what the United States can produce. It was far from it; but, from the interest it has created in this city of three hundred thousand inhabitants, from the independent approbatory remarks of the daily press, and from the desires which come from all quarters that the exhibition should continue, I think that a favorable impression has been made, and I also believe that, from this little affair, we may legitimately argue that there is a most favorable opening here for the various manufactures, &c. of our country. It would require patience and capital, and perhaps the hazarding of something at first; but I believe that the end would more than recompense the adventurers. One or two Americans, a few years ago, commenced the importation of American agricultural implements, &c., and now there is quite a commerce in this line. If importation should be extended, and this people could know what we produce, our commerce would be most rapidly increased. Speculators are not wanted, but moral, sound, enterprising business-men, who will furnish the best articles at the lowest price.

"In conclusion, without wishing to excite expectations which will not be realized, or without desiring to overestimate any thing which has been done in this Exposition, I can only say that, however far short I may have come in my efforts, my intentions have been good, and, when I shall leave Brazil to return to the work of my Master in my own land, I shall have at least the consolation of having endeavored to bring about a closer relation between the strongest Government of South America and the great Republic of the North.

"I remain, gentlemen, very respectfully,

"Your obedient servant,

"J. C. FLETCHER."

A pleasing incident connected with this affair grew out of the late arrival at Rio of one of the presents destined for the Emperor. After the "Exposition," I departed from the city and became engaged in my legitimate labors in another part of the Empire. In the month of July I returned from the Southern provinces, and found that the Messrs. Merriam, of Springfield, Massachusetts, had sent out a superb edition of Webster's unabridged quarto Dictionary. I had also a few more books which were to be placed in the Emperor's own library. An account of the presentation of these volumes was given in a private letter to Mr. J. P. Blanchard, of Boston, from which I extract the following :—

"The gift of Messrs. Merriam arrived during my absence in the Southern provinces; but so soon as I returned I procured it from the custom-house, and in due time conveyed it to the palace. Of course it was too late for the Exposition in the National Museum ; but, as your State had been very poorly represented in May, I was glad to have this specimen of Massachusetts publication, and this monument of the patient and faithful labors of a man who has done so much to define and classify our mother-tongue.

"It was within two days of my departure for Bahia and Pernambuco that I stole a few hours to go out to the Imperial Quinta of Boa Vista,—the Palace of S. Christovão. It is usual to go thither in a coach drawn by *at least* two horses; but, finding a nice new tilbury and a bright mulatto driver, I entered his vehicle, and, with 'Webster's Dictionary,' Hawthorne's 'Mosses from an Old Manse,' and Longfellow's 'Hyperion,' I was soon whirling, through the garden-lined streets of Engenho Velho, to the palace. The Palace of S. Christovão is situated in one of the most picturesque environs of Rio de Janeiro. It stands in bold relief against the lofty green mountains of Tijuca, and is surrounded by the beautifully-foliaged

IMPERIAL PALACE OF BOA VISTA, AT S. CHRISTOVÃO.

trees of the tropics. It has every adjunct that can make it a delightful residence. As we rolled through the long avenue of mango-trees, I saw the coach of one of the Ministers bowling along with the servants in livery. My establishment looked small in comparison with this brilliant equipage; but I felt that the three books which I bore with me would delight His Majesty more than all the carriages of the court.

"I descended after the Minister had entered, and was conducted to an ante-room by a chamberlain, to whom I made known the purport of my visit and the nature of my volumes. Not wishing to trust my precious load to any servant, I carried the three tomes (no light burden) before me. After passing many corridors, I came to a large, wide gallery, which overlooked a courtyard where bright fountains were playing and the choicest and most fragrant flowers were blooming.

"I had supposed that it was a day for private audience; but the long gallery was filled with gentlemen in waiting,—noblemen, Judges of the Supreme Court, Ministers, Chargés, and officers *en grande tenue*, and some of them covered with decorations. I then learned from Senhor Leal, and from the Neapolitan Chargé d'Affaires, that the 13th of July was the anniversary of the Imperial Princess Leopoldina, and these gentlemen had come to felicitate their Majesties on the return of this anniversary. I took my stand at the extreme end of the waiting train, thinking that I had better have chosen a day when His Majesty was less occupied. Presently Dom Pedro II. appeared, his fine manly form towering above every other. He was dressed in black; and, with the exception of a star which sparkled upon his left breast, his costume was simple, and its good taste was most apparent when contrasted with the brilliant uniforms of the court.

"I conjectured that His Majesty would first receive the congratulations of the glittering throng that stood between him and the plainly-dressed clergyman. Judge, then, of my surprise when, merely bowing, he passed by the many titled gentlemen and representatives of foreign courts, and came directly to the 'Webster,' 'Hawthorne,' and 'Longfellow.' With a pleasant smile, he addressed me, and led me to an open arch that overlooked the flowers and the limpid fountain. There he examined the books and bestowed high eulogium upon the Dictionary,—not only for the beautiful style in which it had been prepared by the publishers, but for the almost encyclopedic character of the work. He spoke of Mr. Hawthorne as an author of whom he had heard, and was glad to possess the 'Mosses from an Old Manse.' I called his attention particularly to the 'Celestial Railroad,' which caused an allusion to Bunyan's 'guide and road-book to the Celestial City.' Since the month of May he had procured all the poetical works of Mr. Longfellow, but had not yet added to his library any of his (Mr. Longfellow's) prose compositions. He therefore considered 'Hyperion' a most interesting acquisition.

"His Majesty conversed for a long time on the objects for which I came to Brazil, and expressed his gratitude for the *souvenirs* which he had received from citizens of the United States. I stated to him that I would visit the Northern provinces and then return to my native land. He expressed the customary wishes of a *bon voyage*, &c., but, with great earnestness, said to me, in conclusion, 'Mr. Fletcher, when you return to your country, have the kindness to say to Mr. Longfellow how much pleasure he has given me, and be pleased to tell him *combien je l'estime, combien je l'aime!* —how much I esteem him, how much I love him.'"

Thus ended, so far as my own personal effort is concerned, that which I undertook to do.

Note for 1866.—In many ways we have since learned of the good effect of this effort at Rio de Janeiro in 1855.

The junior author has visited Brazil four times since the last-named year, and had many occasions for long and intimate conversations with D. Pedro II., and can testify to that monarch's continued interest in works that treat of morals, literature, and art. With Mr. Longfellow and the Quaker poet Whittier he has really lived on terms of intimacy, rendering himself familiar with their poetic effusions, and on more than one occasion making felicitous translations of their poems, of which he sent autograph copies to those authors. By his deep interest in science, and by his correspondence with Professor Agassiz, that eminent *savant* came to Brazil, and is now engaged in extensive explorations of the rich treasures of nature in this empire. That the most pleasant relations, literary as well as political, will continue to exist between North America is evident, when the emperor has been represented by such affable gentlemen as the Baron of Penedo, M. M. Lisboa, late Brazilian Minister to the United States, and Sr. Azambuja, the present representative of H. I. M. at Washington.

CHAPTER XIV.

BRAZILIAN LITERATURE—THE JOURNALS OF RIO DE JANEIRO—ADVERTISEMENTS—
THE FREEDOM OF THE PRESS—EFFORT TO PUT DOWN BIBLE-DISTRIBUTION—ITS
FAILURE—NATIONAL LIBRARY—MUSEUM—IMPERIAL ACADEMIES OF FINE ARTS—
SOCIETIES—BRAZILIAN HISTORICAL AND GEOGRAPHICAL INSTITUTE—ADMINISTRA-
TION OF BRAZILIAN LAW—CURIOUS TRIAL.

THE Brazilians, having a ruler with such literary and scientific tastes, will assuredly make more progress in this direction than formerly.

On account of the restrictive policy of Portugal, no printing-press was introduced into this country until 1808. The general taste for reading is mostly confined to the newspapers and the translations of French novels. Authors are by no means numerous in the Empire; but there have been within the last few years a number of very creditable provincial histories, scientific disquisitions, and one or two attempts at the general history of Brazil. The bookstores abound with French works on science, history, and (too often) infidel philosophy.

There is, however, a Government bookmaking which is prolific in the most interesting details. I refer to the annual *Relatorios* or Reports of the Ministers of the Empire, Finance, Justice, Foreign Affairs, War, and the Navy. These are well written and well printed, and contain the most valuable matter for the statesman, the statistician, or the general reader. The Relatorio of the Minister of Justice must demand an amount of labor unknown to officials in the United States or in England; for every case that goes before a jury in each of the twenty provinces must come under his revision and must be placed in its proper table. The crime, age, sex, and nationality of the criminal are given, together with the punishment. In addition to this, matters of prison-discipline and the varied interests of ecclesiastical affairs are not forgotten.

The periodical literature of Rio has, within a few years, been improved in character by the establishment of a Medical Review and also of a Brazilian and Foreign Quarterly. The last-mentioned periodical has been conducted with great spirit and literary enterprise, and promises to be of utility to the country; yet even in this there is a too frequent resort to translations. If Brazilians would only take the time to write, and make the effort to think for themselves, foreigners would soon find their productions to be interesting and valuable, and would prize them accordingly.

The press being free, I doubt whether any journals in the United States, England, or the Continent, contain so many communications from subscribers as those of Rio de Janeiro. As all of these *communicações* must be accompanied with the cash, journalism in Brazil is a lucrative "institution." Some of the editorials of the *Jornal do Commercio*, the *Correio*, and the *Diario* will compare favorably with those of New York or London. The *Correio* has an able corps-editorial, and is an exceedingly readable paper. In the Appendix will be found a leader from the *Jornal do Commercio* which was elicited by a most provoking and uncalled-for note on the African slave-trade, which was sent by the British Minister at Rio de Janeiro to the Brazilian Secretary of State.

The appearance of the newspapers of Rio is like that of the Parisian journals, only the Brazilian dailies are larger, in clearer type, and upon superior paper. The bottom of each sheet contains the light reading, in what is called the *folhetim*; and each Sunday the *Correio Mercantil* has several columns of *pacotilha*, (gossip.) The *Jornal do Commercio*, the *Mercantil*, the *Diario do Rio de Janeiro*, and the *Diario Official* are better printed than French dailies.

The newspaper-press in Rio is quite prolific. It issues four dailies, several tri-weeklies, and a varying number of from six to ten weeklies and irregular sheets. During the session of the National Assembly, verbatim reports of the proceedings and debates of that body are published at length—like those of the English Parliament and the American Congress—on the morning after their occurrence.

The *Jornal do Commercio*, already referred to, has the largest circulation, and is under the direction of Sr. Castro, who translated into Portuguese Southey's "History of Brazil," and Mr. Addé. The *Correio Mercantil* is the property of

Muniz Barreto & Co., and is under the editor-in-chief Sr. Octaviano, one of the finest writers in Brazil.   Mr. Barreto deserves well of his country for the efforts he has made to give a high tone to its press.   The *Diario do Rio de Janeiro* is edited by Saldanha Marinha, (an eminent politician in the Liberal party,) aided by an efficient staff.   The *Diario Official* was created for government purposes in 1862.   Its present editor is Sr. Tito Franco de Almeida, who formerly edited with so much ability the *Jornal do Pará*.   All of the leading editors of the last three journals are members of the Chamber of Deputies.   The *Anglo-Brazilian Times* is a semi-monthly English sheet, edited by Mr. Scully, with more sprightliness than the Brazilian papers, and is a valuable addition to the journals of Rio. There is an interesting agricultural journal published under the auspices of the Imperial Agricultural Society, which merits a wide circulation.   The first paper of this kind, *O Agricultor Brazileiro*, was started by Mr. N. Sands, in 1854.   Only one volume was published.

The most enterprising *typographia* is that of the Brothers Laemmert, in the Rua dos Invalidos.   The press of the Typographia "Universal" turns out fine specimens of work.   The matter of the advertising-columns of the various newspapers is renewed almost daily, and is perused by great numbers of general readers for the sake of its piquancy and its variety.   Several peculiar customs may be noticed, growing out of the Church and Brotherhood advertisements mentioned in a previous chapter, and the patronage of the numerous lotteries authorized by Government.   Persons frequently form companies for the purchase of tickets, and those at a distance order their correspondents to purchase for them.   In order to avoid any subsequent transfer or dispute, the purchaser announces, through the newspaper, the number of the ticket bought and for whose account,—as, for example :—"M. F. S. purchased, by order of J. T. Pinto, two half-tickets, Nos. 1513 and 4817, of the lottery in behalf of the theatre of Itaborahy."   "The treasurer of the company entitled 'The Friends of Good Luck' has purchased, on the company's account, half-tickets Nos. 3885 and 5430, of the lottery of the cathedral of Goyaz."   Following this custom, individuals who wish to publish some pert thing usually announce it as the name of a company for the purchase of lottery-tickets, although that name extends sometimes through a dozen lines of rhyme.

The Brazilians have a most effectual way of collecting debts, which ought to be made known for the benefit of creditors in other portions of the world.   The recipe is found in the following advertisement:—

"Senhor José Domingos da Costa is requested to pay, at No. 35 Rua de S. José, the sum of six hundred milreis; and in case he shall not do so in three days, his conduct will be exposed in this journal, together with the manner in which this debt was contracted."

Another will show that the clergy are not always spared:—

"Mr. Editor:—Since the vicar of a certain parish, on the 8th instant, having said mass with all his accustomed affectation, turned round to the people and said, with an air of mockery, 'As we have no festival to-day, let us say over the Litany,' &c., I would respond, that the reverend vicar knows well the reason why there was no festival. Let him be assured, however, that when intrigue shall disappear the festival will take place; but, if he is in a hurry, let him undertake it at his own expense, since whosoever says the paternoster gets the benefit.*

"(Signed)   An Enemy to Hypocrites."

A school-teacher, after announcing his terms for tuition, thus continues and concludes,—the italics being his own:—

"The first-class day-scholars are instructed in the different branches of science and literature, including the English, French, Portuguese, and Latin languages. Second-class pupils receive a plain education, consisting of reading, writing, grammar, arithmetic, and Christian doctrine.

"The director, not being in the habit of making *splendid advertisements or puffs* in the daily papers, or of throwing dust in the eyes of the public, can only promise that, being the father of a large family and knowing what care and attention children require as to their morals and education, he will do his duty toward them accordingly."

The last specimen which I give illustrates the early marriages which frequently take place in Brazil; but I defy any other country to furnish the like of the following advertisement, which appeared in the *Jornal do Commercio* of Rio de Janeiro in 1852. It is so unique that I furnish the original as well as the translation:—

"Precisa-se de uma senhora branca de afiançada conducta, e com intelligencia bastante para fazer companhia a uma menina casada

* "Quem rese o Pater noster come o pão."

de menor idade, aqual precisa de algumas instrucções proprias de seu estado. Quem estiver nestas circumstancias annuncie por esta folha para ser procurada."

"*Wanted.*—A white lady of faithful character and with sufficient intelligence to be the companion [or, literally, "to make the company"] of a young bride who is a minor, and who is in need of some instructions appropriate to her state. Whoever possesses these qualifications may make known her address in the columns of this journal."

Various allusions to the entire freedom of the press have already been made; and it may be mentioned, in this connection, that there was an interesting example of its use for advertisements for promoting the Bible in Brazil, and also its employment to put down an effort for the diffusion of the Sacred Scriptures. My co-author, (Dr. Kidder,) in the early part of his religious labors in Brazil, commenced by circulating the Bible. I prefer to give his experience in his own words. After speaking of the general influence of the mother-country upon Brazil, he says,—

"Portugal has never published the Bible or countenanced its circulation save in connection with notes and comments that had been approved by inquisitorial censorship. The Bible was not enumerated among the books that might be admitted to her colonies when under the absolute dominion. Yet the Brazilians, on their political disenthralment, adopted a liberal and tolerant Constitution. Although it made the Roman Catholic apostolic religion that of the State, yet it allowed all other forms of religion to be held and practised, save in buildings 'having the exterior form of a temple.' It also forbade persecution on the ground of religious opinions. By degrees, enlightened views of the great subjects of toleration and religious liberty became widely disseminated among the people, and hence many were prepared to hail any movement which promised to give them what had so long been systematically withheld,—the Scriptures of truth for their own perusal. Copies exposed for sale and advertised in the newspapers found many purchasers, not only from the city, but also from the distant provinces.

"At the mission-house many copies were distributed gratuitously; and on several occasions there was what might be called a rush of applicants for the sacred volume. One of these occurred soon after my arrival. It was known that a supply of books had been received, and our house was literally thronged with persons of all ages and conditions of life,—from the gray-headed man to the prattling child,—from the gentleman in high life to the poor slave. Most of the children and servants came as messengers, bringing notes from their parents or masters. These notes were invariably couched in respectful, and often in beseeching, language. Several were from poor widows who had no money to buy books for their children, but who desired Testaments for them to read at school. Another was from one of the Ministers of the Imperial Government, asking for a supply for an entire school out of the city.

"Among the gentlemen who called in person were several principals of collegios, and many students of different grades. Versions in French, and also in English, as well as Portuguese, were sometimes desired by amateur linguists. We dealt out the precious volumes according to our best judgment, with joy and with trembling. This being the first general movement of the kind, we were at times inclined to fear that some plan had been concerted for getting the books destroyed, or for involving us in some species of difficulty. These apprehensions were contradicted, however, by all the circumstances within our observation; and all who came made their errand on the ground of its intrinsic importance, and listened with deep attention to whatever we had time or ability to address to them concerning Christ and the Bible.

"It was not to be presumed, however, that so great an amount of scriptural truth could at once be scattered among the people without exciting great jealousy and commotion among certain of the padres. Nevertheless, others of this class were among the applicants themselves. One aged priest, who called in person, and received by special request copies in Portuguese, French, and English, on retiring, said, 'The like was never before done in this country.' Another sent a note in French, asking for *L'Ancien et le Nouveau Testament*. In three days two hundred copies were distributed, and our stock was exhausted; but applicants continued to come, till it was estimated that four times that number had been called for. All we could respond to these persons was to inform them where Bibles were kept on sale, and that we anticipated a fresh supply at some future day.

"We were not disappointed in the opposition which was likely to be called forth by this manifestation of the popular desire for the Scriptures. A series of low and vile attacks were made upon us in a certain newspaper, corresponding in style with the well-known spirit and character of their authors. Indeed, in immediate connection with this interesting movement a periodical was started, under the title of *O Catholico*, with the avowed object of combating us and our evangelical operations. It was an insignificant weekly, of anonymous editorship. After extravagant promises, and repeated efforts to secure permanent subscribers, it made out to struggle against public contempt for the space of an entire month. Yielding to the stress of circumstances, it then came to a pause. An effort was made to revive it some time after, with the more imposing title of *O Catholico Fluminense*. Thus its proprietors appealed as strongly as possible to the sympathy and patriotism of the people, by the use of a term of which the citizens of Rio de Janeiro are particularly proud. Under this heading it barely succeeded in surviving four additional numbers, in only one of which was the least mention made of the parties whose efforts to spread the pure word of God had given it origin.

"This species of opposition almost always had the effect to awaken greater inquiry after the Bible; and many were the individuals who, on coming to procure the Scriptures, said their attention was first called to the subject by the unreasonable and fanatical attempts of certain priests to hinder their circulation. They contemned the idea, as absurd and ridiculous, that these men should attempt to dictate to them what they should not read, or set up an inquisitorial crusade against the Bible. They wished it, and if for no other reason, that they might show that they possessed religious liberty, and were determined to enjoy it. They poured inexpressible contempt upon the ignorance, fanaticism, and even the immorality, which characterized some of the pretended ministers of religion, who dreaded to have their lives brought into comparison with the requirements of God's word.

"Those of our friends who were consulted on the subject almost invariably counselled us to take no notice of the low and virulent

17

attacks made upon us, with which the people at large had no sympathy, and of which every intelligent man would perceive the unworthy object. Such articles would refute themselves, and injure their authors rather than us.

"The results justified such an opinion. One gentleman (a Portuguese) in particular said to us, with emphasis, 'Taking no notice of these things, you ought to continue your holy mission, and scatter truth among the people.' With this advice we complied, and it is now a pleasing reflection that our energies and time were devoted to vastly higher and nobler objects than the refutation of the baseless but rancorous falsehoods which were put forth against us. We knew full well that this opposition was not so much against us as against the cause of the Bible, with which we were identified, and we were content to 'stand still and see the salvation of the Lord.' And most delightful it was to witness the results of that overruling Providence which can make the wrath of man tributary to the divine praise.

"The malignity of this worse than infidel opposition to the truth excited the curiosity of numbers to examine whether indeed the word of God was not 'profitable for instruction and for doctrine.' The results of such an examination upon every candid mind may be easily conjectured. Thus the truths of inspiration found free course to hundreds of families and scores of schools, where they might be safely left to do their own office upon the minds and hearts of the people.

"Some instances of the happy and immediate effects of circulating the Bible came to our knowledge; but it is reserved for eternity to reveal the full extent of the benefit. While subsequently travelling in distant provinces, I found that the sacred volumes put in circulation at Rio de Janeiro had sometimes gone before me, and wherever they went an interest had been awakened which led the people to seek for more."

There are other means than newspapers for the progress of the Brazilians in knowledge and *belles-lettres*.

In addition to the various colleges and academies described in another chapter, there are a number of public institutions and associations whose object is the cultivation of literature and science, and the diffusion of knowledge.

The *Bibliotheca Nacional* contains 100,000 volumes. These consist chiefly of the books originally belonging to the Royal Library of Portugal, which were brought over by Dom John VI. The collection is annually augmented by donations and Government aid. It was thrown open to the public by the Portuguese monarch, and has ever since remained under suitable regulations, free of access to all who choose to enter its saloon and read. This library is open daily from nine A.M. till two P.M., and was formerly entered from the Rua detraz do Carmo; but the Government has recently purchased the commodious private residence of Sr. Vianna, which is beautifully situated in the vicinity of the Passeio Publico, where the accommodations will doubtless be superior to those which it has hitherto possessed. When it was located in the old library-buildings, it presented an interesting sight to the visitor. Tables covered with cloth, on which were arranged writing-materials, and frames designed to support large volumes, extended through the room from end to end. The shelves, rising from the floor to the lofty ceiling, were covered with books of every language and date. You might here call for any volume the library contained, and sit down to read and take notes at your pleasure. The newspapers of the city and various European magazines were always ready for the reader. Not only this apartment, but also various alcoves and rooms adjoining it on either hand, were filled all around with books. This collection has also been increased by valuable private donations, among which that of the books of the late José Bonifacio de Andrada deserves especial mention.

The publicity of such a library cannot fail to have a beneficial influence upon the literary taste and acquirements of the students of the metropolis,—which, by degrees, will extend itself to the whole community. While the student at Rio may find in the National Library nearly all that he can desire in the field of ancient literature, he may also easily gain access to more modern works in the subscription-libraries.

The English, the German, and the Portuguese residents have severally established such libraries for their respective use. That of the English is somewhat extensive and valuable.

Among the Government institutions must be classed the National

Museum, on the Campo de Santa Anna, which is gratuitously thrown open to visitors; and great numbers avail themselves of this pleasant and instructive resort. The collection of minerals has been much augmented in value by a donation from the heirs of José Bonifacio de Andrada. They presented to the Museum the entire cabinet of their father, who in his long public career had rare opportunities for making a most valuable collection. At an early period of his life he was Professor of Mineralogy in the University of Coimbra, Portugal, where he published several works that gained him a reputation among the scientific men of Europe. Through his life he had been industrious in gathering together models of machines and mechanical improvements, together with choice engravings and coins; and his heirs certainly could not have made a more magnanimous disposal of the whole than to confer them upon the nation. The department of mineralogy is well arranged, but contains many more foreign than native specimens. The same lack of Brazilian curiosities formerly prevailed in other departments, although in that of aboriginal relics there has been from the establishment of the Museum a rich collection of ornaments

THE HARPY EAGLE.

and feather-dresses from Pará and Matto Grosso. There is a constant enlargement and improvement in every respect. Still, it may be said that while the cabinets of Munich and Vienna, Paris, St. Petersburg, London, and Edinburgh have been enriched by splendid collections from Brazil, in various departments of natural history, yet in the Imperial Museum of Rio de Janeiro but a meagre idea can be formed of the interesting productions—mineral, vegetable, and animal—in which the Empire abounds.

It was here that I saw a very fine living specimen of the great harpy eagle, from the forests of the Amazon.

There is an Imperial Academy of the Fine Arts, which was founded in 1824, by a decree of the National Assembly. It is at present organized with a Director and four Professors,—viz.: of painting and landscape, of architecture, of sculpture and of design, and a corresponding number of substitutes. This institution is open to all who wish to be instructed in either department, and about seventy students are annually matriculated,—the greater proportion in the department of design. This Academy also provides funds for the support of a certain number of its most meritorious *alumni* at Rome, where they have ample opportunity for studying the classic productions of ancient and modern art.

The *Conservatorio de Musica* is a State Academy where instruction in instrumental and vocal music is given to both sexes by competent professors. There is also a *Conservatorio Dramatico*, to whose censorship were submitted, in 1854, two hundred and fifty plays, of which one hundred and seventy were approved, fifty-four were amended or suppressed, and thirty-three were of such a character as not only to be suppressed but to merit unqualified rebuke.

The Imperial Agricultural, the Statistical, and the *Auxiliadora* Societies enroll many public-spirited men and good writers. But the association which in its character, dignity, and numbers is the first in all South America is the Brazilian Historical and Geographical Institute, organized at Rio de Janeiro in 1838, which has done more than any other society to awaken the spirit of Brazilian literary enterprise. This association adopted as its fundamental plan the design of collecting, arranging, and publishing or preserving documents illustrative of the history and geography of Brazil. Several distinguished persons took a deep interest in it from the first. The Government also lent a fostering hand. The General Assembly voted an annual subsidy in aid of its objects, and the Department of Foreign Affairs instructed the attachés of the Brazilian embassies in Europe to procure and to copy papers of interest that exist in the archives of different courts, relative to the early history of Brazil. By this movement individual exertions were aroused, and the spirit of inquiry was excited in different parts of the Empire as well as abroad, and interesting results have already been accomplished.

During the first year of its existence, this Institute numbered near four hundred members and correspondents, and had collected over three hundred manuscripts, of various length and value. The most important of these it has already given to the world, together with some valuable discourses and essays furnished by its members. Two Fridays of each month are devoted to the sittings of this association; and none of its members and patrons are so punctual or take so deep an interest in all its proceedings as Dom Pedro II. Its organ is a Quarterly Review and Journal, which publishes the proceedings of the society at length, together with all the more important documents read before it. We have been particularly interested in the articles it has contained upon the aboriginal tribes of South America, and also in its biographical sketches of distinguished Brazilians. The President is the Viscount de Sapucahy.

On the whole, it may be questioned whether the Portuguese language contains a more valuable collection of miscellany than is thrown together in the pages of the *Revista Trimensal ou Jornal do Instituto Historico Brazileiro*.

Almost all the leading men of Brazil belong to the learned professions. Such a thing as an eminent mechanic or merchant holding high position in the State I believe to be unknown. There are certain officers who hold their appointment and receive pay under Government, in accordance with a rule which deserves particular mention. The professors of some of the public institutions, and perhaps the attachés of some of the Government bureaux, receive a certain annual salary. It may not be large; but, after holding office for a stipulated number of years, the employee, if his conduct has been without reproach, can retire, and is paid from the Imperial Treasury a sum equal to the added salaries of his whole term of service. This is a strong inducement to the faithful discharge of duty, and perhaps operates to keep unscrupulous demagogues from seeking office as a reward for party exertions. It is thus that the under-officers in the Brazilian Government acquire a full knowledge of the difficult routine of the various Departments; and the changes of ministry leave no difficulties for the new Cabinet to surmount in carrying on the machinery of government. The Brazilian mode certainly seems more in accordance with common sense than the rotation-in-office principle which prevails in the United States.

In another chapter will be found the course of study pursued in the chief law-school of the Empire. The administration of justice is much simpler than in England or the United States. There are almost the same magistrates and judges, under different names. The *delegado* or *subdelegado* is the justice of the peace; the *juiz municipal* answers to the circuit judge or the presiding officer of the Court of Common Pleas; the *Juiz dos Orphãos* is the Judge of Probate; the *Juiz de Direito* is the Judge of the Supreme Court. There are district supreme judges in all the provinces, and there is a *Supremo Tribunal de Justicia*, which corresponds to the Supreme Court of the United States.

From the experience of Governor Kent with the Brazilian tribunals, and from the interesting letters of Rev. Charles N. Stewart, I cull the following facts in regard to the mode of conducting a criminal trial at Rio de Janeiro. The party accused is first brought before the *subdelegado* in whose district the crime has been committed. He is verbally examined, and his replies, as well as the questions, are all recorded. The accused is asked his age, profession, &c. as minutely as the magistrate thinks proper. He is not compelled to answer, but his silence may lead to unfavorable inferences. The examination of the prisoner is followed by that of the witnesses, who are sworn by placing the hand upon the Bible. The administration of the oath is of the most solemn and impressive character, and in this respect at least the Brazilians read us a wholesome and a needful lesson. All rise—court, officers, bar, and spectators—and stand in profound silence during the ceremony. When the jury retires there is also a great manifestation of respect, —all standing until the twelve have left the court-room.

The *subdelegado*, after the preliminary examination, decides whether the accused shall be held for trial, and submits the papers with his decision to a superior officer, who usually confirms it, and the accused is imprisoned or released on bail.

In civil cases, unless of very great importance, the jury does not form a part of the judicial administration. The jury consists of twelve men. "Forty-eight are summoned for the term; and the panel for each trial is selected by lot, the names being drawn by a boy, who hands the paper to the presiding judge. In capital cases challenges are allowed without the demand of cause. The jury being sworn and empannelled, the prisoner is again examined by the judge—sometimes at great length and with great minuteness —not only as to his acts, but as to his motives. The record of the former proceedings, including all the testimony, is then read. If either party desire, the witnesses may be again examined, if present; but they are not bound over, as with us, to appear at the trial. Hence, the examination of the accused and the witnesses at the preliminary process is very important and material. In many instances, the case is tried and determined entirely upon the record as it comes up."—*Brazil and La Plata.*

When the record is read, witnesses are produced on the side of the Government, and the prosecuting-attorney addresses the jury. The testimony, or the witnesses of the defendant, are then introduced, and his advocate addresses—sometimes at great length— the twelve on whose decision hangs the destiny of his client. The prosecutor replies if he deem it best; after which the judge briefly charges the jury and gives them a series of questions in writing, the answers to which constitute the verdict; and thus, it will be seen, special pleading and legal skirmishing is in a great measure defeated. The decision in each case is by majority, and not by unanimity, as with us. A case begun is generally finished without an adjournment of the court, though it should continue through the day and the entire night.

The arrangement of the court-room is somewhat different from that in the United States. The judge, with his clerk, sits on one side of the hall, and the prosecuting-attorney on the other. The jury, instead of being in a "box," are seated at two semicircular tables placed at the right and at the left of the judge. The lawyers do not stand when they address the jury, but, like the *professores* on examination-day, the *collegios* always make their speeches *ex cathedra*. The lawyers not engaged in the suit which may be before the court occupy a kind of pew which resembles the box for criminals in English and American halls of justice.

The following verdict of a jury was returned in a case of homicide which occurred in Rio in 1851. The trial came off in the spring of 1852, and the "return" is translated from one of the daily newspapers printed at the capital, and gives a clear and concise notion of the nature of the questions propounded by the judge, and the ease with which a jury can come to a speedy conclusion in regard to the guilt or innocence of any accused individual :—

Questions propounded by the Judge to the Jury, and the Verdict rendered, in the Second Trial of B.

In this case the first jury fully acquitted the respondent. The presiding judge appealed to the Court of Relação, consisting of all the judges, twelve in number. This court, on hearing, sustained the appeal and ordered a new trial.

QUESTIONS.

1. Did the defendant, B., on the 23d of September of the last year, kill, by discharging a pistol, the Italian, C., in D.'s hotel?

Answer. Yes; (by twelve votes.)

2. Did he commit the offence in the night-time?

Ans. Yes; (by eight votes.)

3. Did the defendant commit the offence with superiority of arms, in a manner that C. could not defend himself with a probability of repelling the attack?

Ans. Yes; (by eleven votes.)

4. Did the defendant commit the offence proceeding with concealment or surprise?

Ans. No; (by seven votes.)

5. Are there any circumstances extenuating the offence in favor of the defendant?

Ans. Yes; (by eight votes.) By Act 18, § 3, of the Criminal Code :—"If the defendant commits the crime in defence of his proper person;" and ditto, § 4 of same article :—"If the defendant commits the offence or crime in retaliation or revenge of an injury or dishonor which he has suffered."

6. Do the jury find that the respondent commits the act (or offence) in defence of his person?

Ans. Yes; (by seven votes.)

7. Was the defendant certain of the injury (or evil) which he intended to avoid (or escape from)?

Ans. Yes; (by seven votes.)

8. Was the defendant absolutely without other means less prejudicial?

Ans. No; (by eight votes.)

9. Had the defendant provoked the occasion for the conflict?

Ans. No; (by eight votes.)

10. Had the defendant done any wrong which occasioned the conflict?

Ans. No; (by eight votes.)

11 and 12, (like 9 and 10,) in reference to the family of the defendant, if they had provoked, &c.; and answered, No, (by twelve votes each.)

Upon this verdict the court adjudged B. guilty, and sentenced him to twelve years' imprisonment at hard labor and the costs.

An appeal was again taken to the same Court of the Relacão. He was pardoned by the Emperor, October, 1852, upon application of the Minister-Plenipotentiary of his (B.'s) country and by the petition of others.

The following is a curious case of some legal interest:—In February, 1853, a black man was put on trial before the jury on charge of having a pocket-knife, (jack-knife, as we call it.) It did not appear that the black had done or threatened any injury; but the crime was, having a prohibited article. During the trial, a white man appeared and claimed the negro as his slave. This claim was made part of the case on trial, and the jury were directed to determine whether he was free or the slave of the claimant. They found, by the judge giving the casting vote, that he was free, and, by ten votes, that he was guilty of the crime. He was sentenced to one month's imprisonment *as a freeman.* Thus, he obtained a judicial sentence which secured his freedom and had to stay one month as a lodger in jail. A lucky jack-knife to him!

It is impossible, in a work like this, to enter fully into the merits and demerits of the mode of administering law in Brazil. From time to time many charges of corruption have been brought, by rumor, against those who administer it, and doubtless, in some cases, corruption has existed. Those who have had property awaiting certain decisions of the Juizes dos Orphãos have complained that it was much reduced before judgment was rendered. Foreigners have also murmured at what they termed unfairness, and have hinted that some of the magistrates have not been above bribery.

It would not be altogether just to compare the administration of law in Brazil to that of England; but I hazard nothing in saying that in no country of South America is there greater personal security and a fairer dispensation of justice than in this Empire. Each year the various codes are becoming better digested; and the number of eminent men in the legal profession has placed it, in point of mental ability, in the first rank of the learned vocations.

CHAPTER XV.

THE CLIMATE OF BRAZIL—ITS SUPERIORITY TO OTHER TROPICAL COUNTRIES—COOL RESORTS—TRIP TO ST. ALEXIO—BRAZILIAN JUPITER PLUVIUS—THE MULATTO IMPROVISOR—SYDNEY SMITH'S "IMMORTAL" SURPASSED—A LADY'S IMPRESSIONS OF TRAVEL—AN AMERICAN FACTORY—A YANKEE HOUSE—THE RIDE UP THE ORGAN MOUNTAINS—FORESTS, FLOWERS, AND SCENERY—SPECULATION IN TOWN-LOTS—BOA VISTA—HEIGHT OF THE SERRA DOS ORGŌES—CONSTANCIA—THE "HAPPY VALLEY"—THE TWO SWISS BACHELORS—YOUTH RENEWED—PROSAIC CONCLUSION—TODD'S "STUDENT'S MANUAL"—THE TAPIR—THE TOUCAN—THE FIRE-FLIES—EXPENSES OF TRAVELLING—NOVA FRIBOURGO—CANTA GALLO—PETROPOLIS.

THOSE whose tropical experience has been in the East Indies or the western coast of Africa can have no just conception of the delightful climate of the greater portion of Brazil. It would seem as if Providence had designed this land as the residence of a great nation. Nature has heaped up her bounties of every description : cool breezes, lofty mountains, vast rivers, and plentiful pluvial irrigation, are treasures far surpassing the sparkling gems and the rich minerals which abound within the borders of this extended territory. No burning sirocco wafts over this fair land to wither and desolate it, and no vast desert, as in Africa, separates its fertile provinces. That awful scourge, the earthquake,—which causes strong men to become weak as infants, and which is constantly devastating the cities of Spanish America,—disturbs no dweller in this Empire. While in a large part of Mexico, and also on the west coast of South America,—from Copiapo to the fifth degree of south latitude,—rain has never been known to fall, Brazil is refreshed by copious showers, and is endowed with broad, flowing rivers, cataracts, and sparkling streams. The Amazon,—or, as the aborigines term it, *Pará*, "the father of waters,"—with his mighty branches, irrigates a surface equal to two-thirds of Europe; and the San Francisco, the Parahiba do Sul, the vast affluents of the

La Plata, under the names of the Paraguay, Paraná, Cuiba, Paranahiba, and a hundred other streams of lesser note, moisten the fertile soil and bear their tributes to the ocean through the southern and eastern portions of the Empire. Let any one glance at the map of Brazil, and he will instantly be convinced that this land is designed by nature for the sustenance of millions.

Now, there must be some reason for this bountiful irrigation, this fertility of soil and salubrity of climate.

Lieutenant Maury—who seems almost literally to have taken "the wings of the morning" and to have flown to the uttermost parts of the sea—has shown conclusively why it is that Brazil is so blessed above corresponding latitudes in other lands. South America is like a great irregular triangle, whose longest side is upon the Pacific. Of the two sides which lie upon the Atlantic, the longest—extending from Cape Horn to Cape St. Roque—is three thousand five hundred miles, and looks out upon the southeast ; while the shortest—looking northeastward—has a length of two thousand five hundred miles. This configuration has a powerful effect upon the temperature and the irrigation of Brazil. The La Plata and the Amazon result from it, and from those wonderful winds, called the trades, which blow upon the two Atlantic sides of the great triangle. These winds, which sweep from the northeast and from the southeast, come laden, in their journey over the ocean, with humidity and with clouds. They bear their vapory burdens over the land, distilling, as they fly, refreshing moisture upon the vast forests and the lesser mountains, until, finally caught up by the lofty Andes, in that rarefied and cool atmosphere they are wholly condensed, and descend in the copious rains which perpetually nourish the sources of two of the mightiest rivers of the world. The prevailing winds on the Pacific coast are north and south. No moisture is borne from the ocean to the huge barrier of mountains within sight of the dashing waves, and hence the aridity of so much of the hypothenuse of the triangle. I have beheld the western and eastern coasts of South America within thirty days of each other, and the former seemed a desert compared with the latter.

No other tropic country is so generally elevated as Brazil. Though there are no very lofty mountains except upon its extreme western border, yet the whole Empire has an average elevation of more than seven hundred feet above the level of the sea.

This great elevation and those strong trade-winds combine to produce a climate much cooler and more healthful than the corresponding latitudes of Africa and Southern Asia. The traveller, the naturalist, the merchant, and the missionary do not have their first months of pleasure or usefulness thrown away, or their constitutions impaired by acclimating fevers.

The mean temperature of Brazil—which extends from nearly the fifth degree of north latitude to the thirty-third of south latitude (almost an intertropical region)—is from 81° to 88° Fahrenheit, according to different seasons of the year. At Rio de Janeiro,—on the authority of Dr. Dundas,—the mean temperature of thirty years was 73°. In December, (which corresponds to June in the Northern Hemisphere,) maximum, 89½°; minimum, 70°; mean, 79°. In July, (coldest month,) maximum, 79°; minimum, 66°; mean, 73½°. I can add, from my own observations for several years, that I never saw 90° attained in the summer-time, and the lowest in the winter (June, July, and August) was 60°, and this was early in the morning.

The heat of summer is never so oppressive as that which I have often experienced, in the hot days of July and August, at New York and Boston, where frequently the high point of 104° or 105° Fahrenheit has been reached. It must, however, be conceded that three months of weather ranging between 73° and 89° would be intolerable if it were not for the cool sea-breeze on the coast which generally sets in at eleven A.M., and the delicious land-breeze which so gently fans the earth until the morning sun has flashed over the mountains. In the interior the nights are always cool; and it may be added that, one hundred miles from the sea-coast, the climate is entirely different.

Rio is happily situated in its accessibility to the elevated regions. An hour's ride leaves you among the cascades and coolness of Tijuca; six hours by steamer, railway, and coach lift you up to the mountain-city of Petropolis; or twelve hours will bring you amid the sublimities of the Serra dos Orgãos and the silent and refreshing shades of Constancia, where, at Heath's, we may be far away from the dust, din, and diplomacy which are the constant concomitants of the commercial and political capital of Brazil. Again, we may vary our route and ascend the mountains to the elevated uplands upon which are situated the prosperous towns of Nova Fribourgo and Canta Gallo, with their adjacent flourishing coffee-plantations. All of these are delightful resorts, and are becoming each summer more and more frequented.

Not far from the usual route to Constancia is the sweet little valley of St. Alexio, where an American has erected a cotton-factory in the midst of the most beautiful tropic scenery. To some it might be a profanation that these wilds should be startled by any other sounds than the leaping streams from the Serra, or the songs of birds and the shrill music of the cicada; but perhaps there are few who would not be content to behold industry taking the place of indolence, though they might yield to none in love for the beautiful.

I visited St. Alexio a number of times, and enjoyed the kind hospitality of its director, who through many obstacles had at last triumphed in establishing the first successful cotton-manufactory in the province of Rio de Janeiro.

My last visit to St. Alexio was made under such circumstances of weather that I am constrained to give it as an instance of what must be expected at certain seasons of the year. Though in the province of Rio de Janeiro there is no "rainy season," properly so called, yet many visitors to the capital will not soon forget the drenching rains, made doubly perceptible by the uncouth water-spouts (see those in the engraving of the "Senate-House") which formerly poured more than a miniature cascade upon the passers-by. But of these spouts it may now be said their "occupation's gone," and by a city ordinance they are really where *Intrudo* is,—among the curiosities of Rio that have only a historical existence.

The usual mode of getting to St. Alexio is by steamer to Piedade, and thence by carriage to the secluded valley some eight or ten miles from the landing-place. On the occasion of the visit referred to, I was accompanied by a number of friends, among whom was Mr. M., the worthy director and one of the owners of the "Fabrica."

We left the Quai dos Mineiros (not far from the Convent of San

Bento) in the little clumsy steamer that plies between Rio and the upper end of the bay. The morning was bright, but we were soon overtaken by a thunder-storm. Such rain! In temperate zones we fancy that we know what is meant by rain. Quite a mistake! It is child's play when compared to the pouring torrents of the tropics. There was no cabin, and the curtains but half performed their office. In rushed the water over our clothes, under our feet, and out at the scuppers, like holy-stone day on board ship.

When we were sufficiently wet, the rain abated and the curtain rose. And well that it did so; for the bad weather had driven in all the motley crowd of *troupeiros* usually occupying, along with their more respectable animals, the forward-deck of the boat; and the hot steam arising from the greasy cattle-drivers, the unkempt muleteers, and the damp darkies, was not the most agreeable to the lady portion of our company.

The time was beguiled in looking at the glorious scenery and in listening to the improvisation of a mulatto who was going to a *festa* in Majé, there to sell his wit and his *doces*. He told long stories in verse, and imitated different voices with admirable skill. When asked to improvise on Paquetá, the lovely insular gem that we were passing, he immediately dashed off in a strain of poetry, describing the beauties of the island, and then descanted on the faults and failings of its inhabitants, and in a satiric strain worthy of Juvenal lashed the proceedings of the people who frequented the religious festas that are annually held on its bright shores. He concluded with a eulogy on José Bonifacio de Andrada, who here ended his days. In short, had Corinne heard him, jealousy would have saved her the trouble of dying for love. Jesting apart, the man's talent was of a high order, and the harmonious and flowing verse showed the adaptation of the Portuguese language to rhythmical composition

After a hasty repast at a rude inn near the landing-place of Piedade, we prepared for the road. Up came our equipage. I must, in justice to our worthy host, say that his nice American vehicle had received some injury, so that he could only send his mules and engage the best conveyance afforded by the village of Majé. We felt some slight remorse at the destruction of life that

our entrance into the venerable vehicle must have caused, as it
seemed to have served as a temporary refuge to some gay, locked-out
rooster. But we ought not to speak ill of the aged. Guiltless
alike of paint and washing, it far outdid Sydney Smith's "Immortal,"
which, doubtless, was kept in perfect cleanliness by his tidy
Yorkshire servants. However, the sight of a good team reconciled
us to the rudeness of the vehicle. Four fine mules plunged along
through mud and water: I then understood how philosophical it
was to avoid the trouble of washing a carriage. The Hyde Park
turn-out of Count D'Orsay or the Earl of Harrington, in one short
mile, would have been on a par with ours. We forded juvenile
rivers and newly-made brooks; we lumbered up hill and down
dale; now the coachman made a skilful *détour* close to a bank to
avoid a deep mud-hole on the other side, and now he was obliged to
pass under some tree whose overhanging branches gave us a capital
douche. After some miles of this travel we stopped at a venda to
give the animals breath and water before the gallop down the slope.
Soon we were off again.

> "On, on we hasten'd, and we drew
> Their gaze of wonder as we flew!"

And there was as black a tempest gathering for us poor Giaours as
ever threatened to wet that uncomfortable, sword-waving rider of
the "blackest steed!" Down came night and Brazilian rain!
What had formerly been the hood of the carriage was transformed
into a sort of a kitchen-sink, with a hole in the middle, through
which poured the water. Luckily, we had an umbrella: this was
inserted in the hole, and thus the stream was averted from our devoted
heads.

In the midst of all this our driver gave a loud whistle, and
thereupon out rushed four dark figures from a hut by the roadside.
A lady of the party afterward described her romantic impressions
of this scene as follows:—

"What my companions felt I know not; but it was quite allowable
for me, a poor, weak woman, to give myself over as robbed,
or, at least, 'murthered!' One man jumped on the box with a
huge stick in his hand, and the others followed behind, uttering a
series of unearthly yells and undesirable epithets, but all addressed
to the mules; and, as I knew that the skins and skulls of those

beasts were thicker than mine, I was consoled. It was a party
sent to push us up a steep hill; for be it known to all who are
ignorant of the idiosyncrasy of these animals, that, when once they
consider the task assigned to them unreasonable, no persuasion
can induce them to set shoulder to the work. No doubt they cry
to Jupiter, but he will not help them; and so they stand still, or
allow the vehicle to draw them backward; and on the edge of a

THE FABRICA AT ST. ALEXIO.

precipice this is not a pleasant way of travelling. So, after each
mule had clearly learned from the yelling quartette the estimation
in which he was held, we gained the summit. How gladly we
rolled down into that beautiful valley where the factory raises its
white walls! We afterward beheld it under a bright sun, and
Southey's remark that 'even nature herself abhors a factory, and
refuses to clothe its walls with climbers,' is here contradicted, for

the lovely glen in whose bosom this building reposes would lend grace to any structure.

"How hearty was our welcome from the pretty Virginia hostess who met us as we entered, all forlorn! Right gayly we recounted our fright and adventures, and it was the old story over again :—·

> " ' She loved us for the dangers we had pass'd,
> And we loved her that she did pity them.'

"Byron could not bear to see a lady eat,—it is so unethereal. Strictly speaking, it *is* a singular process,—throwing sundry morsels into a hole in your face and using your chin as a mill. Of course, it was only the masculine part of the company who partook of the Westphalia ham, broiled chicken, and other dainties prepared by the good hostess. Such proceedings did not agree with the poetical feelings of my more celestial nature!"

The following morning we surveyed the locality. The proprietor's house stands at a short distance from the factory, and both were actually framed in the United States, brought out in *pieces*, and put together in Brazil. The pine used for the house has, in spite of predictions to the contrary, proved superior in durability to Norwegian pine. A meadow of bright green slopes away from the house toward a clear, rapid brook, which, after rains, may well be called a river; but in dry weather it is easily traversed on the stones that strew its bed. Mr. M. had long and painful researches to find a stream that never dries up even in the hottest season. At last he discovered this little river, and here took up his abode. The hills rise around, covered with the most luxuriant growth; here and there a stately palm rises like a chieftain above its fellows; farther on, the mountains stretch away and blend with the clear blue of the heavens. On the branches sing bright-plumaged birds, that seem, in the early morning, to call on the sensitive-plant trees to awake from their night's slumber. It was, indeed, hard for me to realize that the little sensitive-plant which I had looked upon at home as among the most delicate of exotics is here reproduced in almost giant forms. Its family abounds in Brazil, and the grove that surrounds the residence of Mr. M. is actually composed of trees which quietly fold their leaves in repose at vespers, only to be awakened by the morning sun and the sing-

ing-birds. The city-friends of Mrs. M. used to offer their condolence that she was so far removed from society in that retired vale;
but they were always cut short in their proffered sympathy by the
information that no sense of loneliness prevailed in that sweet
spot. There one may find companionship in those majestic mountains "precipitously steep," the flowering woods, the forest-voices,
and the gushing music of brooks and fountains.

A YANKEE HOUSE IN BRAZIL.

The remembrance of St. Alexio is like that of a pleasant dream,
or the sunny memories of the secluded vales and sparkling waters
at the base of the *Dent du Midi*,—not a day's ride from the upper
end of the Lake of Geneva.

Mr. M. deserves the greatest credit for his persevering efforts
which placed here this first successful cotton-manufactory in the
province. Others had endeavored to establish similar *fabricas*, to
be driven by steam-power, in the city; but they were failures. Not
only had Mr. M. to contend with nature, but probably his worst
annoyances came from a dilatory Government. As to operatives,
the factory is supplied from the German colony of Petropolis.
Another has paid a just tribute of merit to Mr. M.; and I can
heartily subscribe to the sentiments therein contained:—"Though
it is only in the more common fabrics in cotton that the manufacturer can yet compete with British and American goods, yet he

[Mr M.] deserves a medal of honor from the Government, and the patronage of the whole Empire, not only for the establishment of the manufactory, but for the living example—set before a whole province of indolent and sluggish natives—of Yankee energy, ingenuity, indefatigable industry, and unyielding perseverance." (Mr. M. died in 1857.)

It is a comfortable day's ride from St. Alexio to Constancia,—though the usual manner of procedure is to start at mid-day from Rio in the steamer, arrive at Piedade at three o'clock, where mules and guides are awaiting those who have been prudent enough to announce by letter to the "jolly Heath" their intention of spending a few days amid the Serra dos Orgãos. A few hours across the lowlands bring us through the town of Majé to Frechal, (or Freixal,) where the weary and the lazy often spend a night in a dirty inn, surrounded by crowds of children, (the proprietor is the father of twenty-three *meninos*,) and by vast troops of mules, which, laden with coffee, are on their way to the steamer at Piedade. But for those who love a dashing ride up the mountains, on a road in some places paved as the old Roman causeways,—those who wish to feel an evening atmosphere which in coolness and chilliness reminds one of the temperate zone,—the Barreira will be the resting-place. Here is the toll-gate of this fine mountain mule-path, which must have been built at an immense cost, as several miles are paved like the streets of a city.

We zigzag up the steep sides of the Serra, looking down upon the tops of majestic forest-trees whose very names are unfamiliar, and whose appearance is as curious as picturesque and beautiful. Dr. Gardner, who made a most thorough investigation of the *flora* of the Organ Mountains, has recorded in his interesting travels the vegetal riches of this lofty range, and those who would revel in descriptions of palms, *Cassiæ*, *Lauri*, *Bignonias*, *Myrtaceæ*, *Orchideæ*, *Bromeliaceæ*, *ferns*, &c. &c. must turn to the pages of a work which, though necessarily deficient in the history, politics, and present condition of Brazil, is the most unassuming and charming book ever written on the natural aspect of the tropic land under consideration.

In the months of April and May, (October and November in Brazil,) only the autumnal tints of our gorgeous North American

woods can compare with the sight of the forest of the Serra dos
Orgãos. Then the various species of the *Laurus* are blooming, and
the atmosphere is loaded with the rich perfume of their tiny snow-
white blossoms. The *Cassiæ* then put forth their millions of golden
flowers, while, at the same time, huge trees—whose native names
would be more unintelligible, though less pedantic, than their
botanic terms of *Lasiandra, Fontanesia,* and others of the *Melas-
toma* tribe—are in full bloom, and, joining rich purple to the
brightest yellow, present, together with gorgeously-clothed shrubs,
"flowers of more mingled hue than her [Iris's] purpled scarf
can show." From time to time a silk-cotton-tree (the *Chorisia
speciosa*) shoots up its lofty hemispherical top, covered with
thousands of beautiful large rose-colored blossoms, which gratefully contrast with the masses of vivid green, purple, and yellow
that clothe the surrounding trees. Floral treasures are heaped
on every side. Wild vines, twisted into most fantastic forms or
hanging in graceful festoons,—passion-flowers, trumpet-flowers,
and fuchsias in their native glory,—tree-ferns, whose elegance of
form is only surpassed by the tall, gently-curved *palmito,* which
is the very embodiment of the line of beauty,—orchids, whose
flowers are of as soft a tint as the blossom of the peach-tree, or as
brilliant as red spikes of fire,—curious and eccentric epiphytes
draping naked rocks or the decaying branches of old forest-mon-
archs,—all form a scene enrapturing to the naturalist, and bewilder-
ing with its richness to the uninitiated, who still appreciate the
beauty and the splendor that is scattered on every side by the
Hand Divine. The overpowering sensation which one experiences
when entering an extensive conservatory filled with the choicest
plants, exotics of the rarest description, and odor-laden flowers,
is that (multiplied a thousandfold) which filled my mind as I gazed
for the first time upon the landscape, with its tiers of mountains
robed in such drapery as that described above; and yet there was
such a feeling of liberty, incompatible with the sensation expressed
by the word "overpowering," that it is impossible to define it. In
the province of *Minas-Geraes,* from a commanding point, I once
beheld the magnificent forest in bloom; and, as the hills and undulating plains stretched far away to the horizon, they seemed to be
enveloped in a fairy-mist of purple and of gold.

The Barreira is situated in a spot of great wildness and sublimity; for the Organ-peaks, that rise thousands of feet above, seem like the *aiguilles* which start fantastically from the glaciers of Mont Blanc; and the rushing, leaping, thundering cascades are comparable to the five wild mountain-torrents, "fiercely glad," that pour into the Vale of Chamouny. I was once at the Barreira during a tropic storm, and the foaming, roaring rivers, which hurried down with fearful leap from the very region of dread lightning and clouds, madly dashed against the huge masses of granite, as if they would have hurled them from their mighty fastenings, and tore their way into the deep valley beneath with sounds that reverberated among the giant peaks above, giving me a new commentary on the sublime description in the Apocalypse :—"And I heard a voice from heaven, as the voice of many waters and as the voice of a great thunder."

From the Barreira we ascend by zigzags to the uplands, where is situated the former fazenda of Mr. March. His residence— so often visited by Langsdorf, the celebrated Russian voyager, Burchell, the African traveller, and Gardner, the botanist—is now to be numbered among the things that were; for the spirit of enterprise and money-making has laid out in this elevated valley a new resort for the Fluminenses, and speculation in town-lots among the Organ Mountains was rife as I left the shores of Brazil. I hope that it may prove a successful enterprise; for here the wearied and jaded from the city will find coolness, salubrity, and quiet in the midst of the most imposing scenery.

Before reaching March's and the former mountain-home of Mr. H——n, (whose hospitality many a visitor to Brazil will have occasion to remember,) we climb along the very sides of one of the most precipitous of the Organ-pipes. Hence is a view of commanding extent,—of mountain, plain, bay, and ocean,—embracing, it is said, a panorama of more than two hundred miles in circumference, in the midst of which, though distant, the capital of the Empire is seen gleaming amid its verdant and lofty environs. The point for beholding this landscape is appropriately called *Boa Vista*, ("beautiful view.") So enraptured was the Rev. Charles N. Stewart with the grandeur of the scene, that he doubts if—in its combination of mountain, valley, and water—it has a rival; and adds that, in his wide experience in various continents, he only remembers one other prospect that approximates to it,—viz.: the pass "through the mountains of Granada, followed by the first view of the 'Vega,' with the city, the walls, and the towers of the Alhambra, and the snow-covered heights of the Nevada above all, gloriously lighted by the glowing hues of the setting sun."

At the elevation of Boa Vista the climate is very much cooler than at Rio. In the month of June the thermometer has been known to fall as low as 32° Fahrenheit just before daybreak; but this is rare: 40° in the morning and 70° in the warmest portion of the day is the winter *régime;* and, in the summer, 60° and 80° are the two extremes. In January and February, (the July and August of the Southern tropics,) violent thunder-storms often occur,—generally in the afternoon,—and then pass over, leaving the evening delightfully cool.

Here and at Constancia nearly all the European fruits and vegetables thrive; and, as at Madeira and Teneriffe, the apple and the orange, the pear and the banana, the vine and the coffee-plant, may be seen growing side by side. Mr. Heath receives quite an income from the productions of his vegetable-gardens; and, at Rio, the fine cauliflower, (so difficult of cultivation in the tropics,) the best asparagus, and most of the artichokes, peas, carrots, &c. come from Constancia, and are esteemed as rare in that land as the carefully-cultivated hothouse pineapple in England. Two English shillings per head are given for the largest Constancia cauliflower at Rio. This kind of garden, it has seemed to me, might be increased in number in the upper region of the Serra, where are many fertile little valleys, all well irrigated by small streams of cool and limpid water. If they could be managed with the care, industry, and perseverance which Mr. Heath has brought to bear upon such cultivation, they could not but bring a lucrative return to their proprietors, and would confer a great benefit upon the growing city of Rio de Janeiro.

Like the mountains of Tijuca and the curious elevations around Rio, the whole of the Organ range consists of granite. The alluvial soil is very deep and rich in the valleys, and underneath it exists the same red-colored, slaty, ferrugineous clay which is so common throughout Brazil.

The scenery becomes more tame as we leave Boa Vista, and we seem to be far removed from the climate of the plains, though around us the palms, ferns, cacti, tillandsias, &c. tell us that we are not beyond the limits of Capricorn. Creeping and drooping plants, bright flowers and foliage, still abound. Occasionally, howling monkeys hold a noisy caucus over your head, or a flock of bright

THE ORGAN MOUNTAINS.

parrots glides swiftly over the tall and gracefully-bending bamboos, which are a distinctive feature in the landscape. This giant of the grass-tribe has frequently been found in these mountains from eighty to one hundred feet in height and eighteen inches in diameter. They do not, however, grow perpendicularly, nor often singly, but, in vast groups, shoot up fifty and sixty feet, and then curve gently downward, forming most cool and beautiful domes.

As we look back, we have a view of the Organ-pipes, and the aspect which they present is entirely different from that ragged, pointed, and diminutive appearance which they show when seen from the bay. From our nearness and our altitude they seem like sharp naked mountains rising above a sea of foliage. The range from which they are detached is still more lofty, and is most massive in its character. Few persons have ascended these mountains, and those have either been naturalists or daring hunters. Dr. Gardner made probably the most thorough scientific exploration, and up these heights Heath has often pursued the clumsy tapir or the lithe jaguar. The sloth, howling monkeys, the Brazilian otter, a little deer, (*Cervus nemorivagus,*) and two kinds of peccari, may still prove attractions to the naturalist and the sportsman; but every year they are becoming more rare. Of birds there are many varieties, remarkable for their brilliant plumage, and a few are much sought after for their delicacy, the *jacú* and *jacutinga* being the most esteemed.

The difficulties of the ascension of these mountains consist of the thickets of underwood, the serried ranks of great ferns and trailing bamboos, in addition to the steepness of the Serra. The paths of the tapir, however, render the undertaking much more feasible than it otherwise would be. Dr. Gardner, after two attempts,—the latter made several years after the first,—attained the highest summit of the range. These mountains—known in geographies as a portion of the Brazilian Andes, the *Serra do Mar*, and the Organ Mountains—have been variously estimated to possess an altitude ranging from five thousand seven hundred feet up to eight thousand feet. The naturalist mentioned above made the only calculations of their height that have come under my observation; and, though they are only approximate, I give them, in this note, as interesting from the manner in which he reached his conclusions. According to him, the elevation of the highest peak is seven thousand five hundred feet above the level of the sea.*

---

* In the first ascent, Dr. Gardner accidentally broke his barometer before he had made a single observation; but, when on his last excursion he attained the highest summit, with the aid of the thermometer he made the estimate in the manner thus recorded:—"At mid-day the thermometer indicated 64° in the shade, and I found that water boiled at a heat of 198°; from which I estimate the height of the mountain above the sea-level to be 7800 feet. A register of the thermometer—kept

From March's an hour's brisk trotting will bring us within sight of Constancia. Mr. Heath, when expecting guests, is almost

HEATH'S, (CONSTANCIA.)

sure to meet them at an inner gate of his estate, about a half-mile from his residence, the main building of which rises from the midst

---

during our stay in the upper regions of the Serra and observed on the level of Mr. March's *fazenda*—gave a mean difference of temperature between the two places of 12° 5′. Baron Humboldt estimates the mean decrement of heat within the tropics at 1° for every 344 feet of elevation, and considers this ratio as uniform up to the height of 8000 feet, beyond which it is reduced to three-fifths of that quantity, as far as the elevation of 20,000 feet. It has, however, since been found that, in general, the effect of elevation above the level of the sea, in diminishing temperature, is, in all latitudes, nearly in proportion to the height, the decrement being 1° of heat for every 352 feet of altitude: this would give 4400 feet for the elevation of the highest peak of the Organ Mountains above Mr. March's *fazenda;* and, as this is 3100 feet above the level of the sea, we have for the total greatest elevation 7500 feet."—*Gardner's Travels in Brazil*, second edition, p. 405.

of the little cottages like a huge Bernese *châlet*. The smaller buildings are filled, in the summer-time, with boarders who come up to enjoy the cool air of Constancia and the bracing douche of the cascade which rushes down from the hill opposite. In this quiet *cul-de-sac* the Northerner is reminded, by the moss-roses and violets, of his own far-off land in springtime. Not far from the front-door, as we approach the main edifice, is a large clump of roses of a diminutive kind, growing in wild profusion. The tuberose, the Cape jessamine, and the delicate heliotrope, fill the air with sweets; and these and the arbors, with their honeysuckles, attract the tiny humming-birds, who sparkle in the sunshine like winged emeralds of richest hue.

Who that has been to Constancia will forget the material comforts with which Heath surrounds one? It is one of the few resorts for health and recreation that I have visited where the proprietor seems more like a host entertaining his friends than a landlord fleecing his boarders. His anecdotes keep up a constant cheerfulness, while his adventures among the forests and the mountains of Brazil are full of instruction. He accompanied Gardner on many of his excursions, and has been a perfect Nimrod. When the *felis-onça* abounded, the neighbors were sure to send for Heath to avenge depredations upon their folds; and many a well-sent bullet from his rifle has brought the beautiful jaguar—the monarch of the feline tribe in the Western World—to terms, which no troops of hounds or Brazilian guns could have effected. He informed me that many years ago his first visit to Constancia was in hunting the tapir which had made such havoc in the fields of Indian corn belonging to March's *fazenda*, of which he was then the *major-domo*. The number of these huge animals that he has in former years killed in one season at Constancia has been thirty-two. This was merely in the line of duty; for, if he had made a business of it, he could have "bagged" more tapirs, jaguars, peccari, &c. in one year than ever Gordon Cumming or Gerard did of their giant game in the wilds of Kaffraria or Algeria. (Heath died in 1864.)

It has often been a subject of wonder to me that of the tapir, the largest animal of South America, so little should be known. It also derives an interest from the fact that, though one of its species exists in the Old World, it was not discovered until long after the

Tapir Americanus; for the Malay tapir, differing but little from its Occidental congener, was never described until the governorship of Sir Stamford Raffles in Java.

The tapir forms one of the connecting-links between the elephant and the hog. Its snout is lengthened into a kind of proboscis, and, with the exception of the trunk of the elephant, which it resembles, is the longest nasal appendage belonging to any quadruped. It is, however, devoid of that clever little-finger with which nature has enriched the trunk of the land-leviathan. This curious animal has many fossil relatives, but only three living species (two of them belonging to South America) have as yet been discovered.

THE TAPIR.

The tapir is extensively distributed over South America east of the Andes, but especially abounds in the tropical portions. It seems to be a nocturnal vegetarian,—sleeping during the day, and, sallying forth at night, feeds upon the young shoots of trees, buds, wild fruits, maize, &c. &c. It is of a deep-brown color throughout, approaching to black, between three and four feet in height, and from five to six in length. The hair of the body, with the exception of the mane, is scanty, and so closely depressed to the surface that it is scarcely perceived at a short distance. Its muscular power is enormous; and this, with the tough, thick hide (almost impervious to musket-ball) which defends its body, enables it to tear through thickets in whatever direction it chooses. The jaguar frequently springs upon it, but is often dislodged by the activity of the tapir, who rushes through the bushes and underwood and endeavors to brush off his enemy against the thick branches. Its ordinary pace is a sort of trot; but it sometimes gallops, though awkwardly and with the head down. It is very fond of the water, and high up on the Organ Mountains are pools where it delights to wallow. Its disposition is peaceful, and, if not attacked, it will neither molest man nor beast; but, when set upon by the hunter's dogs, it can inflict terrible bites. Mr. Heath informed me that each time it seizes a dog with its teeth the flesh is cut completely from the bone of the canine intruder. The flesh of the tapir is dry, and is often eaten by the Indians of the interior, by whom it is hunted with spears and poisoned arrows. It takes to the water, and is not only a good swimmer, but appears almost amphibious, being enabled to sustain itself a long time beneath the surface: hence it has sometimes been called *Hippopotamus terrestris*. The largest which Mr. Heath ever shot weighed fourteen Portuguese arrobas, (about four hundred and fifty pounds,) though doubtless much larger exist in the Amazonian regions. Naturalists divide the American tapir into two species,—that of the lowlands and that of the mountains,—the latter, found on eastern slopes of the Andes, differing but little from the one already depicted and described.

The peccari is often met with in the woods of Brazil; and this little native swine is the most pugnacious fellow imaginable. Neither men nor dogs inspire reverence; for he will attack both with impunity. It is gregarious in its habits, and will, with its companions, charge most vehemently, no matter how great the odds. It is, I believe, one of the very few animals that has no fear of the detonation of fire-arms.

There are many beautiful and secluded walks and rides in the vicinity of Constancia, and frequently Mr. Heath accompanies his guests in the wild and romantic spots which here abound. I once climbed with two companions to the top of the mountain seen on the right in the sketch of Constancia, (page 283;) and, though I have made many ascensions among the Alps and the Apennines, I have never experienced so much fatigue and difficulty as on that occasion. We were the first, with one exception, to stand upon that height and behold the wondrous view around. I afterward made a sketch of the Organ Mountains at a point some miles distant from Heath's, and where the peaks presented the appearance of irregular saw-teeth; and I could then appreciate better than before the Spanish and Portuguese terms (*Serra* and *Sierra*,—a saw) for mountains.

The sketch alluded to (though not engraved) was made on the fly-leaf of a book which I reread in the Serra dos Orgãos, and which has since circumnavigated with me the Continent of South America. That book was an English edition of Todd's "Student's Manual,"—a work which delighted my boyhood, which gave me new resolution in college, and whose cheerful style, beautiful illustrations, and healthy thought cause it to be a most agreeable companion when no longer under tutors and governors.

Mr. Heath once took our company, through a little belt of forest, to a valley not more than two miles distant from Constancia. From the edge of the woods we looked down upon a dell whose narrow end was next to us. Beyond, on either side of the mountain-spurs which formed the valley, were the dark-green coffee-trees and the pretty shrubs of the Chinese tea-plant. Far beneath us, almost embowered amid giant *bananeiras* and orange-trees, we perceived the red tiles of a cottage. We descended by a little path to this half-hidden habitation, and were introduced to the proprietors, two Swiss brothers, who, after having served in the English army, retired upon a good pension, and here, in quiet, were enjoying life in one of the healthiest and most delightful places upon the earth. The elder brother had not been to the city for eighteen years. He had visited the United States when a younger man, but only that portion which constitutes the northern border of New York. While we were conversing with them, a flock of wild parrots came swooping into the open windows, screaming with delight as they ate the sunflower-seeds which these benevolent old bachelors had scattered for them. The edges of the coffee-*terreno* (where the berries are spread out to dry) were lined with large orange-trees, whose boughs bent downward with their golden burden; running roses had festooned themselves upon shrubs, trees, and outhouses, diffusing grateful fragrance from the thick clusters of buds and blossoms; purling brooks mingled their noisy, gleesome music with the more softened cadence of a distant waterfall, and the whole scene had so much of peace and felicity pervading it, that the "Happy Valley" of Dr. Johnson's imagination seemed here to find its counterpart in reality.

I paid many pleasant visits to this pretty spot, and the lovely valley grew upon me by the hour. In the cottage of the two Swiss I found the best current periodicals, in French, German, English, and Portuguese, all of which languages they speak with fluency. The contrast was, however, most striking, as we conversed about Grindenwald, Martigny, the Riga, and the shores of Lake Leman, (accurate paintings of which hung on the walls,) and then looked forth upon a landscape of perennial bloom and of unchanging verdure. They took me to their garden, where they were, for their pleasure, cultivating moss-roses (which grow with difficulty in Brazil) and vines brought from the warmer parts of their native Switzerland.

During one of my visits they informed me that they had purchased this plantation from a gentleman now residing in the State of Indiana, and they were equally surprised when I informed them that that State was my *terre natale*. They had kept up an active correspondence with the former proprietor, whom they represented as a lover of music and Goethe, but that since 1849 they had received no intelligence from him, and they feared that he had fallen victim to the cholera, which had swept through the Mississippi Valley during the year mentioned. They desired me to write to a friend to see if Mr. R. were dead or alive: accordingly, I wrote to one of the professors of South Hanover College, Indiana; and my correspondent ascertained that Mr. R. was still in the land of the living. Professor T. visited him, and found Mr. R. a venerable German of more than threescore years and ten; but his love for music had not abated, and he was ready to battle for Goethe at a moment's notice. He had not forgotten his friends in Brazil, but, from some cause unknown, had not written to them; and hence their apprehensions. When, however, he heard the description of the "Happy Valley" in the sunny land of the Southern Cross, the vision of its roses, golden fruits, fadeless green, and murmuring brooks came so vividly before him, that, aged as he was, his youth seemed renewed, and he resolved to return once more to that which was his first and beautiful home in the New World. I know not if he carried his resolution into effect, but I can readily imagine how powerfully one may be stirred up by the memory of beauty which is inseparable from that peaceful dale in the Serra dos Orgões.

In July, 1865, I again visited the "Happy Valley," at the invitation of the elder brother, whom I found a cheerful hale man of seventy-three. The younger brother spent the last year (1857) of his life in an attempt to plant a colony near Theresopolis, a town built since 1855 on March's old plantation. Mr. Rinke never returned to Brazil: in 1860 he visited the haunts of Goethe and Schiller, and died at Lucerne, Switzerland, from a cold caught while making a pilgrimage to the scenes of Schiller's "Wilhelm Tell." The "Happy Valley" has lost none of its loveliness. Long may Sr. Fischer live to enjoy it!

In one of my early walks on Heath's plantation, I was very much struck with a tall tree that shot up near the pathway. Its trunk was a little inclined,—otherwise remarkably straight; but its chief attraction was the long and venerable moss which hung from the wide-spreading branches and was gently swayed by the perfume-laden morning-breeze. I sat down to sketch it, and while thus engaged I was startled by a loud chattering; and in an instant a flock of brilliantly-colored birds, in curious flight, came from the neighboring wood and alighted upon the solitary tree. Though their motion on the wing was exceedingly clumsy, they were most nimble as they leaped from limb to limb. They kept up a continual chattering, as if they had met together to arrange their plans for the day. I soon perceived that, notwithstanding their brilliant plumage, which made the lofty tree seem laden with large golden oranges, they were as uncouth in appearance as they had been awkward in flight. Their bill was apparently of most disproportionate length, which did not, however, hinder their active movements among the gnarled branches and pendent moss. Presently, having settled upon their arrangements for the day, they took a unanimous vote, which was uttered in such a *viva voce* scream that the very mountains resounded with wild, unearthly notes

19

This was my first acquaintance with the *toucan,* which in its appearance is one of the most eccentric members of the feathered tribe. The feathers of the breast of the *ramphastos dicolorus* are of the most brilliant orange, chrome, and deep-rose colors, and form a prominent feature in the feather-dresses and ornaments of the wild Indians of the interior. In the sixteenth century the "high-born" dames of the courts of Europe esteemed as their most

THE MOSS-COVERED TREE.

gorgeous and picturesque robes those trimmed with the breast-feathers of the toucan. Its tongue is long, stiff, and is tipped and edged with little, hairlike feathers. It has a singular manner of taking its food. I have watched one in a tame state eating Indian corn; and it would take one grain in its huge bill, throw up its head, elevating its long appendage, and by a series of quick jerks the grain would be tossed along the stiff tongue into the throat.

The toucan belongs to climbing-birds, and is classed with parrots, woodpeckers, and cuckoos. Its foot, provided with two toes in front and two behind, is admirably adapted to the purposes of climbing and clinging. Its bill is by no means solid, and might be termed honey-combed in its structure, and hence is light. This long and heavy-looking instrument seems to be very sensitive and well supplied with nerves, as its owner may be often seen scratching the curious organ with its foot.

Waterton speaks of one species of the toucan in Northern Brazil (the toucans are only found in Tropical America) which "seems to suppose that its beauty can be increased by trimming his tail, which undergoes the same operation as our hair in a barber's shop; only with this difference,—that it uses its own beak (which is serrated) in lieu of a pair of scissors. As soon as his tail is full-grown, he begins about an inch from the extremity of the two longest feathers in it, and cuts away the web on both sides of the shaft, making a gap about an inch long: both male and female adorn their tails in this manner, which gives them a remarkable appearance amongst all other birds."

THE TOUCAN.

The toucan is a most grotesque specimen of ornithology, and the Aracari, (*Pteroglossus*,) with his huge bill and goggle-eyes, appears like a melancholy Jaques, or a spectacled German idealist, who has banished himself far from the haunts of men, to speculate on the miseries of human nature and the exalted excellence of the

"populous solitude of bees and birds
And fairy-form'd and many-color'd things."

The student of natural history can find much to gratify him in the Organ Mountains. There are many beautifully-colored snakes, (only a few of which are very venomous,) a vast variety of lizards, curious frogs and toads,—as some one has remarked,—from the small tree-kind, not more than an inch long, to those marsh ones which are nearly large enough to fill a hat. Beautiful butterflies vie with the flowers which from time to time they taste, or their brilliant wings are reflected from the small pools on whose banks they alight in countless numbers. Large wasp-nests as well as tropical leaves adorn the branches of trees. In some places, beetles like gems attach themselves to the foliage and flowers of low shrubs, and at night the air is lighted up with fire-flies which Gardner compares, in brilliancy, to "stars that have fallen from the firmament and are floating about without a resting-place."

One evening I walked from Heath's toward the "Happy Valley," but, not prolonging my promenade far in that direction, I entered a forest and pursued my way to the edge of a precipice, or rather a crater-like hollow whose centre was a thousand feet below me and whose sides were covered with trees. The night was dark, and it had fallen so suddenly after the brief twilight, that, so far as anticipation was concerned, I was unprepared for it. Before re-tracing my steps I stood for a few moments looking down into the Cimmerian blackness of the gulf beneath me; and, while thus gazing, a luminous mass seemed to start from the very centre. I watched it as it floated up, revealing, in its slow flight, the long leaves of the *Euterpe edulis* and the minuter foliage of other trees. It came directly toward me, lighting up the gloom around with its three luminosities, which I could now distinctly see. This was the *pyrophorus noctilucus*, so well known to every traveller in the Antilles and in Tropical America. It is of an obscure, blackish brown, and the body is everywhere covered with a short, light-brown pubescence. When it walks or is at rest, the principal light it emits issues from the two yellow tubercles; but, when the wings are expanded in the act of flight, another luminous spot is dis-closed in the hinder part of the abdomen. These luminosities—sup-posed to be phosphoric in their composition—are so considerable that the fire-fly is often employed in the countries where it prevails as a substitute for artificial light.

In the mountains of Tijuca I have read the finest print of "Har-per's Magazine" by the light of one of these natural lamps placed under a common glass tumbler, and with distinctness I could tell the hour of the night, and discern the very small figures which marked the seconds of a little Swiss watch. The Indians formerly used them instead of flambeaux in their hunting and fishing expeditions; and when travelling in the night they are accustomed to fasten them to their feet and hands. In some parts of the tropics they are used by the senhoritas for adorning their tresses, or their robes, by fastening them within a thin gauze-work; and through them their bearers become indeed "bright particular stars." It was of this fire-fly (which resembles, in every thing but color, the "snapping-bug" of the Mississippi Valley) that Mr. Prescott, in his "Conquest of Mexico," narrates the terror which they inspired in the Spaniards in 1520. "The air was filled with 'cocuyos,' (*pyrophorus noctilucus*,) a species of large beetle which emits an intense phosphoric light from its body, strong enough to enable one to read by it. These wandering fires, seen in the darkness of the night, were converted by the besieged into an army with matchlocks." Such is the report of an eye-witness,—old Bernal Diaz.

THE BRILLIANT FIRE-FLY.

In one of my rides toward Canta Gallo, I saw in the road the large lizard called the *iguana*. There is nothing to me disgusting in this clean-looking reptile, whose skin, composed of bright, small scales, resembles the finest bead-work. I had often seen them at Rio spitted and hawked about the city; for the flesh is esteemed a great delicacy,—resembling in its appearance and taste that *bonne bouche* for epicures, a frog's hind-leg. The usual pictures of the iguana do not render it full justice; they represent it as horrid in appearance as the imaginary baleful-breathed, javelin-tongued dragon from which good St. George delivered so many devoted virgins. The iguana is from three to five feet in length, and is oviparous. A lady member of my family possessed one which was a great favorite, and she has kindly furnished me with some notes on her pet. I insert them *verbatim.*

"Pedro [the iguana] afforded me much amusement. From his close resemblance to the snake-tribe, it was difficult for strangers to rid their mind of the impression that he was venomous. Such is not the case with iguanas. Their only means of defence is their very powerful tail; and a sportsman told me that he has had a dog's ribs laid bare by a stroke of an iguana's tail.   My poor pet, however, was not warlike, having been long in captivity.   He was given me as a 'Christmas-box' by a friend, and soon became tame enough to go at liberty.   He was about three feet long, and subsisted upon raw meat, milk, and bananas.   He had a basket in my room, and when he felt the weather cool would take refuge between the mattresses of my bed.   There, in the morning, he would be found in all possible comfort.   One evening we missed him from all his usual hiding-places, and reluctantly made up our minds that he was lost; but, on rising in the morning, two inches of his tail hanging out of the pillow-case told where he had passed a snug night!   My little Spanish poodle and he were sworn foes. The moment Chico made his appearance, he would dash forward to bite Pedro; but Chico thought, with many others, that 'the better part of valor is discretion.'   So he made off from the iguana as fast as his funny legs could carry him.   Then Pedro waddled slowly back to the sunny spot on the floor and closed his eyes for a nap. When the winter (a winter like the latter part of a Northern May) began, he became nearly torpid, and remained without eating for four months.   He would now and then sun himself, but soon returned to his blanket.

"I frequently took him out on my arm, and he was often specially invited; but I cannot say that he was much caressed.   It was in vain that I expatiated on his beautiful bead-like spots of black and white, on his bright jewel eyes and elegant claws.

THE IGUANA.

They admired, but kept their distance. I had a sort of malicious pleasure in putting him suddenly down at the feet of the stronger sex, and I have seen him elicit from naval officers more symptoms of terror than would have been drawn forth by an enemy's broadside or a lee shore.   But, alas for the 'duration of lovely things!'   During the summer-months he felt his old forest-spirit strong within him, and he often sallied forth in the beautiful paths of the Gloria. On one of these occasions he met a marauding Frenchman. Pedro, the caressed by me and the feared by others, knew no terror. The ruffian struck him to the earth. It was in vain that a little daughter of Consul B. tried to save him by crying, '*Il est à Madame:*' another blow fractured his skull! My servant ran up only in time to save his body from an ignominious stew-pan; but life was extinct. The assassin fled, and Rose came back with my poor pet's corpse. On my return he was presented to view with his long forked tongue depending from his mouth. He was sent, wrapped in black crape, to a neighbor who delighted in fricasseed lizards, but who, having seen him petted and caressed, could not find appetite to eat him!

"Thus ended the career of poor Pedro, after a life of pleasant captivity; and perhaps it might be said of him, as of many others, 'He was more feared than loved!'"

From Constancia to Nova Fribourgo, or *Morro Queimado*, is a mountain and forest path, which is sometimes taken by travellers who wish to visit the *villa* named above. The route most frequently traversed is by steamboat from Rio de Janeiro, on the bay as far as the Macacù River, and up this stream to the Engenho de Sampaio. Thence we may go by carriage or mule-back to the flourishing town of Porto das Caixas, which is the general rendezvous for the troops of mules that bring coffee and sugars from the Swiss colonies of Nova Fribourgo and Canta Gallo and a large section of the neighboring country. Here are also debarked the goods which return from the capital in exchange for produce.

In addition to its commercial importance, it is distinguished as the family-residence of the Visconde de Itaborahy, (Senhor Joaquim José Rodrigues Torres.) The traveller will here find a very good *hospedaria*, (inn,) kept by a Frenchman, whose prices, though not so moderate as in the interior of the country, may, with other expenses, be interesting to *voyageurs* who may come after me. I find in my note-book the following entry for myself and companion:—

"*Hospedaria de M. Boulanger.*—Two dinners, two candles, two beds, coffee for two, two breakfasts, and the stabling of two mules, —7$200," (equal to about sixteen English shillings.)

At the excellent boarding-house of Mr. Lowenroth, at Nova

Fribourgo, you pay 2$ (one dollar) per diem for every thing. At Canta Gallo, thirty miles farther in the interior, I paid 6$000 (thirteen and sixpence English) per diem, for myself, guide, and three mules. At Pedro Schott's, (a regular *Tête noire* châlet of rude construction,) situated in a wild, secluded spot half-way between the bay and Nova Fribourgo, for two dinners, two beds, two lights, and the stabling of two mules,—4$500, (ten shillings twopence.) At Constancia and at Petropolis you pay 4$000 (nine shillings) per diem, the price of a first-class hotel in the United States. It must be remarked, however, that wine is never extra, and, as this is obtained at a cheap rate direct from Lisbon and Oporto, it is placed upon every table. On going into the fertile province of Minas-Geraes, I found that for myself and company we were charged at Petropolis 16$000, (nearly nine dollars,) and the next night at a little inn called Ribeirão we paid for the same accommodations 4$000, (two dollars and twenty cents.) Upon the sea-coast I have always found the living expensive to the foreigner. Farther in the interior the prices diminish. At the Ponta do Jundiahi, in the province of S. Paulo, dinner for myself and guide, and feed for three animals, the price was but 1$500 (three shillings and fivepence English.) The common Brazilian travels at a rate one-fourth cheaper than either the North American or the European. He rarely stops at the hospedaria, but, when he considers the day's journey ended, whether at two o'clock P.M. or six P.M., he rides under a *rancho*, gives a few handfuls of *milho* (maize) to his mule, and afterward turns him out to pasture. He then—if he has no servant with him —joins with others occupying the same rancho, and *feijões*, and *carne secca*, greased with a little *toucinho*, and well stiffened with *farinha de mandioca*, form a substantial supper, which has as an adjunct coffee, red Lisbon, or water from the running brook. I have found sleep as sweet on a raw hide spread in the dust of a rancho as in the soft bed of the best New York hotel. The ranchos (mere tile-covered sheds) are found all over the country, and, like the caravanserais of the East, are often erected by the authorities; but in many instances they have been built by some *vendeiro*, who charges nothing for the shelter thus afforded to the *tropeiros* and the thousands of sacks of coffee and sugar on their way to the seaboard marts. The *vendeiro*, however, does not count without his host, for

troupeiros need feijões, carne, farinha, cachaça, and coffee for themselves, and milho for their mules. Then an extra girth, a saddle-
blanket, a pointed knife, and an iron spur, are often wanted; and
the Portuguese *vendeiro* thus accumulates property, and in time
becomes a *fazendeiro*, but does not give up the shop, which always
brings him a good return.

Those who intend travelling long journeys in Brazil would do
well to purchase their own mules. Horses and mules (the latter
are much more serviceable) may be hired at the rate of from 5$000
to 10$000 (eleven to twenty-two English shillings) for each fifty
miles, or for a certain sum the trip.

The coffee-plantations of the elevated uplands of Nova Fribourgo and Canta Gallo rank among the best in the province of Rio
de Janeiro: many of them are owned by Swiss and Frenchmen
who came to Brazil at the invitation of Dom João VI., in 1820; but
the colony of which they formed a part fell through, and the most
energetic men have become proprietors. The Baron of New Fri-
bourg has immense plantations in the vicinity of N. Fribourgo,
where he not only employs slaves, but many emigrants from Portugal, the Azores, and Madeira. His residence in the *villa* whence
he derives his title is a large mansion built in good taste. A Protestant chapel of small dimensions is presided over by an old Lutheran
clergyman who came to Brazil with the early German colonists.
I could, however, perceive that there was but little Christian
vitality among this people. Lutherans of the old Church-and-
State School are among the very last men to propagate the gospel.
There is more hope of some of the new pastors in the more recently-
established German colonies.

At Nova Fribourgo are a number of excellent schools, the chief
of which is the Instituto Collegial of Mr. John H. Freese. This
gentleman has devoted many years to instruction in this cool and
healthful spot, and many hundred young Fluminenses have here
received an education in English and French, as well as in the
Portuguese language. I have met with the scholars of Mr. Freese
in different parts of the Empire, and they always manifested a
general intelligence beyond the alumni of other similar institutions.
His *Noções Geraes ácerca da Educação da Mocidade Brazileira* show
that he has given much attention to the subject of education.

Between N. Fribourgo and Canta Gallo the scenery is remarkably Alpine, and such is the cultivation that one is readily reminded of the sweet valleys of Switzerland. In the neighborhood of Canta Gallo I found a number of intelligent German, Swiss, and French gentlemen, whose coffee-plantations bring them most lucrative incomes. I was not a little surprised at a kind offer of a German,

NEAR THE VILLAGE OF NOVA FRIBOURGO.

who manifested the beginning of his hospitality by asking me if I would not take *ein grog*, and was as astonished at my refusal as I had been at his offering.

At the plantation-house of Mr. D., a Swiss from Zurich, I was surrounded by many reminiscences of his fatherland; and when I gazed upon his finely-cultivated fields, which stretched before his mansion, I could almost believe myself in some of the green vales of the Oberland, large paintings of which graced the walls of the *salon.* The illusion was rendered more complete when night had hidden every palm-tree and flowering cactus, and I heard only the sounds of the French and German languages, or from the piano the simple notes of the *Ranz des Vaches,* sweet *nocturnes,* and the majestic strains of Mendelssohn and Beethoven. I could scarcely· believe myself a hundred miles in the interior of Brazil. I, however, realized that I was not in the land of Tell when I returned to Canta Gallo preceded by a negro in livery, who bore (on horseback) a flaming torch, whose flashes of light revealed overhanging mimosas, bignonias, and long, bending bamboos.

The old hotel-keeper at Canta Gallo is a tall Frenchman who was one of the body-guard of Napoleon I., which fact his mellifluous *Français,* as well as rude fresco-paintings, soon inform you.

In returning from this excursion, there is a magnificent view of the whole bay, extending as it does within its mountain-walls one hundred miles in circumference. The most important ports upon the borders of this bay are Majé, Piedade, Porto da Estrella, and Iguassú. At these several places great quantities of produce are delivered by troops from the interior and embarked in steamers and *falluas* for the capital.

A glance at the map shows the Bay of Rio de Janeiro to contain numerous islands, of various form and extent. Ilha do Governador, or Governor's Island, is much the largest, measuring twelve miles from east to west. Most of these islands are inhabited, and under tolerable cultivation. If any thing can add to the imposing scenery of this magnificent bay, it is the vast number of small vessels that are seen constantly traversing it, dotting the green surface of the water with their whitened sails. From morning to evening may be seen, plying in every direction, open and covered boats, canoes, lanchas, falluas, and smacks.

One of the most attractive residences for the people of Rio during the hot season is the newly-formed colony of Petropolis, situated about three thousand feet above the level of the sea. An agreeable steamboat-transit amid the picturesque islands brings you to Mauá, the terminus of the first railroad formed in Brazil, and for which the Empire is indebted to the enterprise of that enlightened and patriotic Brazilian, Evangelista Ireneo da Souza, who, on the opening of this railway was created Baron of Mauá by the Emperor. The road is about ten miles long, and leads to the foot of the mountains, where carriages, each drawn by four mules, receive the travellers. The ascent is by an excellent road, which was built by the Government at an enormous expense, and reminds one of the Simplon route. In some parts the side of the mountain is so steep that three windings are compressed into a space small enough to allow of your being heard as you speak to the persons in the carriages going the opposite direction. When you reach the summit, before descending into the valley in which stands the town, a magnificent prospect opens before you. All the bay and city of Rio, with the plains of Mauá, across which lies the diminutive railroad, are mapped out below.

In the year 1837, Dr. Gardner writes, "We passed through the small, miserable village of Corrego Secco." This is now Petropolis. All the neighboring land was at an earlier date obtained by the Emperor D. Pedro I. with a view to forming a German colony. This design was interrupted by his abdication, but has been carried out by his son, the present Emperor. It now contains ten thousand inhabitants, and on every side are beautiful residences of wealthy Rio families who resort thither during the summer. Nothing can exceed the beauty of the vicinity. Roads, bordered by villas, stretch away from the centre, between hills still covered with virgin forest. Many of these, inhabited by the German colonists, bear the name of places in Fatherland, and the mind is pleasantly transported to scenes in the Old World. The highroad to the mining-district is through Petropolis, and troops of mules, laden with coffee, sugar, and sometimes gold, are perpetually passing down to the head of the bay, where their loads are transferred to falluas and steamers to be transported to the city.

The palace of the Emperor stands in the centre of the town, and when finished and surrounded by cultivated grounds, will present a beautiful appearance. Small streams intersect the streets and are crossed by bridges, adding much to the singular aspect of the place.

There are Roman Catholic and Lutheran churches, large hotels, and many shops. Here the Baron de Mauá had a mansion pleasantly situated at the meeting of two mountain-brooks. Several of the diplomatic corps and other foreigners have villas here and there,—the English generally seeking the heights

The colonists belong to a low class of Germans, and brought with them few arts and but little education. It seems difficult in any tropical climate to prevent the morals and industry of emigrants from deteriorating, and this is particularly to be observed in slave-countries. The degraded colonist, while setting himself above the African, engrafts the vices of the latter upon the European stock, and thus sinks to a lower grade than the negro. The German in Brazil has the want of a sound moral people surrounding him, to sustain and elevate him: therefore it is no marvel if he sink lower and lower in the scale of civilization. Much, however, is being done for the Germans of Petropolis. The clergyman, as the pastor of the church and superintendent of the schools, takes a deep interest in the welfare of his countrymen both spiritually and intellectually.

SWISS VALLEY, NEAR PETROPOLIS.

It is not possible to obtain a view of the entire town of Petropolis at one glance, because it is scattered in various valleys among the hills. More rain falls here than in Rio, and the tiny rivulets often become rushing streams, and the mule-troops labor on through miles of mud. This moisture keeps the air cool and freshens the flowers that cluster round the white-walled cottages which gleam from their dark-green background. The accompanying view is taken in the Swiss valley, where, as you listen to the German accents of the villagers, fancy might induce you to believe yourself in Europe, did not the waving palm and rustling banana remind you that you dwelt under a tropic sun.

Petropolis is annually becoming of greater importance. Its salubrious and delightful climate will make it a large and fashionable resort for the Capital of the Empire, and perhaps the day is not distant when it will become the second city in the province. It stands at the entrance to the fertile province of Minas-Geraes, and, should some plan be devised for constructing a railway up the mountains, its growth will be most rapid. If the Baron of Mauá would pay a visit to the United States and examine the Pennsylvania railways, or the Baltimore and Ohio Railroad, he may be encouraged to persevere. Professor Agassiz considers the engineering triumphs of the Pedro II. Railway of the first rank in the world. The road under President Ottoni is still being pushed into the interior. The União and Industria turnpike is unique in South America. It begins at Petropolis, and extends to Juiz de Fora in Minas-Geraes, and is traversed by stage-coaches and ordinary freight-wagons. Srs. Lage and João Baptista Fonseca are the chief officials of the União and Industria road.

Note for 1866.—Since this chapter was written, some of the most important measures for developing the resources of Brazil have been carried out, at an enormous expense, but which day by day are showing their results. During the cabinet of the late Marquis de Paraná, Sr. Pedreira do Coutto Ferraz, Minister of the Empire, contracted for the construction of the Pedro II. Railroad; for that of the great macadamized road called the *União e Industria;* and for the Canta-Gallo Railway, now twenty-five miles long. The first section of the Pedro II. Railway was opened in 1857. The contract then passed from English into American hands, Messrs. Roberts, Harvey, and Harrah. Major Ellison, of Massachusetts, was the chief engineer of the road. The trains now run to the Ponto Desengano, on the Parahyba River. The great tunnel which terminates near Mendes was opened in December, 1865. The Pedro II. Railway has now passed into the hands of the Brazilian Government, and Mr. William Ellison is engineer-in-chief. Mr. Jacob Humbird and Sr. Carneiro Leão are principal contractors.

CHAPTER XVI.

PREPARATIONS FOR THE VOYAGE TO THE SOUTHERN PROVINCES—THE PASSEN-
GERS—UBATUBA—EAGERNESS TO OBTAIN BIBLES—THE ROUTINE ON BOARD—
ABORIGINAL NAMES—SAN SEBASTIAN AND MIDSHIPMAN WILBERFORCE—SANTOS—
BRAZILIANS AT DINNER—INCORRECT JUDGMENT OF FOREIGNERS—S. VINCENTE—
ORDER OF EXERCISES—MY CIGAR—PARANAGUÁ—H.B.M. "CORMORANT" AND THE
SLAVERS—MUTABILITY OF MAPS—RUSSIAN VESSELS IN LIMBO—THE PRIMA DONNA
—AN ENGLISH ENGINEER—ARRIVE AT SAN FRANCISCO DO SUL.

ALTHOUGH I had resided several years in the Empire, I had never visited its Southern provinces. In June, 1855, duty as well as inclination gave me the privilege which I had so long desired.

Having been kindly provided by Brazilian, German, and English friends at Rio with letters of introduction, and being particularly fortified by a strong *carta de recommendação* from the venerable Senator Vergueiro, (one of the last of the constitutional patriots,) I had every facility for seeing Southern Brazil to advantage.

Wishing to have ample leisure, I procured my passport, several days before my departure, at the proper bureau. One of the first lessons learned by the traveller in Brazil is patience and conformity to all existing formalities. No matter how absurd the regulation, as, for instance, that which requires one to obtain a passport in leaving the city of Rio de Janeiro for the provinces, (where it is never demanded,) you must submit. Protestations only bring a shrug of the shoulders from the snuff-taking official, and woe be to you if the hour for closing the bureau slips around before you have obtained the necessary document. To be perfectly *en règle*, the departing citizen or stranger must have his name registered either in the custom-house or printed in some public journal three days before his passport is granted, in order that his creditors may have an opportunity of knowing his movements. But the passport system, as well as quarantines, never prevented the adit or exit of rogues or pestilence.

In addition to this, I had prepared, the day before, my baggage, consisting of a trunk and a number of large boxes of books, and I had made arrangements with an under-clerk of a mercantile house to have these put on the steamer at an early hour. Believing myself perfectly secure, I was busily engaged in writing up to within half an hour of the time of departure. On entering the mercantile establishment referred to, I found that my baggage was still quietly resting where I had left it the day previous. There was just time to hurry it down to the Consulado in a cart. Off we started, and, on reaching this place, we went through a set of formalities in shipping the boxes; then, taking a boat, (for vessels there do not lie in docks,) we arrived at the steamer, and had the mortification to be informed by the Brazilian second mate that the objects of our haste could not be received on board at that hour without a special permit from the office of the steamer, which was in a street one mile distant from the Consulado.

The blacks rowed me quickly to the shore, where I jumped into a tilbury and rattled through the streets to the much-coveted bureau of the Southern Steam-Packet Company. I obtained the permit, and, with as great celerity in returning as in coming, I was soon on board. I leave to the reader to judge how much easier and more reasonable the whole matter would have been in England or the United States, even if blame were to be attached to me for not attending to my own luggage and seeing it fairly on the steamer the day before.

Once on board, I found that there had been no need of my great fretting, for the engine snorted and hissed more than an hour before we left the moorings. Our passports were all examined by the police-officer, and our personal identities were verified by the agent of the packet, in order to discover if all the passengers had paid their fare: the captain took his stand upon the wheel-house, and to his "Small turn ahead" we moved through the assembled shipping of the loading, discharging, and man-of-war anchorages, until a "Stop her" brought us under the guns of Villegagnon. Here we received the last visit of the agent, and then the Government officials boarded us to see that all was right and——you imagine that we steamed out of the bay, in which imagination you would be egregiously mistaken; for we lay before Villegagnon for two mortal hours, tossing up and down in a delightful swell which rolled in directly from the blue Atlantic. Something had been left behind by the captain's wife, which (of more value than a band-box) proved to have been a large package of money "expressed" to the South; and hence our delay.

It was after five o'clock when we passed the giant sentinels of the Sugar-Loaf and Santa Cruz. The passengers, with the exception of myself, a Frenchman, and a Lombard, were either Brazilians or Portuguese. The captain, though a Baltimorean, had renounced his allegiance to the United States, and had been naturalized in Brazil. Night soon came on, and a heavy rolling sea compelled me to take to my berth,—not, however, before I had seen the Brazilians horribly sea-sick; and all of them have such a bilious look that one would anticipate for them an unusual degree of suffering upon the "vasty deep."

Early the next morning I could see from my cabin-window the mountains of the coast. The same magnificent scenery which so delights the traveller in the vicinity of Rio de Janeiro is reproduced all the way to Rio Grande do Sul, only the mountains vary in form, and in some places the palm-trees are more luxuriant. When I came upon deck, we were just entering the beautiful Bay of Ubatuba. Two vessels were riding at anchor; and, for a small place, there is considerable trade in coffee, which is brought down from the interior and thence shipped to Rio.

The village of Ubatuba stretches along a circular beach, and its bright houses are thrown out in prominent relief by the verdant mountains that lift themselves in the background. The storm had ceased; and I rarely have witnessed a lovelier scene than was presented by this Southern landscape. The captain, seeing the calmness of the water, had the good sense, at this juncture, to invite the passengers to a most substantial breakfast, for which each one on board had been fully prepared by his night's tribute paid to the angry waves.

Every eye beamed with pleasure (doubtless the breakfast had had something to do with it) as the vision of beauty before us came in review. Good-nature and kindness is a predominant characteristic of the Brazilian; but even a churl would have been *alegre* under our present circumstances.

20

We only exchanged mails and took in oranges, (a hundred of the most luscious could be purchased for an English threepence,) and, bidding farewell to Ubatuba, in a short time we were sailing close to woody islands or the green shore. The sea was smooth, the passengers were all upon deck, and the best of feeling pervaded the whole company. Wishing to profit by the occasion, I descended to my trunk and brought up a Portuguese Bible, which I offered to a passenger on the conditions laid down in the rules of the American Bible Society. Only a few moments elapsed ere I had disposed of all the volumes of the Sacred Word which were at my convenience, and on every side my fellow-voyagers were reading with eagerness a book they had never seen before. From time to time I was called on for explanations, and I was renewedly convinced of the freedom from bigotry which is a distinguishing negative quality of the Brazilians. An officer of the Imperial navy had just returned from the Brazilian squadron at the river Plate, and, in seeking the bosom of his family at Santos, wished the Scriptures as a present for his children, and, when purchasing them, he remarked, "Though I am a man forty-five years of age, I have never before seen *A Santa Biblia* in a language which I could understand."

Ubatuba differs in a certain respect from a number of neighboring towns, inasmuch as it rejoices in one of the euphonious aboriginal terms which were found throughout the country at its discovery. Not many leagues from this village is the large town of *Angra dos Reis* and the island denominated *Ilha Grande dos Magos*, which names were given by Martin Affonso de Souza. Although several of these harbors and islands had been previously discovered and probably named, yet—owing to the circumstance that Souza became an actual settler, combined with the fact that in following the Roman calendar he flattered the peculiar prejudices of his countrymen—the names imposed by him have alone remained to posterity. The 6th day of January, designated in English as that of the Epiphany, is termed, in Portuguese, *Dia dos Reis Magos*, (Day of the Kings or Royal Magi.) The island of S. Sebastian and the port of S. Vincente were named, in like manner, on the 20th and 22d days of the same month. The Indian names of Brazilian towns are among some of the most flowing and fine-sounding found in any language:—as *Itaparica, Pindamonhangaba, Inhomerim, Guaratingetá, Parahiba* and its diminutive *Parahibuna,* &c.,—the *h* in each case *non est litera.*

It was only a few hours' run from Ubatuba to our next stopping-place. We were constantly passing one of the boldest and most picturesque coasts that I have ever seen. Near the island and the town of San Sebastian, (the latter on *terra firma,*) I was continually reminded of the banks of the Rhine and of the lake and mountain scenery of Switzerland, though here perpetual verdure crowns cliff and crag, and the valleys were covered with plantations of coffee and sugar, and the orange-groves were prodigal of their golden fruit. The shore was steep and high, and well-wooded promontories stood out with minute distinctness in the bright, pure atmosphere. The island of San Sebastian is only separated by a narrow strait from the mainland, and it seemed to me, as I gazed upon it, like one of the fabled Hesperides. The steep rocky sides of its mountain-ridge are interspersed with belts of forest, from whose thick-foliaged bosom cascades of Alpine magnitude dashed their foaming treasures hundreds of feet below.

It was in a hamlet on this romantic island that Wilberforce—a rollicking, fun-loving English midshipman—says he saw the traces of Portuguese hands in a neat white church which rose from the midst of mud houses. "The antiquity of the building," he writes, "was not the sole proof of its origin. The presence of a church is in itself sufficient to show whether Portuguese or Brazilians have founded the village. It is said that the first building that Portuguese settlers erect is a church : the first

THE ROADSIDE VENDA.

that Brazilians build is a grog-shop." And then he significantly adds, "We order these things better in England, and build both at the same time." I cannot say that the remarks of Midshipman Wilberforce are altogether exact; for it is a fact that the Brazilians already have too many churches for the priests, and also that they do commence the nucleus of their village by a *venda*, which not

only serves as a drinking-house, but as a place for lodging and eating. The Brazilians are a temperate people, as I have already observed, and are not given to drunkenness as the Northern nations; therefore "*grog-shop*" is not the correct term to express the foundation of a Brazilian settlement. Religion and the *venda* are not always inseparable; for you will frequently find a little cross erected near its entrance, and sometimes an alms-box affixed to the door, on which is painted "white souls and black" lifting up from the flames of purgatory hands of supplication; and hard must be the heart that can resist the piteous spectacle.

THE ALMS-BOX.

The midshipman is, however, entirely just in his observations on mosquitos and the very vicious sand-flies called *borruchudos*. Both his indignation and poetry arise at the trouble they gave him; for he eloquently bursts forth in the following :—"Any one who should write an ode to Brazilian scenery [near San Sebastian] would probably begin,—

> " 'Ye mountains, on whose woody heights
> The greedy borruchudo bites;
> Ye forests, in whose tangled mazes
> The dire mosquitos sting like blazes !'—

and so on to the end of the canto. Things that would be poetical in themselves are sadly spoiled by the introduction of such utilitarian adjuncts as mosquitos. Greedy animals! I am ashamed of you. Cannot you once forego your dinner and feast your mind with the poetry of the landscape?"

San Sebastian is twelve or fourteen miles long, and of nearly equal width. It is well cultivated and somewhat populous. Like Ilha Grande, it was a rendezvous for vessels engaged in the slave-trade. Such craft had great facilities for landing their cargoes of human beings at these and contiguous points; and if they did not choose to go into the harbor of Rio to refit, they could be furnished at this place with the requisite papers for another voyage. For no other object was the vice-consulate of Portugal established in the villa opposite.

The sun was setting as our little steamer issued from the Bay of S. Sebastian, and before daylight was gone we neared the Alcatrazes, two rocky islands of curious shape, conspicuous objects well known to all travelled Paulistas.

Before retiring to my cabin I had an interesting conversation with a Portuguese who was proud of his little native peninsular kingdom, and boasted of her great deeds and past prowess, but spoke not of her present glory. The Lombard passenger entertained me with sketches of the Milanese revolt of 1848, and with warlike *chansons*, in which the name of Carlo Alberto Il Ré di Sardegna was ever prominent.

The next morning we arrived at *Santos*, situated a few miles up a river of the same name, which is the chief port of the flourishing province of St. Paul's. Here I landed my two boxes intended for the interior, and which I hoped would reach their destination before I returned to Santos, so that I could ride swiftly after them and not be delayed as I had been in similar excursions in the rural part of the province of Rio de Janeiro. I had some difficulty with the custom-house; and no one but strangers who have gone through this experience in Brazil can imagine the various annoyances to which every species of goods is subjected. There are no objections to the books because they are Bibles, but you must pay duty (small, it is true) a second time upon them. I thought because I had paid duties once at Rio that that was sufficient; but here they have a provincial tariff from which no one is exempt. I had letters from Senator Vergueiro to his two sons, who have a mercantile house here, and also the father and the sons have immense plantations in the interior; and it was to one of these plantations that I determined to go, and, while doing good, be enabled to see for myself the condition of the thousand European colonists which the enterprising Vergueiros have under their charge.

Senhor José Vergueiro, the principal of the Santos house, (Vergueiro & Filhos,) was absent, and his brother, the fourth son of the Senator, was indisposed. But at his order every kindness was shown me by the clerks of the establishment; and through one of them my books were soon liberated from the custom-house. I declined their invitation to dine at the Trapiche, for I had already accepted the kind offer of my Brazilian *compagnons de voyage* at the hotel of Senhor Francisco. Senhor F. was said to be a perfect polyglot; but I found, by trying him in three languages, that he only spoke a smattering of each. The dinner was plentiful and excellent. I found that the convivial qualities of the Brazilians were as remarkable as those of John Bull,—not that there was drinking to any excess, but they ate heartily, and cheered most lustily at every toast or sentiment, with which it seemed our feast was as plentifully provided as with substantial food and *doces.* The Brazilians are great toasters; and I have seen a table at which twenty or more persons were assembled, and each proposed at least one sentiment, while some proposed during the sitting the health of as many as six different individuals. Some of these toasts would be concluded by a song vociferated by the whole company as loudly as if German students had been the performers.

The company at Senhor Francisco's consisted of merchants, physicians, a number of Government civil officers, and one colonel of the regular army. Wine in abundance was placed upon the table; yet it was used in great moderation by those who did partake of it, while others seemed to abstain from it altogether. In settling the bill, ($1 each,) not one of them would allow me to share a penny of the expense; and throughout the whole repast, it being known that I was a Protestant clergyman, they were most respectful in their bearing, and all approved of the work in which I was engaged. I have been thus particular in mentioning this little incident, because some writers and visitors in Brazil, but who certainly have never seen beyond a ship-chandlery, hotel, or at furthest some coast-city, have complained that Brazilians are inhospitable, selfish, and altogether distrustful of strangers. As to inhospitality, away from the great towns it cannot be predicated of them; and even in Rio and Bahia, the largest cities of Brazil, I have met with the very warmest welcomes from Brazilians whom

I had never seen until I handed them my letters of introduction. Among the pleasantest memories of my life will be the recollection of the kind hospitality manifested towards me by Brazilians at the metropolis, where more than elsewhere coldness is said to abound. As to selfishness and distrust of strangers, they possess the one in common with human nature, and of the other they do not possess more than is manifested by Englishmen or Americans when approached by the newly-arrived foreigner without letters of recommendation.

From the hotel of Senhor Francisco we went on board of our steamer. That evening a knot of our passengers, together with the captain and his mate, sat up to a late hour conversing in regard to the demoralizing literature which floods the land from France. They listened with great attention to remarks which were in favor of laying the axe at the root of the tree; and a corrupt religion was measured by the only true standard,—that great Rule of Faith given to us by God in His word.

The next day our steamer did not leave Santos until noon, so that I had an opportunity of going again to the warehouse of Senhor Vergueiro & Filhos. I was glad to find that the youngest Vergueiro was able to be in his counting-room, though Senhor José had not yet returned from the interior. He regretted much that I could not then accept the hospitality of their house, stating that his father had written to them requesting that they would pay me every possible attention, but hoped that on my return from San Francisco do Sul I would give them a long visit. All this was said in a manner so unaffected and cordial as to preclude all idea of formality or insincerity.

At twelve o'clock the "vapor" left Santos, and we were soon steaming down the river.

Santos is situated upon the northern portion of the island of S. Vicente, which is detached from the continent merely by the two mouths of the Cubatão River. The principal stream affords entrance at high-water to large vessels, and is usually called Rio de Santos up as far as that town. At its mouth, upon the northern bank, stands the fortress of S. Amaro. This relic of olden time is occupied by a handful of soldiers, whose principal employment is to go on board the vessels as they pass up and down, to serve as a guard against smuggling. The course of the river is winding and its bottom muddy. Its banks are low and covered with mangroves, so that the foreground is not very inviting; but from the wheel-house a fine prospect of back-country and distant mountains presented themselves on the north. The captain pointed out the site of St. Vincent,—the first regularly-established colony in Brazil. How Martin Affonso de Souza could have chosen this place in preference to the present situation of Rio is indeed hard to account for, except on the ground that the Tamoyo Indians were too numerous around the Bay of Nictherohy.

The sea becoming rough, I took to my old and sovereign remedy against nausea,—viz.: a good berth,—and did not rise until I found that the sun was high above the mountains, and that we were entering the intricate harbor of Paranaguá. Before crossing the bar, we saw outside a Brazilian schooner tossing up and down at anchor. The captain, with his glass, perceived that it was one chartered by the Steam-Packet Company, and was loaded with coals from which he was to obtain his fuel for the remainder of the voyage. It was of the utmost importance, then, that the schooner should cross the bar. With the present wind it would be impossible. The steamer's head was put for the schooner. It was with difficulty that any one became aroused, and then the utmost indifference was manifested by the captain of the little sailing-vessel at a proposition which would have made an English or a Yankee skipper dance with joy,—i.e. to be towed in. His drawling reply was, "Se o Senhor quizer," (If the gentleman wishes it.) This was perfectly in accordance with the general want of energy which characterizes a certain class of Brazilians. The vessel was attached to the P——, and we were soon over the bar, steering up the difficult channel.

A number of letters which I wrote to a friend during this voyage were preserved and afterward returned to me; and I have thought it best from time to time to introduce portions of them which possess at least the interest of being penned amid the scenes which they describe. The following was written from the next port south of Paranaguá.

"SAN FRANCISCO DO SUL,
"PROVINCE OF SANTA CATHARINA.

"This is not that San Francisco of wonderful growth, of adventurers, and of golden dreams. As to gold, there is none; as to adventurers, only two runaway sailors; and as to rapid grow*h, that is reversed, for here there are plenty of houses to let,—plenty 'hurrying [the only haste to be discovered] on to indistinct decay.'

"But I will go back for a day or two in my journey.

"I left Santos on the 15th. It is delightful to travel on a Brazilian steamer, provided that you are not in a hurry. They take things so easy: I mean both steamers and people. And let me say that, of all the travellers with whom I have ever voyaged, the Brazilians are the most good-natured and agreeable after you have made their acquaintance. They are very obliging, yet from time to time can display as much selfishness as other 'humans' on a vessel,—that little world in miniature, where all that is bad is easily brought to light. *Pacienza* is the motto of these steamers. When you arrive at a town, after having been 'terribly' pitched about and sea-sick, you may now count upon a good twenty-four or thirty-six hours on land. It is a great luxury. The passengers desert the vessel, (although good dinners are provided on shipboard,) and off they rush to the hotels; or, in default of this, they seek the Casas de Pasto, and feast to such an extent that you would deem them half famished.

"The 'order of exercises' on board the steamer at sea may be easily stated. Each morning at six o'clock the cabin-boy wakes you up by giving you a cup of coffee, (*noir*,) and thirty or forty minutes afterward a large bowl of *mingau*, (arrowroot, or maize-mush,) well sprinkled with cinnamon and sugar, is placed on the table, and a strapping big fellow, fortified with a ladle, is ready to serve you with all the grace and celerity which appertains to the same kind of presiding genii that you meet with at the Faubourg du Temple in Paris. At ten o'clock a huge breakfast consisting of roast and boiled beef, pork, fresh fish, *pirão*, (a dish of mandioca,) &c. &c., is placed before you. Fall to, help yourself, and your neighbors will do the same without any *ritardo;* and, when satisfied or fatigued with this operation, vary the business by imbibing the *tea* which the steward has just brought simmering in. Now mount the deck. If the sea is not heavy, pipes, cigars, and promenades are the next in the programme. The scenery on shore is *my* cigar; and up to the present time there has been no diminution of my enjoyment in this respect. If any thing, the mountains are still more fantastic and varied than at Rio, and the bays and islets are perfectly picturesque. The passengers are full of pranks and jokes for an hour or so, and then they take a nap or read. I will venture to assert there never was before so much Bible-reading on board of a Brazilian vessel. On account of the warmth of the climate, each of these coast-steamers have, all around the upper deck, little cabins, or, more properly, respectable dog-houses, with a sliding-door. Although there are comfortable berths below, these upper apartments are the choicest to be had; for, night or day, you are always sure of fresh, pure air. My fellow-passengers were stretched around in these little cabins with the sliding-doors pushed back, and

VIEW OF PARANAGUA.

I thus had an opportunity of seeing them as I walked the deck. I was often called upon to explain the Scriptures, and rejoiced in the opportunity of scattering the seed, which, though sown in apparently unpropitious ground, the Master can cause to spring up an hundredfold.

"We arrived at Paranaguá on the Saturday morning after leaving Rio, and now I can say that I have been in the newest Brazilian province,—that of Paraná. The entrance of the bay is a perfect puzzle, and the mountains beyond the city are both lofty and picturesque. While the sun was streaming down upon the deck of our steamer, I took a rough sketch of a portion of the outer harbor, which I herewith enclose to you, premising the impossibility to do justice to this whole coast without the power of a Constable, a Turner, or a Calame.

"Paranaguá was formerly a celebrated rendezvous for scoundrels of all nations engaged in the slave-trade; and when the British Government, a few years ago, ordered its cruisers to make a vigorous demonstration on the Brazilian coast, the 'Cormorant,' of the Royal Navy, steamed up these sinuosities, entered the harbor, and cut out a whole nest of slavers. The fort was well situated near the bar, and H. B. M. 'Cormorant' must pass that point. After a slight resistance before yielding their vessels, the pirate captains and crews ran around by land to the fort and manned the guns, anxiously awaiting the 'Cormorant' as she should proceed to sea, dragging her trophies after her. Proudly she again ploughed through the winding approach to the ocean. The guns of the fort were well pointed, but H. B. M. 'Cormorant' proved to be as much of a sagacious fox as a rapacious bird, for, perceiving the trap laid for her, she prepared a most 'artful dodge.' Her crew very adroitly placed the largest slaver between herself (the man-of-war) and the fort, and then onward steamed the 'Cormorant.' Bang went the cannon of the fortress: the balls touched not the bird of prey; but, in the twinkling of an eye, she slipped beyond the slaver, discharged the heavy guns from her bows, and the dislodged cannon of the fort told how capital had been the aim of H. B. M.'s gunners. The slavers, however, prepared to respond; but the discreet 'Cormorant' cunningly retired behind the big vessel, though but for an instant. She sailed once more onward, and discharged her farewell shot with such telling effect upon the old fort that the inmates made no further attempt to hinder the 'Cormorant,' which soon gained the open sea, and in a few moments, by skilful scuttling, put the slave vessels beyond the reach of *o trafico*, as you know the Brazilians call the accursed slave-trade.

"Most of our passengers went ashore here, many of them bound for Curitiba, the capital of this new province. Their great kindness I shall not soon forget; and I am happy to think that they will carry the Bible, perhaps for the first time, where probably few have ever seen the records of salvation.

"I also went ashore. Paranaguá is a pretty and a clean town,—a little in decay I thought at first; but a second inspection told me that I had not done justice to the only port of Paraná. This town contains about three thousand inhabitants, and annually exports maté to the amount of one million of dollars. Maté is the dried leaves and young stems of a species of oak which is gathered in the interior and brought down in raw-hide cases, exceedingly tightly packed, and is hence shipped for the Spanish-American Republics.

"I found a number of large wholesale stores doing a good business with those who brought hither the products of the back-country. One of these merchants invited me to go to the house of his brother for the purpose of examining a map of the province, which I had in vain sought for in the metropolis, the boundaries not having as yet been definitely fixed. Fancy my feelings when, after threading a number of streets, I entered a house where a recent floor-scrubbing made every thing appear damp, and a large map was brought forth which seemed to have imbibed as much of humidity as possible without being wet; and, though it was perfect in every part save one, that part was just what I wished to see,—viz.: the boundary between Paraná and S. Paulo. Moisture, mildew, and mice had carefully eradicated every design of the engineer and every scratch of the engraver, so that I was left to return, mourning over the mutability of maps and the carelessness of man in Paranaguá.

"In one of the streets the ruins of a church attracted my attention; and I was informed that it was an edifice nearly completed by the Jesuits when they were expelled. You can scarcely travel a hundred miles along the Brazilian sea-coast (which stretches, with its bays and inlets, nearly four thousand miles) without encountering, in some rich valley or upon some picturesque emi nence, the immense churches, chapels, and convents of this order, whose members entered Brazil when its prosperity was at its height and when its ambition was hindered by no external circumstances. I have been more surprised at the hugeness of some of the conventual edifices in Brazil than at any thing of the kind I have ever seen in France, Germany, or Italy.

"As the little canoe in which we went from the steamer to the town neared the inner harbor, where vessels were moored close to the shore, I perceived two which looked remarkably desolate and forlorn. They were Russian vessels which were found near this port at the commencement of hostilities, and, fearing to be *nabbed* by some H.B.M. 'Bulldog,' 'Grabber,' or 'Jowler,' slid into this out-of-the-way place. It appears very singular to see these Northern birds of the ocean clipped of their wings *here*. They are truly out of place; for their yards are taken off, the topmasts are down, and, with their stiff hulks, awnings of canvas in the house-roof style, and with their general want of rigging, they seem like the 'Fury' and 'Hecla' in their Greenland clothes, or rather as if the winter-bound Bay of Archangel were their resting-place, and it and the surrounding shores were suddenly clad by the 'Hand divine' with the warmth and flowers and verdure of this perpetual-summer land.

"When, on my return, I reached the steamer, I found that a lady whose peculiar taste in dress had attracted the attention of all on board was attended by a number of 'spruce gentlemen' whose well-trimmed moustaches and highly-polished patent-leather shoes indicated that they belonged to a class very different from the poncho-clad passengers bound to Curitiba and the Sertões. It was not long before I ascertained that the lady in question was the 'bright particular star' of a theatrical company then travelling the provinces, and that the gentlemen were from the same establishment, they having arrived some days previous to their prima donna assoluta.

"The passengers who were destined for Santa Catharina remained that night upon the steamer; but the next day, (Sunday,) at an early hour, all left, with the exception of myself, to pass the hours of sacred time at Paranaguá, where a grand festa was to take place in honor of some saint. One of the greatest inducements was to attend the theatrical performances of the strolling actors, who were to give dignity and honor to the occasion by stupid and vulgar comedies. You will think, perhaps, 'What is the use of disseminating the word of God among such a people?' I will reply, 'Be not weary in well-doing;' and it is God's own word. My duty is to scatter it far and wide, to preach it by precept whenever I can, and by example always, and then leave the rest to Him. I have already found more than one notable instance in Brazil, where a Bible, left under circumstances just as untoward, has produced its fruits.

"I spent my day on board, but had very little quiet while the steamer was receiving her cargo of coals from the schooner along-side, from which—in some manner very unaccountable to the skipper—there were many tons short. I had all to myself, a large table well spread with viands; but, being of a social nature, I invited the engineer (a common-sense and wide-awake fellow of the Manchester machine-shop stripe) and the Brazilian second mate to join me. I find out from the Englishman that there are many of his countrymen and their children at the Saude, [a division of the municipality of Rio de Janeiro,] uncared-for either morally or intellectually. They are too far from the English church to attend service: but this plea of distance perhaps is only put forward to hide the real one of indifference. Now, can you not put something in train for them? They are workmen, and he says that both adults and children are not doing what they ought, one class running to cachaça and the other to ignorance, and 'Sunday is no Sunday.' Next year there are a thousand English and Irish laborers coming out for the Pedro Segundo Railway, and, on account of the distance and the pulpit-duties of Mr. ———, a clergyman, he cannot have facilities for attending to their minds or souls.

[In regard to the matter here referred to, some English ladies and an American theological student (then on a visit to Brazil) took it up, and interested both English and American merchants in the plan. They furnished the means, and, just as all was well organized, a competent man was found in an English mate, then on his homeward voyage from Australia, and intending to devote the remainder of his days to God in some other employment than that of following the ocean, and was persuaded to take charge of the new school, which in a short time was in full operation, and disseminating its ameliorating influences upon both parents and children.] This school in 1865 is still a great success.

"The next day (Monday) we left Paranaguá. After a fine run of eight hours along a coast abounding in repetitions of Corcovados and Peaks of Tijuca, we entered the safe Bay of San Francisco do Sul.

"Letters of introduction are great things in Brazil. They have smoothed the way for me everywhere previous to arriving at this port, and I here find no exception to the general rule expressed in the line above. Mr. V., the agent of the steamer, received me very kindly, and my boxes were soon despatched and landed upon the beach, which was filled with fishermen, mulatto women, half-naked children, and an indescribable lot of sundries in the shape of timber, rice spread out to dry, canoes drawn up, &c. &c. In another hour the steamer had rounded the promontory, and was soon out of sight on its way to Desterro. So, for the present, I will say,—Adeos."

Note for 1866.—The Saude School, referred to in this chapter, has been the means of great good at Rio; and, though its chief patroness, Mrs. Jane S. D. Garrett, has returned to England, thus leaving a void not easily filled, it is steadily accomplishing its good work. An acknowledgment for hospitality received is here due Mrs. Garrett, the recollection of whose home in the Larangeiras will long be one of the "pleasures of memory." The junior author, during his visits of 1862, '63, '64, and '65, received in Rio much kindness from the Swiss family of Mr. Gustave Lutz, and in the American homes of Mr. George N. Davis, Mr. Henry E. Milford, and Mr. John Hayes. It gives us pleasure also to recognize the courtesies of Admiral Tamandaré, Senhora Andrade e Pinto, (of the Rua St. Ignacio,) the Baron de Mauá, Sr. Militão Maximo de Souza, Mr. W. G. Ginty, and Mr. Bennett.

CHAPTER XVII.

THE PROVINCE OF PARANÁ—MESSAGE OF ITS FIRST PRESIDENT—MATÉ, OR PARA-
GUAY TEA—ITS CULTURE AND PREPARATION—GROWS IN NORTH CAROLINA—SAN
FRANCISCO DO SUL—EXPECTATIONS NOT FULFILLED—CANOE-VOYAGE—MY COM-
PANIONS NOT WHOLLY CARNIVOROUS—A TRAVELLED TRUNK—THE TOLLING-BELL
BIRD—ARRIVAL AT JOINVILLE—A NEW SETTLEMENT—CIRCULAR ON EMIGRATION
TO BRAZIL.

THE province of Paraná, whose chief port, Paranaguá, I had
just left, merits a still further mention. It commenced its full
provincial career about the year 1853, though for a number of
years previously projects had been entertained in the General
Assembly at Rio to set off the *comarca* of Curitiba from San Paulo
as a distinct province. As to its limits, they are essentially
those of the old district of Curitiba. Its first President, Zacarias
de Goes e Vasconcellos, was Minister of Marine in 1852–53. and is
one of the instances so frequent in Brazil of a young man who.
rising rapidly by his talents, attains the highest positions of State.
He is probably the youngest person ever called to take a seat in
the Imperial Cabinet, where by his eloquence and by his readiness
at response (for the ministers are interpellated as formerly in
France and as now in England) he rose to an eminent place among
the statesmen of Brazil.

In 1854, he opened for the first time the Provincial Assembly of
Paraná, and his Relatorios (messages) of that year and the following, now both before me, display ability and research.

He places the population at 62,000, only one-sixth of which is
composed of slaves; and, if his statistics be correct, the province of
Paraná must enjoy a salubrity beyond any other portion of the
world,—the births exceeding the deaths between two and three
hundred per cent.

He enforces upon the legislators the duty of making the common-school education far more obligatory than it is. "Primary

320

instruction," he urges, "is more than a mere right of the child, a duty discharged toward him; it is a rigorous obligation. It is thus that you (the representatives) should consider and dispose of the subject in the legislation of the new province.

"The people oblige themselves to be vaccinated. They respond to this without fail, for vaccination is a preservative from fatal pestilence.

"Now, primary instruction is, so to speak, a moral vaccine, which preserves the people from that worst of pestilences,—ignorance,— from those crude notions which bring man to the level of the brute, and which change him into the fit and facile instrument for robbery, assassination, revolution, and, in fine, for all evil.

"Primary education is more: it is a kind of baptism with which man is regenerated from the dark ignorance in which he is born, and truly effects his entrance into civil society and into the enjoyment of those rights and privileges which are his heritage."

When we consider what are the views of Roman Catholics in regard to baptism, we can see the force of the remarks of Senhor Zacarias.

The President does not merely confine his attention to the early training of the youth of his provincial charge, but his remarks in reference to the various branches of agriculture show him to be a man of enlarged views, and that he is as ready to combat indolence as ignorance. He alludes to the fact that wheat was formerly not only an article of cultivation in the fertile comarca of Curitiba, but that it was exported. This branch of agriculture is now almost abandoned, and, according to his statements, because a large portion of the population, eschewing the labor required in the production of the cereals, rush to the virgin forests, and there, stripping the evergreen leaves and the tender branches of the *Ilex Paraguayensis*, easily convert them into the popular South American beverage known as the *yerba maté* or *herva Paraguaya*, and thus amass fortunes or obtain a livelihood without the intervention of persevering industry or great exertion.

Large quantities of this kind of tea are annually exported from the province of Paraná. Senhor Zacarias would not have the tea-bearing Ilex uprooted to produce the same effect as the vigorous Marquis de Pombal brought about by the destruction, in the last

21

century, of the vineyards of Portugal; but he wishes to control its gathering, to moderate the inclinations and the causes that push the people into this branch of labor for a few months and then leave them indolent for the remainder of the year.

The *maté* of Paraguay, doubtless from prejudice, is considered superior in quality to that of Paraná; but the inhabitants of the interior neighboring Spanish provinces prefer the former to the latter, as they are accustomed to use the beverage without sugar; while in the cities of Buenos Ayres and Montevideo the former is the favorite, and is almost always sweetened before consumption.

In the interior of the province of San Paulo, after my visit to Santa Catharina, I met with an American physician, a man of great scientific tastes and acquirements, who has taken up his residence in South America for the purpose of research in his favorite study of botany. In the course of many interesting conversations with him in regard to the various vegetable riches and wonders of the surrounding regions, I was not a little pleased to find that he was perfectly acquainted with the mode of preparation, as well as the class and family, of the plant in question. Maté, as I have already mentioned, is the name of the prepared article of the tree or shrub which is commonly known to botanists as the *Ilex Paraguayensis*. It is classified by Von Martius as belonging to the *Rhamnée* family, and he gives it the scientific name of *Cassine Gongonha*. The Spaniards usually denominate it *Yerba de Paraguay*, or *maté*.

While in Paranaguá, I observed many raw-hide cases which the blacks were unloading from mules or conveying to the ships riding at anchor in the beautiful bay. Upon inquiry, I learned that these packages, weighing about one hundred and twenty pounds each, consisted of maté. This substance, so little known out of South America, forms truly the principal refreshing beverage of the Spanish Americans south of the Equator, and millions of dollars are annually expended in Buenos Ayres, Bolivia, Peru, and Chili in its consumption. This town of Paranaguá, containing about three thousand inhabitants, exports every year nearly a million of dollars' worth of maté.

In Brazil and in Paraguay it can be gathered during the whole year. Parties go into the forest, or places where it abounds, and break off the branches with the leaves. A process of kiln-drying is resorted to in the woods, and afterward the branches and leaves are transported to some rude mill, and there they are by water-power pounded in mortars.

The substance, after this operation, is almost a powder, though small stems denuded of their bark are always permitted to remain. By this simple process the maté is prepared for market. Its preparation for drinking is equally simple. A small quantity of the leaf, either with or without sugar, is placed in a common bowl, upon which cold water is poured. After standing a short time, boiling water is added, and it is at once ready for use. Americans who have visited Buenos Ayres or Montevideo may remember to have seen, on a fine summer evening, the denizens of that portion of the world engaged in sipping, through long tubes inserted into highly-ornamented cocoanut bowls, a liquid which, though not so palatable as iced juleps, is certainly far less harmful. These citizens of Montevideo and Buenos Ayres were enjoying with their *bombilhas* a refreshing draught of maté. It must be imbibed through a tube, on account of the particles of leaf and stem which float upon the surface of the liquid. This tube has a fine globular strainer at the end.

Great virtues are ascribed to this tea. It supplies the place of meat and drink. Indians who have been laboring at the oar all day feel immediately refreshed by a cup of the herb mixed simply with river-water. In Chili and Peru the people believe that they could not exist without it, and many persons take it every hour of the day. Its use was learned from the natives; but, having been adopted, it spread among the Spaniards and Portuguese, until the demand became so great as to render the herb of Paraguay almost as fatal to the Indians of this part of America as mines and pearl-fisheries had been elsewhere.

It grows wild, and never has been successfully cultivated, although attempts were made by the Jesuits of Paraguay to transplant it from the forests to their plantations. These attempts have been considered by many without result; still, there are others who consider that the experiment justifies further efforts, and are urging this day the *domestication*, so to speak, and the cultivation, of maté under a regular system.

But that which astonished me most in the doctor's conversation was the statement that a shrub similar to the *Ilex Paraguayensis* was indigenous to the United States, and that a decoction of its leaves and branches was actually used as a beverage in the region where it grew.

His life had been full of adventure in every portion of the globe; and, when he was a younger man, he roamed over each Southern and Western State, hunting for the weed which was vulgarly supposed to cause the "milk-sickness." Although he did not find the cause of that disease, which has so damaged many a speculation in Western towns and villages, yet he made the acquaintance of a little tree in North Carolina, from the leaves of which many of the country-people of the old North State "make tea." If I remember rightly, he informed me that it was the *Ilex euponia;* but scientific readers must not hold me responsible for this name, as my note-book may probably mislead me. A few years afterward, Dr. —— was in this, the most glorious field for a botanist in the world,—this Southern Brazil, whose magnificent *flora* has been the wild delight of every favored follower of Linnæus who has been permitted to enter it. In the course of his rambles he encountered the *Ilex Paraguayensis*, and immediately saluted it as his old acquaintance (under features but little different) of North Carolina. Some months elapsed, and he visited Paranaguá; and he was almost as much surprised at another discovery, which was not, however, in the botanical line. He found, in this out-of-the-way part of Brazil, an American woman engaged in the delightful art of preparing *feijões* and *toucinho* (pork and beans) for natives and foreigners who might patronize her establishment. In conversation with Dr. —— in regard to the maté, she exclaimed, "Why, doctor, this is the same *truck* we use in *Caroliner* to make tea." Here was a most striking confirmation of the true conclusion of science.

Now, if this tree or bush really abounds in North Carolina, why may not the enterprise of some of her citizens add to the exports (laid down in every geography as tar, tobacco, turpentine, and lumber) maté? Brazil and Paraguay are reaping their millions from a shrub which grows spontaneously, and the subject is really worth investigation in the United States.

Returning from the new province of Paraná, attention will be now directed to the province of Santa Catharina.

San Francisco is an ancient town which has evidently seen better days. The arrival of a stranger with such a peculiar cargo as mine created quite a sensation in the usually-stagnant society of this northern portion of the province of Santa Catharina. All the idlers, gossipers, men of business, and even the *Padre*, came to see the new books. The priest found no objection to them, and two hours had not elapsed before they were all disposed of, and I made my arrangements to ascend the river San Francisco do Sul to the German and French colonies founded on the lands once belonging to the Prince de Joinville.

In the mean time, with Mr. V. and two new acquaintances, both Germans, I strolled around the town, which is finely situated on an island separated from the mainland only by a very small stream. Before us stretched a bay three miles in width and six in length. It is well protected from the ocean, and in it is discharged the river San Francisco do Sul, which flows from the mountains that rear their green summits far in the distance. That lofty ridge, in its highest elevation, is more than four thousand feet above the level of the sea, and from its inland base to the rich plain where Curitiba is situated there is a gradual ascent of twenty miles. With an energetic people, this district—which in regard to fertility and climate is one of the finest in the world—would bloom with a cultivation not surpassed by the rich fields of Lombardy or the model farms of Midlothian.

Great hopes were entertained at the beginning of this century that San Francisco do Sul would become a flourishing mart, on account of the road which would open the high plains to the commerce of the bay. Furthermore, there was great activity at that time, the chief occupation of the inhabitants consisting in ship-building and in the cutting of timber. Vessels of large dimensions were formerly built here, as well as coasters, at the order of merchants from Rio, Bahia, and Pernambuco. The wood used was so strong, holding the iron so firmly, that ships built of it were of the most durable quality, and were in greater esteem with the Portuguese and Spaniards than those built in Europe. In 1808, Mr. Mawe, one of the earliest English voyagers in Brazil, wrote that, on account of its ship-building, "the harbor of San Francisco do Sul is likely to become of considerable value to Brazil; and if it be connected with Curitiba, the cattle of which have been found superior to those of Rio Grande, there is every probability that at no distant day the Portuguese navy will touch here to be supplied with salt provisions."

As I looked upon the silent streets of San Francisco,—as I beheld its bay innocent of any vessel except the smallest coasters, and its once-busy shipyards containing but two small mandioca sloops upon the stocks,—I thought how wide a difference there was between the reality of the present and the speculations of half a century ago in regard to the commercial activity and future growth of the town, situated upon the waters of *Babitonga*, by which name the natives called the bay. It was thought that the establishment of a colony of Europeans in the vicinity of the decaying town would resuscitate it; but thus far there has been no such result, and I fear that many a year will elapse before this can be accomplished.

I determined to start for the colony at an early hour the next morning, and to this end Mr. V. kindly sought for a canoe belonging to a gigantic slave who rejoiced in the appropriate name of José Grande. After nightfall the African made his appearance, and it was settled that we should commence our trip at three and a half o'clock in the morning.

Mr. V. regretted that the circumstance of his boarding prevented his offering me his hospitality, but recommended me to a hotel, or, more properly speaking, a regular country-inn, which had just been opened by a German from the colony of Donna Francisca. My experience in that establishment was at the time detailed in a letter to a friend at Rio:—

"Herr Sneider, mine host, and all his family, spoke scarcely any thing but German, and as much of English and Portuguese as can be compressed into 'yes' and 'Sim, Senhor.' By-the-way, I have picked up a certain quantum of that same jaw-breaking language of Goethe and Schiller, which I have neglected since my university days for the tongues of Southern Europe. My supper was perfectly German; for it closed with beer, which, in default of barley, had been made from rice, that abounds in this vicinity.

Having finished my repast, I gave orders that, as they had prepared supper enough for three men, the remainder should be arranged for my breakfast in the canoe, as it would be entirely too early to partake of that meal before embarking.

"We then had a mutual-instruction society,—an exchange of English and German. How many children there were I cannot say; but there was any quantity of blooming fresh frauleins from nineteen years and downward, together with a number of healthy, rosy boys. It had been so long since I had looked upon blue-eyed and fair-haired children that they were quite a curiosity. Having occasion to see Mr. V. before retiring, I said to them, 'I go now to Mr. V.'s: when I return, I wish to have a large room and a good clean bed.' A patron of the inn informed me that I should be thus accommodated in every particular.

"When I again entered Herr Sneider's, I was told that my room was ready, and, upon my signifying my intention to go to bed, the whole family,—Herr S., Frau S., Frauleins S., and the boys,—to my astonishment, followed me to the apartment, which proceeding I did not fancy, because it did not seem quite *convenable*, taking into view the feminine portion of the procession. I, however, concluded to be led to my quarters, of which I entertained the highest expectations. These expectations were realized so far as the size of the chamber was concerned; but, unfortunately, mine was not the only bed in it, for there were four or five others, filled with snoring occupants. I determined to be gracious and make no complaint, for assuredly my clean sheets would make up for a little too much of society. So, pulling down the supposed coverlet, I found that it was a feather-bed for a regular Prussian winter. These Germans, when they left Fatherland, could conceive of no country where winter and snow could not even be exotic. I discovered also that, instead of the good, healthy, and hard Brazilian mattress, there was a second huge feather-bed; and I must thrust myself between these. When my eyes got beyond the first, I found my clean sheets to be of the color of the dirty Minas cotton which so plentifully (or scantily, as the case may be) clothes the slaves throughout the Empire. A closer inspection informed me that they had seen whiter days, and had also made the acquaintance of many other lodgers, which fact I roundly asserted, and to which they partly assented. I, however, resolved to make the best of it, when they would let me,—for they hung around as if they would never give me the opportunity of going to rest. A young German ship-chandler had his bed in the same room, and, without ceremony, commenced to divest himself before the company preparatory to sleep. This I could hardly do, and seated myself and began to read. Finally the family left me, with many *schlafen Sie wohl.* Having read as long as I wished, I determined to enter my bed, fortified with a pair of pantaloons, (I had not forgotten the sheets,) which after a time, proving rather uncomfortable with feather-beds, I threw to one side. But this operation caused the young ship-chandler much concern; for, hearing me moving around in the dark, and supposing me ill, he screamed for the family, and the scene which ensued is indescribable with pen: only the pencil of Rembrandt could depict the depth of shadow and the rich chiaro-oscuro, and that of Teniers the ruddy, jolly features of the group of young Germans thus aroused to see what was the matter with the American, who by this time was snugly ensconced in his bed and almost bursting with laughter.

"I slept badly, and at half-past three o'clock heard the ponderous step of José Grande. Following him through the deep gloom that hung around, we (for I had given a bright German lad permission to go with me) entered the canoe, which was soon shoved from the shore, and were propelled by José toward Donna Francisca. Young Germany and myself lay down in the bottom of the narrow 'dug-out.'

"The morning was dark and drizzly, and a feeling of loneliness crept over me as I lay listening to the pattering raindrops and the dripping oar disturbing the oppressive silence. I thought of those so dear to me, but who now were separated from me by thousands of miles of ocean; but I was less lonely when I breathed a prayer for them and felt in my heart the ever-cheering sentiment of poor Pringle:—

> "'A still small voice comes through the wild,
> (Like a father consoling his fretful child,)
> Which banishes bitterness, wrath, and fear,—
> Saying, "Man is distant, but God is near!"'

"I tried to sleep, but it was impossible; so, after three hours, I said to José, 'We will breakfast.' On opening the budget, I found two plates, four pieces of meat, and—nothing else,—not even a knife and fork; but, as I am neither a lion, a vulture, nor even a Guacho of Corrientes, I could not breakfast on flesh alone. The rain had now ceased, and I proposed to José to land and to purchase something from one of the farm-houses on shore. *'Não tem nada, senhor,'* ('They have nothing,') was José's sage reply. Nevertheless, at my request, he put into a pretty cove at the foot of a mountain, and sallied forth for a bargain. He soon returned, accompanied by a sickly-looking boy, bringing oranges, bananas, and enough farinha for four men. Young Germany and myself fell to work while José's strong arm was sending us over the glassy waters. At Rio de Janeiro I had often looked with admiration upon the slaves in the boats stuffing and throwing farinha into their mouths; but I never then dreamed that I should employ my digits for the same purposes. I must admit, however, that there was neither gracefulness nor dexterity on my part; for my face became powdered with the effort to 'pitch in' the farinha *à la Brazilienne.* We had one other *compagnon de voyage,* but not an eating one. Faithful old trunk! What sketches thou mightest give of Europe, America, (North and South,) and of the African Isles!—what scenes thou hast witnessed in three zones, on the Atlantic and the Pacific Oceans, in the Straits of Majellan and on the Isthmus of Panama, in the Mexican Gulf, and, lastly, on the Rio San Francisco do Sul! Each time that I open thee, and see there imprinted 'W. S. Chase, trunk and harness maker, Providence, R.I.,' my thoughts run over the past, and I recall the bright summer-day that I bought thee, when on the eve of my first voyage 'over the seas and far away.' Thou callest up a host of memories,—

> 'the fond recollections of former years,—
> And the shadows of things that have long since fled
> Flit over the brain like the ghosts of the dead.'

"Speaking of sketches, I send you one which I took of myself and fellow-voyagers. They are after (a very long way, indeed) a compound of Gainsborough and Turner, with a slight addition of Wilkie and Kenny Meadows thrown in."

The river became narrower, and every moment some large
aquatic bird would be startled by our voices or by the dash of the
oar. Now it would be a beautiful white ibis, then a blue heron or
a band of dancing cranes. From the mangrove-bushes and the

ASCENDING THE RIO S. FRANCISCO DO SUL.

more distant woods we could hear the sometimes harsh and sometimes musically-solemn sound of the uruponga, or tolling-bell bird,
making the air resonant with its peculiar and solitary note. I had
listened again and again to these birds in my journeys in different
parts of Brazil, but I never had the good fortune to see but one,
and that was in the province of San Paulo. The sound which the
uruponga (what a sweet aboriginal onomatope!) sends forth varies
little, but it can always be said to be *metallic*. To hear it from
afar, it is not unlike the tolling of a bell; but, when distance does
not mellow the cadence, it is more like striking an anvil or the
filing of a large piece of iron. To listen to it in a Brazilian forest
at mid-day, ringing forth its mournful knell when every other
songster is mute, powerfully disposes one

         " To musing and dark melancholy."

Wallace says, in his account of the Amazonian regions, "We
had the good fortune one day to fall in with a small flock of

the rare and curious bell-bird, (*Casmarhynchos carunculata,*) but they were on a very thick, lofty tree, and took flight before we could get a shot at them. Though it was about four miles off in the forest, we went again the next day, and found them feeding on the same tree, but had no better success. On the third day we went to the same spot, but from that time saw them no more.

The bird is of a pure white color, the size of a blackbird, has a broad bill, and feeds on fruits. From the base of the bill above grows a fleshy tubercle, two or three inches long and as thick as a quill, sparingly clothed with minute feathers: it is quite lax, and hangs down on one side of the bird's head. The bird is remarkable for its loud, clear, ringing note,—like a bell,— which it utters at mid-day, when most other birds are silent."

URUPONGA, OR TOLLING-BELL BIRD.

Waterton, in his wanderings in Demerara, often alludes to the campanero, (uruponga.) In one passage he says, "It never fails to

attract the attention of the passenger: at a distance of nearly three miles you may hear this snow-white bird tolling every four or five minutes, like the distant convent-bell. From six to nine A.M. the forests resound with the mingled strains of the feathered race; after this they gradually die away. From eleven to three all nature is hushed in midnight silence, and scarce a note is heard saving that of the campanero."

No bird has been more misrepresented by artists than the uruponga. The mistake has been in copying stuffed specimens. The accompanying illustration is one of many that represents the uruponga with a stiff horn in the unicorn style. The body is well enough, but the rhinoceros-appendage is utterly at variance with nature. The little engraving is a correct likeness of this singular bird, whose small, flexible, and drooping appendage is very similar to that which is a part and parcel of every turkeycock.

I was struck by the fact that, though the aquatic birds were at first startled by us, they did not seem to have much fear. They flapped their great wings and moved slowly from us a few paces, and then speedily resumed their former position.

On, on sped our canoe under the sturdy strokes of José. The scenery was still more striking and beautiful. A background of high mountains was prefaced by gentle eminences and by a woody margin of bright-green trees. Even the tall African, whom no one would have suspected of a taste for these glorious views, exclaimed, from time to time, "*É muito bonito, senhor!*" ("It is very beautiful, sir.") By the way, José gave me his idea of Protestants, —viz.: people who were not baptized, and were destined to *inferno.*

After some hours' rowing, the river became exceedingly narrow, so that the trees, with their rich parasites, completely overarched us. This was near the new village of Joinville, in the colony of Donna Francisca. We jumped ashore, tied our canoe to the stump of a recently-fallen tree, and tramped over—or, rather, through—a road which was like a sponge soaked with water. Here, indeed, was the beginning of a new town in the wilderness,—houses stuck down in the woods, and plenty of mud and children: but for the difference of the *flora,* I would have believed myself beyond the Missouri, on the borders of Kansas. On every side the forest was to be seen, and here and there an opening, in the centre of which was the cabin of the colonist. The smallness and newness of the houses, the deadened trees, the muddy streets, and the general appearance of every thing, reminded me of a pioneer settlement in the West. It was curious to see men from the Rhine, and some from the environs of Berlin, here *planted* amid wild woods, in cottages of the rudest construction, thatched with palm-leaves.

The "Hotel" of Herr Palma was my goal, and a hearty welcome

awaited me; for the letters of Mr. V., in addition to the prospect of gain from the stranger, prompted it. The German cannot forget his native land; and one glance showed me that, though hard work must necessarily be the morning, noon, and night *regime* of the colonist in these woods, yet here were all the appliances for amusement,—a ballroom, a gallery for the orchestra, and a ten-pin alley. Mine host sent immediately for the schoolmaster, so that I might receive every mark of honor and distinguished village-consideration.

Note for 1866.—In regard to emigration from the South of the United States, I here insert the circular issued by Sr. Galvão, the official agent at Rio, and countersigned by the Brazilian Consul-General at New York:—

"EMIGRATION TO BRAZIL.

"The Imperial Government looks with sympathy and interest on American emigration to Brazil, and is resolved to give it the most favorable consideration. Emigrants will find an abundance of fertile land, suitable for the culture of cotton, sugar-cane, coffee, tobacco, rice, etc. These lands are situated in the provinces of Rio Grande do Sul, Santa Catharina, Paraná, São Paulo, Espirito Santo, and Rio de Janeiro; and each emigrant may select his own lands. As soon as the emigrant has chosen his land, it will be measured by the Government, and possession given on payment of the price stipulated. Unoccupied lands will be sold at the rate of 23, 46, 70, or 90 cents per acre, to be paid before taking possession, or sold for terms of five years, the emigrants paying six per cent. interest yearly, and receiving the title of property only after having paid for the land sold. The laws in force grant many favors to emigrants, such as exemption from import duties on all objects of personal use, implements of trade, and agricultural implements and machinery. Emigrants will enjoy under the Constitution of the Empire all civil rights and liberties which belong to native-born Brazilians. They will enjoy liberty of conscience in religious matters, and will not be persecuted for their religious belief. Emigrants may become naturalized citizens after two years' residence in the Empire, and will be exempt from all military duties except the National Guard (militia) in the municipality. No slaves can be imported into Brazil from any country whatever. Emigration of agriculturists and mechanics is particularly. desired. Good engineers are in demand in the Empire. Some railroads are in construction and others in project: besides there are many roads to be built and rivers to be navigated. On sale, at the disposal of emigrants, lands of the best qualities, belonging to private persons. These lands, varying in price from $1.40 to $7.00 per acre, are suitable for the growth of coffee, sugar-cane, cotton, tobacco, rice, Indian corn, etc., and may be obtained in every condition, from virgin forest to that in a complete state of cultivation."

CHAPTER XVIII.

COLONIA DONNA FRANCISCA—THE SCHOOL-TEACHER—THE CLERGYMAN—A TURK—
BIBLE-DISTRIBUTION—SUSPECTED—A B C—THE FALLEN FOREST—THE HOUSE OF
THE DIRECTOR—A RUNAWAY—THE VILLAGE CEMETERY—MORAL WANTS—
ORCHIDACEOUS PLANTS—CHARLATANISM—SAN FRANCISCO JAIL—THE BURIAL OF
THE INNOCENT, AND THE MONEY-MAKING PADRE—THE PROVINCE OF STA. CATHA-
RINA—DESTERRO—BEAUTIFUL SCENERY—SHELLS AND BUTTERFLIES—COAL-MINES
—PROVINCE OF RIO GRANDE DO SUL—HERDS AND HERDSMEN—THE LASSO—
INDIANS—FORMER PROVINCIAL REVOLTS—PRESENT TRANQUILLITY ASSURED BY
THE OVERTHROW OF ROSAS, AND OF THE PARAGUAYAN LOPEZ JR.

THE Colonia Donna Francisca is a new enterprise, whose origin may be stated in a few words. In 1843, Prince de Joinville married Donna Francisca, the sister of the Emperor of Brazil. With her hand he received, as a dower, a large forest-estate in the province of Santa Catharina. A few years ago, at some of the watering-places of Germany, the Prince met with Senator Schrœder, of Hamburg, who proposed to him a plan for making his dower profitable,—viz.: to grant a certain portion of land to a company, who should form a colony upon it. The Prince granted nine square leagues, reserving a certain number of acres for himself in the most desirable situations. The company was formed, and agreed to bring out some sixteen hundred colonists within a given time. From March, 1851, to March, 1855, the number, according to contract, had arrived. The greater portion of the colonists are from German Switzerland, though France and Germany are represented by a respectable minority. The village of Joinville contains about sixty houses; in the surrounding country there are one hundred and twenty buildings, and others in construction. After deducting deaths, there are something like fifteen hundred inhabitants in this colony; while there are a considerable number of French, and French Swiss, in an adjoining colony founded by Prince de Joinville

334

on his own lands. Two-thirds of all the colonists are doubtless Protestants, while the other third are Romanists.

What will be the success of the colony remains to be seen. The colonists, with few exceptions, are not of the first class who seek the New World; and doubtless the company, wishing to fulfil their contract as to numbers, were not by any means careful in the selection of the emigrants. They are obliged to pay for their land, which is much dearer than in the United States, and, having the thick forests to fell, are soon out of funds. Their distance from any market, and the impossibility of obtaining remunerating crops until the hard labors of the pioneer are performed in the unbroken wild wood, operate powerfully against all but the most courageous hearts. With lands, however, (which the company has now obtained,) away from the low district bordering the river, the prospect will be brighter. I am nevertheless convinced that the best means of colonizing Brazil is not by private speculation in village-lots and farming-grounds.

Herr Palma returned, accompanied by the school-teacher. The latter was a dandified-looking gentleman, dressed in the latest Parisian fashion, but withal a person not wanting in ability or in acquirements; for at his rooms I found chemical apparatus, with which he was constantly experimenting, and I also ascertained that he was an engineer and an artist of no ordinary merit. He offered his services to go with me to the Lutheran clergyman, and to be at my disposition generally. To the clergyman I had no letters. In a few moments I was at his house, which was most scantily furnished: indeed, I have rarely seen in the backwoods of the United States a minister surrounded with so little comfort, or so few of the necessaries of life. He spoke neither French nor Portuguese, and his stock of English exceeded very little my stock of German; so that I had great difficulty in making him comprehend my mission. I attempted to be more explicit through the teacher, to whom I spoke in French, which he translated into German. Still he did not seem to comprehend, and I left his house feeling somewhat discouraged at my reception, especially when I contrasted it with the warm co-operation which I had received from the Lutheran clergyman at Petropolis.

In the mean time a rumor ran through the village that a

stranger with Bibles had arrived, and when I returned to the little inn I had as much as I could do to attend to the visitors. Among them was an accomplished and refined lady, the daughter of an LL.D. of Hamburg, and wife of the head-director of Prince de Joinville's colony, which must not be confounded with the Hamburg colony in Joinville. My German Bibles and Portuguese Testaments were soon exhausted, but I had some still left at San Francisco, for which they paid me the money, and I sent them the next day after my return.

The clergyman now joined us. He was a little more cordial this time. I invited him and the school-teacher to take tea with me. During the repast, the latter left us a few moments, and then returned; but while he was absent, the clergyman said to me, "How did you become acquainted with the teacher? He is a *turncoat.*" I then understood his reserve, and non-comprehension of my remarks which I had made in the presence of the pedagogue at the parsonage. The teacher was born in Bulgaria,—was a Mohammedan: he afterward went to Germany, and finally came to Brazil with some Belgian *savants* whose object was scientific exploration. The young man became attached to a Brazilian girl twelve years of age, renounced his religion, became a Romanist, and married her. I could still further appreciate the cautious movements of the clergyman, when he informed me that he himself was a Bohemian by birth, was educated in Vienna, and was the means of turning some seventy Papists to Protestantism, and on this account he was expelled from Austria. Although I received the kindest of treatment from the schoolmaster, truth compels me to say that among the people of the village he has the reputation of being Roman Catholic only in *theory*, for in *practice* he was as much of a Turk as if he resided in the heart of the Ottoman Empire.

The company around me was a mixed one, some being Romanists, others Protestants. In the course of the evening an honest-looking Bernese Swiss came into the room. I saluted him, and spoke of the Bible, but observed that he viewed me with a cautious eye. Soon I saw him and the pastor go out together. They returned in a few minutes; and a short time after the Bernese took me aside and said, "I am convinced that you have a good object in view. I was afraid you were a Jesuit," (he had not forgotten the Sonder-bund in his own country;) "but the pastor assures me that you are not. I wish to do good. I once hoped to be a missionary, but early circumstances prevented, and therefore I must be content to work through others: so please accept this small sum of money, and all that I wish you to do is to spread the good news of the blessed Saviour." After he went away, the pastor handed me another small sum, which the same Bernese had given him for me. The total was only nine francs; but that sum is equal to one hundred francs in the United States. I afterward sent him, from San

A GERMAN EMIGRANT'S CABIN AT DONNA FRANCISCA.

Francisco do Sul, sufficient Bibles in return for his gift, and hope that he will thus be more immediately made the instrument of spreading "the good news of the blessed Saviour."

It was late when my visitors retired. The next morning, at

22

an early hour, mounted upon a wild-looking horse, and dashing through mud and mire, I went to breakfast with the director of the Hamburgese (the Joinville, not the Prince's) colony. As I rode along, I saw on either hand the small cottages of the colonists, (distinguished from Brazilian houses· by their chimneys,) reared amid the overshadowing, broad-leafed banana-trees, in this land of no winter. But they have a hard lot, for the forest-land is difficult to clear; the soil is not so rich for cereals and other productions which they have been accustomed to cultivate, and, above all, the people are poor, and, many of them being from the lowest classes in Germany, quite a number give themselves up to drink. It was on this latter account that the pastor solicited German temperance-tracts.

As I passed one house, in the midst of hundreds of palms and other magnificent trees, I heard the sweet sound of a mother teaching her little one to lisp its A B C.

It was a new sight for me to behold the primeval forest of the tropics being prostrated under the fell swoop of the woodman's axe. On every side, noble palms and rare and gigantic parasites were hurled in wild confusion to the ground. Near the house of Mr. H., I saw one of these wood-kings lifting his solitary head amid his fallen companions. The monarch was crowned and festooned with magnificent orchidæ and clambering wild vines. His own bright-green foliage spoke of life and vigor; but the dripping dew-drops seemed like falling tears mourning the desolation around. But, to make this world a fit habitation for man, nature, as well as man, must make her sacrifices: so utility reconciled me.

The little long-tailed birds (closely resembling the whidah-birds of Africa) that I had often seen pining in cages were here in glorious freedom, playing before me, gracefully floating from fern to fern, or swinging in fearless glee upon the pendent parasitic vanilla which loaded the morning air with its rich perfume.

The house of Mr. H. was prettily situated, and, in this remote corner of the world, it was as interesting as it was strange to con over, in his little parlor, the last London "Illustrated News," "La Presse," and the Paris "Illustration." Madame H., from La Belle France, demonstrated that others besides American women could enter the backwoods and undergo with contentment the hardships and the excitements of a pioneer life.

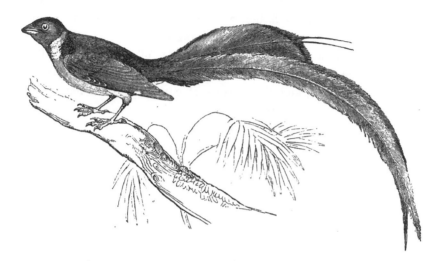

When Mr. H. and myself were ready to return to the village, our horses were brought to the door; but mine had the bad taste to break his halter, and, snorting a loud adieu, away he went, careering along the road toward Joinville. His free movement, crested mane, and distended nostril, made him look for all the world like one of the steeds on the Elgin marbles; only he was *minus* his rider. As he disappeared from sight, he flung his heels high in the air, and gave a series of farewell kicks and other antics which were enough to provoke laughter from even brooding melancholy. Mr. H. kindly furnished me with another horse, and the last that I saw of my steed was just as we reached Joinville. He had entered a small sugar-plantation, and was enjoying a most delightful repast of the tender young cane.

Before entering the village, we turned aside from the road, ascended a forest-crowned hill, upon whose sides was the rural cemetery where were buried the colonists of the Hamburg settlement. It was a sad yet beautiful spot. The morning sun had risen high above the forests, yet the dense foliage was still sparkling with matinal freshness. Each day and each year the sun will shine upon that remote little cemetery; but those who there sleep will never again behold the morning glories of this bright land. The earth was yet fresh that covered the remains of one of the finest men of the colony : a few wreaths *immortelles* had been hung with rustic taste by some kindly hand near the humble grave; but no father or mother or gentle sister would ever shed the silent tear over the sleeping dead.

From the same hill we had a fine view of the village. The living and the dead are thus brought near each other; but man is a forgetful creature, and the lessons of cemeteries and new-made graves are as easily forgotten in this retired nook as amid the busy hum of the vast city.

Before leaving the colony, I visited the school, which is sustained by the common-school fund of the province, and I found that the Bulgarian had not been neglectful of his little charge, which he instructed in both German and Portuguese.

In wandering through Joinville, I called upon a colonist who has a brother in New York, and, while in his house, a gentle-manly-looking man entered. By his conversation I ascertained that he was a physician. So soon as he knew who I was, and in what capacity I had visited the colony, he took me warmly by the hand, and I learned that he was one of those physicians who care for the souls as well as for the bodies of their patients. My inter-course with him was very pleasant; for, in addition to his piety, I found him a gentleman of cultivated mind, having been educated at the University of Halle; and that which particularly interested me was that he had, apart from his professional studies, attended the lectures of Tholuck.

He, as well as the Lutheran clergyman, highly approved of the proposition of another German pastor in the Empire, which is to have an ordained missionary colporteur to go from colony to colony throughout Brazil, with Bibles and tracts, encouraging such communities as have pastors; by the printed Word and reli-gious works rallying those who are without a clergyman; and performing the rites of marriage where, for want of a minister, this—so essential to the purity of a community—has been to a great extent neglected.

There are German colonies scattered here and there throughout the whole length of the Brazilian sea-coast, and there is, from the nature of the case, a loud call upon the evangelical Germans of our land to care for the spiritual welfare of their countrymen in Brazil. I believe that such a work, carried on by a few of the Lutheran churches of the United States, would redound in great good. They could thus direct the operations of the man who should be called to this labor better than a large benevolent society that has fifty other lands in view. Such an enterprise is of the most imperious necessity, not only for keeping alive evangelical piety, but the knowledge of Protestant Christianity.

On returning to the hotel, I found that a large basket of orchidaceous plants of the rarest species had been prepared according to my order, which I sent as a present to a kind friend at Rio de Janeiro. The lot, with the basket, cost but three dollars: in England they would have brought a fabulous price, considering the rage that now exists among royal and noble horticulturalists for these curious subjects of Flora's kingdom. They can be easily transported over the ocean, if care be taken that all contact with salt water be avoided. I found that there was a naturalist not far from Rio who often sent orchidæ to England. Brazil is exceedingly rich in parasites and air-plants; but none among the vast variety is more graceful than the vanilla, which is found in greater or less abundance from the northern limit of the Empire to the province of St. Catharine's. Its little star-like flower, its pretty leaf, and its delicious fragrance,

THE VANILLA.

make it an object of beauty and of admiration. I, however, could never understand why the vanilla-bean should be imported into

Rio from Mexico and Central America *viá* New York, when the plant itself abounded in Brazil.

I left the colony with sincere regret that I could not remain longer and see more of the people; but, according to the announcement, the steamer which was to take me back to Santos was to arrive the next morning. So I bade farewell to my newly-made friends, and, after several hours' hard rowing in the cramped-up, narrow canoe, arrived at San Francisco do Sul.

The steam-packet was not in the harbor on the appointed day, and I passed the time very agreeably with Mr. V. and a number of Germans, one of whom was a young physician educated at Breslau, but was about to retire in disgust from the colony and from Brazil. He was certainly more adapted to a formed than to a forming society. He alleged, as his principal reason, that Brazil was a great field for charlatanism; that pretenders and quacks could always succeed better than the regular scientifically educated. He instanced the case of a barber of the Schleswig-Holstein army, who emigrated to the new province of Paraná and is now the physician in highest repute in that region. I was further informed that this *ci-devant* knight of the razor had recently appeared in the theatre at Paranaguá with a decoration bespangling his breast, pretending that it was conferred in Europe for his distinguished surgical services! My Breslau friend was evidently a cultivated man, and well read in his profession, but home-sickness was doubtless the disease that made him look at every thing with distorted vision; for I doubt if there can be found on the Western Continent a country where the Government and the medical faculty are more strict than in Brazil. There are successful charlatans under the very eyes of the medical schools in Paris, and it is not therefore strange that examples occur in a vast, thinly-populated country.

Often, leaving my companions, I would stray alone into the foliaged walks which are found on every side, and there I could be as retired as if a thousand miles from the haunts of man. A favorite p ace was the ruins of an old convent on the summit of a vine-clad hill, near which were the new foundations of an hospital erected as an expiatory offering by some rich lady of San Francisco: she having died, her pious work, I fear, will soon be in the same condition as that of the Jesuits.

In one of my rambles I paid a visit to the jail, the only occupant of which was a German who, in a fit of anger, had struck the director of the Hamburg colony. Now, it is perfectly allowable in Brazil to call a man very hard names and cheat him as much as you please with impunity; but to strike a man is beyond all bounds of decency, and the jail or some other punishment is sure to follow. The prisoner seemed very happy under the circumstances, having a finer room than that which I occupied at Herr Sneider's, and perfect freedom to go where he pleased at certain hours of the day.

From the jail I entered the large church, situated near the centre of the village. The floor was so constructed of wood that it could be lifted up in sections, which was always done when interments took place. Here for nearly two centuries people had been buried who died with the fond hope of being brought nearer to heaven by having their bodies within these precincts made by man's hands. An old negro was digging a grave, and each time his heavy hoe (the spade is rarely used) went down, it ruthlessly crunched and smashed through skulls and ribs and whatever else is fragile in our poor human frame. The fragments were pitched up as common clay.

I was disturbed in my meditations of this scene by the fat, jolly, round *padre*, who, with a giggling face, gave orders, in a loud and any thing but solemn voice, to an assistant who was bearing a coffin to the centre of the church. It was a small coffin, yet it was large enough. It was uncovered, and in it lay, in the slumber of death, a little girl of twelve months. A sweet smile was upon her features; her tiny hands were clasped together, and her eyes were open and beaming with such a lovely expression that they seemed to be gazing into heaven. The tinsel and the ornaments with which the body was bedecked I scarcely saw. Three women, clad in deep mourning, and with mantillas of richest broadcloth trailing from their heads to the ground, swept noiselessly through the church, giving one lingering look at the innocent dead. The priest approached and saluted me. I had seen him upon my arrival, and made bold to make a few inquiries in regard to the child. He informed me that he was just preparing to say mass for it: I, however, took up the words of our Saviour, and said, "Of such is the kingdom of heaven," and that the little one redeemed by the Saviour was already an angel in the realms of light, and that there was no need of saying mass for such, even waiving the question of right to say mass for any one. He replied with an *é verdade, senhor*, but, notwithstanding, went on to his work,—because he made by it money,—because the church is corrupt, and man seeks out new inventions rather than follow the plain precepts of truth.

After speaking with him against intermural burials, I espied a pulpit, and asked him if he preached: he answered, "Sometimes, especially at the festas." To all my remarks on preaching the righteousness of Christ only, he bowed, grinned, uttered many *é verdades* and *muito obrigados*, (it is very true; I am much obliged to you;) and I left, profoundly convinced that a moral earthquake will be necessary to shake off the indifference of the Brazilian priesthood before their minds will be directed aright.

The steamer entered the bay, and I turned my face northward.

The province of St. Catharine, in which the colony of Donna Francisca is situated, is the smallest in the southern part of the Empire. In fertility and salubrity it is second to none. Its resources, however, have been developed only fifty or sixty miles from the coast: beyond this, the aborigines still abound, and farther in the interior they are warlike, and cherish a deadly hatred to the white man. Yet I would not convey, through this statement, the impression that the province is a howling wilderness; for the towns on the sea-coast, the villages, and the flourishing small plantations, more remote from the littoral, and the numerous colonies founded by the Imperial and provincial governments, by private companies and by single individuals, on the belt of land stretching from the Rio San Francisco do Sul to the Mampituba, all speak of a certain amount of civilization and progress. The population is estimated at ninety thousand.

The capital of the province is often called Santa Catharina, though its proper and full name is *Nossa Senhora do Desterro*, which may be translated either "Our Lady of the Desert" or of "Banishment." It is situated upon the island which gives the name to the province, and its harbor, though small, is compared with that of Rio de

Janeiro for excellence and beauty. Desterro is the seat of a considerable trade; yet the planters are not engaged in grand agricultural operations, as in the provinces farther north. The coffee exported thence enjoys a high reputation, and is of a superior quality.

The island of Santa Catharina is mountainous and finely wooded, and the scenery with which the city of Desterro is surrounded has been the admiration of every traveller who has been privileged to visit this picturesque region. A friend who resided many years ago in the islands of the Pacific, on visiting St. Catharine's wrote home his impressions, stating that the general aspect of all around him was so like the South Seas that he felt as if he were suddenly transported thither and were again amid the scenes of bygone years. He added, "The palm-tree tossing its plumed branches in the wind, the broad leaves of the banana rustling in the breeze, the perfume of the orange-blossoms and Cape jessamine, the sugar-cane, the coffee-plant and cotton-bush, the palma Christi and guava, the light canoe upon the water, and the rude huts dotting the shore,—all hurried me in imagination to the Marquesas, the Society, and the Sandwich Islands."

There is a commerce here in artificial flowers made from beetles' wings, fish-scales, sea-shells, and feathers, which attract the attention of every visitor. These are made by the *mulheres* (women) of almost every class, and thus they obtain not only pin-money, but some amass wealth in the traffic. The wreaths, necklaces, and bracelets made from the scales of a large fish are not only curious, but are exceedingly beautiful. Their effect at night is that of the most brilliant set of pearls, and they are as much superior in splendor to the small specimens of fish-scale flowers manufactured in Ireland, and exposed in the Sydenham Palace, London, as the diamond surpasses the glisten of cut-glass.

Not only tropic fruits and flowers are here to be found in profusion, but the choicest horticultural productions of Europe can be cultivated to perfection; and such is the salubrity of the air, that Desterro is often visited by invalids from the more northern provinces, and even from more distant countries.

The natural history of Santa Catharina is peculiarly interesting. Among the shells abounding on the coast there is a species of *Murex*, from the animal of which a beautiful crimson color may be extracted. It is, however, the department of entomology which has excited the most lively admiration of the naturalists who have visited the province. The butterflies are the most splendid in the world. Langsdorff says they are not like the tame and puny lepidopters of Europe, which can be caught by means of a small piece of silk. On the contrary, they rise high in the air, with a brisk and rapid flight. Sometimes they light and repose on flowers at the tops of trees, and rarely risk themselves within reach of the hand. They appear to be constantly on their guard, and, if caught at all, it must be when on the wing, by means of a net at the extremity of a long rod of cane. Some species are observed to live in society, hundreds and thousands of them being sometimes found together. These generally prefer the lower districts and the banks of streams.   When one of them is caught and fastened by a pin on the surface of the sand, swarms of the same species will gather round him, and may be caught at pleasure.

It has been rumored for many years that mines of coal exist within the bounds of the province; but, notwithstanding some examinations by order of Government, no satisfactory discoveries have yet been made.   Doctor Parigot, who was employed to make surveys in the province in 1841, reported the existence of a carboniferous stratum, from twenty to thirty miles in width and about three hundred in length, running from north to south through the province.   The best vein of coal he opened he pronounced half bituminous, and situated between thick strata of the hydrous oxide of iron and bituminous schist; but hitherto there has been no very encouraging result from these explorations.   In the neighboring province of Rio Grande do Sul, coal of a better kind, though somewhat argillaceous, was found about the same time at a place called Herval, not far from S. Leopoldo.   But in 1861 the most important mineral discovery ever made in Brazil was made by Mr. Nathaniel Plant, in Rio Grande do Sul; and the name of Candiota, in connection with coal, will be as famous in Brazil as Cardiff in England. For full accounts of this great discovery, see Appendix H.

The province of *São Pedro do Rio Grande do Sul* (more commonly known as simply *Rio Grande do Sul*) constitutes the extreme southern portion of the Empire of Brazil.   It is so called from the first parochial Church of St. Peter, (S. Pedro,) and the river called Grande, (see on the map Barra do Rio Grande,) near whose margins it was erected.   In many of the official papers of the Empire, this province occurs as S. Pedro, to distinguish it from Rio Grande do Norte.   In the salubrity of its climate and the fertility of its soil it resembles the Republic of Uruguay, upon which it borders.   It is admirably adapted for European immigration, and the most successful of all the colonies established by the Imperial Government is that of S. Leopoldo, founded in 1825, which to-day numbers a busy and prosperous population of more than eleven thousand souls.

All the cereals and fruits of Central Europe can be cultivated in this province, and formerly immense quantities of wheat were grown, so that not only was there sufficient for home-supply, but for exportation. This branch of agriculture has now so dwindled that flour is, to some extent, imported from the United States.

The great wealth of Rio Grande do Sul consists of that which constituted the riches of the patriarchs,—flocks and herds. The *Guachos* of Buenos Ayres are not more expert on horseback or more skilful in the use of the lasso than are the Rio Grandenses, whose occupation from childhood is the care and culture of the herds of cattle which roam the vast campinas or prairies. It has been estimated that in the province of Rio Grande do Sul, not mentioning parts of Santa Catharina and S. Paulo which are devoted to the same purposes, five hundred thousand cattle are slaughtered annually for the sake of preserving their hides and flesh, while as many more are driven northward for ordinary consumption. Most of the *carne secca*, or jerked beef, in common use throughout Brazil, is prepared here. After the hide is taken from the ox, the flesh is skinned off in a similar manner from the whole side, in strips about half an inch in thickness. The meat, in this form, is stretched in the sun to dry. But very little salt is used in its preservation, and, when sufficiently cured, it is shipped to all the maritime provinces, and is the only kind of preserved beef used in the country. Stacks of this meat (emitting no very agreeable odor) lie piled up, like cords of wood, in the provision-houses of Rio de Janeiro.

In the financial year 1853–54, Rio Grande do Sul exported the value of near $3,000,000 in hides, horns, hair, and wool, $1,000,000 of which were imported into the United States.

The character of the people is somewhat peculiar, owing to their circumstances and mode of life. They are generally tall, of an active and energetic appearance, with handsome features, and of a lighter skin than prevails among the inhabitants of the northern portions of the Empire. Both sexes are accustomed, from childhood, to ride on horseback, and consequently acquire great skill in the management of those noble animals upon which they take their amusements as well as perform their journeys and pursue the wild cattle of their plains.

The use of the lasso is learned among the earliest sports of boyhood, and is continued until an almost inconceivable dexterity is acquired. Little children, armed with their *lasso* or *bolas*, make

war upon the chickens, ducks, and geese of the farmyard, until
their ambition and strength lead them into a wider field.

For the pursuit of wild cattle the horses are admirably trained,
so that, when the lasso is thrown, they know precisely what to do.
Sometimes, in the case of a furious animal, the rider checks the
horse and dismounts, while the bull is running out the length of
his raw-hide rope. The horse wheels round and braces himself to
sustain the shock which the momentum of the captured animal
must inevitably give. The bull, not expecting to be brought up so

THE LASSO.

suddenly, is thrown sprawling to the ground. Rising to his feet,
he rushes upon the horse to gore him; but the latter keeps at a
distance, until the bull, finding that nothing is to be accomplished
in this way, again attempts to flee, when the rope a second time
brings him to the ground. Thus the poor animal is worried, until
he is wholly within the power of his captors.

Nor is it only in Rio Grande do Sul or San Paulo that scenes of
this kind may be observed. They were formerly witnessed in Rio
de Janeiro itself. At the *Matadouro publico*, situated on the Praya
d'Ajuda, before the municipal butcheries were removed to the spacious *abattoirs* at San Christovão, vast numbers of cattle were daily
slaughtered. Among the droves that reached the capital from the

distant sertões was occasionally an ox so wild and powerful that he was not disposed to surrender life without a desperate struggle. He would break from his enclosure and dash into the streets of the city, threatening destruction to whoever opposed his course. A horse, accoutred with saddle and bridle, and with a lasso fastened to him by a strong girth, stood ready for the emergency, and was mounted in an instant to give pursuit. The chase was widely different in its circumstances from that which occurs in the open *campos;* but perhaps no interest was lost in the rapid turning of corners of streets, the heavy clatter of hoofs upon the pavement, and the hasty accumulation of spectators. In a short time, usually, the noose of the lasso whirled around the horns of the fugitive, an area was cleared, and the scene already described was enacted, until the runaway ox was killed on the spot or led away in triumph to the slaughter. The lasso is, moreover, in frequent use in the Campo de Santa Anna, in the same city, where vast herds of mules

THE GUAYCURUS.

are frequently congregated for sale. The purchaser has only to indicate which animal out of the untamed multitude he would like to examine, and the *tropeiro* soon has him "slippernoosed" at the end of his long rope, by which he holds or leads him at will.

This portion of Brazil was inhabited at the period of the settlement by two peculiar tribes of savages. On the eastern part of

Rio Grande do Sul and in St. Catharine's were the *Carijos*, who were said to be the most humane of all the aborigines, and were the most accessible to European manners and cultivation.   North of the province under consideration were the *Guaycurus*,—Indian ca·valry,—so called because the Portuguese found them ready to give battle on horseback.   Where they obtained these horses is an unexplained mystery, but doubtless they were procured either through the Spaniards on the Pacific coast, or from some of the earliest settlements on the La Plata.   I have in my possession an old picture of Guaycurus charging regulars, and their position reminds one of that resorted to by the wild Camanches of New Mexico.

Rio Grande do Sul is in population and commerce the fifth or sixth province in the Empire.   Until the rapid augmentation of exports from Pará, she occupied with certainty the fifth place.

For a series of years Rio Grande was in open rebellion against the Imperial Government, to which fact allusion has already been made.   The effect of this struggle was the proclamation of freedom to the slaves by both parties, so that the number of those in bondage was greatly diminished.   The proximity of this province to the Spanish-American Governments doubtless did much, before the Empire of Brazil was fully established in strength, to incline it to republican notions, and it was thought at one time that Rio Grande would sever itself from the Empire, and, like the Banda Oriental, or Uruguay, (once a province of Brazil,) become an independent State.   But, between generous concessions and vigorous measures, Rio Grande was brought back to allegiance, and to day none of her sister-provinces excel her in loyalty to the existing *régime*.   Brazil, however, has taken effectual means and preventives that her southern border be no longer disturbed.   The tyrant Rosas* was overthrown through the aid of the Brazilian

---

* Allusion having been made to the part which Brazil took in the overthrow of the Nero-Borgia of the New World, the following note from Mr. Hadfield's work will give an outline of the history of affairs in the Argentine Confederation :—

"In January, 1831, the provinces of Buenos Ayres, Entre Rios, Corrientes, and Santa Fé, entered into a federal compact, to which all the other provinces at subsequent periods became parties.   The union was a voluntary alliance.   No general Constitution was promulgated, and the adhesion of the several members

army and navy, and Brazil is now (1866) endeavoring to conquer
a peace by overthrowing a new despot,—Lopez Jr. Brazil is en-

---

was left to be secured by the resources of the person who might obtain the direction of affairs. This Argentine Confederation, like the Republic which it had
succeeded, soon fell into a state of anarchy; and it was not till the election of
General Rosas as governor or captain-general, with almost absolute power, in
1835, that even temporary quiet was secured. But cruelty and despotism marked
his sway at home, and his ambition, which continually prompted him to endeavors
to extend his power over the whole country watered by the La Plata and the Paraná,
led him into disputes with foreign powers; and these ultimately brought about
his downfall.

"On the death of Francia, Dictator of Paraguay, Rosas refused to acknowledge
the independence of that power, insisting that it should join the Argentine Confederation. At the same time he refused to allow the navigation of the Paraná by
vessels bound to Paraguay. Lopez, the new Dictator of Paraguay, therefore entered
into alliance with the Banda Oriental, now called Uruguay, with which Rosas was
at war. These powers applied for assistance to Brazil. The war was prolonged
until the whole country on both sides of the Plata and the Paraná was in a state
of confusion. Great Britain volunteered her mediation, but it was rejected by
Rosas. England and France tried various measures with Rosas from 1845 to 1849,
but in vain. On the final withdrawal of the two great powers in 1850, Brazil
determined on active interference. Accordingly, toward the close of 1850, Brazil,
Uruguay, and Paraguay entered into a treaty, to which Corrientes and Entre Rios,
as represented by General Urquiza, became parties, by which they bound themselves to continue hostilities until they had effected the deposition of Rosas,
'whose power and tyranny' they declared to be 'incompatible with the peace
and happiness of this part of the world.' Early in the spring of 1851, a Brazilian
fleet blockaded Buenos Ayres, and soon after an Argentine force commanded by
Urquiza crossed the Uruguay. General Oribe, who commanded the army of Rosas
at Montevideo, capitulated. His soldiers for the most part joined the army of
Urquiza, who—at the head of a force amounting, it is said, to seventy thousand
men—crossed into Buenos Ayres. A general engagement took place on the plains
of Moron, February 2, 1852, when the army of Rosas was entirely defeated. Rosas,
who had commanded in person, succeeded in escaping from the field; and, in the
dress of a peasant, he reached in safety the house of the British minister at Buenos
Ayres. From thence, with his daughter, he proceeded on board H.B.M. steamer
Locust, and on the 10th of February sailed in the Conflict steamer for England."

Note for 1866.—THE PARAGUAY WAR OF 1865-66.—In 1862, Lopez Sr., the
second Dictator of Paraguay, died. In 1859 he had given the Brazilian Government difficulty by his non-compliance with the solemn treaties made in 1850,
which granted the right of way for steamers going to Matto Grosso up the river
Paraguay, and also by his refusal to settle the boundary-line question between
Paraguay and Brazil. He thus treated the power which had saved Paraguay from
the tyrant Rosas. However, matters were bridged over, because Brazil made
earnest diplomatic efforts, accompanied by a strong show of force. In 1862,
Lopez Sr. died. Lopez Jr. assumed the reins of government, and became third
Dictator of Paraguay. He then sent for mechanics from Europe, imported vast
quantities of machinery and iron, nominally for the railway from Asuncion to

gaged in a just war, though not one of her seeking, and the downfall of Lopez the younger will be promotive of as great a benefit as the victory over the former disturber of South American peace,—Rosas.

Villa Rica, but in reality, as subsequent events have shown, for war purposes, as he seems to have been filled with hatred of Brazil for presuming, through the Brazilian statesman Paranhos, to interfere in Paraguayan affairs. In 1863–64, the Banda Oriental, or Republic of Uruguay, became torn by internal faction: the *Blancos* (the *ins*) were opposed by the *Colorados* (the *outs*), led by General Flores. Brazilian citizens in Uruguay suffered at the hands of the *Blancos*, and Brazil was compelled, after long and peaceable protestation, to send down Vice-Admiral Visconde de Tamanderé with the Brazilian fleet to protect her citizens She did this effectually by aiding Flores, and the government of Uruguay fell into the hands of the *Colorados*. Forced to take up arms to protect her subjects in Uruguay from ill treatment and extortion, she showed by her moderation in the hour of triumph that her practice was consistent with her profession, and that no ideas of conquest or oppression had mingled with the exaction of the reparation she had so long and vainly sought by peaceful means. But Lopez, before the *Blanco* party fell, had said to Brazil, "If you attack Uruguay, I will attack you." This was a mere pretext, as his ample preparations showed. On the 13th of November, 1864, without declaration of war, Lopez caused the Brazilian mail-steamer Marquis de Olinda, on her way to Matto Grosso, to be seized and brought back to Asuncion, and her passengers, including the President of Matto Grosso and a number of Brazilian army and navy officers, to be put into prison, where they are to this day, (March, 1866.) The Brazilian Minister, Vianna de Lima, could not obtain his passports without the intervention of the United States Minister, Mr. Washburn. Paraguay steamers then went up the river, bombarded and seized Coimbra, took Albuquerque, Corumbá, and other points in Brazil, and committed great outrages upon an almost defenceless people. Of course Brazil had no other resource than to fly to arms. But Brazil, like all large bodies, moved slowly, and in the mean time Lopez, (whose object should have been to secure the neutrality of the Argentine Confederation,) without judgment and without knowledge of international law, demanded from the Argentine Confederation passage for the Paraguayan arms across the Argentine State of Corrientes. The President (Mitre) of the Confederation replied, in effect, "We are at peace with Brazil: it cannot be done." At this Lopez seized, without declaration of war, steamers belonging to the Confederation. Uruguay, the Argentine Confederation, and Brazil were then all brought into alliance, through the good offices of the Brazilian envoy, Sr. Octaviano. On June 11, 1865, at a place on the river Paraná, not far from Corrientes, the first naval conflict took place between Paraguayans and Brazilians Barroso commanded the Brazilian fleet. The odds (in number of Paraguay steamers and land-batteries) were against the Brazilians; but the victory achieved was the most brilliant in the annals of South America. Lopez's troops had invaded Corrientes and Rio Grande do Sul, but were defeated at the Yatay (or Ytati) on the 17th of August, 1865, and at Uruguayana (Rio Grande do Sul) on the 18th of September, 1865, the Emperor commanding in person. The great conflict at the mouth of the Paraná and Paraguay will doubtless settle forever that despotism in Paraguay, which has kept one of the finest countries of our globe as an inland Japan, without progress or development.

23

CHAPTER XIX.

JOURNEY TO SAN PAULO—NIGHT-TRAVELLING—SERRA DO CUBATAO—THE HEAVEN OF THE MOON — FRADE VASCONCELLOS — ANT-HILLS — TROPEIROS — CURIOUS ITEMS OF TRADE—YPIRANGA—CITY OF SAN PAULO—LAW-STUDENTS AND CONVENTS—MR. MAWE'S EXPERIENCE CONTRASTED—DESCRIPTION OF THE CITY—RESPECT FOR S. PAULO—THE VISIONARY HOTEL-KEEPER.

On my return from the province of Santa Catharina I again touched at Paranaguá, and, with the usual slowness which characterized Brazilian coast-travelling a few years ago, I came leisurely to Santos, and thence proceeded to the city of San Paulo. A young Brazilian had the intention of accompanying me to the capital of the province; but when I informed him that it was my determination to start for the interior the day of my arrival at Santos, he at first laughed at me, considering it an impossibility, and intimated that I would gladly accept the proffered hospitality of friends. When he found me unmoved in my resolution, he dropped his smiles, and looked at me with that pity which is bestowed upon the hopelessly insane.

At half-past five o'clock in the evening I set out alone. I have often heard exclamations of surprise, from those who have never been in Brazil, at the very idea of journeying without a companion in a land which their imaginations have pictured as the abode of brigands and wild beasts. Though I have compassed many leagues *solus*, I have never met with the former, and the latter have been quite harmless. My horse, in size, in his trappings, and in general appearance, was befitting a Calmuck Tartar. He had never made the acquaintance of a curry-comb, but got over the fine road which leads to Cubitão with a speed worthy of a better-looking animal. It was dark before I reached the bridge which spans the Rio do Cubitão; and, not feeling exactly sure of a *hospedaria*, I rode up to a little way-side venda, and my inquiries were answered very satisfactorily in French. The same man I saw upon my return, and learned from him that he came to Brazil twenty years ago under the impression that gold was as plentiful as paving-stones. He directed me to an inn kept by a German beyond the bridge. Having given my name at the *Registro*, and having paid a slight toll, I clattered over, and was soon at the house of the German. I felt half inclined to push onward over the mountains, so as to make San Paulo before mid-day of the morrow. I however concluded to refresh myself and horse, and gave orders for supper. The refreshment, so far as sleep was concerned, was a minus quantity, and at an early hour I was astride my steed and on my way up the Serra. The road which traverses this range of mountains is probably the finest in Brazil, with the exception of the Imperial highway to Petropolis. When Dr. Kidder visited this portion of the Empire, there existed a very excellent road, made at great expense; yet, owing to its steepness, it was perfectly impassable for carriages. His description of that route is as follows:—

"It embraces about four miles of solid pavement and upward of one hundred and eighty angles in its zigzag course. The accomplishment of this great work of internal improvement was esteemed worthy of commemoration as a distinguished event in the colonial history of Portugal. This appears from a discovery made on my return. Halting on the peak of the Serra, my attention was drawn to four wrought stones, apparently imported. They corresponded in size and form to the mile-stones of the United States, and had fallen prostrate. One lay with its face downward, so embedded in the earth as to be—to me at least—immovable. From the others, having removed with the point of my hammer the moss and rubbish by which the tracery of the letters was obscured, I deciphered as follows:—

"MARIA I. REGINA,
NESTE ANNO, 1790.

———

OMNIA VINCIT AMOR SVBDITORVM.

———

FES SE ESTE CAMINHO NO FELIS GOVER-
NO DO ILL° E EX° BERNARDO JOSE DE
LORENO, GENERAL DESTA
CAPITANIA.

"A solid pavement up this mountain-pass was rendered essential from the liability of the road to injury by the continued tread of animals, and also from torrents of water which are frequently precipitated down and across it in heavy rains. Notwithstanding the original excellence of the work, maintained as it had been by frequent repairs, we were obliged to encounter some gullies and slides of earth, which would have been thought of fearful magnitude had they not been rendered insignificant in comparison with the heights above and the deep ravines which ever and anon yawned beneath precipitous embankments. At these points a few false steps of the passing animal would have plunged both him and his rider beyond the hope of rescue. Our ascent was rendered more exciting by meeting successive troops of mules. There would first be heard the harsh voice of the tropeiros urging along their beasts, and sounding so directly above as to seem issuing from the very clouds: presently the clattering of hoofs would be distinguished, and at length would be seen the animals, *erectis auribus*, as they came borne almost irresistibly down by their heavy burdens. It was necessary to seek some halting-place while the several divisions of the troop passed by, and soon their resounding tread and the echoing voice of the guides would be lost in the thickets beneath."

The above description of the road was strictly true fifteen years ago; but now, by judicious engineering, the grades are not nearly so steep, and at a vast expense the whole is finely macadamized. Still, the ascent is too precipitous for heavily-laden carriages. But this will soon be remedied. English engineers are surveying a route into the interior which may extend as far as the province of Goyaz; and it is the fond hope of the Vergueiros that the time is not distant when the coffee of Campinas, Limeira, and Itú will be brought upon wheels to Santos. In the engraving the present comparatively greatly-winding highway is in strong contrast with the almost perpendicular road made by the early Jesuits before the one of which Dr. Kidder speaks. The Jesuits' Road is the dark line seeming to divide the conical mountain into equal parts.

As I pushed up with my sorry-looking steed, the Serra became enveloped in mist, so that I could scarcely see a rod before me; but upon my return the mountains were not only bathed in glorious sunlight, but the plains beneath and the distant ocean seemed brought near, as by magic. There was a wildness and sublimity in the landscape which I have not seen surpassed in the vicinity of Rio de Janeiro. From the summit of the mountain the dark and rugged gorges were not even clothed with the abundant foliage which is found everywhere else. Streams burst forth from some of the loftiest peaks, and thundered down into the deep ravines beneath.

THE BRIDGE AND SERRA DO CUBITÃO.

The Jesuit Vasconcellos made the ascent of this Serra two hundred years ago, and his description of the scenery is sketched with a masterly hand; but his estimate of the altitude was certainly extraordinary:—

"The greater part of the way you have not to travel, but to get on with hands and feet, and by the roots of trees; and this among such crags and precipices, that I confess my flesh trembled when I looked down. The depth of the valleys is tremendous, and the number of mountains, one above another, seems to leave no hope of reaching the end. When you fancy you are at the summit of one, you find yourself at the bottom of another of no less magnitude. True it is, that the labor of ascent is recompensed from time

to time; for when I seated myself upon one of these rocks, and
cast my eyes below, it seemed as though I was looking down from
the heaven of the moon, and that the whole globe of earth lay
beneath my feet.   A sight of rare beauty for the diversity of
prospect, of sea and land, plains, forests, and mountain-tracks, all
various, and beyond measure delightful.   This ascent, broken with
shelves of level, continues till you reach the plains of Piratininga,
in the second region of the air, where it is so thin that it seems as
if those who newly arrive could never breathe their fill."

Dr. Kidder thus criticizes Vasconcellos:—

"The last sentence is as erroneous as the preceding are graphic
and beautiful.   I should not, however, deem it necessary to correct
the statement, had not Southey, upon its authority, represented
this ascent to continue eight leagues to the very site of S. Paulo,
which is upon the plains of Piratininga.   The truth is, that from
the summit of the Serra, before stated to be three thousand feet
above the sea, the distance to S. Paulo is about thirty miles, over a
country diversified with undulations, of which the prevailing
declination, as shown by the course of streams, is inland.   Nevertheless, so slight is the variation from a general level, that the
highest point within the city of S. Paulo is estimated to be precisely the same altitude with the summit mentioned.   What inconvenience would be experienced from rarefaction of the atmosphere
at such an elevation may be easily determined."

It however appears to me that the estimated altitude of the
Serra, made by the good *frade* Vasconcellos, was a just one according to his standard; for, even considering that he did not have the
asthma, to go up a steep mountain, ("the heaven of the moon"
in elevation,) not by travelling, "but to get on with hands and
feet, and by the roots of trees, and this among such crags and
precipices," was assuredly sufficient to make one pant and feel
as if he were "in the second region of the air" and "could never
breathe his fill."   I once encountered a tall, lank Californian on
the Isthmus of Panama.   It was at the end of a hot and sultry
day: the pedestrian gold-digger had set his face toward the Pacific, while I was seeking the port of Aspinwall.   I accosted him,
and inquired the distance to Obispo, (at that time the terminus of
the Panama Railway.)   "Stranger," said he, "they call it five

miles; but I can assure you that it is about *five hundred,* for I never have been so tired in all my life." He estimated distance as Frade Vasconcellos estimated the altitude of the Serra de Cubatão.

Having once attained the summit of the mountain, I galloped over the upland plains, feeling more uncomfortable from the cold than ever before in Brazil. At ten o'clock I reached the hotel of M. Lefevre, a Frenchman from Roussillon, at whose well-provided table my chilliness was soon removed.

The plains between this and San Paulo, where there was no cultivation, were dotted by termite-ant-hills of such a size and form as to remind one of the pictures of a Hottentot village. In some places the industrious little creatures had literally ploughed up the ground for many yards around. The earth composing the outer shell of these insect-habitations becomes so indurated by the action of the sun that they retain their original erect position and oval form for scores of years.

The country over which I passed, save that the earth has a marked ferruginous appearance, resembles what are called the "oak-openings" of the western parts of the United States. In the vicinity of the village of San Bernardo there are considerable plantations of coffee and Chinese tea

I was constantly meeting with troops of mules laden with coffee, on their way to Santos, or passing others returning from the seaboard to the interior. It may be here remarked, that ordinary transportation to and from the coast is accomplished with no inconsiderable regularity and system, notwithstanding the manner. Many planters keep a sufficient number of beasts to convey their entire produce to market; others do not, but depend more or less upon professional carriers. Among these, each troop is under charge of a conductor, who superintends its movements and transacts its business. They generally load down with sugar and other agricultural products, conveying, in return, salt, flour, and every variety of imported merchandise. I was informed that two hundred thousand mules annually arrived with their burdens at Santos. A gentleman who had for many years employed these conductors in the transmission of goods stated that he had seldom or never known an article fail to reach its destination.

The Paulista tropeiros, as a class, differ very much from the

Mineiros and conductors that visit Rio de Janeiro. They have a certain wildness in their look, which, mingled with intelligence and sometimes benignity, gives to their countenance altogether a peculiar expression. They universally wear a large pointed knife, twisted into their girdle behind. This *faca de ponta* is perhaps more essential to them than the knife of the sailor is to him. It serves to cut wood, to mend harnesses, to kill and dress an animal, to carve food, and, in case of necessity, to defend or to assault. Its blade has a curve peculiar to itself, and, in order to be approved, must have a temper that will enable it to be struck through a thick piece of copper without bending or breaking. This, being a favorite companion, is often mounted with a silver handle, and sometimes encased in a silver sheath, although it is generally worn naked. Many foreigners (among them Englishmen) have purchased these knives to take home as curiosities, not knowing that they were manufactured in Great Britain or in the northeastern part of France. Lady Emeline Stewart Wortley, in her interesting gossiping letters from the New World, states that she procured in Peru, as a great curiosity, a poncho of the country, so that she might show to her friends in England the peculiar costume and the manufactures of the people who are descended from Castilian adventurers and the subjects of Atahualpa. Before leaving South America, some kind friend engaged in commerce, not wishing Lady Emeline to be duped, broke her pleasant delusion by informing her that the poncho in question was from the looms of Scotland. It might also be mentioned that many of the beautiful water-vases seen by foreigners at Rio de Janeiro are manufactured at the potteries in Staffordshire, and are sent out in large quantities to South America. The mysteries of the supply of distant countries with the productions considered as peculiar to those lands would form a curious book, far more interesting than the "blue-books" of Old England, or the annual "Commerce and Navigation" issued from the United States financial department.*

---

* Paper manufactured in New England bears the stamp "Bath Post" and "Paris." Large establishments near New York import labels and wrapping-paper from France, to put in and around hats which go over the Union as made on the banks of the Seine. Staffordshire not only makes water-vases supposed in South America to

VIEW OF THE CITY OF S. PAULO.

(By Richards, after a Sketch by Mrs. Elliot, of S. Paulo.)

Before the sun had set, I saw in the distance the city of San Paulo. Its elevated position on a small table-land that springs up from the plain, and its many towers and steeples and old conventual buildings, give it an appearance far more imposing than a town of greater population. Before ascending the hill, I passed the pavilion erected on the margin of the Ypiranga to commemorate the declaration of Brazilian independence which was emphatically made by Dom Pedro I. when (September 7, 1822) in this place he exclaimed "*Independencia ou Morte!*" Such a spot should be hallowed in the thought of every Brazilian, as well as memorable throughout the world; and it is therefore not much to the credit of Brazil or to the province of San Paulo, fertile in patriots, that a more fitting monument, of "enduring brass or marble," has not hitherto been erected commemorative of an event of such vast national interest.

Eventide was setting in as I splashed through the Tiete, the first of the La Platan affluents that I had crossed; and I soon ascended to the city. When I entered the first street, I felt more convinced than ever that I was south of the tropic of Capricorn; for, though verdure unchanging can be seen everywhere, yet in the nights of June (which answers to December in the northern hemisphere) there is experienced a chilliness which renders overcoats comfortable. Mine had been left behind by accident, and not only my feelings told me of its absence, but, beholding several law-students well cloaked, I was forcibly reminded of my carelessness and my consequent suffering. I fell into conversation with the young "limbs of the law," and found them exceedingly affable and communicative, as they kindly guided me to the hotel of Senhor C. Observing a large convent near at hand, I remarked that a new country like Brazil had little need of a body of monks and friars. I was somewhat surprised at the earnest and ready reply of one, who, apparently uttering the sentiments of the party, said, "No, Senhor, we need none of them: they are a lazy set; and we approve of what the King of Sardinia has recently done in regard

have been manufactured on the spot, but drives a good trade with statues of the Virgin, supposed to be the production of Italy and France, where they adorn so many houses of the peasantry.

to convents." Brazil has few monks in her splendid conventual buildings, and those few, with the exception of the Italian Capuchins, are indolent, luxurious, and licentious. The many edifices already secularized are used for state arsenals, provincial palaces, libraries, hospitals, &c.

I could not but contrast my introduction to S. Paulo with the entrance of Mr. Mawe, who nearly half a century ago made the acquaintance of the same city. In my case I rode into town and went to the hotel in the same manner as I would have done in Boston, Liverpool, or Geneva. But Mr. Mawe's experience with Brazil was immediately succeeding the opening of the country by royal decree in 1808. In his very readable "Travels" he says, "Our appearance at S. Paulo excited considerable curiosity among all descriptions of people, who seemed by their manner never to have seen an Englishman before. The very children testified their astonishment,—some by running away, others by counting our fingers and exclaiming that we had the same number as they. Many of the good citizens invited us to their houses, and sent for their friends to come and look on us. As the dwelling we occupied was very large, we were frequently entertained by crowds of young persons of both sexes who came to see us eat and drink. It was gratifying to us to perceive that this general wonder subsided into a more social feeling: we met with civil treatment everywhere, and found great pleasure in a more refined and polished company than we had seen in the Spanish settlements."

Though San Paulo is still distinguished for its "refined and polished" society, it is hard at this day to conceive of the curiosity at seeing strangers which must have been one of the direct consequences of Portugal's Japanese policy toward the colony of Brazil.

S. Paulo is situated between two small streams upon an elevation of ground, the surface of which is very uneven. Its streets are narrow, and not laid out with regard to system or general regularity. They have narrow side-walks, and are paved with a ferruginous conglomerate closely resembling old red sandstone, but differing from that formation by containing larger fragments of quartz,—thus approaching breccia.

Some of the buildings are constructed of this stone; but the material more generally used in the construction of houses is the

common soil, which, being slightly moistened, can be laid up into
a solid wall. The method is to dig down several feet, as would be
done for the foundation of a stone house, then to commence filling
in with the moistened earth, which is beaten as hard as possible.
As the wall rises above ground, a frame of boards or planks is made
to keep it in the proper dimensions, which curbing is moved upward as fast as may be necessary, until the whole is completed.
These walls are generally very thick, especially in large buildings.
They are capable of receiving a handsome finish within and without, and are usually covered by projecting roofs, which preserve
them from the effect of rains. Although this is a reasonable pre-
caution, yet such walls have been known to stand more than a
hundred years without the least protection. Under the influence
of the sun they become indurated, and are like one massive brick,
impervious to water, while the absence of frost promotes their
stability.

From San Paulo I wrote to one of my friends at Rio a letter,
from which I take the following extracts:—

"June 26, 1855.

"I am in a cold room,—such cold as I have not before experienced in Brazil. The moon is shining coldly; men creep
about in cloaks, (I wish I had one,) and the only thing that
possesses caloric is the candle which throws its dim light upon this
paper. I ought, however, to except the stirring strain of a distant
bugle, that really fills the night-air with a warming melody.

"Here I am stopped, because people do nothing *d'appressado*
(in a hurry) in Brazil. I put my two boxes ashore at Santos on
the 14th, and they were not sent forward until the 23d; and
to-day I passed the rancho where the *troop* encamped last night.
This evening they have reached a point two miles beyond San
Paulo,—at which rate they will attain their destination—Limeira
—about the 14th of July, the day on which I hope to sail from
Rio for the northern provinces. But if possible I shall hire extra
mules, overtake my boxes, transfer them to my animals, and push
on so as to reach the colony of Vergueiro (more than one hundred
miles from here) by Saturday night.

"Tell Senhor Fernando Rocha that his friend, Senhor Seraphim,
has been most useful and kind to me, running over the whole town

to procure for me the requisite animals. Do you think that an American or an English merchant would have done as much, late at night, for a stranger three hours after his arrival?

"I fear you will find me quite complaining, and place me in the category of those travellers who, like Smollett, were always scolding and grumbling about the inconveniences of the country in which they were 'voyaging.' I assure you that I take things as much like a philosopher as possible,—eating all kinds of food in all sorts of places, and sleeping where I would have scruples about making a daylight examination. Fancy, I slept, or at least attempted it, last night in a dirty German *hospedaria*, with a wild parrot overhead and my Calmuck horse haltered just the other side of a thin partition: so, between the music of one biting his chain, and the other crunching his *milho*, (Indian corn,) I got a very small share of 'nature's sweet restorer.'

"Yesterday I left Santos, although I was informed that it was impossible to start for the interior the same day that I arrived; yet my kind friends, the Vergueiros, enabled me to keep my word which I gave on board the steamer, to the effect that night should see me on my way. To-day I rode thirty-two miles, which you know, as Paulistas travel, is a good day's journey. As I drew near to San Paulo and gazed upon the green prairies dotted by herds, the white houses surrounded by trees, and in the background the distant mountains, I seemed to behold, as in years gone by, the like scenes of Burgundy, Piedmont, and Northumberland.

"I felt a more profound respect for San Paulo than for any South American city that I have yet visited. It was larger than I anticipated, and its houses, with their overhanging eaves, give it an appearance not unlike that of Vevay, on the Lake of Geneva. These eaves, I should say, extend over the streets five or six feet, protecting the passers-by from the rain and sun, and giving a Swiss picturesqueness to the whole.

"My feelings of respect, however, arose not from the size of the city, nor from its picturesqueness, but because there is a more intellectual and a less commercial air about the people than you see elsewhere in Brazil. You do not hear the word *dinheiro* constantly ringing in your ear, as at Rio de Janeiro. There are no less than five hundred law-students in the legal college here established, and their appearance really recalls the Dane law-school of Harvard University and the students of Heidelberg. The *genus* student is the same the world over,—full of pranks, fun, and mischief. The week of my arrival, several scores of these fellows had 'kicked up a row' (as one of them elegantly expressed it) at the theatre, so that the President of the province ordered a strong police-force to be present at the next representation, and it was not without difficulty that order was preserved.

"In entering the city, I fell in with a number of these young legalists, who conducted me to the hotel where many of their classmates were whiling away their time at billiards; and, judging from the sound of rolling balls and 'lucky hits' at this late hour, one would suppose they will have little opportunity for preparing their morning lesson. The hotel-keeper is a young Brazilian, educated at ——'s, in Nova Fribourgo, and speaks very good English. He has too many projects, however, to succeed. His last plan is to establish a sort of Surrey Zoological Gardens, for concerts, exhibitions, and recreation generally, at Rio de Janeiro. His chosen spot for this purpose is on the Praia Vermelha, not far from the Sugar-Loaf. Speaking of gardens, I am reminded of plantations, and will only say that to-day I saw immense plantations of what I had first supposed to be coffee, but which proved to be genuine Chinese 'green tea.'

"But now to bed: if rolling billiard-balls will let me sleep, I will be refreshed for the journey of to-morrow.

"P.S. Wednesday morning.—I have a horse, a conductor, and two mules, and shall be off in a few moments. You will next hear from me at Limeira."

Note for 1866.—The São Paulo and Jundiahy Railway is now nearly completed from Santos to Jundiahy (mentioned on page 399). The effect of these various railways is becoming apparent, and though some of them in the Empire are not very paying stock, there must be in the end a great gain to the country. The São Paulo road is rightly located; for it penetrates the interior of one of the most fertile of the Southern provinces.

CHAPTER XX.

HISTORY OF SAN PAULO—TERRESTRIAL PARADISE—REVERSES OF THE JESUITS— ENSLAVEMENT OF THE INDIANS—HISTORICAL DATA—THE ACADEMY OF LAWS— COURSE OF STUDY—DISTINGUISHED MEN—THE ANDRADAS—JOSÉ BONIFACIO— ANTONIO CARLOS—ALVARES MACHADO—VERGUEIRO—BISHOP MOURA—A VISIT TO FEIJO—PROPOSITION TO ABOLISH CELIBACY—AN INTERESTING BOOK—THE DEATH OF ANTONIO CARLOS DE ANDRADA—HIGH EULOGIUM—MISSIONARY EFFORTS IN SAN PAULO—EARLY AND PRESENT CONDITION OF THE PROVINCE— HOSPITALITIES OF A PADRE—ENCOURAGEMENTS—THE PEOPLE—PROPOSITION TO THE PROVINCIAL ASSEMBLY—RESPONSE—RESULT—ADDENDA—PRESENT ENCOU-RAGEMENTS.

THE history of San Paulo takes us back to an early period in the settlement of the New World by Europeans. It has already been remarked that, in 1531, Martin Affonso de Souza founded S. Vicente, the first town in the captaincy, which for a long time bore the same appellation. There had previously been shipwrecked on the coast an individual by the name of João Ramalho, who had acquired the language of the native tribes and secured influence among them by marrying a daughter of one of their principal caciques. Through his interposition, peace was secured with the savages and the interests of the colony were fostered. By degrees the settlement extended itself inland, and in 1553 some of the Jesuits who accompanied Thomé de Souza, the first captain-general, found their way to the region styled the plains of Piratininga, and selected the elevated locality on which the city now stands, as the site of a village, in which they commenced to gather together and instruct the Indians.

Having erected a small mud cottage on the spot where their college was subsequently built, they proceeded to consecrate it by a mass, recited on the 25th of January, 1554. That, being the day on which the conversion of St. Paul is celebrated by the Roman Church, gave the name of the apostle to the town, and subsequently to the province. St. Paul is still considered the patron saint of both. A confidential letter, written by one of these Jesuits to his brethren in Portugal, in addition to many interesting particulars on other subjects, contains the following passage, which may serve to show how the country appeared to those who saw it nearly three hundred years ago. This letter exists in a manuscript book taken from the Jesuits at the time of their expulsion from Brazil, and still preserved in the National Library at Rio de Janeiro. Its date is 1560. No part of it is known to have been hitherto rendered into English previous to the translation made by Rev. Dr. Kidder.

"For Christ's sake, dearest brethren, I beseech you to get rid of the bad idea you have hitherto entertained of Brazil: to speak the truth, if there were a paradise on earth, I would say it now existed here. And if I think so, I am unable to conceive who will not. Respecting spiritual matters and the service of God, they are prospering, as I have before told you; and as to temporal affairs, there is nothing to be desired. Melancholy cannot be found here, unless you dig deeper for it than were the foundations of the palace of S. Roque. There is not a more healthy place in the world, nor a more pleasant country, abounding as it does in all kinds of fruit and food, so as to leave me no desire for those of Europe. If in Portugal you have fowls, so do we in abundance, and very cheap; if you have mutton, we here have wild animals, whose flesh is decidedly superior; if you have wine there, I aver that I find myself better off with such water as we have here than with the wines of Portugal. Do you have bread, so do I *sometimes*, and always what is better, since there is no doubt but that the flour of this country (mandioca) is more healthy than your bread. As to fruits, we have a great variety; and, having these, I say let any one eat those of the old country who likes them. What is more, in addition to yielding all the year, vegetable productions are so easily cultivated (it being hardly necessary to plant them) that nobody can be so poor as to be in want. As to recreations, yours are in no way to be compared with what we have here.

"Now, I am desirous that some of you should come out and put these matters to the test; since I do not hesitate to give my opinion, that, if any one wishes to live in a terrestrial paradise, he should not stop short of Brazil. Let him that doubts my word come and see. Some will say, What sort of a life can that man lead who sleeps in a hammock swung up in the air? Let me tell them, they have no idea what a fine arrangement this is. I had a bed with mattresses, but, my physician advising me to sleep in a hammock, I found the latter so much preferable, that I never have been able to take the least satisfaction, or rest a single night, upon a bed since. Others may have their opinions, but these are mine, founded upon experience."

The Jesuits, unhappily, did not find this paradise to be perennial. Their benevolence, and their philanthropic devotedness to the Indians, brought down upon them the hatred of their countrymen, the Portuguese, and of the Mamalucos, as the half-breeds were denominated. These two classes commenced at an early day the enslavement of the aboriginals, and they continued it through successive generations, with a ferocious and bloodthirsty perseverance that has seldom found parallel. As the Jesuits steadfastly opposed their cruelties, the Portuguese resorted to every means of annoyance against them. They ridiculed the savages for any compliance with the religious formalities in which they were so diligently instructed,—encouraging them to continue in their heathen vices, and even in the abominations of cannibalism. Nevertheless, these missionaries did not labor without considerable success. The Government was on their side, but was unable to protect them from the persecutions of their brethren, who, although calling themselves Christians, were as insensible to the fear of God as they were regardless of the rights of men. From the pursuit of their imagined interest, nothing could deter them but positive force. As the Indians were driven back into the wilds of the interior, through fear of the slave-hunters, the Jesuits sought them out, and carried to them the opportunities of Christian worship and instruction. It was thus that a commencement was made to the celebrated Reductions of Paraguay, which occupy so wide a space in the early history of South America. Sometimes the Paulistas would disguise themselves in the garb of the Jesuits, in order to decoy the natives whom they wished to capture. At other times they assaulted the Reductions, or villages of neophytes, boasting that the priests were very serviceable in thus gathering together their prey.

Voluntary expeditions of these slave-hunters, styled *bandeiras*, spent months, and sometimes years, in the most cruel and desolating wars against the native tribes. Instigated by the lust of human plunder, some penetrated into what is now the interior of Bolivia on the west; while others reached the very Amazon on the north. As the Indians became thinned off by these remorseless aggressions, another enterprise presented itself as a stimulant to their avarice. It was that of hunting for gold. Success in the latter enterprise created new motives for the prosecution of the former. Slaves must be had to work the mines. Thus, the extermination of the native tribes of Brazil progressed, for scores of years, with fearful rapidity. One result of these expeditions was an enlargement of the territories of Portugal and an extension of settlements. By the growth of these settlements four large provinces were populated. They have since been set off from that of S. Paulo, in the following order:—Minas-Geraes, in 1720; Rio Grande do Sul, in 1738; Goyaz and Matto Grosso, in 1748.

During the period when Portugal and her colonies were under the dominion of Spain, a considerable number of Spanish families became inhabitants of the captaincy of S. Paulo; and when, in 1640, that dominion came to an end, a numerous party disposed itself to resist the Government of Portugal. They proceeded to proclaim one Amador Bueno, king; but this individual had the sagacity and patriotism peremptorily to decline the dignity his friends were anxious to confer upon him. The Paulistas have been subsequently second to none in their loyalty to the legitimate Government of the country; unless, indeed, the unhappy disturbances that occurred among them in the years 1841–42 be considered as forming an exception to this remark. It is now one of the most prosperous provinces of the Empire.

My colleague remained many days in the provincial capital, and gives the following account of its institutions and great men:—

"The Academy of Laws, or, as it is frequently denominated, the University of S. Paulo, ranks first among all the literary institutions of the Empire. I enjoyed an excellent opportunity for visiting it, being introduced by the secretary and acting president, Dr. Brotero. This gentleman—whose lady is a native of the United States—deserves honorable mention, not only for the zeal and

24

ability with which he administers the affairs of the institution of which he has since become the president, but also as an author. He has published a standard work on the Principles of Natural Law, and a treatise upon Maritime Prizes.

"The edifice of the Curso Juridico was originally constructed as a convent by the Franciscan monks, whom the Government compelled to abandon it for its present more profitable use. Being larger and well built, a few alterations rendered it quite suitable to the purposes for which it was required. The lecture and recitation rooms are on the first floor, the professors' rooms and library on the second; these, together with an ample court-yard, compose the whole establishment, save two immense chapels still devoted to their original design. In one of these I found several very decent paintings, and also an immense staging, upon which workmen were engaged finishing the stucco-work upon the principal arch of the vaulted roof. Both chapels abounded with mythological representations of the patron saint, both in images and colors. The library of the institution, containing seven thousand volumes, is composed of the collection formerly belonging to the Franciscans, a part of which was bequeathed to the convent by the Bishop of Madeira; the library of a deceased bishop of S. Paulo; a donation of seven hundred volumes from the first director; and some additions ordered by the Government. It was not overstocked with books upon law or belles-lettres, and was quite deficient in the department of science. The only compensation for such deficiencies was a superabundance of unread and unreadable tomes on theology. Among all these, however, there was not to be found a single copy of the Bible—the fountain of all correct theology—in the vernacular language of the country; a rarer volume than which, at least in former years, could scarcely have been mentioned at S. Paulo. This particular deficiency I had the happiness of supplying by the donation of Pereira's Portuguese translation, bearing this inscription :—

"AO BIBLIOTHECA DA ACADEMIA JURIDICA DE S. PAULO

DA SOCIEDADE BIBLICA AMERICANA

PELO SEU CORRESPONDENTE

D. P. KIDDER.

CIDADE DE S. PAULO,
15 de Fev'o de 1839.

"The history and statistics of the institution were kindly communicated to me by the secretary, in a paper, from which the following abstract is translated :—

"The Academy of the Legal and Social Sciences of the city of S. Paulo was created by a law dated August 11, 1827. It was formally opened, by the first professor, Dr. José Maria de Avellar Brotero, on the 1st day of March, 1828,—Lieutenant-General José Arouche de Toledo Rendon being first director.

"The statutes by which it is governed were approved by law, November 7, 1831.

"The studies of the preparatory course are—Latin, French, English, Rhetoric, Rational and Moral Philosophy, Geometry, History, and Geography.

"The regular course extends through five years. The several professorships are thus designated :—

"*First Year.*—1st professorship, Philosophy of Law, Public Law, Analysis of the Constitution of the Empire, and Roman Law.

"*Second Year.*—1st professorship, Continuation of the above subjects, International Law, and Diplomacy; 2d professorship, Public Ecclesiastical Law.

"*Third Year.*—1st professorship, Civil Laws of the Empire; 2d professorship, Criminal Laws, Theory of the Criminal Process.

"*Fourth Year.*—1st professorship, Continuation of Civil Law; 2d professorship, Mercantile and Maritime Law.

"*Fifth Year.*—1st professorship, Political Economy; 2d professorship, Theory and Practice of General Law, adapted to the Code of the Empire.

"The age of sixteen years and an acquaintance with all the preparatory studies are requisite in order to enter the regular course. No student can advance without having passed a satisfactory examination on the studies of the preceding year. When the examinations of the fifth year are passed acceptably, the Academy confers the degree of Bachelor of Arts; and every Bachelor is entitled to present theses on which to be examined as a candidate for the degree of Bachelor of Laws.

"In examinations on the course, students are interrogated by three professors for the space of twenty minutes each. Competitors for the Doctorate are required to argue upon their theses with nine professors successively, each discussion lasting half an hour. At the end of each examination, the professors, by secret ballot, determine the approval or rejection of the candidate.

"In order to explain the peculiarities of the above course of study, it should be remarked that, in its arrangement, the University of Coimbra was followed as a model. The education imparted by it may be formal and exact in its way, but can never be popular. The Brazilian people look more to utility than to the antiquated forms of a Portuguese university; and I apprehend it will be found necessary, ere long, in order to secure students at the University of S. Paulo, to condense and modernize the course of instruction."

In 1855, the prosperity of the Law-Academy was no longer a matter of doubt, as at that time there were two hundred and ninety-six students in the five classes, and three hundred more in the preparatory course, which, by recurring to their list of studies, I find (*minus* the Greek language) to be very similar to the studies in most colleges in the United States. Under Senhor Brotero, the institution at San Paulo has become exceedingly popular, and, doubtless, is far more practical than in the first years of its existence. It is here and at the Pernambuco Law-School (which contains three hundred and twenty students in the *regular* course) that the statesmen of Brazil receive that education which so much better fits them for the Imperial Parliament and the various legislative assemblies of their land than any preparatives that exist in the Spanish-American countries.

"My sojourn at S. Paulo," continues Dr. Kidder, "was rendered increasingly interesting by repeated interviews with several distinguished citizens of the province. One evening, while walking in company with several gentlemen in the extensive gardens of Senhor Raphael Tobias d'Aguiar, a popular ex-president of the province and one of its largest land-proprietors, the conversation turned upon the different foreign travellers in Brazil. Mawe was recollected by some; but St. Hilaire, the French botanist, enjoyed the highest consideration of all, as having accomplished his task in the most thorough manner.

"Senhor Raphael related a very interesting anecdote, communicated to him by St. Hilaire. A poor man in England, in reading the work of Mr. Mawe, had become so enthusiastic with the idea of the vegetable and mineral riches of Brazil, that, in order to get to the country, he actually came out in the capacity of a servant. After reaching Rio de Janeiro, he had by some means found his way up the Serras into the interior, where his industrious exertions had been rewarded with success, and where the botanist found him actually possessed of a fortune.

"Among the distinguished men of S. Paulo, I will first mention the Andradas,—three brothers, whose family residence is Santos. These brothers were all educated at the University of Coimbra, in Portugal, and received the degrees of Doctors in Jurisprudence and Philosophy, and the younger that of Mathematics.

"José Bonifacio, the eldest, after his graduation, travelled several years in the northern countries of Europe,—devoting himself meanwhile to scientific researches, the results of which it was his intention to publish in Brazil. On his return to Portugal he was created Professor of Metallurgy in Coimbra, and of Medicine in Lisbon. While engaged in these professorships, he published several treatises of much merit, among which was a dissertation on 'The Necessity of Planting New Forests in Portugal, and particularly of Fir-Trees along the Sandy Coasts of the Sea-Shore.' His valor was called out by the invasion of Portugal, when he organized and headed a body of students who determined to do what they could toward repelling the army of Napoleon. In 1819 he returned to Brazil in time to take a leading part in the revolution of independence. (He died at Praia Grande in 1838.)

"Antonio Carlos returned to Brazil soon after having completed his education. In the year 1817, while executing the office of Ouvidor in Pernambuco, he was arrested as an accomplice of the conspirators in a revolt which broke out at that time. He was sent to Bahia and thrown into prison, where he remained four years. As a proof of his philanthropy as well as of his indomitable energy of mind, it must be mentioned that he spent this long period almost exclusively in instructing a number of his fellow-prisoners in rhetoric, foreign languages, and the elements of science. Being at length liberated, he returned to San Paulo, where he was shortly afterward elected deputy for that province in the Cortes of Lisbon. He assumed his duties in that body, and remained in it until the increasing insults and aggravations which were heaped upon the Brazilians, without the hope of redress, forced him and several of his colleagues, among whom was Feijo, to withdraw and embark secretly for England. Having arrived at Falmouth, they published a solemn declaration of the motives which induced them to desert the Cortes and to quit Lisbon. Thence they returned to their native country.

"Martin Francisco, the younger brother, had won high distinctions as a scholar, and, from early life, was the frequent subject of political honor. At the first organization of the Imperial Government he was created Minister of Finance, and in this capacity did the country important service,—his elder brother being at the same time Minister of State and of Foreign Affairs. At this period the three brothers were all elected members of the Assembly which convened to prepare a Constitution for the Empire.

"Before the discussions of that body were brought to a close, the Emperor was induced, by the coalition of two minor parties, to dismiss the Andrada Ministry and to appoint Royalists as their successors. The powerful opposition which the brothers immediately arrayed against those by whom they had been supplanted made the position of the new Ministry and that of the Emperor also extremely embarrassing. Attacks produced recrimination, until the Emperor at length resolved upon the rash and desperate expedient of dissolving the Assembly by force, which he succeeded in accomplishing, and then apprehended the three brothers Andrada and a few others who were leaders of the opposition. They were all, without the least examination or trial, conveyed on board a vessel nearly ready for sea, and transported to France.

"Their time in Europe was not idly spent. Already acquainted with all the more important modern languages, they devoted themselves to literary pursuits and the society of the learned with all the enthusiasm of students.

"In the year 1828, the two younger brothers returned to Rio, and, after a short detention in the prison of the Ilha das Cobras, received a full pardon from the Emperor. José Bonifacio came out in 1829 from France.

"The French admiral, who had known him in Europe, sent immediately to offer him every attention; but Andrada requested him to make no demonstration, as he was very uncertain how he might be received. But as soon as the arrival of the ship was known, Calmon, the Minister of Finance, went immediately on board to offer his congratulations and every kind civility. On Andrada's interview with the Emperor, it is said that the latter proposed an embrace, and that all the past should be forgotten. Andrada replied, with Roman firmness, that the embrace he would most cheerfully give, but to forget the past was impossible.

"The Emperor then proposed to him to enter into the Ministry, but he declined, assuring His Majesty that he only returned to Brazil to live in retirement. Nevertheless, José Bonifacio, in his old age, was the individual to whom the Emperor, on his abdication, confided the guardianship of his children. He had then proved the faithlessness of many of those officious partisans who had urged him forward in his attempted overthrow of the men who were his earliest and most devoted friends. The Emperor had learned, by painful experience, how to appreciate real patriotism.

"Antonio Carlos and Martin Francisco had no sooner returned to their native province, than they were immediately restored by their countrymen to important offices, and have ever since retained a prominent position in the national councils. They have, moreover, continued the same ardent and fearless advocates of their principles that they were in early life.

"It has been said, and perhaps justly, that 'the Andradas, when in power, were arbitrary, and, when out of place, factious; but their views were ever great, and their probity unimpeachable.' Their disinterestedness was manifest, and is deserving of eulogy. Title and wealth were within their reach; but they retired from office undecorated, and in honorable poverty. In many of their acts they were doubtless censurable; yet, when the critical circumstances of Brazil at the period are taken into consideration, surely some apology may be made for their errors. When old age required José Bonifacio to withdraw from public business, he retired to the beautiful island of Paqueta, in the Bay of Rio de Janeiro. He died in 1838; and, if there is any one fact that more loudly than another upbraids the lack of literary enterprise in Brazil, it is that no memoir of so distinguished an individual has made its appearance, or, so far as I could learn from his brothers, was ever contemplated.

"Both Antonio Carlos and Martin Francisco are distinguished, powerful orators. The latter is clear, expressive, and chaste in his diction; the former is fluent, impetuous, and sometimes extravagant. Antonio Carlos is particularly fond of the arena of debate, and few questions come before the Provincial or National Assembly which are not subjected to the searching analysis of his acute mind and to the often-dreaded ordeal of his flaming rhetoric. His speeches abound in beautiful illustrations from the French, Spanish, Italian, and English poets; and, when discussing questions of jurisprudence and diplomacy, his references display a critical acquaintance with standard English authors upon those subjects. As a random specimen of his style of eloquence, I will translate a paragraph from his speech in the General Assembly at Rio de Janeiro, in 1839, on the much-debated question whether foreign troops should be hired to compose the standing army of the Empire.

"After having gone through with an elaborate argument, he says, 'I am unwilling to weary the house. I have proved that the measure is anti-constitutional, that it is injurious to the dignity of Brazil, that it is useless, that it is impolitic, and that it will be oppressive to the nation.

"'Now I must close. It pains me to think that such a measure can possibly be approved. Such is the aversion I cherish toward it, that I am caused to fear that, if it should pass, some of our citizens will wish themselves alienated from the land of their birth; alienated, I was about to say, from a degraded nation. But this tongue cannot utter such a reproach, nor this heart anticipate such an injury, to the Brazilian people.

"'Every night, when I seek rest upon my humble couch, the first act of devotion I render to God is a thanksgiving that I was born upon this blessed soil,—in a country in which innocence and liberty were natives, but from which they temporarily fled away on the approach of those iron fetters of social bondage which Cabral, the accidental discoverer, imported in connection with the limited civilization of Portugal.

" ' Eis, descobreis Cabral os Brazis não buscados,
    C' os salgados vestidos gotejando, .
    Pesado beijas as douradas prayas,
    E ás Gentes que te hospedão, ignaras
    Do Vindouro, os grilhões lanças,
    Miserandos !   Então a liberdade,
    As azas não manchadas de baixa tyrannia
    Soltou isenta pelos ares livres.

" ' So it was an infamous series of oppressive laws and shameful proscriptions was imposed upon our poor ancestors, and would have rested upon *us* to-day, had not the grand achievement of our national independence set us free!   Allow me to remark a startling coincidence.   To-morrow will be the anniversary of that independence,—an event ever to be remembered.   To-day an effort is made, which, if successful, will throw clouds and gloom over it, and thus efface the brightest picture in our history.

" 'How is it that we, who were able to shake off the yoke of foreign bondage without the aid of mercenary troops, are supposed to be incompetent to crush rebellion within our own borders? Shameful reflection !   Is Bento Gonsalves some European adventurer?   No! he is a Brazilian, like us; and least of all can he withstand Brazilians.

" 'My heart is overflowing, but my tongue fails to express my thoughts.   If this measure pass, I shall have nothing left me to do but to hide my head, and to weep and sigh, in the language of Moore,—

" 'Alas for my country ! her pride is gone by,
    And that spirit is broken which never would bend :
    O'er the ruin her children in secret must sigh,—
    For 'tis treason to love her, 'tis death to defend.'

" An intimate friend and political associate of Antonio Carlos is Senhor Alvares Machado, another aged Paulista, also celebrated for his prompt and often passionate eloquence.   A brief extract from one of his speeches in the Chamber of Deputies forcibly expresses the provincial pride which the Paulistas cherish together with their sentiments of independence.   'How,' said he, 'can the present administration expect to intimidate *us*, who never succumbed to the founder of the Empire?   We spoke the language of liberty, of justice, and of truth, to a king and the descendant of kings.

"'On one occasion it was proposed to construct our constitution after the monarchial model, and to accomplish this intrigues were set on foot in all the provinces. What then was our language? "Sire," said we to the monarch, "despotism may be planted in the province of S. Paulo, but it will be upon the bones of the last of her inhabitants."'

"Another prominent member of the provincial legislature of S. Paulo was Vergueiro, a Senator of the Empire. This gentleman, a Portuguese by birth, has long been conspicuous in Brazil. Previous to the independence of the colony, he was one of the deputies to the Cortes of Lisbon, and had there distinguished himself above most of his colleagues for the open and explicit manner in which he defended the interests and privileges of the land of his adoption. Subsequently, while in the Brazilian Senate, he maintained his reputation as a skilful debater and a sincere friend of liberal institutions. During the scenes connected with the abdication of the first Emperor, he acted an important part, and, as has already been stated, was appointed at the head of the provisional Regency.

"During one of my visits to the Provincial Assembly of S. Paulo, this gentleman made a long and interesting speech on the subject of the outbreak and disorders at Villa Franca.

"The sessions of this legislative body are held in an apartment of the old College of the Jesuits, which has long since been appropriated to the uses of the Government. My attendance upon its deliberations was not very frequent, although several of my visits were quite interesting. Probably no provincial legislature in the Empire presented a greater array of learning, of experience, and of talent, than did this. At the period of which I am speaking, Martin Francisco de Andrada occupied the Presidential chair, while Senhores Antonio Carlos, Vergueiro, Alvares Machado, Raphael Tobias, the Bishops of S. Paulo, of Cuyabá, and Moura, the Bishop-elect of Rio de Janeiro, with various other gentlemen of distinction, took part in the proceedings.

"At the close of one of the sessions, I had the pleasure of meeting several of these gentlemen in a saloon adjoining the hall of debates, and of hearing from them the warmest expressions of American feeling and of a generous interest in the affairs of the United States.

"Antonio Maria de Moura was considered the special representative of the ecclesiastical interests in this legislature. This individual had gained a great degree of notoriety during a few years previous. He had been nominated by the Imperial Government to fill the vacant bishopric of Rio de Janeiro. The Pope of Rome was, for some reasons, displeased with the nomination, and accordingly refused to consecrate him. This circumstance gave occasion for long diplomatic negotiations, and for a time threatened to interrupt friendly relations between Brazil and the Holy See. For several years questions relating to this subject were frequently and freely discussed before the National Assembly. During these debates expressions were often used not the most complimentary to His Holiness, and facts of a startling character were brought to view. For example, a reverend padre, in speaking on the subject, alluded to a canonical objection to this candidate, which, he said, was very generally known,—viz.: the illegitimacy of his birth: 'that, however, was a trifling matter, it having been dispensed with in the case of two of the actual bishops of the Empire. But this gentleman had signed a report declaring against the forced celibacy of the clergy, and, when interrogated by His Holiness on the subject, had refused to give explanations.'*

"The longer this subject was discussed, the wider the difference seemed to grow. The Pope was unwilling to recede from his position, and the Brazilians resolved not to brook dictation from the Pope.

"The proposition to make the Brazilian church independent of His Holiness was more than once started, and it was finding increased favor with the people. But the question was regarded solely in its political bearings. Consequently, it became an object for the Government to settle it in the easiest way practicable. On the accession of a new ministry, measures were adopted to satisfy Moura and to induce him to step out of the way. Accordingly,

---

* See Jornal do Commercio, June 30, 1839.

he was at length persuaded to waive his claim, and to resign aL office which he could not be permitted to fill peaceably. The question was then easily disposed of. The Government made another nomination, which the Pope approved,—at the same time complimenting the rejected candidate with the title and dignities of bishop *in partibus infidelium*. At the time I met him, Padre Moura did not appear to be over thirty-five years of age. His demeanor was affable and his conversation interesting. He was understood to be the confidential adviser and assistant of the old Bishop of S. Paulo. He had been for a series of years engaged in political life, and will probably continue in similar engagements, since they will be in no wise inconsistent with the obligations of his office of bishop *in partibus*.

"I had the honor of more than one interview with the ex-Regent Feijo. The first was in company with an intimate friend of his, in the lower room of a large house, where he was staying as a guest, in the city of S. Paulo. There were no ceremonies. His reverence appeared to have been lying down in an adjoining alcove, and had hastily risen. His dress was not clerical. In fact, his garments were composed of light striped cotton, and appeared by no means new; while his beard was apparently quite too long for comfort in so warm a day. He was short and corpulent, about sixty years of age, but of a robust and healthful appearance. His countenance and cranium bore an intellectual stamp and conveyed a benevolent expression, although there might have been something peculiar in the look of his eyes, which gave rise to a remark made to me before I saw him, that he had 'the physiognomy of a cat.' His conversation was free and very interesting. My friend mentioned to him that I had made several inquiries respecting the customs of the clergy and the state of education and religion in the country. He proceeded to comment upon these several topics, and expressed no little dissatisfaction with the actual state of things, particularly among the clergy. He said 'there was scarcely a priest in the whole province that did his duty as the Church prescribed it, and especially with reference to catechizing children on the Lord's day.'

"He was on the eve of a journey to Itú and Campinas, and, being asked when he would set out, replied, *Dizem no Domingo*, ('Sunday is talked of;') thus indicating that even he himself had not too high a respect for the institution of the Sabbath-day. On another occasion I called on him at his own house in Rio de Janeiro, while he was in attendance on the Senate, of which he was a member, and for a long time president. It was in the morning, and I found him alone in his parlor, occupied with his breviary; while at the same time there lay on the table by which he was sitting a *faca de ponta*, or pointed knife, of the species already described, enclosed in a silver sheath. I presented him with copies of some tracts that we had just published in the Portuguese language for circulation in the country. He received them courteously, and again entered into conversation respecting various plans for the religious amelioration of Brazil. He, however, seemed to have little faith, and less spirit, for making further exertions, having been repeatedly baffled in his cherished projects for improvement. So little encouragement, indeed, had he met with from his brethren the clergy, that he was inclined to compare some of them to the dog in the manger, since they would neither do good themselves, nor allow others to do it.

"Feijo is a remarkable man. Like many others among the Brazilian clergy, he entered upon a political career in early life, and laid aside the practical duties of the priesthood. His abandonment of the Cortes of Portugal, to which he had been elected in the reign of Dom John VI., has already been mentioned.

"After the establishment of the independent Government of Brazil, he became a prominent member of the House of Deputies. During a debate in that body he listened to what seems at first to have struck him as a very strange proposition,—viz.: 'that the clergy of Brazil were not bound by the law of celibacy.' Coming, however, as the statement did, from a gentleman of great learning and probity, it secured his candid attention. Subsequent reflection, while meditating upon the means of reforming the clergy, and examining the annals of Christianity, convinced him not only that the proposition was correct, but also that the most fruitful source of all the evils that affected this important class of men was a forced celibacy. Whereupon, as a member of the Committee on Ecclesiastical Affairs, he offered to the House his views on the subject in the form of a minority report.

"In this report he proposed, 'that since celibacy was neither en-joined upon the clergy by divine law nor apostolical institutions, but, on the contrary, was the source of immorality among them; therefore, the Assembly should revoke the laws that constrained it, and notify the Pope of Rome of the necessity of revoking the ecclesiastical penalties against clerical matrimony; and, in case these were not revoked within a given time, that they should be nullified.'

"As a matter of course, such a report, coming from an ecclesias-tic of high standing, excited a great deal of attention. To the surprise of many, it was received with great favor by both priests and people. This circumstance, together with his own convictions of duty, prompted the author to develop his opinions at length and in a systematical treatise. Thus originated his celebrated work on Clerical Celibacy. From the remarks of a competent critic on that work, we select the following :—'It is really a novelty in the literary world. We can, in truth, say no less than this:—that the book contains unquestionably the best argument ever advanced, in any Papal or Protestant country, against the constrained celibacy of priests and nuns. It sets forth all that a Protestant can say, and what a Roman Catholic priest, in spite of every early prejudice, is constrained to say, against a cruel and unnatural law, enacted against the immovable law of the almighty Creator.'

"The author is master in ancient as well as in modern Catholic lore,—in canon law, and in the writings of the fathers; and we should be no less amazed than instructed by seeing any one of his brother-prelates in America or in Europe come out with any thing like a rational answer to 'FEIJO'S DEMONSTRATION OF THE NECES-SITY OF ABOLISHING CLERICAL CELIBACY.'

"Notwithstanding the violent attacks made upon him in con-nection with this startling attempt at innovation, yet he was sub-sequently elevated to the highest offices in the gift of the nation. He was, successively, appointed Minister of State, Regent of the Empire, and Senator for life.

"He was, moreover, elected by the Imperial Government as Bishop of Mariana, a diocese which included the rich and important province of Minas-Geraes. He, however, did not see fit to accept this dignity, but, on resigning his Regency, returned to his plantation, a few miles from the city of S. Paulo, where he resided during my visit to that part of Brazil.

"After that period his health declined, and a pension of four thousand milreis per annum was conceded to him, in consideration of his distinguished services in the past. In 1843 he died."

Since the above was written by my co-laborer in this work, many of the leading men whom he met at San Paulo have gone to their rest. Antonio Carlos, Martin Francisco de Andrada, and Alvares Machado, are no more. The constitutional Empire which, with self-sacrificing toil, they aided in erecting, and for which they suffered in the crucible of political persecution, exists on a firm foundation, and their labors are not forgotten, though as yet no lofty monument rears its form to tell of their true patriotism.

Antonio Carlos de Andrada expired on the 5th of December, 1845, and from the *Necrologia* in the *Annuario do Brazil* for 1846 I extract the following testimonial to his talent, worth, and statesmanship. It may be remarked that, if every foreigner who investigates the character of the deceased finds so much to command his admiration, we should pardon the high strain of eulogium pronounced by his countrymen upon one who, for so many years, nobly filled the first places in the gift of the monarch and the people.

"'The *Assemblea Geral* of 1844 being dissolved, Antonio Carlos de Andrada was, in 1845, newly elected Deputy for his native province of San Paulo. But he had scarcely been informed of his election by the Paulistas, when he heard that he had been chosen Senator for Pernambuco, after having also received the popular votes of the provinces of Pará, Minas, Ceara, and Rio de Janeiro. He took his seat thus late in life in the Senate-chamber,—a tardy recompense for his great merit.

"In literature, in Parliament, and in the whole Empire, his death left a great void, which will long be felt by all his compatriots.

"With no other ambition save that of serving his country,—the sole glory desired by his generous heart,—he neither desired nor sought for honors.

"The Councillor Carlos Antonio de Andrada was of medium height and of a robust constitution: every feature of his face expressed genius, feeling, and energy of mind. Of easy and graceful manners, mild and jovial in familiar conversation, he rendered himself agreeable to every one who approached him. Severe for himself, he was indulgent to others, and ready to pardon an offence or an injustice done to him. He was a devoted friend, and a generous adversary to his competitors in public life: he never employed his power to injure others, but always to protect the weak. An excellent father, a loving husband, the best of brothers,—there was not a single domestic virtue which was not found in Antonio Carlos!"

What matters it if to such a man no monumental stone be erected?—

> " The fame is lost which it imparts :
> Who for his dust a tear would claim
> Must write his name on living hearts."

The conclusion of the eulogy to the deceased statesman is the highest encomium that could be pronounced upon a public man in a government where, too often, those in power have not scrupled to enrich themselves at the expense of the State.

There is the noblest and most eloquent praise in the simple fact and statement,—viz.: "Such was the Councillor Antonio Carlos de Andrada: *he lived and died poor!*"

The following details of the missionary efforts of my colleague and predecessor will be found, I doubt not, deeply interesting:—

"Although two hundred years had elapsed since the discovery and first settlement of the province of San Paulo, it is not known that a Protestant minister of the gospel had ever visited it before. Although colonized with the ostensible purpose of converting the natives, and subsequently inhabited by scores of monks and priests, there is no probability that ever before a person had entered its domains, carrying copies of the word of life in the vernacular tongue, with the express intent of putting them in the hands of the people.

"It is necessary to remind the reader, that, throughout the entire continent to which reference is now made, public assemblies for the purpose of addresses and instruction are wholly unknown. The people often assemble at mass and at religious festivals, and nearly as often at the theatre; but in neither place do they hear principles discussed or truth developed. The sermons in the former case are seldom much more than eulogiums on the virtues of a saint, with exhortations to follow his or her example. Indeed, the whole system of means by which, in Protestant countries, access is had to the public mind, is unpractised and unknown. The stranger, therefore, and especially the supposed heretic, who would labor for the promotion of true religion, must expect to avail himself of providential openings rather than to rely on previously-concerted plans. The missionary, in such circumstances, learns a lesson of great practical importance to himself,—to wit, that he should be grateful for any occasion, however small, of attempting to do good in the name of his Master. The romantic notions which some entertain of a mission-field may become chastened and humbled by contact with the cold reality of facts; but the Christian heart will not be rendered harder, nor genuine faith less susceptible of an entire reliance on God.

"The unexpected friendship and aid of mine aged host at San Bernardo, already mentioned, was not a circumstance to be lightly esteemed. Scarcely less expected was the provision made for me, at the city of S. Paulo, of letters of introduction to gentlemen of the first respectability in the various places of the interior which I wished to visit. At one of those places, the individual to whom I was thus addressed, and by whom I was entertained, was a Roman Catholic priest; and it affords me unfeigned satisfaction to say, that the hospitality which I received under his roof was just what the stranger in a strange land would desire.

"When on reaching the town where he lived I first called at his house, the padre had been absent about two weeks, but was then hourly expected to return. His nephew, a young gentleman in charge of the premises, insisted on my remaining, and directed my guide to a pasture for his mules. In a country where riding upon the saddle is almost the only way of travelling, it has become an act of politeness to invite the traveller, on his first arrival, to rest upon a bed or a sofa. This kindness, having been accepted in the present instance, was in due time followed by a warm bath, and afterward by an excellent but a solitary dinner. Before my repast was ended, a party of horsemen passed by the window, among whom was the padre for whom I was waiting. After reading the letter which I brought, he entered the room and bade me a cordial welcome. He had arrived in company with the ex-Regent Feijo,

25

with whom I had previously enjoyed an interview at the city of S. Paulo, and from whom he had received notices of me, as inquiring into the religious state of the country. My way was thus made easy to introduce the special topic of my mission. On showing me his library,—a very respectable collection of books,—he distinguished, as his favorite work, Calmet's Bible, in French, in twenty-six volumes. He had no Bible or Testament in Portuguese. I told him I had heard that an edition was about to be published at Rio, with notes and comments, under the patronage and sanction of the Archbishop. This project had been set on foot in order to counteract the circulation of the editions of the Bible-societies, but was never carried into effect. He knew nothing of it. He had heard, however, that Bibles in the vulgar tongue had been sent to Rio de Janeiro, as to other parts of the world, which could be procured gratis, or for a trifling consideration. Judge of the happy surprise with which I heard from his lips that some of these Bibles had already appeared in this neighborhood, three hundred miles distant from our depository at Rio. His first remark was, that he did not know how much good would come from their perusal, on account of the bad example of their bishops and priests. I informed him frankly that I was one of the persons engaged in distributing these Bibles, and endeavored to explain the motives of our enterprise, which he seemed to appreciate.

"He said Catholicism was nearly abandoned here and all the world over. I assured him that I saw abundant proofs of its existence and influence; but he seemed to consider these 'the form without the power.' Our conversation was here interrupted; but, having an opportunity to renew it in the evening, I remarked that, knowing me to be a minister of religion, he had reason to suppose I would have more pleasure in conversing on that subject than upon any other.

"I then told him I did not comprehend what he meant by saying that Catholicism was nearly abandoned. He proceeded to explain that there was scarcely any thing of the spirit of religion among either priests or people. He, being only a *diacono*, had the privilege of criticizing others. He was strong in the opinion that the laws enjoining clerical celibacy should be abolished, since the clergy were almost all *de facto* much worse than married, to the infinite scandal of religion; that such was their ignorance that many of them ought to sit at the feet of their own people to be instructed in the common doctrines of Christianity; that the spirit of infidelity had been of late rapidly spreading, and infecting the young, to the destruction of that external respect for religion and the fear of God which used to be hereditary. Infidel books were common, especially Volney's 'Ruins.' I asked whether things were growing better or worse. 'Worse,' he replied; 'worse continually!' 'What means are taken to render them better?' 'None! We are waiting the interference of Providence.' I told him there were many pious persons who would gladly come to their aid if it were certain they would be permitted to do the work of the Lord. He thought they would be well received if they brought the truth; meaning, probably, if they were Roman Catholics.

"I asked him what report I should give to the religious world respecting Brazil. 'Say that we are in darkness, behind the age, and almost abandoned.' 'But that you wish for light?' 'That we wish for nothing. We are hoping in God, the Father of lights.'

"I proceeded to ask him what was better calculated to counteract the influence of those infidel and demoralizing works he had referred to than the word of God. 'Nothing,' was the reply. 'How much good, then, is it possible you yourself might do, both to your country and to immortal souls, by devoting yourself to the true work of an evangelist!' He assented, and hoped that some day he should be engaged in it.

"I had before placed in his hands two or three copies of the New Testament, to be given to persons who would receive profit from them, and which he had received with the greatest satisfaction. I now told him that whenever he was disposed to enter upon the work of distributing the Scriptures we could forward them to him in any quantity needed. He assured me that he would at any time be happy to take such a charge upon himself; that when the books were received he would circulate them throughout all the neighboring country, and write an account of the manner of their disposal. We accordingly closed an arrangement, which subsequently proved highly efficient and interesting. When I showed him some tracts in Portuguese, he requested that a quantity of them should accompany the remission of Bibles. On my asking how the ex-Regent and others like him would regard the circulation of the Scriptures among the people, he said they would rejoice in it, and that the propriety of the enterprise would scarcely admit of discussion. 'Then,' said I, 'when we are engaged in this work we can have the satisfaction to know that we are doing what the better part of your own clergy approve.' 'Certainly,' he replied: 'you are doing what we ought to be doing ourselves.'

"Seldom have I spent a night more happily than the one which followed, although sleep was disposed to flee from my eyelids. I was overwhelmed with a sense of the goodness and providence of God, in thus directing my way to the very person out of hundreds best qualified, both in circumstances and disposition, to aid in promoting our great work. This fact was illustrated in the circumstance that, although I had a most cordial letter of introduction to the vigario of the same village, which I left at his house, I did not see him at all, he happening to be out when I called. To use the expression of a gentleman acquainted with the circumstances, 'he hid himself,' as though fearing the consequences of an interview, and, by not showing at least the customary civilities to a stranger, greatly offended the gentleman who had given me the letter. The padre whose kindness I experienced had paused in his clerical course some years before, and was engaged in the legal profession, although he retained his title and character as a priest. In correspondence with this circumstance, there is scarcely any department of civil or political life in which priests are not often found. After the second night I was under the necessity of taking leave of him in order to pursue my journey.

"At another village, a young gentleman who had been educated in Germany was often in my room, and rendered himself very agreeable by his frank and intelligent conversation. He represented this to be one of the most religious places in the country, having a large number of churches and priests in proportion to the population. In one church particularly the priests were unusually strict, and, in the judgment of my informant, quite fanatical. They always wore their distinguishing habit, were correct in their moral deportment, required persons belonging to their circle to commune very often, and, moreover, discountenanced theatres. This latter circumstance was unusual; for, in addition to the clergy being often present at such amusements, there was even in that place the instance of a theatre attached to a church.

"I introduced to this young gentleman the subject of circulating the Bible. He at once acknowledged the importance of the enterprise, and expressed great desires that it should go forward; saying that the Brazilians, once understanding the objects of the friends of the Bible, could not but appreciate them in the most grateful manner. He proposed to converse with his friends, to see what could be done toward distributing copies among them. I put two Testaments in his hands as specimens. The next morning he told me that, having exhibited them the evening previous to a company of young persons, there had arisen a universal demand for them, and many became highly urgent not to be overlooked in the distribution. He consequently repeated his assurance that the sacred books would be received with universal delight, and requested a number of copies to be sent to his address. I was told that here also many of the rising generation had very little respect for religion, through the influence of infidel writings and of other causes. The apology for almost any license was, 'I am a bad Catholic.' The people generally assented to the dogmas of the Church, but seldom complied with its requirements, except when obliged to do so by their parents or prompted by the immediate fear of death. The rules requiring abstinence from meats on Wednesdays and Fridays, also during Lent, had been abolished by a dispensation from the diocesan bishop for the last six years, and the Provincial Assembly had just asked a repetition of the same favor. The decision of the bishop had not then transpired, but many of the people were expressing a disposition to live as they should list, be it either way.

"Just previous to my visit to this place, a young man of a respectable family, having sunk his fortune in an attempted speculation on a newly-arrived cargo of African slaves, had committed suicide. It was said to be the first instance of that crime ever known in the vicinity, and the result was an unusual excitement among all classes. I may here observe, that suicide is exceedingly rare throughout the whole of Brazil; and there can be but little question that the rules of the Church, depriving its victim of Christian burial, have exerted a good influence in investing the subject with a suitable horror and detestation. Would to Heaven a similar influence had been exerted against other sins equally damning but more insidious! The very abomination of moral desolation could exist in the same community almost unrebuked.

"At a third village I was entertained by a merchant of truly liberal ideas and of unbounded hospitality. He also offered to co-operate with me in the circulation of the sacred volumes, not only in his own town, but also in the regions beyond.

"Having accomplished a journey of about two hundred miles under very favorable circumstances, I again reached the city of S. Paulo. 1 had not stayed so long in various places as I should have been interested and happy to do, in compliance with urgent invitations. I had, however, important reasons for not indulging my pleasure in this respect. My mind had dwelt intensely upon the state of the country, as shown by facts communicated to me from various and unexceptionable sources. I had anxiously inquired how something for its good might be accomplished; whether there was any possibility of exceeding the slow and circumscribed limits of private personal communication of the truth. Hope, in answer, had sprung up in my mind, and was beginning to be cherished with fond expectation.

"From the idea of distributing a couple of dozens of Testaments in several schools of the city, I was led to think of the practicability of introducing the same as reading-books in the schools of the whole province. This seemed to be more desirable from the fact, universally affirmed, that there then prevailed an almost entire destitution of any books for such use in the schools. The Montpellier Catechism was more used for this purpose than any other book; but it had little efficacy in fixing religious principles upon a proper basis, to resist the undermining process of infidelity.

"Encouraged by the uniform thankfulness of those individuals to whom I presented copies, and also by the judgment of all to whom I had thought proper to suggest the idea, I had finally resolved to offer to the Government, in some approved form, a donation of Testaments corresponding in magnitude to the wants of the province. Fortunately I had, in the secretary and senior professor of the university, a friend fully competent to counsel and aid in the prosecution of this enterprise. I laid the whole subject before him.

He informed me that the proper method of securing the object would be by means of an order from the Provincial Assembly, (if that body should see fit to pass one,) directing the teachers of schools to receive said books for use.

"Early next morning he called with me to propose the subject to various prominent members of the Legislative Assembly. We visited gentlemen belonging to both political parties : two priests, one a doctor in medicine and the other a professor in the Academy of Laws; the Bishop-elect of Rio de Janeiro, who was confidential adviser of the old Bishop of S. Paulo,—the latter also belonging to the Assembly; and at length the Andradas. Each of these gentlemen entertained the proposition in the most respectful manner, and expressed the opinion that it could not fail to be well received by the Assembly. The bishop, who was chairman of one of the committees to which it would naturally be referred, said he would spare no effort on his part to carry so laudable a design into effect. He, together with one of the padres referred to, had purchased copies of the Bible, at the depository in Rio, for their own use, and highly approved of the edition we circulated.

"Our visit to the Andradas was peculiarly interesting. These venerable men, both crowned with hoary hairs and almost worn out in the service of their country, received me with gratifying expressions of regard toward the United States, and assurances of entire reciprocity of feeling toward Christians who might not be of the Roman Church. They were acquainted with, and appreciated the efforts of, the Bible Societies : they, moreover, highly approved of the universal use of the Scriptures, especially of the New Testament. They pronounced the offer I was about to make to be not only unexceptionable, but truly generous, and said that nothing in their power should be wanting to carry it into full effect. Indeed, Martin Francisco, the president of the Assembly, on parting, said that it gave him happiness to reflect that their province might be the first to set the example of introducing the word of God to its public schools. Senhor Antonio Carlos, at the same time, received some copies of the Testament as specimens of the translation, which, with the following document, as chairman of the Committee on Public Instruction, he presented in course of the session for that day :—

"'*Proposition to the Honorable Legislàture, the Provincial Assembly of the Imperial Province of S. Paulo.*

"'Whereas, having visited this province as a stranger, and having received high satisfaction, not only in the observation of those natural advantages of climate, soil, and productions with which a benignant Providence has so eminently distinguished it, but also in the generous hospitality and esteemed acquaintance of various citizens; and,

"'Whereas, in making some inquiries upon the subject of education, having been repeatedly informed of a great want of reading-books in the primary schools, especially in the interior; and,

"'Whereas, having relations with the American Bible Society, located in New York, the fundamental object of which is to distribute the Word of God, without note or comment, in different parts of the world; and, whereas the New Testament of our Lord and Saviour Jesus Christ is a choice specimen of style, as well on subjects historical as moral and religious, in addition to embodying the pure and sacred truths of our holy Christianity, the knowledge of which is of so high importance to every individual, both as a human being and as a member of society; and,

"'Whereas, having the most unlimited confidence in the philanthropic benevolence of said Society, and in its willingness to co-operate for the good of this country, in common with all others, and especially in view of the happy relations existing between two prominent nations of the New World: therefore I propose to guarantee, on the part of the said American Bible Society, the free donation of copies of the New Testament, translated into Portuguese by the Padre Antonio Pereira de Figueiredo, in sufficient number to furnish every primary school in the province with a library of one dozen,—on the simple condition that said copies shall be received as delivered at the Alfandega (Custom-House) of Rio de Janeiro, and caused to be distributed among, preserved in, and used by, the said several schools, as books of general reading and instruction for the pupils of the same.

"'With the most sincere desires for the moral and civil prosperity of the Imperial province of San Paulo, the above proposition is humbly and respectfully submitted.   "'D. P. KIDDER.

"'CITY OF SAN PAULO, Feb. 15, 1839.'

"The same day I received a verbal message, saying that the Assembly had received the proposition with peculiar satisfaction, and referred it to the two committees on ecclesiastical affairs and on public instruction. The following official communication was subsequently received:—

TRANSLATION.

" 'To MR. KIDDER :—I inform you that the Legislative Assembly has received with especial satisfaction your offer of copies of the New Testament, translated by the Padre Antonio Pereira de Figueiredo, and that the Legislature will enter into a deliberation upon the subject, the result of which will be communicated to you.

" 'God preserve you!

" 'MIGUEL EUFRAZIO DE AZEVEDO MARQUEZ, *Sec.*

" 'PALACE OF THE PROVINCIÁL ASSEMBLY,
S. PAULO, Feb. 20, 1839.'

"Among other acquaintances formed at S. Paulo was that of a clergyman, another professor in the Law University. His conversation was frank and interesting, and his views unusually liberal. He gave as emphatic an account as I have heard from any one of the unhappy abandonment of all vital godliness and of the unworthiness of many of the clergy. He approved of the enterprise of the Bible Societies, and cheerfully consented to promote it within the circle of his influence by distributing Bibles and tracts, and reporting their utility. Exchanging addresses with this gentleman, I left him, entertaining a high estimation of his good intentions, and with ardent hopes that he might yet be greatly useful in the regeneration of his Church and in the salvation of his countrymen.

"Thus were happily completed arrangements with persons of the first respectability and influence, in each principal place of the interior which I had visited, that they should distribute the word of God among their fellow-citizens. All the copies that I brought were already disposed of, and there was a prospect that the day was not distant when it could be said that a Roman Catholic Legislature had fully sanctioned the use of the Holy Scriptures in the public schools of their entire territory. I was told, on the best authority, that the committees of the Assembly were drafting a joint report, recommending compliance with the offer by means of an order on the treasury for the funds needed in payment of the duties and the expense of distribution.

"Such circumstances as the results of this short visit were so far beyond the most sanguine anticipation, that, on leaving, I found it difficult to restrain my feelings of gratitude and delight for what mine eyes had seen and mine ears had heard.

"In conclusion, it becomes necessary to add that, owing to the agitations and intrigues common to most political bodies, action in reference to my proposition was delayed beyond the expectation of its friends. The last direct intelligence I had from the subject was received in conversation with the president of the Assembly. I met this gentleman on his subsequent arrival at Rio de Janeiro to discharge his duties as a member of the House of Deputies. He informed me that such were the political animosities existing between the two parties into which the Assembly was divided that very little business of any kind had been done during the session. The minority as a party, and individuals of the majority, favored the project, but, under the circumstances, did not wish to urge immediate action upon it. Meantime, through some slanders circulated by an English Catholic priest residing at Rio, the suspicions of the old bishop were excited lest the translation was not actually what it purported to be, but had suffered alterations.

"An examination was proposed, but, either through inability or wilful neglect, was not attempted; and thus the superstitious humor of the old diocesan was counted among other things which caused delay. The president expressed a hope that on the next organization of the Assembly the proposal would be fully accepted.

"I subsequently saw in a newspaper that the committee to whom the subject had been referred, or probably its chairman, in direct contravention of his voluntary promise to me, but in obedience to the old bishop's idle fears, had filed in the secretary's office a report unfavorable to the proposal. The proposition was probably never acted upon. To the credit of the province, it certainly was never formally rejected."

The dissemination of the truth, however, does not depend upon legislative acts or the aid of statesmen, though we may hail with pleasure every move of the "powers that be" for the advancement of knowledge and religion. The circulation of the Scriptures is not a matter of sectarianism; and all should rejoice in the diffusion of that "which" (as the barbarian chieftain in Northumberland said to his compeers when the first monk visited Britannia) "teaches us the origin and the destiny of our souls."

I visited the province of S. Paulo more than sixteen years after the events narrated above, and I found the same willingness manifested by all ranks of society in the reception of the word which my companion in authorship experienced among the Paulistas, and I was thus enabled to diffuse very many copies of Holy Writ. From time to time, in this pleasant portion of Brazil, I found much to encourage my labors among the humble and ignorant as well as among the more elevated and intelligent. It was not less pleasing occasionally to trace the workings of the seeds of truth sown so many years before by Dr. Kidder. I found that an eminent Brazilian had been won, by the perusal of *A Santa Biblia*, to "wisdom's ways," and to become the earnest advocate of its circulation. Far in the interior of this province I met with two gentlemen who did not profess to be Christians, but who, as philanthropists, took a deep interest in the Bible cause. One of them told me that a Brazilian came to him a few days before with a Portuguese Bible, saying that he was "so rejoiced to have the Bible in his own vernacular." My informant thinks this *Biblia* must have come either from my predecessor or from the Bibles left at the house of an American merchant in Rio de Janeiro. I was also informed by an English watchmaker at Campinas that he had met with a Brazilian who had in his possession a Portuguese Bible, and that he took great pleasure in carrying it with him to the Roman Catholic church each Sunday.

In a most fertile and densely-populated portion of the province I made the acquaintance of a physician who had resided in Brazil eleven years,—had travelled, for scientific purposes, through much of the Empire,—had won the respect and esteem of the Brazilians by his affability as well as his professional ability. He therefore has a great influence. It is his opinion that Brazil, in a certain sense, is ready for a reformation; but that the inhabitants have had such immoral priests, and are themselves so low in a moral point of view, that it would not be a vigorous breaking away from the trammels of Romanism. They are, however, not bigoted, and are willing to read. He it was that gave me the instance of the padre who, by reading some of the works of Luther that had strayed from Germany into Brazil, preached such Protestant sermons that he was attacked by the bishop, and finally driven away from his parish, but not from his sentiments. It seemed to me, when hearing of this incident, that the old German Reformer was still hurling his inkstand.

CHAPTER XXI.

AGREEABLE ACQUAINTANCE—OLD CONGO'S SPURS—LODGING AND SLEEPING—COM-
PANY—CAMPINAS—ILLUMINATIONS—A NIGHT AMONG THE LOWLY—ARRIVAL AT
LIMEIRA—A PENNSYLVANIAN—A NIGHT WITH A BOA CONSTRICTOR—EVENTFUL
AND ROMANTIC LIFE OF A NATURALIST—THE BIRD-COLONY DESTINED TO THE
PHILADELPHIA ACADEMY OF NATURAL SCIENCES — YBECABA—SKETCH OF THE
VERGUEIROS — PLAN OF COLONIZATION — BRIDGE OF NOVEL CONSTRUCTION —
FUTURE PROSPECTS.

ON the morning of the 21st of June, I left the city of San Paulo
for Limeira. Before starting, I called upon Messrs. E. and C., two
English engineers who had come out to make the surveys for a carriage-road into the interior. In the bookcase of Madam E. I
found many an old friend. How curious it was to see Cheever's
"Windings by the Waters of the River of Life," Hamilton's "Life
in Earnest," and other good books, in this distant city, whose very
existence was perhaps unknown to the authors mentioned! I was
loath to leave the agreeable company at Mr. E.'s; but my mules, horse,
and conductor were all ready, and now, with this cavalcade, *vamos*.

My conductor was an old darkey of sixty, whose vestments consisted of a roundabout, a pair of pantaloons, and an old straw
hat. His naked, bony heels were ungarnished by the slightest
sign of a spur. As I was to ride fast, in order to accomplish my
journey in a given time, I saw that it would never do to have old
Congo go unarmed as to his pedal ex-
tremities; so, reining up at a hard-
ware-store, I furnished the ancient with
a pair of iron spurs, each spike of
which was large enough for the gaff of
a fighting-cock. With a bit of whipcord he fastened them to his skinny
ankles, and, mounting, we were soon

en route, and in a few minutes cleared the city of San Paulo.

At ten o'clock in this climate the sun is by no means cold. The

extra animals, once outside of the streets, had a great disposition
to roam over the plains of Piratininga, and much of our time was
lost in changing from one side of the road to the other in search
of the fugitives.   Under the influence of his unusual exercise and
the warmth of the day, the juice of youth seemed to be oozing out
of old Congo.   He uttered prayers, at a most vociferous rate, to
Santa Maria and *Diabo*.   And I am sorry to record that most of
his pious ejaculations were to the latter character, whose name,
though not in the calendar, is more frequently used in Brazil than
those of all the saints put together.   Hearing the clatter of hoofs
behind us, I turned round, and beheld two Paulistas galloping in
the same direction with ourselves.   In passing us, they both burst
into a fit of immoderate laughter.   I could not at first divine what
so excited their cachinnatory powers, until one of them exclaimed,
" *Olha as esporas.*"   Upon looking down, I perceived that the whipcord which fastened the iron spikes to the heels of old Congo had

slipped around, and the spur was standing
out prominently in front of his instep.   The
old fellow, in his arduous chase after the
wandering mules, had not perceived this, and
went on belaboring and thumping the sides
of his animal with his blunt, bony heels.

After the ride of a league, I found my
boxes; but Joachim Antonio da Silva, the muleteer who had them
in charge, would not give them up until I made many assurances
that all was right.   And now once more forward !

Previous to to-day, I had always had young negroes or German
boys for my conductors, and I feared that the ambition of old
Congo was dead, and that no hope of reward would resurrect it.
He went very slow: the journey must be accomplished with those
boxes in four days, or I could not come off victor.   The trip was
considered, by muleteers, one of eight days; so, in order to accelerate the speed of my animals, I determined not to leave old Congo.

We pushed on, as rapidly as possible, through a fine region of
country, abounding in coffee and sugar plantations.   I had much
conversation with the old negro, who could remember when, more
than half a century ago, he was stolen on the coast of Africa, but
did not recollect ever having heard the story of the Creation and

Redemption; so I employed myself in endeavoring to pour into his mind some light on that greatest of all subjects to man. He found it very interesting, and pronounced it *"muito bonito,"* (very beautiful.)

With all our pushing, driving, and changing animals, we only got over twenty-four miles,—which is a good day's work for Brazilians, but did not satisfy me. By a bright moon we arrived at a house where we could find no "entertainment for man or beast." We rode on to a mere road-side hovel, and to our question, *Tem lugar?* we received the response, "We cannot receive you: we have no room." This was from a slatternly-looking mulattress. Every thing was against us; but it was impossible for us to go farther. Old Congo, however, made a speech with such eloquence that the desired quarters were obtained. And such a room! No cabin in Old Ireland, or clapboard shed in the "Far West," could surpass it in ugliness and narrowness, to say nothing of dirt. The floor was mud, and the walls were of dried mud, ornamented with the marks of the "daubing" fingers. It was six feet by eight, and here were stowed self, saddles, sacks, and Congo. No wonder that they said they had "no room." We supped off of beans, uncooked corn-meal, and eggs, whose durable qualities were not to be questioned. We (that is, I first and Congo afterward) stood up (for there was no chair in the house) to a table something like a horse-trough. I am capable of any thing. My bed was a mat spread on a board and graced by a pillow and a sheet. Such an article as a coverlet did not exist in that *casa.* The African had more sense than I had, for his poncho was large and heavy. By a dim light stuck into the mud wall, I read to poor old Congo the first passage of the Holy Word that he, doubtless, had ever heard in a language which he understood; then, praying in Portuguese, I lay down upon my board, and he upon the ground, which I think must have been a softer couch than mine. In a letter to a friend I thus detailed my experience:—"I piled on to me, in lieu of coverlet, my saddle-cloth and mackintosh. I was more sensitive to the cold than the night before, and sleep would not be wooed. I then put on my coat; but that did not keep off the cold nor the fleas, which were 'still so gently o'er me' creeping. I kicked away until I could stand it no longer, and then (I scarcely dare write it to you) I aroused old Congo from a sound sleep, and made him get into—no —on to my board, to warm me. It was not exactly the case of the aged monarch of Israel; for it was cruel to transfer the ancient darky from the comfortable bosom of mother-earth to the hard realities of a soft board and a cold young man. I profited nothing by it, for slumber came not to my eyelids, and the thought of certain *bixos* rendered me still more wakeful, if such a thing were possible."

Before cock-crowing I ordered the mules to be saddled, and at daylight we were again on our way. I rode on, far in advance of my muleteer, and, passing a mile beyond the village of Jundiahy, I arrived at the hotel of Senhor José Pinto. I found a large party at a twelve-o'clock breakfast, which repast was perfectly *à la Brazilienne.* They supposed that I would wish matters in a different style, but I made them all at ease by sitting down, telling them that I was not a stranger, and manifesting my "at-homeness" by eating as heartily of their dishes as if I had been accustomed to them all my life. This opened their hearts, and thus gave me, both then and afterward, an opportunity of speaking of those higher interests which concern man here below.

In two hours or more my baggage-mules came up. I perceived that, at this rate, it would be impossible for me to get on as I wished, or to complete all my arrangements at Limeira and Ybecaba and get back to Rio de Janeiro for my northern trip. Fortunately for me, I found at José Pinto's the two Paulistas whose mirth had been so excited at the revolution of the old African's spurs. They were going far into the interior, and had an extra animal, which I hired, and pushed on, accompanied by them, leaving my old Congo to come up *sem duvida* (without fail) two days after me.

I had now a better opportunity of knowing something more of the *moradores*, or road-side dwellers, of which class my companions were specimens. They sang for me fandango melodies, Ethiopian airs in bad Portuguese, and entertained me in various ways. In return, I gave them some information about the world outside of Brazil, not leaving out, in the end, a mention of the "Happy Land."

Our resting-place was to be the important town of Campinas,

(or San Carlos,) more than one hundred miles in the interior. As we approached this town, I was struck by the beauty and fertility of the surrounding country. The grand old mountains had been left far behind us, and around, as far as I could see, were extensive plains, or rather rolling prairies, and almost every acre occupied. There were most highly-cultivated coffee-plantations, from whose deep green could be seen, peeping here and there, the large white residences of the planters. It was on the evening of the 28th of June that we drew near Campinas. The clear beauty of the tropic night was made even more beautiful by the illumination of the city, by the huge bonfires spread over the plains, and by the most brilliant fireworks sent up from every street and from all the surrounding plantations. The sight and sounds were such that one, without any stretch of imagination, would have believed himself near some besieged city during a fierce bombardment. It was "St. Peter's Eve;" and every man who had a *Pedro* attached to his name felt himself obligated to burn a huge heap of combustibles before his door, and to send up any quantity of sky-rockets and fire off innumerable pistols, muskets, and cannon. Under such a storm we entered Campinas. My two Paulistas led me through the narrow streets, and we finally arrived before a row of small whitewashed houses. These were the residences of the friends of my Paulistas; but I could not think of stopping there, and desired that some one would lead the way to an inn. They were all very kind, but were so occupied with our tired animals that no one could be spared for the purpose. The hotel, if one can call it such, was at a great distance, and it was suggested that I had better stop with them, though it was *muito mal*, (very bad fare.) I thought that it could not be harder than the night before. I entered: this was the residence of Senhor *Theobardo o Carpinteiro;* or, in plain English, Theobald the carpenter. Senhor Theobardo, however, had not expended any of his skill upon his own house, for the floors and the walls were composed of the same substance as the street. The night before I had only been in the outer court. I now had an opportunity of seeing the inner temple. Senhor Theobardo was half Indian, half mulatto, and I think that, if he could have had an extra half, it would have been yellow Portuguese. He and his children had formed such a close alliance with the substance of which his floors were made, that one could literally say that all (judging from their complexion) were of the "dust of the earth." The kitchen, which served the purpose of parlor and dining-room, was without chimney, chairs, or any of the appliances of civilized life. A few earthen pots were the culinary utensils, and a fire in one corner of the room, in the style of the Patagonians, (indeed, I have seen the same kind among the Terra del Fuegians,) served for cooking, the smoke the meanwhile escaping as best it could. When I saw Mr. Theobardo, Mrs. T., and all the little T.s squatting around the fire, and the mellow light of the embers not softening their sallow features, which, excepting their flashing eyes, were un-relieved by a single trace of cleanliness or grace, I thought that Borrow, in his wildest adventures among the gypsies of Spain, could not have witnessed a group more wild, more dirty, or more picturesque. But I soon found that, although they had dirty faces, they had large hearts, and I reflected that my mission was to them as well as to the more elevated; so I made myself at home, and also put them at their ease. We talked about the United States, and finally I got out a Portuguese New Testament, and, collecting whites, and those who had all sorts of mixtures, from the white, through the red, down to the negro, I commenced read-ing the Holy Book. I had a most interested audience, who proba-bly for the first time heard the message of salvation. I shall never forget that night, and the kindness of the most lowly people I ever met with,—lowly, at least, as to this world's goods; and it is my earnest hope and prayer that the truth may reach and enrich their souls.

The room which they assigned to me was not quite so large as the one I had occupied the night before, and was shared between boards, planes, chisels, saws, harness, saddles, a Paulista, and my-self. Just as I was retiring, a huge wooden bowl, as large as a bath-tub, was brought to me filled with water. This was of their own accord: but who would have thought it, among these people who apparently never performed any ablutions?

That night slumber was sweet indeed; and the next morning I departed at an early hour, leaving my blessing and one milreis with the kind Theobardo. The former he accepted, but the latter he declined, until I forced it upon him as a *lembrança*.

Our route was still more picturesque than that of yesterday The fine road was overshadowed by trees and wild vines; and the carolling birds and singing Paulistas made the ten leagues appear short. Our party was enlarged by two young Germans on their way to Ybecaba. All the houses by the road-sides, and even the huge churches, are built of (or, rather, rammed down with) mud or clay. The large conventual buildings of S. Paulo and the immense church of Campinas (whose walls are five feet in diameter) are composed of beaten earth.

The whole feature of the country had changed: the sublime scenery of the coast was not here to be found, but, in its stead, that which reminded me of the United States. In the newness of the settlements and plantations, I could have easily believed myself in the northern part of Ohio. We were now constantly fording and passing over streams, which were the head-waters of the River Plate. We pushed on until night, illumined by a full moon in an unclouded sky, brought us to the town of Limeira. Here I had before been informed I should find an American physician, Dr. ——, formerly of Pennsylvania. I rode up to his house, and had a most welcome reception. I desired to journey on by moonlight to the plantation of Senator Vergueiro; but the doctor would take no refusal, and stated as a further inducement that another American had arrived that very day, and that we together would compose such a trio as had never before been seen in the distant villa of Limeira.

Limeira is situated in a most fertile region, watered by streams that send their tribute to the mighty Paraná. If Dr. —— was surprised at my unexpected arrival, I was no less astonished to learn that another American had arrived that day, who was perambulating the province, practising his profession of dentist. In what nation pretending to civilization will you not find the American dentist? I may be permitted to indulge a little patriotic pride when speaking of this profession, whose members more than any other of my compatriots may be found in almost any portion of the world. Their superior merits have been repeatedly acknowledged by Englishmen and Frenchmen of the same profession. The secret of their perfection and success has been owing to various causes, not the least of which is the regular dental colleges which exist in the United States, being the first institutions of the kind ever founded, and until recently the only ones in the world. I have met with American dentists at Rio de Janeiro, Valparaiso, and in New Granada. At Paris the dentists *à la mode* are Americans.* A sickly schoolmate, with whom in years gone by I had dug out many a page of hard Latin, is now the most popular dentist in Berlin. On the continent, in interior cities, you will meet with Yankee teeth-replacers and teeth-extractors; and, if the professor or doctor has not the advantage of being a citizen of the great Republic, he publishes in emphatic characters in his advertisements that he has studied his profession in the United States, or fills molars *à la mode Americaine.*

But to return to Dr. ——. He gave me a hearty Pennsylvania welcome, and, as it was late, soon conducted me to my chamber. Now, this chamber was adjacent to a medicine-room, where were not only plenty of the bottled doses which flesh in Brazil is frequently "heir to," but also the apartment was adorned with many specimens of the rich floral and animal kingdoms of Brazil. There being no door to close the aperture that existed between this room and mine, I was frequently disturbed during the night by a strange noise, which could not proceed from unemployed physic or from the dried and stuffed specimens which were hung around in profusion. When daylight returned, I ascertained that the singular noise had arisen from the rustling of a very fine *boa-constrictor*, that had slept (or rather attempted to sleep) within about eight feet of my bed.

---

* American Dentists.—Mr. Walsh, the Paris correspondent of the Journal of Commerce, in a late letter, says:—

"A few days ago I had occasion to apply to the principal Paris bookseller in the department of medicine for some recent comprehensive and elegant work on Dentistry. He wrote to me at once the following reply:—'I regret that it is not in my power to meet your wishes: there is nothing recent nor good in France on the art and science of dentistry. Our surgeons are obliged to borrow from the Americans their proficiency and treatises on this subject, acknowledging that your countrymen are much further advanced than they themselves are in this important branch of the medical art. It is unnecessary for me to mention to you works published fifteen years ago.' Your dentists may be gratified by this testimony. The success of the Americans of the profession who have settled in this capital is strong evidence of the justness of appreciation."

This room-mate of mine had been presented to the dtctor, and was one of the chief occupants of the medical apartment.

The doctor's life had been of that romantic kind which from time to time we find coupled with devoted study and hard reality. A great lover of nature, he early turned his attention to botany and geology. He roamed over the whole United States, and finally came with a few others to Brazil, many years ago, to explore the *flora* and mineralogy of this Empire. Being an enthusiastic naturalist, he fairly revelled in the glorious field of his favorite studies; but the sickness of one of the expedition brought him back to Rio de Janeiro, where he was induced by the American minister to fill the place of mineralogist on board of an American frigate which was on its way to examine the coal-fields of Borneo. I shall not soon forget the interesting account which he gave me of this expedition, during which he visited Madagascar, the coasts of Zanzibar, China, Tonquin, Manilla, &c. &c. His reports adorn the publications of the Smithsonian Institute. After he had filled his accepted time of service on board the frigate, he returned to Brazil, penetrated the forest, and resumed, on his own account, further explorations; but, in order to obtain the necessary means, he first practised his profession as a physician.

From other lips I learned the sequel of the doctor's adventures in a field widely different from that of botany. He opened his office on the plaza of an important town in the interior of San Paulo. On the opposite side of the square was a young Brazilian widow, endowed with the double attraction of wealth and beauty. It was not long before the doctor was approached by *empenhos*,* and became duly informed that the bereaved *Brazilienne* thought that she could find in him a solace for all her afflictions. The doctor replied that he was already married to the virgin forests, and, not contemplating another marriage, ran away to his beautiful woods.

---

* *Empenho:* this word is used in Brazil to express the idea, in politics, commerce, &c. &c., of soliciting aid, promotion, and favors not by direct approaches. Thus, A wishes a favor from D: A ascertains that B is very well acquainted with C, who is a most influential friend of D, and to whom D is under obligations. B goes to C and C in turn to D, and thus the favor is obtained through intermediates. The verb *empenhar* means to lay, to pawn, to pledge, to persuade. *Dinheiro, Diabo,* and *Empenho* are most frequently used in Brazil.

On his return, however, a more powerful *empenho* was brought to bear upon him. The doctor yielded,—was led to the church, and the fair Paulista married him. Their union was blessed by a fine, chubby boy, whom the patriotic physician named George Washington, fondly hoping that this was the first child born in Brazil who bore the illustrious name. "But," said he, "fancy my disgust when, the other day, I learned that some yellow *Sertanejo* had anticipated me, and had his clay-bank urchin baptized also George Washington!"

At the earnest request of influential persons, he took up his residence at Limeira; but his plans for botanical researches, foiled for a time, have not been given up, and it is his intention at some future day to explore the dense *sylva* of the interior, where nature so luxuriantly abounds in the gigantic, the wonderful, and the beautiful.

On the following morning after my arrival at Limeira, accompanied by Dr. ——, I went to the Fazenda de Ybecaba, the plantation of the Vergueiros. It was a clear and lovely day, and we rode along under an archway of forest-trees, many of them clad with the most curious epiphytes and orchidaceous plants. From time to time the doctor would point out some very remarkable subjects of this portion of Flora's kingdom, and delineate their peculiarities and qualities as only one can whose heart is bound up in the beauties of nature. We halted in an open space, and my companion indicated with his finger one of the common palms of this region. In the tree itself there was nothing to render it worthy of attention above its fellows to those accustomed to its graceful form; but there was an accidental interest given to it which called forth the doctor's enthusiastic admiration. He was not only a thoroughly-educated botanist and mineralogist, but was an amateur ornithologist, and loved to watch every trait of the gaudy and brilliant birds of Brazil. From the tufted crown of the palm there hung twenty nests of the large oriole called the *Iguash;* and the feathery inhabitants of this swinging town were hovering around and chattering like "children just let loose from school." The doctor informed me that, though so many leagues intervened between Limeira and the sea-coast, he would cause the tree to be carefully cut down, sawed into sections, and trunk, top, and

nest transported to Santos, and there shipped for Philadelphia. Its
destiny, after it arrived at the City of Brotherly Love, was to be the
Academy of Natural Sciences.   The nests would also be sent, with
several specimens of the *Iguash.*   This whole project, however,
was to be coupled with one condition, which was a *sine quâ non; i.e.*
the Directors of the said Academy of Natural Sciences were to re-
erect the palm-tree, with its long nest-adornments, in the centre or
in some conspicuous part of their edifice; for, unless this was
guaranteed, the doctor added, " palm-tree, birds, and all would soon
be consigned to oblivion."   It was a grand idea—and I doubt if it
were ever before entertained by a naturalist—to transport a lofty
nest-covered tree on the shoulders of men for more than two hundred miles, in order that it might be sent thousands of leagues
over the ocean as a specimen of the wonders of vegetation and of
the bird-architecture of this Southern Hemisphere.

We resumed our route, and
in a few minutes we over-
took old Congo, who, true to
his word, had driven and
ridden well, and had got over
more ground in forty-eight
hours than he had on any
previous occasion in five days.
We emerged from the forest-
bordered road, and saw in the
distance the celebrated plantation of Senator Vergueiro.

Though I had heard more of this establishment than of any
similar one in Brazil, it did not fall behind my anticipation.
We passed through the great gateway, and were welcomed by the
screams of a flock of gayly-painted parrots, which were at times
alighting, and at times whirling around the tops of a group of lofty
trees.   Two pairs of them rested upon different branches, and
seemed to be in amiable confab in regard to the newly-arrived.
Between Campinas and Limeira, and also at Ybecaba, I beheld the
loftiest trees that I met with in any portion of the country.   Three
noble denizens of the forest have been left not far from the residence of Senhor Vergueiro, and form a conspicuous object in the

landscape. In the distance could be seen the manor and the chapel, and on either side of them various out-buildings, which served as shops, store-rooms for coffee, and sheds for machinery. On our left were neat little cottages belonging to the colonists. The peculiarity of Ybecaba consists in the fact that free labor is employed in carrying on its vast operations; and those whom Senator Vergueiro and his sons have brought to displace the Africans are men of the working-classes from Germany and Switzerland. With enlarged views and true economy, we shall see in the sequel that they have adopted that plan which has not only been productive of great and profitable results to themselves, but that they have helped to elevate and greatly benefit the condition of those who were in narrow circumstances at home. The Vergueiros have solved the question, so often asked, " What is the true mode for colonization in Brazil?"

As we drew near the mansion we saw on every side of us evidences of thrift. For the first time away from Rio de Janeiro I saw carts whose wheels were not of the old primitive Roman kind, but turned upon their axles like civilized cartwheels. And it may be mentioned that these, and all the agricultural implements and machinery, are manufactured on the plantation. When subsequently examining the workmanship of carpenters, cabinet-makers, blacksmiths, and wheelwrights, from the Cantons de Vaud and Valais, and from interior villages of Prussia, I perceived that not only had they not lost their skilfulness, but had actually gained under the supervision of their enlightened proprietors.

Senhor Luiz Vergueiro received us with marked attention. The doctor was, of course, an old favorite; but Senhor V. soon made me feel at home, and I afterward discovered that he took a deep interest in my visit to Brazil, from the account which he had read in the *Correio Mercantil* of my presentation, at Rio de Janeiro, of the various specimens of American arts and manufactures to the Emperor and to the different scientific societies of the metropolis.

Every facility was given me for full investigation of the books of the plantation and the condition of the colony, which enabled me to make a just and fair comparison between this system of colonization and those of Petropolis and Donna Francisca, and also to see more clearly the results of contrasted free and slave labor. The whole of the day was thus occupied; but, before detailing any account of that examination, it will be best to give a more full account of the family Vergueiro, whose venerable head has been mentioned several times in previous pages of this work.

Nicoláo de Pereira de Campos Vergueiro is a native of Portugal, and of noble descent. He arrived in Brazil before the King, Dom John VI. By profession a lawyer, he is a man of cultivated and disciplined mind. He early settled in the province of San Paulo, and took a conspicuous part in the political affairs of the country. From the very commencement of agitations for extending the rights of his adopted land, he stood in the foremost rank of patriots, shoulder to shoulder with the Andradas, Feijo, and others eminent in the struggle for Brazilian independence. His private virtues, his moderate and enlightened views, and his great firmness, made him an object of confidence on the part of the people. He was deputed to the Cortes of Portugal, having for his colleagues José Bonifacio de Andrada, and Feijo. He did not, however, escape to England with them when they were threatened by the Cortes, but demanded, fearlessly and firmly, his passport, and succeeded in obtaining it. He returned to Rio de Janeiro, and from that time to this has been a leader on the liberal side of politics, and is to-day called a *Santa Lusia.* From the era of Brazilian liberty until now, he has either been Deputy or Senator. On that trying night when the people in the Campo Santa Anna clamored for the reinstatement of the Ministry dismissed the previous day, Dom Pedro I., before resorting to the last expedient left to him by the Constitution, sent for Vergueiro, knowing that he was one who possessed the confidence of the populace, to desire him to form a ministry in accordance with their wishes. Vergueiro was not found, or the revolution would have either been stayed or put off to a more distant period. He has been repeatedly Minister of the Empire, has received eminent places from the people, but has steadfastly refused all title of nobility, and every honor from the Imperial Executive, except the Grand Cross of Santa Cruz.

Before leaving for Southern Brazil, I called upon Senator Vergueiro at Rio de Janeiro. He was at that time present in the capital during the session of the Assemblea Geral, and resided in

FAZENDA OF SENATOR VERGUEIRO, AT YBECABA.

the beautiful suburb of Botafogo. It was in the evening that I entered his residence, and was received by his daughters, whom I found intelligent and possessing one accomplishment so often lacking in a Brazilian lady: they could converse. Not many moments elapsed before the venerable Senator entered. His hair was white, and his form was bowed under the weight of fourscore years; yet in the glance of his eye there was something which told that the soul was neither slumbering nor decrepid. His smiling countenance also proclaimed that neither the burdens of age nor of past and present public and private service had affected in the least degree the cheerfulness of his nature. Whether conversing about the copies of the sacred truth, or of my contemplated visit to Ybecaba,—whether addressing a playful remark to his family, or a word of information to me,—he was a most pleasant picture of a hale and happy old man, with his mental powers unimpaired, and with the hopefulness of youth. The aged statesman stands almost alone in the Brazilian Senate-Chamber; for the patriotic yet impetuous Andradas are gone; the eloquent, the irresistible, but unsafe Vasconcellos has long since been laid in the tomb; the old Marquis of Valença has recently been followed to his "long home;" a new generation of Brazilians fill their places: nevertheless, Nicoláo de Pereira de Campos Vergueiro still represents an admiring constituency, no longer, as in stormier times, battling for right, but as the advocate of every measure for the advancement of his beloved country.

Few men in Brazil have been blessed with such sons; few, we may add, have taken such pains to have their children properly educated. Co-operating with their father, they have presented in their colony a model to their compatriots. His four sons were educated in Brazil, Germany, and England. The oldest, Senhor Luiz, studied law at the University of Göttingen. Senhor José (head of the Santos house) was trained in the military school of Prussia, and rose to the position of first lieutenant of the thirty-seventh Prussian infantry during the troubles between Belgium and Holland.

The third son (who had charge of the Rio house of Vergueiro & Filhos) was educated as a commercial man in London and Hamburg, and the younger brother had a thorough mercantile training

in the same cities.   By their European education they have been
enabled to carry out more easily the plans of their father concerning emigration.

In 1841, Senhor Vergueiro, in the teeth of public opinion, sent
to Germany for forty families as colonists; but the General Government was so opposed to the old Senator during the troubles of 1842,
in the province of San Paulo, that the colony was broken up.   In
1846, he again commenced carrying out his project; and, in so
doing, he has been completely successful.   The Government itself
through official organs, has commended the system of Vergueiro
as the system worthy of imitation.

That system may be stated in few words.   Sr. Vergueiro has in
Europe an agent who communicates with cantonal and communal
authorities, and with private individuals, offering inducements to
the able-bodied poor who wish to emigrate with their families to
the New World.   The emigrant, at his option, can defray his own
expenses to Brazil, or, permitting Sr. Vergueiro to transport him,
he (the emigrant) agrees in such case to refund at his own time
and convenience the price of his passage at a small rate of interest.
The agent at Hamburg charters a vessel, and thus a large number
of colonists are enabled to seek a new home at a very moderate
outlay.

Sr. V. guarantees on his part to defray all the expenses of the
colonists from the sea-coast to his plantations, and, on their arrival
at their final destination, to furnish each head of a family with a
house, so many thousand coffee-trees, proportioned to the number
of each family, and to supply all with provisions, articles of
clothing, &c. at wholesale prices.   The colonist, on his part, agrees
to tend faithfully his allotted portion of coffee-trees, to share the
profits and expenses of the crop, and not to leave without giving
one year's notice and paying his indebtedness (if any exist) for
passage-money advanced.

This contract is very simple, and is a safe investment for both
contracting parties.

During the year 1854, the result of the coffee-culture on the
plantation of Ybecaba was one million six hundred thousand pounds,
of which one-half of the expenses and profits belong to the
laborers.

I visited the cottages of the colonists, about one mile from the manor. As I passed along, I was constantly saluted by cheerful Swiss and German workmen, some of whom were surrounded by noisy and joyous fair-headed children, who capered about with as much life and glee as if at the foot of the Hartz or in the valleys of the Oberland.

Before reaching the hamlet, (of which I present a sketch drawn by a young German at Ybecaba,) I crossed a small stream upon a

COLONIA VERGUEIRO.

bridge of a novel and cheap construction, which in its simplicity commends itself to every settler in Australia or Western America, where proprietors are many but laborers are few. It may be styled a "self-made" bridge. A number of logs are fastened longitudinally in the water, leaving, of course, spaces between them. On the top of these are thrown large branches, and then finer brush; and on the surface is placed a certain quantity of clay and loose earth. A portion of the brook higher up is turned aside by a ditch through the light soil, and conducted over the log and brush-heap. In a few days this little side-stream has borne down an immense burden of red soil across the bridge, and has rendered the superstructure as firm as the road, while beneath, through branches and logs, the "river runs merrily by." The ditch, the water through it having finished its work, is closed, and a solid passage-way is thus obtained.*

At the hamlet I found an intelligent head-agent, who kept the books of the colonists, and gave to the latter orders for every pound of bacon, yard of cloth, &c. Without his signature they could not obtain these articles at the manor storehouse.

The larger portion of the colonists were Roman Catholics; but I did not leave before every opportunity was afforded for their obtaining the Scriptures, both in Portuguese and German.

Some of the colonists have thriven remarkably, having in five years' time gained five and seven thousand milreis, ($2500 and $3500.) The state of morals was certainly most creditable when comparing it with that of the countries whence they came. From 1847 to '55, (the period of my visit,) among several hundred laborers of the humblest classes of German and Swiss, not an illegitimate child had been born. The Vergueiros encourage the marriage-institution as not only essential to purity, but for the interest of both planter and colonist. There are now about one thousand European workmen, including children.

Ybecaba is a small plantation, containing but five or six square miles; but near by the V.s possess a *fazenda* not so well cultivated, but three times as large. At Angelica they own a new plantation, well adapted to the culture of coffee, which is twelve leagues in circumference. Hitherto blacks have been employed upon this large estate; but it is the intention of the proprietor to introduce,

---

* In some of the mining-districts there is a simple and philosophical mode of splitting off the side of clayey mountains. Wells are dug into them, and, during the heavy rains, these, by means of gutters, become filled with water. The hydrostatic pressure of the liquid columns forces off masses from the faces of mountains which would require hundreds of men for months to accomplish with the mattock and shovel.

as soon as possible, free white laborers. I demanded of Sr. Luiz Vergueiro if it were mere philanthropy which prompted their efforts to introduce free labor: he replied, most promptly and decidedly, "We find the labor of a man who has a will of his own, and interests at stake, vastly more profitable than slave-labor."

I could not but contrast the happy and cheerful condition of these colonists with the discouraged residents in the colony Donna Francisca. Though the Germans of Petropolis have every advantage of a nearness to market, and a growing city which has many wants to be supplied, yet the condition of the colonists at Ybecaba is infinitely superior if we consider the prosperity of the individual. The settlement at Leopoldina in Rio Grande do Sul has been the only truly successful Imperial colony, that of Petropolis being under the *Governo Provincial*. By the report of the Minister of the Empire for 1854–55, I ascertain that, out of seventeen colonies founded by the Imperial Government and by the provincial authorities, only four can be called prosperous; and of but two can it be said, "*muito prospera.*" The remainder are placed under the heads "not prosperous," "confounded with the population," "in decay," or "no information of its condition." Of the twenty-four private efforts at colonization, twenty-one are reported prosperous, nearly all of which have been founded since 1852, and more or less on the Vergueiro system. These colonies are in five provinces, and the excellence of the "plan-Vergueiro" consists in this,—viz.: its applicability throughout the Empire on a great or small scale. Nine of the twenty-one senhors have each less than one hundred and twenty colonists, thus enabling the small proprietors to have, to a certain extent, the advantages of the larger landholders. Slavery (since the vigorous measures of 1850 were adopted against the slave-trade) has been doomed in Brazil. The Emperor and his Government are against this inhuman traffic, and the popular voice sustains them. The comparative ease with which a slave may obtain his freedom, and, by the possession of property, the rights of citizenship, will probably in twenty years put an end to servitude in this South American Empire. There must then be a supply of laborers from some other source than Africa. The mother-country, the Portuguese islands, Germany, and Switzerland will furnish that supply. Individual emigration as it exists from Europe to the United States can never succeed in Brazil on a large scale, owing to the peculiar structure of the Government; but the system inaugurated by Sr. Vergueiro & Sons is capable of indefinite extension, while it protects the interests of both employer and employee. Though there may be individual instances of oppression under a powerful and unjust proprietor, yet, as a whole, this plan will in the end prove a great blessing to Brazil and to the poorer classes of Europe. Already the Swabian, the Fribourgeois, the Vaudois, the Valaisan, the Portuguese, and the Ilheo, look up like men in their new homes: they have no longer that appearance—too common in their native districts—of the crushed and cringing peasant who has no thought beyond the pinching want of to-day. As we look upon their joyous faces, we can readily believe what Sr. José Vergueiro said to me at Santos:—"They breathe here the air of freedom, sir,—such as they never snuffed in their native land."

Under such a system, they have not the pressing cares of the pioneer; they are not the victims of speculating land-companies, and, at the same time, though enjoying comparative ease, their own interest keeps them from indolence. At a year's notice, when they have learned, under the tuition and protection of a powerful Brazilian, the cultivation of tropical productions, they can leave and "set up" for themselves if they choose. They can easily become naturalized; their children grow up as citizens attached to the soil; and, if nothing untoward occurs, Brazil, in half a century, will have a host of small proprietors infusing a new life-blood into the body politic. Under her mild Government there will spring up a more hardy people, who will be the subduers of the virgin forests and the pioneers in the vast, fertile, healthy, but almost unexplored regions of Paraná, Goyaz, Mato Grosso, and Minas-Geraes, where the head-waters of the Amazon and the La Plata are interlaced or separated by a narrow dividing-ridge.

To the speedy and sure accomplishment of this desired consummation, Brazil should still more modify her laws, so that there may be every facility for the introduction of colonists. Already the Empire has done away with some of the most objectionable features; but much remains to be done. Every obstacle should be removed, and the Government, by a general act, should proclaim its policy as liberal in all the initiatory steps for the newly arrived as it is generous in regard to the holding of property by foreigners. Such measures would promote immigration, and in time a new population would grow up in this beautiful country, worthy of its vast resources. Let a pure gospel be in the hearts of such a people, and Brazil, in the future, will be a land in every respect unsurpassed on the face of the earth.

Sr. Vergueiro and his sons are making constant improvements in modes of cultivation, and are studying the best manner of applying Northern labor and skill to tropical agriculture. I before mentioned the workshops of the mechanics, where agricultural implements in wood and iron are turned out in a style equal to any thing of the kind made in Europe or North America. Among the various machines for facilitating the preparation of coffee for market was one—the invention of Senator Vergueiro—which cleans no less than thirty-two thousand pounds of coffee per diem.

We had been kindly invited to dine at the mansion-house, and it is unnecessary that I should particularize the component parts of a most sumptuous dinner. Suffice it to say that the "fat of the land" was there in profusion, and that the "feast of reason," &c. was well supplied by Sr. Luiz V., Dr. ——, and the intelligent padre, who conversed fluently in both French and German.

The doctor and myself left Ybecaba at a late hour, and, after a pleasant ride by moonlight, reached Limeira.

Note for 1866.—Senator Vergueiro died in 1860. On account of financial and other difficulties, it is said that Ybecaba, though still kept up, is not in so flourishing a condition as formerly. The conclusion of the long intestine struggle in the United States has caused many Southern planters to look to Brazil. The Imperial Government, as has been mentioned, is determined to receive them on most liberal terms; and "colonization," which it must be confessed has not fulfilled the expectation of its friends, will give place to "immigration," which has done so much for the United States, and, if the Government of Brazil will only be liberal and will fulfil its promises, viz., to sell land at a fair rate and cut the *red tape* of public offices and petty *fiscaes* and *subdelegados*, (interior supervisors and justices of the peace,) a population will be introduced which will add a thousand-fold to the well-being and honor of the Empire.

CHAPTER XXII.

A NEW DISEASE—THE CULTURE OF CHINESE TEA IN BRAZIL—MODUS OPERANDI— THE DECEIVED CUSTOM-HOUSE OFFICIALS—PROBABLE EXTENSION OF TEA-CULTURE IN SOUTH AMERICA—HOMEWARD BOUND—MY COMPANION—SENHOR JOSÉ AND A LITTLE DIFFICULTY WITH HIM—CALIFORNIA AND THE MUSICAL INNKEEPER— EARLY START AND THE STAR-SPANGLED BANNER—THE SENHORES BROTERO OF S. PAULO—FOURTH OF JULY INAUGURATED IN AN ENGLISH FAMILY—"YANKEE DOODLE" ON THE PLAINS OF YPIRANGA—LAME AND IMPOTENT CONCLUSION— ASTRONOMY UNDER DIFFICULTIES—DELIVERANCE—RETURN TO RIO DE JANEIRO.

THE next day after my visit to Ybecaba, I was employed in obtaining such information from Dr. —— as one would be sure to find in a man of intelligence and observation who had long resided in the country. I made many inquiries in regard to the various diseases of Brazil, and the remarks of this experienced physician confirmed my own oft-repeated opinion that few portions of the world could boast of so great a salubrity as this Empire.

Probably no tropical country has been so exempt from a general disease as Brazil. It has only been within the last five years that the yellow fever invaded these healthy realms, and not until 1855 has that dreadful scourge, the cholera, touched these shores. The ravages of these two devouring pestilences—both of which were confined to a narrow belt of the sea-coast—have been greatly over-estimated. During the prevalence of the cholera in the vicinity of Bahia, I was in that city of one hundred and twenty thousand inhabitants. I have seen it gravely stated in American and English journals that so great was the mortality and the panic there that there were not enough people left to bury the dead! Now, if the perpetrators of this horrible fiction had given the truth, they would have described a great deal of sickness among the blacks and much panic among the whites; that, out of a provincial population of nearly a million, 9,490 died from all diseases in the

416

political year 1855–6, the majority of cases being cholera, but that business went on as usual. I was in Rio de Janeiro during several yellow-fever seasons, and though—from personal knowledge, by visiting the hospitals and examining the list of the deceased—I ascertained that a truly large proportion of the foreigners in the city did fall before the terrific disease, yet, as a general thing, there were about as many natives that died of consumption each day as of the yellow fever.

Though no general pestilence has swept through the land, yet there are peculiar diseases in different parts of the Empire. In some of the mountainous districts there exists the same swelling of the throat and neck which is known in Switzerland as *goître*. The Brazilians call it *papos;* and Von Martius says that he found in the valley of the Parahiba River instances of this swelling larger than are seen in Europe, but not accompanied with the melancholy and idiotic appearance so often combined with the *goître* in Switzerland, Germany, and Northern Italy.

At Limeira I became aware of a new disease, which, like the *goître*, seems to be confined to certain localities. I was sitting in the office of Dr. ——, conversing with him in regard to Brazil, when I observed a Portuguese, about sixty years of age, enter, and demand, with great earnestness, if he thought that he could live. Soon after, a middle-aged Brazilian came, and, seeming to cling to the words of the physician as tenaciously as to a divine oracle, made nearly the same interrogatory. Neither of these men appeared in ill health, and, if I had not heard them state that they had great difficulty in swallowing, I would have considered them in a perfect sanitary condition. Upon inquiry, I ascertained from the doctor that these men had a disease which is widely prevalent in some portions of Interior Brazil, but he has never seen a notice of it in any medical work whatever. The Brazilians call it *mal de engasgo*. The first indication of its existence is a difficulty in swallowing. The patient can swallow dry substances better than fluids. Wine or milk can be drunken with more facility than water; still, both are attended with difficulty. To take thin broth is an impossibility. In some cases fluids have been conveyed to the stomach in connection with some solid. The person thus affected appears to be in good health, but in five or six years death

27

ensues from actual starvation. The sufferings of such a one was described to me as most horrible.

Some physicians in the province of San Paulo think it a paralysis of the œsophagus; but Dr. ——, who has seen many cases of *mal de engasgo*, inclines to the belief that it is a thickening of the mucous membrane. As the œsophagus is in general the least affected by disease of any part of the body, and is very rarely paralyzed, he cannot believe that so wide-spread a disease as the *mal de engasgo* can proceed from paralysis. Living as he does in the interior, it is difficult to obtain a subject for dissection, or permission to make a post-mortem examination, and therefore he has had no opportunity for a thorough investigation of the disease; but it is his intention to do so as soon as facilities present themselves, and then to lay the result before the medical world. He informed me that he was called to visit a man suffering from this malady eighty miles from Limeira, and to his astonishment he found in the same room no less than nine persons similarly affected. As yet no remedy has been found. The full extent of country over which the *mal de engasgo* prevails is not known; but to this physician's certain knowledge it exists from Limeira (two hundred miles from the sea-coast) to Goyaz,—a distance of four hundred miles. It is not found upon the coast; and a journey to the sea-board always benefits the afflicted patient. In 1855 I communicated the above facts in regard to the *mal de engasgo* to the New York "Journal of Commerce." A few days after its publication, a physician of Brooklyn suggested, in the columns of the same journal, that there might be *erysipelas* at the bottom of the disease. He gave as an instance one of his own patients who suffered from symptoms like those described, and which finally resulted in the discovery of erysipelas. I know that one case of similarity in a disease does not prove a general rule: still, the subject is worthy of investigation.

One topic of our conversation possesses a far more general interest than the nature of a new disease: it was the cultivation of the Chinese tea in Brazil.

There is probably no other country where the culture of this Asiatic shrub has been so successful away from its native region. The Portuguese language is the only European tongue which has

preserved the Chinese name (*cha*) for tea; and as the stranger at
Rio de Janeiro and other towns of the Empire passes the *vendas*,
he is always sure to see a printed card suspended, announcing *Cha
da India* and *Cha Naçional*: the former is the designation given to
tea from China, and the latter to the same production grown in
Brazil.

In 1810, the first plants of this exotic were introduced at Rio de
Janeiro, and its cultivation for a time was chiefly confined to the
Botanical Garden near the capital and to the royal farm at Santa
Cruz.   In order to secure the best possible treatment for the tea,
which it was anticipated would soon flourish so as to supply the
European market, the Count of Linhares, Prime Minister of Portugal, procured the immigration of several hundred colonists, not
from the mingled population of the coast of China, but from the
interior of the Celestial Empire,—persons acquainted with the
whole process of training the tea-plant and of preparing tea.

This was probably the first colony from Asia that ever settled in
the New World, of which we have authentic records. The colonists,
however, were not contented with their expatriation: they did not
prosper, and they have now disappeared.   Owing in part, doubtless, to characteristic differences in the soil of Brazil from that of
China, and perhaps as much to imperfect means of preparing the
leaf when grown, the Chinese themselves did not succeed in producing the most approved specimens of tea.   The enthusiasm of
anticipation, being unsustained by experiment, soon died away;
and near the city of Rio de Janeiro the cultivation of tea has
dwindled down to be little more than an exotic grown on a large
scale at the Botanical Gardens.

As a Government matter it was a failure; but several Paulista
planters took up the culture, and, though they encountered years
of discouragement, they have lived to see it, though as yet in its
infancy, one of the most flourishing and remunerative branches of
Brazilian agriculture.

Between Santos and San Paulo, near San Bernardo, I saw large
and productive tea-plantations.   The manner of its culture differs
but little from that adopted in China.   Tea is raised from the seed,
which, being preserved in brown sugar, can be transported to any
portion of the country.   These little tea-balls are planted in beds,

and then, in the manner of cabbage-plants, are transported to the field and placed five feet apart. The shrubs are kept very clean by the hoe, or by the plough, which, though a recent introduction, has on some plantations been eminently successful for this purpose.

The shrubs are never allowed to attain a height of more than four feet; and the leaves are considered ready for picking the third year after planting. The culture, the gathering, and the preparation of tea are not difficult, and children are profitably and efficiently employed in the various modes of arranging it for market. The apparatus used is very simple, consisting of—1. Baskets, in which the leaves are deposited when collected; 2. Carved framework, on which they are rolled, one by one; 3. Open ovens, or large metallic pans, in which the tea is dried by means of a fire beneath. Women and children gather the leaves and carry them to the ovens, where slave-men are engaged in keeping up the fire, stirring, squeezing, and rolling the tea,—which operations are all that it requires before packing it in boxes for home-sale or for exportation to the neighboring provinces.

The tea-plant is a hardy shrub, and can be cultivated in almost any portion of Brazil, though it is perhaps better adapted to the South, where frosts prevail, and which it resists. If left to itself in the tropics, it will soon run up to a tree. The coffee-tree requires rich and new soil, and a warm climate unknown to frosts; but the tea-plant will flourish in any soil. Dr. ——, who visited various portions of China, is of the opinion that the *cha* can be grown in any part of the United States from Pennsylvania to the Mexican Gulf. There are not many varieties of the plant, as is often supposed, black and green teas being merely the leaves of the same tree obtained at different seasons of the year. The flavor is sometimes varied, as that of wines from the same species of grape grown on different soils. The plant is not deciduous, as in China, and in Brazil is gathered from March to July, which in the Northern hemisphere would correspond to the interval between September and January.

I was informed that several million pounds are now annually prepared in the provinces of San Paulo and Minas-Geraes, and its culture is on the increase.

Some years ago the tea-planters were greatly discouraged; for the *cha* was badly prepared, was sold too new, and hence the demand did not increase.   But, since a greater experience in its culture and preparation, a better article for this favorite beverage has met with corresponding encouragement.   Formerly the cultivators said that, if they could obtain sixteen cents per pound wholesale, it would be as remunerative as coffee.   In 1855, twenty cents for the poorer article could be obtained; and for superior qualities—the greater portion of the crop—forty cents per pound wholesale was readily commanded.   The demand for it is constantly increasing.   When rightly prepared, it is not inferior to that imported from China.   Much, indeed, of the tea sold in the province of San Paulo as *cha da India* has merely made the sea-voyage from Santos to Rio de Janeiro, and there, after being packed in Chinese boxes, is sent back to the Paulistas as the genuine aromatic leaf from the Celestial Empire.   I have seen foreigners in Brazil who esteemed themselves connoisseurs in tea deceived by the best *cha naçional*.

A few years ago, Mr. John Rudge, of the province of San Paulo, sent some tea from his plantation as a present to his relatives in Rio de Janeiro.   This was prepared very nicely, each separate leaf having been rolled by the slaves between the thumb and forefinger until it looked like small shot.   It was thus invested with a foreign appearance, packed in small Chinese tea-caddies, and shipped at Santos for the capital.   When the caddies arrived, they were seized at the custom-house as an attempt to defraud the revenue.   It was on the other hand insisted that the boxes contained *cha naçional*, although, by some neglect, they did not appear upon the manifest. The parties to whom the tea had been sent offered to have it submitted to inspection.   The caddies were opened, and the custom-house officials screamed with triumph, adding to their former suspicions the evidence of their senses, for the sight, the taste, the smell of the nicely-prepared tea proclaimed emphatically that it was *cha da India*, and that this was an attempt to defraud His Imperial Majesty's customs.   It was not until letters were sent to Santos, and in reply the certificates of that provincial custom-house had been received, that the collectors at Rio were satisfied that there was no fraud, and that the province of San Paulo could produce as good tea as that brought around the Cape of Good Hope.

A few years may suffice to show on the pages of the "Commerce and Navigation" of Great Britain and the United States that tea enters largely into the articles of importation from Brazil. Fifty years only have elapsed since the first cargo of coffee was shipped from Rio de Janeiro, and now Brazil supplies two-thirds of the coffee of the world. The revolution in Hayti was the commencement of a new era for the coffee of Brazil.

In 1846, Dr. —— learned that several planters were about to root up their tea-shrubs. He besought them not to carry out their intention; "for," said he, "there is to be a great revolution in China, [in 1845 he had been informed in the Celestial Empire of the existence of the *Triad* Society,] and the price of teas will be sure to go up in a few years." The disheartened planters were encouraged to go on; and, only a short time before my visit to Limeira, one of these *fazendeiros* sent to Dr. —— several pounds of most excellent tea, at the same time assuring him (the doctor) of his deep gratitude for having been prevented from the destruction of his plantation. He had found it exceedingly remunerative, and next year he intended to enter into more extensive operations.

Throughout the world the use of tea is becoming as universal as that of coffee, and any continued disturbance in China must bring into prominent notice the tea-culture of Brazil. The *récolte* is now almost entirely used within the Empire; but the adaptability of the culture to almost any portion of the immense territory, and the ease by which it can be carried on, will doubtless, in a very brief period of time, fully develop this new source of national wealth.

It was on the morning of the 2d of July that I set out on my departure from Limeira. I shall never forget the kindness and attention of my generous host, as well as the welcome reception at the model plantation of Senator Vergueiro. The few days spent there so pleasantly gave me fresh hopes and great encouragement for the future of Brazil.*

---

* At Limeira I met a German engineer, who, with his accomplished Hamburgese wife, (to whom I am indebted for the sketches of the bridge at Cubitão and the German colonist's house) forms an agreeable society for Dr. ——.

The moon was shining brightly as I bade farewell to the two Americans and turned my face, for the first time in months, *homeward*. I rode on in silence for half an hour, and was then overtaken by a "lone horseman" going in the direction of Campinas. We journeyed together, and at noon we halted near a clear, purling brook, and beneath the shade of lofty, overarching trees we shared a palatable dish of *farinha de milho* and fried chicken, which the good *mulher* of the Paulista had thoughtfully provided for his journey. I have often had occasion to speak of the kindness manifested by Brazilians of all classes toward strangers. The casual visitor to Brazil may, in the coast-cities, come in contact with shopkeeping Portuguese, whose fleecing propensities are not excelled by their brethren in London, Paris, or New York; and hence he may grandly generalize, in writing home to some obscure journal, that the Brazilians are the greatest set of rascals in the world.

My travelling-companion was a carpenter, but was an adept in other crafts. My horse having cast two of his shoes, we turned to a road-side venda and purchased the necessary articles, which Sr. Tomaso attached with all the skill of a practised blacksmith.

We arrived at Campinas at four o'clock in the afternoon. I rode immediately to a *hospederia;* but the innkeeper seemed so perfectly indifferent as to custom that I bade him good-day, and sought the house of an English daguerreotypist, to whom I had letters. I had there a warm welcome, and the remainder of daylight was spent in rambling through this mud built city in company with my host and an Italian physician to whom Dr. —— of Limeira had given me a note of introduction. I found much to interest me in the vast cathedral, built wholly of *taipa:* the carved woodwork (reminding one of old European cloisters) was by some mulatto sculptors from Bahia, and would have done credit to the best Italian artists in that line. The physician, who was a fierce Malthusian, entertained me with long-winded speeches in support of his favorite ideas, until I finally obtained a respite by leading him on to some wonderful snake-stories, which, though equalling in length (the stories, not the snakes) his Malthusian arguments, were far more interesting.

I made arrangements at the house of a mule-dealer for an extra

animal, which was to carry me forward on the morrow, as my
Rosinante gave evidence of exhaustion.   My newly-engaged quadruped was to be forthcoming, together with a guide, at sunrise.
The sunrise came, and two succeeding hours; but neither biped
nor quadruped appeared.   Finally, when almost in despair, the
long-expected pair clattered up to the door.   The usual apologies
of "mules in pasture," "difficult to catch," &c., were offered and
accepted.   I soon perceived that my guide, instead of being a mere
employee, was the son of the proprietor of the animals which we
bestrode,—that he was not simply José, but *Senhor* José,—and that
he was musical withal.   I, however, feared that his position as a
gentleman might somewhat interfere with the orders for increased
speed which from time to time I might find it necessary to issue.

We rode on through a finely-cultivated region, large coffee-
plantations stretching on either hand as far as the eye could reach,
variegated with fields of waving sugarcane or groups of umbrageous forest-trees.   My companion enlivened the way by many
songs to the Virgin and "to his mistress's eyebrows;" but, when
the sun had sunk beneath the horizon, Sr. José concluded that we
had journeyed far enough for one day, and proposed that we should
tarry for the night at the house of a planter near by.   To this I
strongly objected, as my contract was that I should be carried for
a specific sum to a specific point, several leagues farther on.   I
found that he was no underling, to be crossed in his wishes; and he
firmly resisted.   I would have left him where he was, without
further ado; but, knowing the difficulty of separating animals that
have travelled in company, I thought best to compromise the
matter by stating that we would remain here for the night, in
which case, however, the compensation would be several milreis
less than if we had accomplished the contemplated number of
leagues.   But he was not the man for a compromise: he demanded
full pay for short work.   I then determined, at all hazards, to push
on without him.   I found my perverse horse as stubborn as Sr. José.
I endeavored to start him in the direction of San Paulo: he, however, was resolved to travel only toward the plantation.   I spurred
the mule, which I rode, after him, endeavoring to head off the
horse: this I found a most difficult task.   Sr. José, meantime, sat
motionless as a statue, secretly and maliciously enjoying my un-

successful efforts. I was fatigued beyond measure; but my will was unbroken, (as well as that of my horse,) and at last victory crowned my struggles, and, shouting to Sr. José "*Boa noite*," and triumphantly exclaiming, "I know how to protect my rights," I trotted off, Rosinante in advance, toward San Paulo.

Glancing over my shoulder, I beheld my guide still statue-like bestriding his mule, and comparable to any thing else than "Patience on a monument smiling at grief." Poetically speaking, he was *planted*.

My way was now over a good road, though the overhanging forest obscured almost every ray of moonlight. My animal went gayly on, leaving, however, time enough for a few reflections. Among them the most prominent was, "Suppose Sr. José rides after me and salutes me in the back with his long knife," (*faca de ponta*,) which looked innocent enough when reposing in its sheath or cutting an orange. In all my travels in Brazil I never carried a weapon of any kind, and this was the first time that I felt the least suspicion that all might not be perfectly safe. In the midst of these reflections and thoughts about that long knife, I had accomplished more than a half-league, when I heard the rapid movement of mule-hoofs. Sr. José came thundering up the hill, and overtook me. Instead, however, of a knife-salutation or loud words, he instantly, in the mildest possible voice, suggested that we should change beasts, as he was very much fatigued, and that the difference in the gait of the two animals would be a relief to both parties. We went on as cosily as if nothing had happened, and at eleven o'clock rode up to the house of one Sr. João Baptista, whose residence was christened with the mellifluous and auriferous name of *California*.

We soon aroused Sr. J. Baptista, who, while we sipped our *cha*, tinkled on his guitar "many a roundelay." I informed Sr. J. B. that the morrow was the *dia da independencia* in the United States, and requested the favor of "Hail Columbia." Sr. J. B. declined, on the ground of not possessing the tune in question; but (like a skilful shopkeeper who, destitute of a certain article, suggests to his customer another which, in his estimation, is equally good if not superior) Sr. J. B. proposed the *Brazileiro*, as being nearer the required national air than any thing else in his musical

treasury.   Its spirit-stirring strains were quivering in my ear
when I thought how difficult it would be to find in the backwoods of Wisconsin or Minnesota ac-
complished musicians such as Sr. J. B.
or Sr. José, who was also skilled in the
art.   The Brazilians, as a whole, are a
musical people, and sometimes, during
a storm, when I have been plodding
on in darkness, I have been cheered
by the sound of a violin, a guitar, or
by human voices singing sweetly in
concert.

HERCULES BEETLE.

I could sleep but little, and that
little was rudely interupted, (whether
by a giant beetle or a stealthy bat I
was unable to ascertain;) and I jumped
from my hard bed at two o'clock on
the morning of the Fourth of July,
and aroused the household of Sr. J.
Baptista and the sleepers in the neighboring rancho by screaming
at the top of my voice the "Star-spangled Banner."

I bade my musical host and Senhor José adeos, mounted my
Rosinante, and accomplished thirty-two miles before breakfast.
My primary object had been to get to Santos, in order to take the
steamer of the 6th for Rio; and a secondary consideration was to
celebrate the Fourth of July at the house of Mr. E., the English
engineer.

I visited Senhor Brotero, the President of the Law-School for
which San Paulo is so justly celebrated.   Madame Brotero I found
to be a countrywoman, from Boston.   I also made the acquaintance
of Senhor Brotero, Jr., to whom Senhor Octaviano, the accomplished editor of the *Correio Mercantil*, of Rio, had given me a letter
of introduction.   This gentleman, who bids fair to be one of the
leading men of S. Paulo, possesses enlarged views, and has had the
advantage of extended travel in Europe and North America.

It was a pleasant forenoon that I spent with Mr. and Mrs. E.
and Mr. C., inaugurating with them the celebration of my nation's
birthday.   Mr. C., however, threw something of a damper upon

my patriotism by dropping in, "By-the-way, it is the birthday of George III.:" but chronology shows that Mr. C. was just four weeks out of the way, and his inappropriate remark in no manner marred the general harmony of the occasion.

These and other friends pressed me not to hasten on at my rapid rate, thinking that thirty-two miles before breakfast was sufficient for one day: but my purpose was to make twenty miles that night before I sought repose.

Senhor Coelho (the *maître-d'hôtel*) had procured for me a fine mule. He was a lithe animal, and when I mounted he bounded

YANKEE DOODLE ON THE PLAINS OF YPIRANGA.

away as though he had wings. He clattered through the streets, descended the hill, splashed through a little affluent of the La Plata, and, just as the sun was setting, went galloping gayly over the plains of Ypiranga. I soon came in sight of the pavilion erected over the spot where Dom Pedro I. exclaimed, *Independencia ou Morte*, and, being animated with Fourth-of-July sentiments, I gave vent to my patriotism in shouting, at a furious rate, "Yankee Doodle" and "Hail Columbia," to the no small amusement and astonishment of the sable passers-by.

I reached San Bernardo and passed through its silent streets. The atmosphere was laden with the perfume of some sweet night-

opening flower, and the sky overhead seemed joyous as my homeward-bound spirits. My mule flagged not, and I was congratulating myself that my journey's end would soon be accomplished,
when, to my surprise, the spirited beast whirled suddenly to the
right, and plunged into the stable-yard adjoining a large white
house. I kicked, and cuffed, and spurred, all to no purpose. The
noise which I made aroused two poncho-clad Brazilians, who came
toward me, thus discoursing in Portuguese:—"Yes, it is he."
"No; let me look again." "Yes, I am certain it is." These little
monosyllables are as brief and as elliptical in the language of Lusitania as in the plainest Saxon, and could give me no clue to the
meaning of the locutors. I was not, however, long left in doubt,
for one of them approached, and thus addressed me:—"*Senhor,
isto é meu animal.*" ("This is my beast, sir.") Supposing that he
was mildly accusing me of theft, I replied that he must be mistaken, for I had hired that mule at S. Paulo. "It may be," he
said; "but still he is mine." I then ascertained that the man was
the proprietor of my long-eared steed, and that he (the proprietor)
had preceded me in company with a number of law-students who
were on their way to Santos. Feeling by this time much fatigued,
and considering the stubbornness that had come over my quadruped,
I asked if I might lodge at the house for the night. The other
personage now turned up his sombrero and informed me that there
was no room in the inn, but possibly I might be accommodated a
mile farther on. I could not make my mule stir; so these two
benevolent individuals aided me in whipping and kicking the
brute until he was fairly under way. I had, however, advanced
only five hundred yards, when master long-ears pulls me up again,
and no dint of beating, pulling, pounding, and tugging could make
him budge a peg on the "forward march." He willingly beat a
retreat, and the next moment I again stood before the white
hospedaria from which I had been politely sent away a short time
before. My two new-made acquaintances were soon by my side,
and I once more begged for a room. One of them gave a negative
answer; but, when I suggested that I was willing to pay a good
price for my accommodations, he left me as if to consult some one.
I then heard an emphatic female voice screech out, "*Não, Senhor.*"
This reply was brought to me, and I sent back word that I had

letters from Senator Vergueiro, showing that I was a respectable person. It was of no avail, for at each fresh attempt to move the tender mercies of the woman to whom belonged that voice, I received a more emphatic "*Não, senhor.*" My last resort was to claim, in "the sacred name of Brazilian hospitality, only room enough upon their floor for a stranger who is here stopped contrary to his own will." The reply was the same, "*Não, senhor.*" "Then," said I, "it is an outrageous shame. I have travelled through a number of your provinces, and have mingled much with the rich and the poor, but this is the first time that I have been unable to obtain shelter. Here I am, compelled before a large house to pass the night in the road." My appeals and denunciations were equally unsuccessful; so I sat down upon a curbstone, holding the bridle of my obstinate and tired animal. Poor fellow! his fatigue was not equal to mine. I had ridden since morning nearly fifty miles, and had spent seven hours in San Paulo. Three or four days had elapsed since I had had a comfortable sleep, and the night-air was keen for Brazil, though it was as balmy as a May evening in the Northern hemisphere. The body, however, was not suffering so much as the mind. I felt

ASTRONOMY UNDER DIFFICULT CIRCUMSTANCES.

this inhospitality to the quick. I sat with my head bowed down upon my left hand, turning my eyes from time to time toward the stars and the waning moon. It was studying astronomy under difficult circumstances, so that I did not make much progress.

While thinking of my condition, and feeling that it was worse, and my treatment more outrageous, than when, a mere innocent student-traveller, I was once taken prisoner on suspicion by the Austrians in Lombardy, and led by an armed soldier through the streets of Pavia, I was aroused from my reflections by an old negress, who said to me, "Come here, senhor." I followed her to a comfortable room, where she left me with a nice cup of tea and *doce* accompaniments. My mule was cared for, as well as myself, and when the morning sun awoke me I found that I was to have as my fellow-travellers the young law-students. I ascertained that this house was kept by a respectable Brazilian widow, who was making a large fortune by letting mules for riding or for the transportation of baggage, and that whoever employed her animals in S. Paulo would be entertained *gratis* at this otherwise inhospitable *hospedaria*. It so happened that the students and myself were not aware of this regulation, and had hired our mules of another man, who had guided them as far as this house. Here the young "legals" insisted on stopping. The Donna da Casa refused them accommodation, and they had taken possession *vi et armis*. It may be that, owing to senhora being somewhat embittered by such proceedings, had refused me when I pleaded the name of Senator Vergueiro and Brazilian hospitality. For assuredly there was plenty of room, when we consider that there were eight unoccupied beds in the house. It may be, also, that the senhora was suspicious of a stranger travelling alone at that hour of the night, as she had been deceived a few weeks before by an individual who pretended to have letters from a nobleman, but who turned out to be an unmitigated scoundrel. I was (justly, as I thought) indignant for a time, and entertained an idea that it would be right that the public should know through the Rio journals of such treatment to an *estrangeiro;* but the more I reflected upon it, I became rather ashamed of my indignation. I had travelled thousands of miles in Brazil, and this was the first experience of the bitter; and how foolish it would be to lay it before the public! The widow had a perfect right to make such regulations as she chose concerning her household, and an Anglo-

American who is firm for the independence of the home-castle is assuredly the last man who ought to complain. So I dismissed the whole subject, and have never recurred to it since, except to indulge in a laugh at my own ludicrous position in the stable-yard, and the tableau of the stubborn mule and the curbstone. Thus ended my Fourth of July, 1855.

The next day I arrived with my student-friends at Santos, and, after enjoying for a few days more the hospitality of *Casa Vergueiro*, I steamed away in the comfortable old Paraense for Rio de Janeiro. From San Sebastian to the Sugar-Loaf we were pitched about in fine style by an angry sea; but the sun shone forth brilliantly as on the following day we lay under the guns of Villegagnon, and the glorious panorama of the magnificent bay, sparkling in the freshness of morning, lost none of its splendor by comparison with the beautiful scenes which I had witnessed in Southern Brazil, and which I afterward found unequalled in the provinces of the North.

Note for 1866.—The province of S. Paulo, like other Southern provinces, by its climate, soil, &c., offers very desirable inducements to emigrants from the United States. The hilly portions of S. Paulo, Paraná, St. Catharines, and Rio Grande do Sul are the best adapted to sheep-farming. A very pleasant offering of four fine merinos to the Emperor of Brazil was made by Dr. George B. Loring, of Salem, Massachusetts, in 1865. The sheep were received at Rio with thankfulness by the Emperor, and were placed in the hands of Mr. John Hayes, the energetic and intelligent American director of the plantations of the Baron of Mauá. These sheep will be the beginning of better things in the ovine race in the Southern provinces. Senhor Marcondes, the Minister of Agriculture in the Cabinet of August, 1864, and Senhor Paulo Souza, filling the same station in the Cabinet of May, 1865, highly praise the gift of Dr. Loring.

CHAPTER XXIII.

THE BRAZILIAN NORTH—EXTENT OF THE EMPIRE—THE FALLS OF ITAMARITY—GIGANTIC FIG-TREE—THE KEEL-BILL—A PLANTATION IN MINAS-GERAES—PETER PARLEY IN BRAZIL—SWEET LEMONS—BARONIAL STYLE—THE PADRE—VESPER-HOURS—THE PLANTATION-ORCHESTRA—THE WHITE ANTS OBEDIENT TO THE CHURCH—THE GREAT ANT-EATER—THE PACA—THE MUSICAL CART—THE MINES AND OTHER RESOURCES OF MINAS-GERAES—COFFEE: ITS HISTORY AND CULTURE—THE PROVINCE OF GOYAZ—STINGLESS BEES AND SOUR HONEY—MATO GROSSO—LONG RIVER-ROUTE TO THE ATLANTIC—A NEW THOROUGHFARE—LIEUTENANT THOMAS J. PAGE—THE SURVEY OF THE LA PLATA AND ITS AFFLUENTS—FIRST AMERICAN STEAMER AT CORUMBA—STEAMBOAT-NAVIGATION ON THE PARAGUAY—OFFICERS OF THE AMERICAN NAVY—DR. KANE AND LIEUTENANT STRAIN—DIAMOND AND GOLD MINES THE HINDERERS OF PROGRESS—THE DIFFERENCE IN THE RESULTS FROM DIAMONDS AND COFFEE.

Now to the North: not the Boreal North, with hoary beard and glistening spears and crunching ice-batteries,—but a genial, sunny, laughing, flowery, Austral North. We on the hither side of the equator are so wedded to experience, that it is difficult to conceive of a North where

"The fields are florid in eternal prime,"

and where mighty rivers, with unabated force, sweep onward,—

"And traverse realms unknown and blooming wilds,
And fruitful deserts, worlds of solitude;
Where the sun smiles and seasons teem in vain."

THE MINEIRO.

CASCADE OF ITAMARITY, NEAR PETROPOLIS.

I could never become accustomed to look for the sun and the equator in the direction which all past experience told me was the region of stern winter. I could not be reconciled to the idea that the southern front of my Brazilian residence was the coldest side, although I knew that reason and geography informed me that that portion of my house looked toward the Falkland Islands and the unexplored snow-continent of the Antarctic zone.

But to the Brazilian North! If by land, it will be many months of painful journeys up mountains and hills, through dense forests and jungles, over wide *campos* and broad rivers, before we reach the *Serra Pacaranua*, which divides Brazil and Venezuela. I have not seen the record of a single traveller who has ever accomplished this long terrestrial route. Eschwege, Rodriguez, Ferreira, Natterer, Mawe, Prince Maximilian, Spix and Von Martius, St. Hilaire, Langsdorf, Pohl, Burchell, Gardner, Lieutenant Strain, the expedition under Castlenau, and Wallace, have traversed large districts of Brazil; while—not to mention earlier fluvial explorations—Mawe, Smyth, Edwards, Herndon, Gibbon, and Wallace (the most thorough explorer) have examined the Amazon, and Lieutenant Page has the honor of being the first scientific investigator of the La Plata and some of its tributaries. Still, it is hazarding nothing to say that the greater portion of this extensive Empire has only been trodden by the foot of the wild Indian, or, at long intervals, by the most adventurous of the Portuguese traders. It is difficult for us to comprehend even the dry tables of distances: how much more inconceivable the toil and the almost insurmountable obstacles to be endured and overcome in a vast country with a sparse population, and, in certain portions, no roads save the paths of cattle and the tracks of the tapir! The distance, on a straight line drawn from the head-waters of the river Parima, on the north, to the southern shores of the Lagoa Mirim, in Rio Grande do Sul, is greater than that from Boston to Liverpool. It is farther from Pernambuco to the western boundary which separates Peru and Brazil, than by a direct route from London, across the Continent, to Egypt. Brazil has neither been explored nor surveyed, and its full extent cannot be accurately ascertained; but, according to the best calculations made in 1845 for the *Diccionario Geographico*

28

Brazileiro, the Empire contains within its borders 3,004,460 square miles. The United States, by the latest computations of the Topographical Bureau at Washington, has an area of 3,002,013 square miles. But by the settlement of different boundary-lines since 1856, Brazil has acquired additional territory: so that we should have to add to the possessions of the United States an area equal to that of the adjacent States of New England, New York, and Pennsylvania, to make it of the same dimensions as the land of the Southern Cross European Russia possesses an area of 2,142,504 square miles, and the remainder of Europe 1,687,626. It is by these figures and comparisons that we may arrive at an approximate idea of the vastness of Brazil.

It is not, however, its extent which should attract our attention so much as the fact that no portion of the globe is so available for cultivation and for the sustentation of man.

It has already been seen that the internal resources of this Empire are commensurate with its favored position and its wide extent. It is neither the gold of its mines nor the diamonds that sparkle in the beds of its inland rivers that constitute the greatest sources of its available wealth. Although nature has bestowed upon Brazil the most precious minerals, yet she has been still more prodigal in the gift of vegetable riches. Embracing nearly five degrees north of the equator, the whole latitude of the southern torrid and ten degrees of the southern temperate zone, and stretching its longitude from Cape St. Augustine, (the easternmost point of the continent,) across the mountains of its own interior, to the very foot of the Andes, its soil and its climate offer an asylum to almost every valuable plant. In addition to numberless varieties of indigenous growth, there is scarcely a production of either India which might not be naturalized in great perfection under or near the equator; while its interior uplands, and its soil in the Far South, welcome many of the fruits, the grains, and the hardier vegetables of Europe.

Every year this Empire is becoming more developed; yet it will require two centuries of its present progress to bring it to an equal position with the United States. The signs of the times are, however, that Brazil will not go on at the snail's-pace which characterized her up to the abolition of the slave-trade; and the internal improvements auspiciously begun under D. Pedro II. will rapidly unfold the resources of the country.

Of the twenty provinces, four only are inland,—viz.: Minas-Geraes, Goyaz, Mato Grosso, and Amazonas, (sometimes called Alto Amazonas.) It is in Mato Grosso ("dense forest") and Goyaz that the head-waters of the Amazon and the La Plata have their origin, within a few miles of each other; while on the borders of Minas-Geraes the sources of the San Francisco, the Tocantins, and the La Plata take their rise from the same mountain-ridge.

The usual route to the fertile province of Minas-Geraes is through Petropolis, and the traveller thither should not fail to make a little *détour* and visit one of the prettiest cascades in Brazil. Following for a few miles the highroad to the Minas, we turn to our right, and there, among the dells formed by the Serra da Estrella, we find the Falls of Itamarity. The name, in the Guarani language, signifies "shining stones," or "the rock which shines;" so called, doubtless, from the glittering appearance of the large mass of rock, the face of which is worn smooth by the water. *Ita* means "stone or rock." This cascade is composed of three distinct falls, formed by a stream of small size unless after heavy rains. The charm of this lovely spot consists in the surrounding woods and the murmuring waters; so that we may truly say that

> " the gush of springs
> And fall of lofty fountains, and the bend
> Of stirring branches, and the bud which brings
> The swiftest thought of beauty, here extend,
> Mingling, and made by Love unto one mighty end."

Garlands of parasites enfold the old trees in their graceful arms, and bands of verdant climbers depend from the highest boughs to the very ground. The torrent has undermined the banks and prostrated the trees that stood near the edges, and they now lie in wild disorder across the bed of the stream, mingled here and there with huge stones brought down by the force of the water.

The bridge represented in the engraving was improvised for the occasion of the visit by Sir W. Gore Ouseley, formerly British Minister to Brazil. Such crossings are easily formed by felling a few trees and binding them together with the supple vines that abound. Nature soon heals her wounds and clothes them with parasites, so that in a few weeks the artificial structure seems like a work of her own hand.

The road from Petropolis to Barbacena is exceedingly picturesque,—sometimes winding along the side of a mountain which

A BRAZILIAN MOUNTAIN-ROAD.

gives extensive views of plains beyond, and sometimes in deep valleys along the banks of babbling streams. Long troops of mules on their way to Estrella are constantly passing; but—to show the wildness of the region notwithstanding frequent villages and fazendas—we were startled every few moments by flocks of wild parrots, and could hear in the trees the chattering of monkeys. Now a fine road for coaches leads to Barbacena.

At a place called Padre Corréas, not far from Petropolis, is a celebrated wild-fig tree, whose branches extend over a circumference of four hundred and eighty feet, and four thousand persons, it is computed, can stand under its shade at noonday. Near by, on the height east of the hamlet, can also be seen two rows of the Brazilian pine, (*Araucaria Braziliana*,) so well known in the large conservatories of Europe and the United States. Very fine specimens of this Brazilian pine-tree are to be found in the Crystal Palace at Sydenham. When one hundred miles farther in the interior, I saw many *jacarandá* (rosewood) trees. Their resemblance to the common locust of the United States is very striking. There are a number of species of the jacarandá, varying in tint from a deep rich brown to a beautiful violet. The latter kind I have never seen north of the equator, save in small specimen-pieces; but, at the Fazenda do Governo, Dr. Joaquim A. P. Da Cunha, the amiable proprietor, showed me, in his establishment for making sugar, a beam, fifty feet long and three feet in diameter, of the violet-tinted jacarandá. It had performed the menial office of a connecting-beam for fifty years, and its exterior was dusty; but, on chipping it, I found it to be of the most beautiful violet. The wood of Dr. Da Cunha's pig-pen consisted of boards and sticks of rosewood: but let none of my readers imagine a highly-polished piano or a splendid centre-table; for exposure to the atmosphere renders the jacarandá as plebeian in appearance as the commonest weather-beaten pine. The rosewood-tree is cut down, deprived of its branches, and conveyed to market generally by floating it to some seaport-town, whence it is shipped to North America and Europe. It is of exceeding hardness and durability,—cog-wheels made of this wood lasting longer than those constructed from any other ligneous substance. The United States annually purchase of Brazil eighty thousand dollars' worth of rosewood.

As I was journeying in the province of Minas, I observed a flock of birds of which I had seen the same species at the foot of the Organ Mountains, and which I then took to be the common blackbirds so well known in North America; but a closer inspection showed them to possess a bill of remarkable thickness. They had a clear and musical whistle, and I afterward discovered them to be the *ani*,— a genus of scansorial birds found only in Tropical America. They are sometimes called the keel-bill. They live in flocks, and it is said that they have practical communism among them, many pairs using the same nest, which is built on the branches of trees, and is of a large size. Here they lay and hatch in concert.

THE KEEL-BILL.

I cannot enter into the details of my journey in Minas-Geraes, but I am reluctant to pass over a visit to one of the finest plantations in the province. The proprietor was a Brazilian, and the whole fazenda, in its minutest details, was carried on in the manner peculiar to the country, without any admixture of foreign modes of government and culture.

Twelve miles beyond the Parahibuna (an affluent of the Parahiba) we turned aside from the highway, and, after riding through a belt of enclosed forest-land, we saw before us the large plantation-house of Soldade, belonging to Senhor Commendador Silva Pinto. The approach to the mansion was between two rows of palm-trees, around whose trunks a beautiful *bignonia.* (the *venusta*) entwined itself, and then threw its climbing branches over the feathery leaves of the *palms*, thus forming a magnificent arch of flowers and foliage. The buildings, in the form of a hollow square, occupied an acre of ground. On two sides of the square was the residence of the Commendador and his family, while the remaining sides consisted of the sugar-establishment and the dwellings of the slaves. We entered the court-yard by a high gateway, and then for the first time we perceived the venerable planter sitting in a second-story veranda, reading. So soon as he saw us he laid down his book, descended into the square, and with great affability bade us a warm welcome. The American party doubtless owed this hospitable reception to one of our companions, Dr. Ildefonso Gomez, a Brazilian whom almost every man of science visiting the Empire has delighted to honor for his intelligence, for his eminent abilities as a naturalist, and for his integrity as a man.

Servants flew about noiselessly at the commands of the Commendador: they gave us rooms, hot coffee, hot baths, &c. &c. Then both they and their master did that which is most grateful to the weary traveller: they let us alone.

When I had performed my ablutions and was recovered from

fatigue, I went to the veranda where the Commendador had been reading. I picked up his book, and to my astonishment I here found that it was *A Historia Universale do Senhor Pedro Parley*, (Peter Parley's Universal History!) Old Peter Parley in the interior of Brazil! I knew that England had availed herself of those books which have delighted Anglo-American childhood, and that hosts of counterfeiters and imitators had arisen, assuming that *nom de plume;* but it was beyond my most sanguine expectations to have ever seen in the Portuguese language, and in an interior province of distant Brazil, the history of the Eastern and Western Continents by Senhor Pedro Parley amusing and instructing youth and old age. It was no imitation. In reading the preface, I perceived that some priest had had to do with the translation, for it roundly asserted that Senhor Pedro Parley was *um bom Catholico Romano!* which will doubtless be an important piece of information to the veritable Puritan-descended Peter.

I looked from the veranda upon a scene of cultivation. Close at hand were one hundred and fifty hives with bees; gently-rounded hills were covered with grazing flocks and herds, cotton and sugar fields were in valleys, while Indian corn and mandioca in large tracts were far to our right. The orange-orchard was the largest that I ever saw in any land: it was computed that there were ten thousand bushels of six different kinds of the luscious fruit. The sweet lemon abounded to such an extent that it was estimated that there were five thousand bushels. A "sweet lemon" seems almost as much of a contradiction in terms as an honest thief; but it is a reality. Dr. Ildefonso Gomez informed me that this fruit, exactly resembling the acid one bearing the same name, was originally a sour lemon, but, by a disease and by grafting, a new species has been produced. The taste is not so rich as that of an orange, but is very quenching to the thirst, and the Brazilians at Rio consume great quantities of them. Near S. Romão, a little place on the head-waters of the San Francisco, the lemon-tree has become naturalized, and the cattle that pasture in the woods are so fond of the fallen fruit that when killed their flesh smells strongly of it.

Of all the articles mentioned above, not one finds its way to market. They are for the sustenance and clothing of the slaves, of whom the Commendador formerly had seven hundred. These are engaged in cultivating coffee, (for this is the great coffee-region,) which is the only crop intended by the proprietor to bring back a pecuniary return. This senhor owns other plantations, but that of Soldade contains an area of sixty-four square miles.

At dinner we were served in a large dining-room. The Commendador sat at the head of the table, while his guests and the various free members of his family sat upon forms, the *feitors* (overseers) and shepherds being at the lower end. He lives in true baronial style, and I was reminded of the description by Mr. J. G. Kohl of castle-life among the noblemen of Courland and Livonia. A pleasant conversation was kept up during the long repast, and at its close three servants came,—one bearing a massive silver bowl a foot and a half in diameter, another a pitcher of the same material containing warm water, while a third carried towels. The newly-arrived guests were thus served in lieu of finger-basins, which are rarely seen outside the capital.

The Commendador had a chapel in his mansion, and each morning mass was performed by an amiable young Portuguese priest, who knew much more about music than the gospel. The padre had many questions to ask concerning the peculiar doctrines of Protestants, and I was surprised to find that he possessed no Bible. I presented him with a New Testament, and before my departure we had many most earnest and serious conversations in regard to vital piety and the solemn responsibility that was upon him to teach the truth as it is in Christ Jesus. With the approval of the Commendador, (which was heartily given,) explanations of the Scriptures were hereafter to constitute a portion of the chapel-service on Sundays. This planter is now the Baron of Berthioga.

On these interior plantations there is a beautiful custom at vespers of offering a short prayer and wishing each other a good-night; not that they then retire, but *boa noite* is the form of a blessing. We were all sitting on the veranda as the last rays of the sun were gilding the hill and the distant forest. The chapel-bell struck the vesper-hour. The conversation was arrested: we all arose to our feet. The hum of the sugar-mill ceased; the shout of the children died away; the slaves that were crossing the court-yard stopped and uncovered the head. All devoutly folded their hands and breathed the evening prayer to the Virgin. I too joined in devotion to the blessed Saviour, the sole Mediator, and when the padre and others wished me the blessing in the name of *Nossa Senhora*, I returned the benediction *em nome de Nosso Senhor Jesus Christo*. The noise of merry voices again rang through the court-yard; the day's labor was finished; and soon night, with its dark-ness, silence, and repose, reigned over Soldade.

Another custom I observed in various parts of Brazil, which, though a mere unmeaning form, is a custom both Christian and beautiful. I doubt, however, if one in a thousand attach any deeper significancy to it than we do to "good-morning." At the close of the day the slaves enter the room where their master is, and, with their hands crossed, each addresses the fazendeiro in a pious salutation, the full form of which is, "I beseech your blessing in the name of our Lord Jesus Christ," and the reply should be, "Our Lord Jesus Christ bless you forever;" but in time this prayer and benediction are abbreviated to the last words of each sentence, which are pronounced in a most rapid and business-like manner by both parties:—*Jesus Christo —— sempre*, (forever.)

In the course of our conversation the Commendador told us that he had his "own music now." He spoke of it very humbly. We desired to hear his musicians, supposing that we should hear a wheezy plantation-fiddle, a fife, and a drum. The Commendador said that we should be gratified in the evening. An hour after vespers I heard the twanging of violins, the tuning of flutes, short voluntaries on sundry bugles, the clattering of trombones, and all those musical symptoms preparatory to a beginning of some march, waltz, or polka. I went to the room whence proceeded these sounds; there I beheld fifteen slave musicians,—a regular band: one presided at an organ, and there was a choir of younger negroes arranged before suitable stands, upon which were sheets of printed or manuscript music. I also observed a respectable colored gentle-man (who sat near me at dinner) giving various directions. He was the *maestro*. Three raps of his violin-bow commanded silence, and then a wave of the same, *à la Julien*, and the orchestra com-menced the execution of an overture to some opera with admirable skill and precision. I was totally unprepared for this. But the next piece overwhelmed me with surprise: the choir, accompanied by the instruments, performed a Latin mass. They sang from their notes, and little darkies from twelve to sixteen years of age read off the words with as much fluency as students in the Freshman year. I could scarcely believe my eyes and ears, and in order to try the accomplishments of the company I asked the maestro for the *Stabat Mater:* he instantly replied, "*Sim, Senhor*," named to the musicians the page, waved his *báton*, and then the wailing and touching strains of *Stabat Mater* sounded through the corridors of Soldade. While at supper we were regaled by waltzes and stirring marches,—among the latter "Lafayette's Grand March," composed in the United States. The maestro regretted that they had it not in their power to play our three national airs; but I promised him that when an opportunity should afford I would take pleasure in adding to his musical library "Yankee Doodle," "Hail Columbia," and the "Star-spangled Banner." One morning at three o'clock I was awakened by a servant, who informed me that the orchestra was about to play the *Brazileiro* in honor of O Senhor Commendador's guests; and in a few minutes the band, with the addition of big drum, little drum, and cymbals, startled the early birds by the national anthem of Brazil, which was succeeded by "Lafayette's Grand March."

Before our departure from Soldade, the hospitable proprietor furnished us horses, and we sallied forth to roam over the immense plantation. A portion of our party carried their guns, hoping to meet with game in our ramble. We rode over hills used as pasture-ground, which were literally dotted with the upright and fallen columns that had been erected by the *termites*, or white ant. These curious edifices and their still more curious architects have always had a great attraction for the naturalist. The hillocks are conical in their shape, but not with a broad base and tapering point as those built by the termites of Africa. Exposure to the sun has rendered them exceedingly hard, and doubtless many that are seen upon the uplands of S. Paulo and Minas-Geraes are more than a century old; for houses whose walls have been built from the same earth are still in existence which were built by early settlers in the seventeenth century. Sometimes the termites' dwelling is overturned by the slaves, the hollow scooped wider, and is then used as a bake-oven to parch Indian corn. In my ride over

Soldade I saw a number of very large vultures, who, during the rain, had taken refuge in the houses that had been vacated by the white ant.

These insects do not, however, always dwell in columnar edifices of three and six feet in height. I have seen, in some portions of Brazil, the ground ploughed up, to the extent of one hundred feet in circumference, by one nest of white ants. Again, they will climb trees, carrying building materials with them, and erecting a small archway (resembling what carpenters call an "inch-bead") over them for protection against their sworn enemy, the black or brown ant, and on the loftiest branches they will construct their nest. In cities they are sometimes very destructive: hence every Brazilian lady keeps her fine robes in tin boxes, and each gentleman who pretends to a library must often look at it

WHITE ANTS IN A TREE.

to see if the *cupim*, or white ant, has not become a most penetrating reader of his volumes. My introduction to the cupim was in the house of our former Consul, ex-Governor Kent. A box of books sent out by the American Tract Society was placed in a lower room, and the next morning it was announced to me that the cupim had entered my property. I hastened to the room, and, turning over the box, beheld a little black hole at the bottom, and white, gelatinous-looking ants pouring out as though very much disturbed in their occupation. I opened the box, and found that a colony of cupim had eaten through the pine wood, and then had pierced through "Baxter's Call," "Doddridge's Rise and Progress," until they had reached the place where Bunyan's Pilgrim lay, when they were rudely deranged in their literary pursuits.

On another occasion I saw a Brussels carpet, under which cupim had insinuated themselves and had eaten out nearly all the canvas before the proprietor made the sad discovery.

Dr. Kidder, at Campinas, witnessed the depredations of the white

ants in the *taipa* (clay-built) houses. They insinuate themselves
into the mud walls, and destroy the entire side of a house by per-
forations. Anon they commence working in the soil, and extend
their operations beneath the foundations of houses and undermine them. The people dig large pits in various places, with the
intent of exterminating tribes of ants which have been discovered
on their march of destruction.

Mr. Southey states, on the authority of Manoel Felix, that some
of these insects, at one time, devoured the cloths of the altar in
the Convent of S. Antonio, at Maranham, and also brought up into
the church pieces of shrouds from the graves beneath its floor;
whereupon the friars prosecuted them according to due form of
ecclesiastical law. What the sentence was in this case, we are
unable to learn. The historian informs us, however, that, having
been convicted in a similar suit at the Franciscan Convent at
Avignon, the ants were not only excommunicated from the Roman
Catholic Apostolic Church, but were sentenced by the friars "to
the pain of removal, within three days, to a place assigned them

GREAT ANT-EATER.

in the centre of the
earth." The canon-
ical account grave-
ly adds that the ants
obeyed, and carried
away all their
young and all their
stores!

The white and
other ants have,
however, enemies
far more tangible
than bulls of ex-
communication, in
the *Myrmecophaga*,
or the great ant-
eater, the Taman-
dua, and the "little
ant-eater," of which the last two have a prehensile tail. The great
ant-eater is a most curious animal, but well adapted to the purposes

THE JAGUAR, OR BRAZILIAN TIGER.

for which it was designed by the Creator. Its short legs and long claws (the latter doubled up when in motion) do not hinder it from running at a good pace; and when the Indians wish to catch it they make a pattering noise upon the leaves as if the rain were falling, upon which the myrmecophaga cocks his huge bushy tail over his body, and, standing perfectly still, soon falls a prey. In the northern part of Minas-Geraes a naturalist once came suddenly upon the great ant-eater, and, knowing the harmless nature of its mouth, seized it by the long snout, by which he tried to hold it, when it immediately rose upon its hind-legs, and, clasping him around the middle with its powerful fore-paws, completely brought him to a stand. It was struck down with a club a number of times, but soon recovered and ran off; and not until a pistol-ball was lodged in its breast was the naturalist able to add it to his collection. It measured six feet in length without the tail, which, together with the long tufts of hair, measured full four feet more.

When the great ant-bear sleeps, it lies on one side, rolls itself up so that its snout rests on its breast, places all its feet together, and covers itself with its bushy tail. When thus curled up, it is so exactly like a bundle of hay that any one might pass it carelessly, imagining it to be a loose heap of that substance.

When it walks or runs, the claws of the fore-feet are doubled up, causing one side only of the foot to rest upon the ground. The proper use of these powerful claws is to obtain the white ant. When the ant-bear wishes a meal, he attacks one of the hard hillocks already described, and with his huge fore-paws furiously tears out a portion of the walls, and, thrusting in his long, slender tongue, which is covered with a viscid saliva, and to which myriads of ants adhere, he opens his little mouth and draws it in: then, shutting his lips, he pushes out his tongue a second time, retaining the ants in his mouth until the tongue has been completely exserted, when he swallows them. Wallace says that the Indians of the Upper Amazon positively assert, that the great ant-eater sometimes kills the jaguar by tightly embracing the latter and thrusting its enormous claws into the jaguar's sides. The aborigines also "declare that these animals are all females, and believe that the male is the 'curupira,' or demon of the forest.

The peculiar organization of the animal has probably led to this error."

As we descended the hills of Soldade on our return to the plantation-house, one of our party fired at two pacas which were feeding near a little stream. Either the aim of the hunter was not good, or the buckshot did not tell upon the hairy side of the animal, and in a few moments he had swum the river and was hidden in the thick copse of bushes and ferns. The paca, the capybara, and agouti abound in Brazil, and are of the same family as marmots

THE PACA.

and beavers. The paca attracts the attention of the hunter both on account of the difficulty of its capture (as it takes the water and swims and dives admirably) and the esculent nature of its flesh. It is about eighteen inches in height and two feet in length, and its color is brown, spotted with white. The hinder limbs (being considerably bent) are longer than the anterior ones, and its claws are well formed for digging and burrowing. They are easily domesticated, and make lively pets, eating readily out of the hand of those it is accustomed to, but hiding from strangers. A friend bound to the United States had one on shipboard, which was a great favorite, and bade fair to weather the voyage and visit the shores of North America; but either the new paint, or some salt water that he drank in a storm, cut short the thread of his existence, and poor paca was consigned to the blue waves of the Atlantic.

After leaving our kind host, we journeyed toward Barbacena, over roads that can be used for vehicles; but the only movable article of that kind which we saw was the Roman cart, unimproved since the days of the Georgics. Indeed, all Roman carriages were of the same simple plan. The wheels did not turn on their axis, but axis and wheels turned together. We could often hear music of a most *fortissime* character, which they ground out as they moved slowly over the plantations. I was informed that the Brazilians construct these carts of a particular wood, having special reference to the musical qualities, which, when put into action under a heavy load and behind three yoke of cattle, resemble the concentrated powwow of a thousand belligerent tomcats. On the day of some

THE MUSICAL CART.

festa, I was travelling near the banks of the Parahiba, and miles away I heard the grinding of a cart. The distance had somewhat mellowed its music, and, after a long ride, I came up with it, and found a gay party of country Brazilians in their holiday attire riding upon the old Roman chariot, which was adorned with bed-covers of a bright pattern. The unbonneted senhoras seemed as much at home in their turn-out, and doubtless as proud of it, as the most dashing lady of the Fifth Avenue in her cushioned coach which sways softly upon the most modern elastic springs.

The province of Minas-Geraes is the most important of all the inland divisions of the Empire, owing to its mineral and vegetal riches, its immense herds, its accessibility to market, and its population. It contains eight hundred thousand inhabitants, and yet is so extensive that there are within its area of one hundred and fifty thousand square miles many forests,—a perfect wilderness, overrun with Indian tribes, and where the jaguar roams in undisturbed independence.

Other portions are among the most improved and eligible parts of the Empire. One writer has remarked, with great emphasis, that, if there be one spot in the world which might be made to surpass all others, Minas is that favored spot. Its climate is mild and healthful; its surface is elevated and undulating; its soil is fertile, and capable of yielding the most valuable productions; its forests abound in choice timber, balsams, drugs, and dye-woods.

But all these circumstances together have not given the province so much celebrity as the 'single fact of its inexhaustible mineral wealth. Its name signifies the general or universal mines, and, accordingly, mines of gold, silver, copper, and iron are found within its borders, besides quantities of precious stones. Several of the most valuable gold-mines not far from Ouro Preto have been wrought by an English mining company for the last twenty years. This enterprise has been unquestionably a source of profit to its stockholders, and has rendered great service to the country generally, by introducing the most approved methods of mining and by giving an impetus to Brazilian industry. This company constantly employs a large number of miners from Cornwall, and has established quite an English village at its principal mine.

The agricultural capacities of the province are very great. It yields coffee, sugar, tobacco, and cotton. It indeed produces some coarse manufactures of cotton. Its soil yields Indian corn in great profusion, and may be made to grow wheat. Upon its campinas, or upland prairies, innumerable herds of cattle, and some flocks of sheep, are pastured. The milk of the cows is converted into a species of soft cheese, known as the *queijo de Minas*. Immense quantities of them may be seen at Rio de Janeiro, and from that port they are scattered along the coast, being very much esteemed as an article of food.

The great staple, however, of Minas-Geraes, and of the whole Empire of Brazil, is coffee. What a history might be written of the voyages, the naturalization, and the uses of this member of the *Rubiaceæ* family! The coffee-tree is not, as is generally supposed, a native of Arabia, but its home is Abyssinia, and particularly that district called Kaffa, whence the name of the beverage-berry. To this day the coffee-plant is found growing as far as the sources of the White Nile. It was not taken to Arabia until the fifteenth century, when, being cultivated extensively, with great success as to quantity and quality, in the province or Kingdom of Yemen, and embarked from Mocha, the coffee of that portion of the world obtained a celebrity which it has never lost. When it was introduced by the Orientals into Europe we know not; but as early as 1538 we find edicts against it, issued by the Mohammedan priests, on the ground that the faithful went more to the coffee-shops than to the mosque. The earliest notice that we have of it in France is in 1643, when a certain adventurer from the Levant established in Paris a coffee-house, which did not succeed. In a few years, however, it became the *mode* among the aristocracy, through its inauguration by Soliman Aga, the Ambassador of the Sublime Porte at the Court of Louis XIV. Several of the high personages of the time resisted its introduction,—among them the celebrated Madame de Sévigné, who had declared that the popularity of coffee would be merely ephemeral; and, in the intensity of her admiration for Corneille, she predicted that *Le Racine passerait comme le café,* (Racine will be forgotten as soon as coffee,) both of which predictions have proved rather detrimental to the prophetic reputation of the renowned lady letter-writer. Before the middle of the seventeenth century it was in vogue in the principal capitals of Europe. An English merchant from Constantinople was the first to introduce it to the Londoners, and his wife, being a young and pretty Greek, was a most attractive saleswoman. It is said that the coffee-houses were greatly multiplied during the Protectorate, and that Cromwell, wishing to protect the interest of the taverns, and doubtless urged on by the publicans, caused them to be closed.

Previous to the eighteenth century, all the coffee consumed in

Europe was brought from Arabia Felix *viâ* the Levant, and the Pachas of Egypt and Syria took good care to increase their coffers by exorbitant transit duties. This exaction was broken up by the vessels of Holland, (first,) England, and France sailing around the Cape of Good Hope to Mocha. In 1699, Van Horn, first President of the Dutch East Indies, obtained coffee-plants and had them cultivated in Batavia, where they wonderfully prospered, and the berries of Java obtained a reputation second only to those of Mocha. One of the Batavian shrubs was transplanted to the Botanical Gardens of Amsterdam in 1710, and by great care succeeded so well that a shoot was sent to Louis XIV. and placed in the Jardin des Plantes. From this last plant, slips were confided to M. Isambert to be taken to Martinique; but M. Isambert died before the arrival of the ship, and consequently the coffee-plants perished. In 1720, Antoine de Jussieu, of the Royal Botanical Gardens, sent, by Captain Declieux, three more coffee-shrubs, also destined to Martinique. The voyage was long, the vessel was short of water: two of the plants died, but Captain Declieux shared his ration of water with the *cafier*, and thus succeeded in introducing it into the West Indies: that plant was the ancestor, it is said, of all the coffee-plantations in America.

The honor of planting the first coffee-tree in Brazil belongs to the Franciscan Friar Villaso, who in 1754 placed one in the garden of the San Antonio Convent at Rio de Janeiro. It was not, however, until after the Haytien insurrection that coffee became an object of great cultivation and commerce in Brazil. In 1809, the first cargo was sent to the United States, and all the coffee raised in the Empire in that year scarcely amounted to 30,000 sacks, while in the Brazilian financial year of 1855 there were exported 3,256,089 sacks, which brought into the country nearly $25,000,000. The United States, during the financial year ending June 30, 1856, imported, from all coffee-producing countries, 235,241,362 pounds of the beverage-berry, 180,243,070 pounds (*i.e.* nearly three-fourths of the whole) of which came from Brazil. The next highest country on the list is Venezuela, which sent us 16,546,166 pounds; and thirdly, Hayti, from which we imported about 13,500,000 pounds. The whole sum paid by the United States for coffee was $21,514,196, of which Brazil received no less than $16,091,714.

The great coffee-region, as has been mentioned, is on the banks of the Rio Parahiba, and in the province of San Paulo; but every year it is more widely cultivated, and a considerable quantity is now grown in provinces farther northward. It can be planted by burying the seeds or berries, (which are double,) or by slips. The trees are placed six or eight feet apart, and those plants which have been taken from the nursery with balls of mould around their roots will bear fruit in two years; those detached from the earth will not produce until the third year, and the majority of such · shrubs die. In the province of S. Paulo, and the richest portions of Minas-Geraes, one thousand trees will yield from 2560 to 3200 pounds, in Rio de Janeiro from 1600 to 2560. In some parts of S. Paulo, one thousand trees have yielded 6400 pounds; but this is extraordinary. In the province of Rio de Janeiro, trees are generally cut down every fifteen years. There are some *cafiers* on the plantation of Senator Vergueiro which are twenty-four years old, and are still bringing forth fruit. As a general rule, they are not allowed to exceed twelve feet in height, so as to be in reach. When the berry is ripe, it is about the size and color of a cherry, and resembles it, or a large cranberry: of these berries a negro can daily collect about thirty-two pounds. There are three gatherings' in the year, and the berries are spread out upon pavements or a level portion of ground, (the *terreno*,) from whence they are taken when dry and denuded of the hull by machinery, and afterward conveyed to market. Nothing is more beautiful than a coffee-plantation in full and virgin bloom. The snowy blossoms all burst forth simultaneously, and the extended fields seem almost in a night to lay aside their robe of verdure, and to replace it by the most delicate mantle of white, which exhales a fragrance not unworthy of Eden. But the beauty is truly ephemeral, for the snow-white flowers and the delightful odor pass away in twenty-four hours.

It is by toilsome journeys on mule-back that the coffee-sacks from Minas-Geraes generally reach a market, and nothing so much hinders the general prosperity of this province as its lack of good roads and some feasible thoroughfare to a market. The province has, of late years, expended considerable sums upon the construction of roads, but as yet it cannot send a single ton of its produce to market upon wheels. The journey from Ouro Preto, the capital, to Rio de Janeiro,—a distance of about two hundred miles,—is performed on the backs of mules and horses only, and ordinarily requires fifteen days.

As to education, it is but just to say that Minas-Geraes, according to official statistics, takes the lead of all the provinces in this praiseworthy enterprise. The provincial Government has made large expenditures for the support of schools, and the people seem to have appreciated the benefit to be derived from them.

Should the long-talked-of enterprise of steam navigation upon the Rio Doce and the Rio de S. Francisco ever prove successful, the interests of Minas-Geraes would be greatly promoted. A most thorough survey of the Rio de S. Francisco was made by Mr. Halfeld

As to the navigation of the Rio San Francisco,—a river as large as the Volga,—a glance at the map will show its importance to Minas and all other provinces watered by it and its tributaries. The San Francisco is the largest river emptying into the Atlantic between the Amazon and the Rio de la Plata. It rises in the province of Minas, and waters the soil of Bahia, Pernambuco, Sergipe, and Alagoas, in its course to the ocean. From the mouth of the Rio das Velhas to the Falls of Paulo Affonso, not many leagues east of Joazeira, a distance of seven hundred miles, its waters are suitable for navigation, although, from the sparseness of population on its banks, and the lack of enterprise, it is but little used for this purpose. The Falls of Paulo Affonso are described by those who have seen them as an immense cataract, over which the river plunges, forming a spectacle of the utmost grandeur. The vapors arising from the ravine may be seen at a great distance. They resemble the smoke of a conflagration in the midst of the forest. The river does not again find a tranquil bed until near its embouchure, but for the space of seventy-five miles dashes with fury over a succession of rapids and smaller cataracts, which effectually interrupt the passage of vessels and forbid the hope of any artificial connection between the upper and lower navigation.

But these difficulties are about to be overcome in another manner: a railway from Pernambuco to Joazeira has already been projected, through the enterprise of the Messrs. de Mornay, who have obtained the concession of the first portion for its construction from the city of Pernambuco to Agoa Preta, on the river Una,

a distance of seventy-four miles. From Bahia also another road
has been projected northward to Joazeira. Now, from the latter
point to the mouth of the Rio das Velhas there is an uninterrupted
steamboat navigation for seven hundred miles, and numerous tributary rivers increase the navigation to nearly two thousand miles.
It is therefore from the Barra das Velhas that a railway will most
probably be made to Rio de Janeiro, about four hundred and thirty
miles in a straight line,—the whole comprising, by rail and by
river, as Mr. Borthwick in his excellent report says, "a grand internal communication between the capital and the most thriving
provinces;" and such is its necessity that it is only a question of
time. When such a system of internal improvements is completed,
no province will be more benefited than Minas-Geraes. The recent
investigations of Mr. Halfeld have been published by the government.

INHABITANTS OF THE FORESTS OF GOYAZ.

Upon the west and north of Minas-Geraes is the large province
of Goyaz. Like most of the interior portions of Brazil, Goyaz was
discovered and overrun at an early day by the Paulistas, in their
search for mines and Indian slaves. It abounds in gold, diamonds,
and precious stones; but its remoteness from the sea-shore, and its
lack of roads, canals, and steamboats upon its navigable rivers, are
great obstacles to the development of its resources.

This province, bounded on the west by the Araguaia, may be
considered as occupying the central portion of Brazil, and is not
generally mountainous, although its surface is elevated and unequal. Some tall virgin forests are seen upon the banks of its
rivers, in which most comical monkeys abound; but the larger

part of the province is covered with that species of low and stunted shrubbery which prevails in large portions of the province of Minas, and is designated by the terms *catingas* and *carasqueiros.* Its soil yields the usual productions of Brazil, together with many of the fruits of Southern Europe. Cultivation has progressed further in Goyaz than in Mato Grosso, though it is still extremely backward.

The name of this province is derived from the *Goyas,* a tribe of Indians formerly inhabiting its territory, but now nearly extinct. Various other tribes still exist within its borders, several of which cherish a deadly hatred to the people who have invaded their domains and disturbed them in their native haunts. Settlements are often laid waste by the hostile incursions of these Indians.

In Goyaz, as well as in other portions of the interior, the traveller will find plenty of honey made by stingless bees. I do not know that it holds true in Brazil, as in North America, that the bee precedes by a few miles the onward march of civilization,— advances as the Indian and the wild beast prepare to take their departure,—and thus is the pioneer of a better state of things; but it gives of its sweets to sustain and cheer the settler and the *voyageur* in those vast and fertile solitudes. I suppose that the bees of Brazil are indigenous, and not like the honey-bee of the United States, which was unknown before the arrival of Europeans, and to which the Indians—having no term for it in their language —gave the name of "English flies." The greater portion of the Brazilian bees possess, in their absence of weapons, a peculiarity which many a stung sufferer would wish the *Apis mellifica* of North America possessed. Some of these bees make *sour* honey, which will compensate for *sweet* lemons.*

---

* Dr. Gardner, in his visit to Goyaz, was entertained at a little place not far from Natividade, near the mountains which form the southwestern boundary of Piauhi. "The owner of the house," he says, "returned from the woods, shortly after our arrival, with a considerable quantity of wild honey, some of which he kindly gave us, and we found it excellent: it was the product of one of the smaller bees so numerous in this part of Brazil. This was the season in which the people go to the woods in search of honey. It is so generally used, that, after leaving Duro, [where Goyaz, Piauhi, and Pernambuco are contiguous,] a portion was presented to us at almost every house where we stopped. These bees mostly belong to the genus *Melipona,* Illig., and I collected a great many, which, with some other zoo-

In some portions of Goyaz society is very backward, but not altogether in the state which existed at the time (1817) of St. Hilaire's visit. There is a powerful class of the inhabitants called *vaqueiros*, or cattle-proprietors. These men possess vast herds of horned cattle, and their principal business is to mark, tend, and fold them. They understand the use of the lasso, and also of the long knife. However, their moral and intellectual condition is by no means perfect.

---

logical specimens, were afterward lost in crossing a river. A list of them, with their native names and a few observations, may not be uninteresting:—

"1. *Jatahy.*—This is a very minute yellowish-colored species, being scarcely two lines long. The honey, which is excellent, very much resembles that of the common hive-bee of Europe.

"2. *Mulher branco.*—About the same size as No. 1, but of a whitish color: the honey is likewise good, but a little acid.

"3. *Tubi.*—A little black bee, smaller than a common house-fly: the honey is good, but has a peculiar bitter flavor.

"4. *Manoel d'Abreu.*—About the size of the *tubi*, but of a yellowish color: its honey is good.

"5. *Atakira.*—Black, and nearly the same size as the *tubi*,—the principal distinction between them consisting in the kind of entrance to their hives: the *tubi* makes it of wax, the *atakira* of clay. Its honey is very good.

"6. *Oariti.*—Of a blackish color, and about the same size as the *tubi*: its honey is rather sour, and not good.

"7. *Tataira.*—About the size of the *tubi*, but with a yellow body and a black head: its honey is excellent.

"8. *Mumbúco.*—Black, and larger than the *tubi*: the honey, after being kept about an hour, becomes as sour as lemon-juice.

"9. *Bejui.*—Very like the *tubi*, but smaller: its honey is excellent.

"10. *Tiubá.*—Of the size of a large house-fly, and of a grayish-black color: its honey is excellent.

"11. *Bord.*—About the size of a house-fly, and of a yellowish color: its honey is acid.

"12. *Urussú.*—About the size of a large humble-bee: the head is black and the body yellowish. It produces good honey.

"13. *Urussú preto.*—Entirely black, and upward of an inch in length: it likewise produces good honey.

"14. *Cuniára.*—Black, and about the same size as No. 13: its honey is too bitter to be eatable. It is said to be a great thief of the honey of other bees.

"15. *Chupé.*—About the size of No. 10, of a black color. It makes its hive of clay on branches of trees, and is often of a very large size. Its honey is good.

"16. *Urapua.*—Very like No. 15, but always builds its hive rounder, flatter, and smaller.

"17. *Enchú.*—This is a kind of wasp about the size of a house-fly: its head is black and the body yellow. It builds its hive in the branches of trees: this is of a papery tissue about three feet in circumference. Its honey is good.

"18. *Enchú pequeno.*—Very similar to the last, but always makes a smaller hive: it also produces good honey.

"The first eleven of these honey-bees construct their cells in the hollow trunks of trees, and the others either in similar situations or beneath the ground. It is only the last three kinds that sting, all the others being harmless. The only attempt I ever saw to domesticate these bees was by a Cornish miner in the Gold District, who cut off those portions of the trunks of the trees which contained the nest, and fastened them up under the eaves of his house. They seemed to thrive very well; but whenever the honey was wanted, it was necessary to destroy the bees. Both the Indians and the other inhabitants of the country are very expert in tracing these insects to the trees in which they hive. They generally mix the honey—which is very fluid—with farinha before they eat it, and of the wax they make a coarse kind of taper about a yard long, which serves in lieu of candles, and which the country-people bring to the villages for sale. We found this very convenient, and always carried a sufficient stock with us: not unfrequently we were obliged to manufacture them ourselves from the wax obtained by my own men." 1865, M. Brunet, of Bahia, has found forty kinds of bees.

But, in the general improvement which is gradually pervading all Brazil, this province receives its share; and, when the railways are completed to Joazeira, Goyaz will be easily brought within a few hours of the great marts on the Atlantic seaboard. The various affluents of the Tocantins and of the Parahiba do Sul water this province, and afford it a certain species of communication with the adjacent provinces; and yet in the middle and southern provinces I have met with travellers and mule-troops taking the long and fatiguing land-route to Rio de Janeiro and Santos. From Goyaz, the capital of the province, to Pará, the distance is more than one thousand miles, and this journey has been performed the whole way by water, with the exception of a few leagues. This long river-route was accomplished as early as 1773, under the governorship of José d'Almeida de Vasconcellos Sobral e Carvalho, and we of the North are filled with wonder that this navigation does not become permanent and reliable. As Brazilian steamers have been running regularly upon the Amazon since 1853, we may hope in time to see the waters of the Tocantins and its tributaries furrowed by suitable *vapores*. The President of this province, Sr. Magalhães, descended the Araguaya to Pará in 1863.

Mato Grosso is an immense province, containing a greater area than the original thirteen States of the Union. It is west of Goyaz, and borders upon Bolivia, the Argentine Confederation, and Paraguay.

Mato Grosso may be reached from Pará by ascending either the Tocantins, the Chingu, the Tapajos, or the Madeira Rivers. A glance at the map would lead one to suppose that the passage of the Madeira was not only the longest, but also that which would be in every way the most difficult. It is, however, better known than either of the others, and is the only one which has, to any extent, been a commercial thoroughfare.

The distance in a right line from Pará to Villa Bella, or Mato Grosso, (one of the principal towns of the province,) is about one thousand miles. Not less than two thousand five hundred miles must be traversed in making the passage by water. Lieutenant Gibbon, U.S.N., has given a very interesting account of his descent (in 1852) of the Mamoré River, from the fort Principe de Beira to the Madeira, and thence to Pará; but the best detailed sketch of this long route and the numerous difficulties it opposes to either the traveller or the merchant is found in a memoir published by the Geographical and Historical Institute of Rio de Janeiro. Brazil established mail-steamers to Cuyabá in 1856.

For the distance of fifteen hundred miles up the Amazon and the Madeira, to the Falls of St. Antonio, there is nothing in the way but a powerful current. Much of the country through which the last-named river flows is very unhealthy. From the Falls of St. Antonio a succession of falls and rapids extend upward more than two hundred miles. Nearly all this distance it is necessary to transport canoes and cargoes overland, by the most tedious and difficult processes imaginable. Precipices must be climbed, roads cut, and huts built from time to time as a temporary shelter against the rains. In short, three or four months are necessarily consumed on this part of the route. Once above this chain of obstacles, there remain about seven hundred miles of good navigation on the Mamoré and Guaporé Rivers. Previous to steam-navigation on the Amazon the entire voyage occupied ten months, when made by traders carrying goods. Vast numbers of Indians and negroes are required as oarsmen and bearers of burdens. It is customary for several companies to associate together, and the supplies which must necessarily be provided beforehand occasion great expense and inconvenience. The downward voyage, as a matter of course, would be much more easily and quickly performed. Notwithstanding the tedium and the toil of this long and dreary passage, it is generally less dreaded than the overland route to Rio de Janeiro. On the latter, an interminable succession of mountains, the lack of any direct or suitable roads, the impossibility of procuring provisions by the way,—at least for great distances,—and the slow pace of loaded mules, are by no means trifling difficulties in the way of either despatch or pleasure.

But by the enterprise and ability of Lieutenant Thomas J. Page, U.S.N., a new route by water to the capital of the Empire has been opened to Brazil and the world. This gentleman, acting under orders of the United States Government, sailed from Norfolk in 1853, in the U.S. steamer "Water-Witch," four hundred tons' burden and nine feet draft. The object of this expedition was the survey of the river La Plata and its tributaries, for the advancement of commerce and the promotion of science. Although some obstacles presented themselves at Rio de Janeiro, the Imperial Government finally granted its consent, and the Water-Witch went on its mission of peace; and no one can read Lieutenant Page's report to the late Secretary of the Navy (Mr. Dobbin) without the deepest interest, and the conviction that the surveys and discoveries of the Commander and those under him are of the greatest importance to North America and Europe, as well as to Brazil and the South American States.

The investigations of Lieutenant Page on the Paraná, Paraguay, and also a number of their tributaries, show conclusively that these rivers can become the richest channels of commerce. Of the Paraguay he says:—

"This river differs from the Paraná in several particulars. Its period of rising is generally the reverse; it contains but few islands, is confined between narrow limits, is more easy of navigation, because less obstructed by shoals, and the course of its channel is less variable; its width from one-eighth to three-fourths of a mile, its velocity two miles per hour, and its rise is from twelve to fifteen feet. In October it attains its maximum and in February its minimum state. From its mouth to Assuncion, a distance of two hundred and fifty miles, there were found no less than twenty feet of water when the river had fallen about two feet. This depth of water remained unchanged for the distance of several hundred miles above Assuncion, and the Water-Witch had ascended the Paraguay seven hundred miles above this place before she found less than twelve feet. At this time the river had fallen several feet.

"The admirable adaptation of these rivers to steam-navigation cannot but forcibly strike the most casual observer.

"There are no obstructions from fallen trees, neither shoals nor rocks, to endanger navigation. At suitable points—in fact, at every point in Paraguay particularly—an abundance of the best wood may be procured immediately on the banks; and, when populated, no difficulty will be found in obtaining a supply of it prepared for immediate use. By experiment carefully made, one cord of the Paraguay wood was ascertained to be equal, in the production of steam, to a ton of the best anthracite coal.

"The left bank of the river, up to the distance of four hundred
and fifty miles from Assuncion, is populated, but more and more
sparsely as the northern frontier is approached.   Between the most
northern Paraguayan and the most southern Brazilian settlements—
a distance of two hundred and fifty miles—there is no habitation of
civilized man.   Various tribes of Indians were met with at different points, with some of whom we 'held a talk,' and parted on
such friendly terms, because of the numerous presents we made
them in trinkets and tobacco, that they became somewhat troublesome, following us along the banks on horseback, desirous that we
should repeat the visit on shore."

This was the first steamer that ever ploughed the upper waters
of the Paraguay.   The arrival of the Water-Witch at Coimbra
(Brazil) was hailed with the liveliest demonstrations of joy, and
Lieutenant Page was received by the authorities with the most
marked attention.   His command, owing to the proper permission
from the Imperial Government arriving too late, did not proceed
higher than Corumba.   Lieutenant Page is, however, of the opinion
that Cuibá, in Mato Grosso, may be reached by small steamers.   It
is hoped that this energetic and intelligent officer may yet prose-
cute his surveys for the benefit of the world.

It is interesting to reflect that while the American navy has
been to a great extent, for nearly fifty years, exempt from the work
of war, her gallant officers have won imperishable laurels in the
nobler pursuits of scientific investigation.   The names of Bache,
Lieut. Strain, Kane, Gillis, Page, and the scores who have been
employed on coast-surveys, have done more to benefit their country
and mankind than all the naval battles of the nineteenth century.
Since these pages were commenced, three whose names are mentioned above have slept the "last sleep."   When scientific attainments, self-sacrifice, and suffering shall be connected together, the
hero of the Arctic regions and the hero of the Isthmus of Darien
will not be forgotten by the thousands who shall come after us.
To both may be applied the language of Mr. George Ripley, of New
York, in regard to Kane :—"The admirable qualities which they
displayed in the discharge of their official duties are a sure pledge
of permanent fame.   Courage, wisdom, fertility of resource, power
of endurance, and devotion to an idea, are stamped upon their

intrepid career." As Dr. Kane, though bent on an errand of mercy, was the first American to attempt "to lift the dead veil of mystery which hangs over the Arctic regions," so Lieutenant Strain, for the benefit of mankind, was the first American to explore the wonderful rivers of that region of fabulous fertility in the South.

While a midshipman, he obtained leave to enter the interior of Brazil, and, accompanied by a small party of brave spirits, (among whom was Dr. Reinhart,) he explored the province of San Paulo, tracing the rivers Tieté and Paranapanema nearly to their confluence with the Paraná. The dangers and hardships he encountered in this expedition were only inferior to those of the more recent and better-known expedition to the Isthmus of Darien. His services as an explorer were suitably acknowledged by the Imperial Government; and in Brazil I have heard high encomiums on Lieutenant Strain, and in his death science has lost a noble son.*

It would be an interesting expedition, and great good would be accomplished, if the Government of Brazil would consent to send out, with England, France, and the United States, a joint scientific commission, to explore thoroughly the whole district of Central Brazil, from Bolivia to Bahia, with particular reference to the navigability of the waters, that here interlace, of those vast rivers which irrigate such a wide extent of territory.

In the northern part of this province are countless hosts of monkeys, mostly of the howling kind. M. de Castelnau, on the

---

* The career of this officer after leaving Brazil may be briefly stated:—From South America he went to California. "In 1849, returning from the Pacific, he crossed the continent from Valparaiso to Buenos Ayres, of which he published a narrative entitled 'The Cordillera and Pampa.' Subsequently, he was attached to the Mexican Boundary-Commission. An African cruise followed his return from Mexico, and not long after he led the fatal expedition across the Isthmus of Darien, which cost so many valuable lives, and undermined the health, and has now caused the death, of the leader. Rallying from the effects of the hardships of that adventure, he accompanied Lieutenant Berryman in the voyage of the steamer *Arctic* to sound the course of the Atlantic telegraph. This was his last public service. But his energetic spirit could not brook inaction, and at the time of his death he was on his way to join the same ship from which he had been detached three years before to examine the Darien route; and on the same spot where he won so high a name among American explorers he yielded up his life."—*Providence* (R.I.) *Journal*

head-waters of the Amazon, found the written authentic account of a padre of very early times, who affirmed that there was here a race of Indians which he had seen, who were dwarfish in size and had tails. He says that one was brought to him whose caudal extremity was "the thickness of a finger, and half a palm long, and covered with a smooth and naked skin;" and also he further sets his seal to the fact that the Indian cut his own tail once a month, as he did not like to have it too long. Was not the padre's dwarf the *Brachyurus calvus*, with the short, ball-like tail, discovered a few years ago in this region by Mr. Deville?

THE BALD-HEADED BRACHYURUS.

Cuiabá, the capital of Mato Grosso, has a healthy location upon a river of the same name. Although called a city, it is, in fact, but a village. Its houses are nearly all built of taipa, with floors of hardened clay or brick. The region immediately surrounding it is said to be so abundant in gold, that some grains of it may be found wherever the earth is excavated. It is about one hundred miles from the diamond-district.

Its soil is fertile, but it almost universally lacks cultivation. In some parts particular attention is given to grazing; but, gene-rally speaking, the inhabitants make no exertions to produce any thing that is not requisite for their own consumption. Indeed, they do not always reach the limit of their own necessities. The

province abounds in gold and diamonds; but, owing to the lack
of skill employed in searching for them, the products of either, for
latter years, have been very small. What is gained by the miners
and the *garimpeiros*, as the diamond-seekers are called, together
with a certain quantity of ipecacuanha, constitute the whole
amount of exports from the province. These articles are generally sent on mule-back to Rio de Janeiro, where manufactured
goods in return are purchased and sent back over the tedious land-
route.

The first printing-press ever seen in Mato Grosso was procured
at the expense of the Government in 1838. In matters of education this province is exceedingly backward. The schools are not
only few in number, but great inconveniences are suffered from the
lack of books, paper, and nearly every other material essential to
elementary education. In addition to this low and unpromising
state of education, that of religion appears, from the reports of
successive presidents of the province, to be still worse. There are
but few churches in existence: not more than half of these are
supplied with priests; and all, without great expenses in repairing,
will ere long be in ruins.

Goyaz and Mato Grosso may be ranked together in the relation
they bear to the other portions of the Empire and of the world.
Both were originally settled by gold-hunters. The lure of treasure
led adventurers to bury themselves in the deep recesses of these
interminable forests. Their search was successful. Their most
eager avarice was satiated. But agriculture was neglected; people could not eat gold, and in many instances those who were able
to count their treasure by *arrobas* were in the greatest want of the
necessities of life. The ground was not cultivated; nothing was
exported; no flourishing towns were built. The gold-fever, abating,
left society in a state so enfeebled that we see its effects even to-
day. Gold and diamonds hindered the progress of Goyaz and
Mato Grosso more than their dense forests and great distance
from the sea-shore. It is instructive to look at the widely-different
results of the mineral and vegetable riches of the Empire. After
Mexico and Peru, (before the discovery of Australian and Californian treasure,) Brazil furnished the largest quantum of hard cur-
rency to the commercial world. Here the diamond, the ruby, the

sapphire, the topaz, and the rainbow-tinted opal sparkle in their native splendor. And yet so much greater are the riches of the agricultural productions of the Empire, that the annual sum received for the single article of coffee surpasses the results of eighty years' yield of the diamond-mines. From 1740 to 1822, (the era of independence,) a period which was the most prosperous in diamond-mining, the number of carats obtained were two hundred and thirty-two thousand, worth not quite three and a half millions pounds sterling. The exports of coffee from Rio alone during the year 1851 amounted to £4,756,794! And when we add the sums obtained for the other great staples of sugar, cotton, seringa, (or the India rubber,) dye-woods, and the productions of the immense herds of the South, we have, it is true, a better idea of the sources of wealth in Brazil, but only a faint conception of the vast resources of this fertile Empire.

Having thus glanced at all the interior provinces except Amazonas, we next turn our attention to the maritime provinces north of Rio de Janeiro.

Note for 1866.—The war with Paraguay (which was a piece of unparalleled barbarity on the part of President Lopez, son of the old Dictator) brought untold misery upon the thinly-settled population. Until November, 1864, steamers plied up to Cuyabá, and the products of the country, chiefly ipecacuanha, descended the river and were thus brought to market. All commerce, though never considerable, has been checked. Probably the largest purchaser in the world of ipecacuanha and of Brazilian sarsaparilla is the well-known Dr. J. C. Ayer, of Lowell, Massachusetts, whose remedies are found over the whole world. In 1863, Brazilian officials descended the rivers Araguaya and Tocantins to Pará in the old style, (see page 456.) Though there are many difficulties, these great rivers may yet be made to serve as highways from an interior almost closed to the outer world.

CHAPTER XXIV.

CAPE FRIO—WRECK OF THE FRIGATE THETIS—CAMPOS—ESPIRITO SANTO—ABORI-
GINES—ORIGIN OF INDIAN CIVILIZATION—THE PALM-TREE AND ITS USES—
THE TUPI-GUARANI—THE LINGOA GERAL—FEROCITY OF THE AYMORES—THE
CITY OF BAHIA—PORTERS—CADEIRAS—HISTORY OF BAHIA—CARAMURU—ATTACK
OF THE HOLLANDERS—MEASURES TAKEN BY SPAIN—THE CITY RETAKEN—THE
DUTCH IN BRAZIL—SLAVE-TRADE—SOCIABILITY OF BAHIA—MR. GILLMER, AME-
RICAN CONSUL—THE HUMMING-BIRD—WHALE-FISHERY—AMERICAN CEMETERY—
HENRY MARTYN—VISIT TO MONTSERRAT—VIEW OF THE CITY—THE EMPEROR'S
BIRTHDAY—MEDICAL SCHOOL—PUBLIC LIBRARY—IMAGE-FACTORY—THE WON-
DERFUL IMAGE OF ST. ANTHONY—NO MIRACLE—ST. ANTHONY A COLONEL—
VISIT TO VALENÇA—DARING NAVIGATION—*IN PURIS NATURALIBUS*—THE FAC-
TORY AND COLONEL CARSON—AMERICAN MACHINERY—SKILFUL NEGROES—
RETURN HOME—COMMERCE WITH THE UNITED STATES.

To reach the Brazilian North by sea has been no difficult task since 1839. At Rio de Janeiro, scarcely three days elapse unless some steamer, either foreign or *nacional*, embarks for the city of Bahia. Entering one of these, in a few hours we will be abreast of Cape Frio, which huge oval mass of granite marks the spot where the line of coast turns to the north and forms nearly a right angle.

Some years ago, the English frigate Thetis, bound homeward at the expiration of a cruise in the Pacific, was wrecked upon Cape Frio. This vessel, on leaving the harbor of Rio, where she had touched, encountered foul weather. After struggling against it till it was presumed she had cleared the coast, she bore away on her course. The darkness of the night was impenetrable, and, the wind being strong, the ship was running eight or ten knots an hour, when, without the slightest warning or apprehension of danger by any one on board, she dashed upon this rocky bulwark. The officers and crew, in the shock and consternation of the moment, had barely time to transfer themselves to contiguous por-

464

tions of the promontory, before the shivered frigate went to the bottom. Most of those on board were saved by drawing themselves up, on shelves of the rock, out of the reach of the waves, where, in the most constrained position, they were forced to remain throughout the dismal night.

A good light-house has since been constructed upon Cape Frio, which at the present time renders the approach of the navigator nearly as safe by night as it is by day.

We pass the Parahiba River, twenty miles from the mouth of which is the flourishing town of Campos, formerly called S. Salvador. The vast region surrounding this town is known as the Campos dos Goyatakazes, or plains of the Goyatakaz Indians, the aboriginal inhabitants. It is a rich tract of country, and has, for beauty, been compared to the Elysian fields. Campos is situated on the western bank of the river. The town has regular and well-paved streets, with some fine houses. Its commerce is extensive, employing a vast number of coasting-smacks to export its sugar, its rum, its coffee, and its rice. The sugars of Campos are said by some to be the best in Brazil.

Not many leagues beyond the disemboguement of the Parahiba we sail along the coast of Espirito Santo. This province embraces the old captaincy of the same name, and part of that of Porto Seguro. Although this portion of the coast was that discovered by Cabral and settled by the first Donataries, yet it is still but thinly inhabited, and has not made the improvements that may be found in most other parts. Its soil is fertile, and especially adapted to the cultivation of sugarcane, together with most of the inter-tropical productions. Its forests furnish precious woods and useful drugs, and its waters abound with valuable fish. But vast regions of its territory are only roamed by savage tribes, who still make occasional plundering incursions upon the settlements. Surveys have recently been instituted upon the rivers Doce and S. Matheus, and it is thought practicable to render those streams navigable to small steamers. Organized companies have had these enterprises in charge, and propose to open new and direct means of transport between the coast and the province of Minas-Geraes. Should this undertaking succeed, it will be of great importance, not only to the provinces of Espirito Santo and Minas-Geraes, but also to the city of Bahia, to which large quantities of the produce exported would be directly conveyed.

The distance from Rio de Janeiro to Bahia is about eight hundred miles. There is no large city or flourishing port on the coast, nor is there a single direct or beaten road through the interior. The only author who has ever travelled over this portion of Brazil by land is Prince Maximilian of Neuwied. Few naturalists have exhibited more enthusiasm, and few travellers more persevering industry, than did His Highness in passing through these wild and uncultivated regions. It is difficult to form an idea of the impediments, annoyances, and dangers which he had to surmount. But such was the interest and cheerfulness with which the Prince performed his journeys, that he described his condition by saying, "Although scratched and maimed by thorns, soaked by the rains, exhausted by incessant perspiration caused by the heat, nevertheless the traveller is transported in view of the magnificent vegetation." His travels in Brazil were accomplished between the years 1815 and 1818, and the rich and interesting work in which he gave their results to the world furnishes up to the present day the best account we have of the scenery and of the people on this section of the coast. No part of Brazil has been less agitated by the revolutions of the last half-century. Under the present *régime*, there has been a gradual improvement; yet, up to 1839, the whole province of Espirito Santo contained not a single printing-press, and many of its churches, built with great expense by the early settlers, are going to decay. But when we look at recent educational statistics, we find that there is progress even in this quiet corner of the world. In 1839, there were but seven primary schools in the province; but in 1855, the Minister of the Empire reports twenty-nine sustained by the Imperial fund, to say nothing of those conducted by provincial and private enterprise. Various internal improvements are contemplated; and we hope the day is not far distant when Espirito Santo shall have her fertile soil, which is so well adapted to the sugar and coffee plants, teeming with cultivation.

Frequent allusion has been made to the aboriginal tribes of Brazil. Their history would fill many volumes. The same interest which attaches to the Incas and their subjects, to the Montezumas and the millions over whom they lorded it, does not belong to the tribes or nations which inhabited Brazil at its discovery. The few remains of antiquity which have been reported in the North are doubtless monuments of the Empire of the Incas east of the Andes.

That erudite and accurate student of Indian antiquities, Mr. Schoolcraft, has, I think, clearly shown that the germ of Mexican civilization was the cultivation of the maize, which, to produce in quantities and in perfection, requires, at least for some months, continued labor. Thus the ancient Mexicans, if they were even for a short time nomadic, would be recalled to the spot whence they drew their principal sustenance. The want of rain either called forth efforts for artificial irrigation, or for the construction of floating gardens on the lakes which gem the great Valley of Azteca. These could not be well abandoned without the greatest sacrifice, and thus there grew up insensibly a community,—a *settlement.* If the early history of the great Peruvian nation, which numbered more than three times the population of Mexico, could be known, we should doubtless find that their civilization originated in endeavoring to procure food by the cultivation of the rainless and arid Pacific sea-coast, by resorting to artificial irrigation. When strength of mind and skill were developed, they could push their way into a more favored region, driving back other tribes. Thus, in time, they extended their conquests, their comparative civilization, and their Sabean religion over a territory comprising the country from the Pacific coast on the west to the eastern slope of the Andes, and from the equator to Valparaiso.

The tribes of Brazil, however, from the natural irrigation, and from the spontaneous products of the forests and plains, had no motives to call forth that mental effort for existence which often results in civilization. They were not settled; neither were they habitually and widely nomadic, each tribe having certain limits, where it remained until driven out by a superior force. The plantain, the banana, the cashew, the yam,—above all, the mandioca, and the more than two hundred species of palms,—furnished them food, drink, and raiment. The little cultivation to which they attended was that of the mandioca-root, which, when planted in burned ground, thrives among the stumps and roots of trees without further husbandry.

JARÁ-ASSÚ PALM (LEOPOLDINA MAJOR.)

But the most generous gift (to which allusion has been made) that bountiful Providence gave Brazil is the palm-tree. The traveller in the interior provinces and upon the sea-coast away from the cities is struck by the very great application of this "Prince of the Vegetable Kingdom" to the wants of man. And if the prince plays so important a part in the domestic economy of Europeans and their descendants, his highness was and is servant for general house and field work among the aborigines of Brazil. To this day it furnishes the Amazonian Indians house, raiment, food, drink, salt; fishing-tackle, hunting-implements, and musical instruments, and almost every necessary of life except flesh. Take the hut of an Uaupé Indian on one of the affluents of the Rio Negro. The rafters are formed by the straight and uniform palm called *Leopoldina pulchra*; the roof is composed of the leaves of the Caraná palm; the doors and framework of the split stems of the *Iriartea exhoriza*. The wide bark which grows beneath the fruit of another species is sometimes used as an apron. The Indian's hammock, his bow-strings, and his fishing-lines are woven and twisted from the fibrous portions of different palms. The comb with which the males of some of the tribes adorn their heads is made from the hard wood of a palm; and the fish-hooks are made from the spines of the same tree. The Indian makes, from the fibrous spathes of the *Manicaria saccifera*, caps for his head, or cloth in which he wraps his most treasured feather-ornaments. From eight species he can obtain intoxicating liquor; from many more (not including the cocoanut-palm, found on the sea-coast) he receives oil and a harvest of fruit; and from one (the *Jará assú*) he procures, by burning the large clusters of small nuts, a substitute for salt. From another he forms a cylinder for squeezing the mandioca-pulp, because it resists for a long time the action of the poisonous juice. The great woody spathes of the *Maximiliana regia* are "used by hunters to cook meat in, as, with water in them, they stand the fire well:" (Wallace.) These spathes are also employed for carrying earth, and sometimes for cradles. Arrows are made from the spinous processes of the *Patawá*, and lances and heavy harpoons are made from the *Iriatea ventricosa;* the long blowpipe through which the Indian sends the poisoned arrow that brings down the bright birds, the fearless peccari, and even the thick-skinned tapir, is furnished by the *Setigera* palm: the great, bassoon-like musical instruments used in the "devil-worship" of the Uaupés are also made from the stems of palm-trees

One would have supposed that a people thus supplied with almost every necessity of life would have exhibited gentleness and docility, and would have been among the most peaceful of the denizens of the New World. On the contrary, the aborigines of Brazil were a warlike, ferocious people, unskilled in the usual arts of peace, and were of the most vengeful and bloody character. Many of these tribes were cannibals: some ate their enemies in grand ceremonial; others made war for the purpose of obtaining human food; and others still devoured their relatives and friends as a mark of honor and distinguished consideration. At this day, in the remote interior, on the upper waters of the Amazon, there exist, in as wild a state as when South America was first discovered, tribes whose anthropophagous propensities are as fully indulged as if the European had never placed foot upon the continent. We would feel inclined to discredit the accounts of all the early navigators who touched upon the Brazilian coasts in regard to the cannibalism of the natives, were it not that it is fully confirmed at the present day : forty days' journey (as travellers travel) from the mouth of the Amazon up the river Purus, are found the Catauixis, and near them other tribes of Indians, who, Mr. Wallace (a thorough· and truthful explorer) says, "are cannibals, killing and eating Indians of other tribes, and they preserve the flesh thus obtained smoked and dried."

So far as can be ascertained, there were more than one hundred different tribes inhabiting Brazil at the discovery of South America.

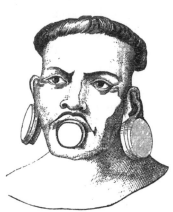

A BOTACUDO DANDY.

The large majority of these belonged to one race, and were called, upon the sea-coast, Tupi Tupinaki, Tupi-nambi, or something similar, in the way of a compound of the root *Tup*. In the South, upon the head-waters of the La Plata, they were called Guarani. They were most curiously situated, dwelling in a narrow belt upon the whole sea-coast from the mouth of the Amazon down to the present province of S. Paulo. Here they extended inland to the Paraguay, and up its waters and across the interlacings of the La Platan and Amazonian sources, where, it is surmised, they had their origin : thence they were found upon the Marmora, the Madeira, the Tapajoz, and other rivers, down the Amazon to the great island of Marajo. This people spoke in effect the same language, called by Dr. Latham, in his treatise on the languages of the Amazon, the Tupi-Guarani. This learned philologist says that as far northward as the equator and as far south as Buenos Ayres the Tupi-Guarani language was to be found. Now, there were, surrounded by this widely-spread race, numerous tribes of other aborigines, who spoke a class of languages totally distinct and different. These different tribes, it was ascertained by the Jesuits and traders, comprehended, to a certain extent, the Tupi-Guarani tongue, though their own languages were so unlike that they scarcely had one word in common. The priests, the traders, and the slave-hunters pushed their way through these tribes, and each, in their widely-different mission, aided in the formation of a remarkable language, called the *Lingoa Geral* or *Lingoa Franca*, which was the common vehicle of communication, from the Orinoco to the La Plata, among people whose languages remain unknown. The trader, the scientific explorer, and the Brazilian Government official, at this day have their intercourse with the savages of the Japura, the Paraná, the Chingu, and the Araguaia, by the Lingoa Geral. The basis of this, as already observed, is the Guarani or Tupi-Guarani tongue.*

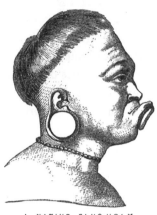

A NATIVE PLUG-UGLY.

These surrounded tribes, so to speak, occasionally, though rarely, succeeded in reaching the coast. Thus, the Aymores—a cannibal tribe who acquired such a terrible celebrity—made their appearance upon the sea-shore a long time after the discovery of Brazil. The coast-tribes regarded them with horror, and considered them as irrational beings, ignorant of the construction of huts and of the art of adorning their persons with the rich plumage of the parrot and the gay-painted macaw. They had a still more distinctive characteristic, that consisting in an unconquerable fear of water, which impeded them from following their enemies when they swam a river or plunged into a lake. They assaulted Porto Seguro and the Ilheos with such ferocity that Bellegarde says that labor ceased on all the plantations for want of workmen who had gone to give them battle. They were afterward routed and nearly all

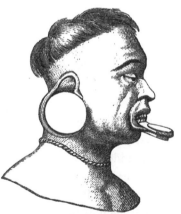

LIP-ORNAMENT OF THE SOUTH AMERICAN INDIAN.

---

* Dr. Latham says, "With two exceptions, the distribution of the numerous dialects and subdialects of the Tupi-Guarani tongue is the most remarkable in the world,—the exceptions being the Malay and the Athabascan tongues."

dispersed, and there only remain as their descendants the Bota-
cudos, a few hundred of whom still—now peacefully—wander upon
the banks of the rivers Doce and Bellemonte. These Indians,
like many of the savages of South America, wear the most absurd
ornaments of light wood, (the aloe,) which they at pleasure insert
and take out from slits in their ears and lips.

But the question naturally arises, What have become of the
numerous tribes once inhabiting the sea-coast and those provinces
where now a civilized population most abound? Where are the
Tupi-Guarani? Many wandered to remote parts of the Empire;

A BOTACUDO FAMILY ON THE MARCH.

European diseases and vices, as well as war and the march of
civilization, swept them from their places. The Guarani of South
Brazil, under the Jesuits, reached a certain degree of advancement; but the inhuman Portuguese slave-hunter, who pushed his
way as far as Bolivia, with ruthless hands broke up the missions
and led them into captivity, and they soon melted away before
cruel taskmasters. Of the Tupinambás and the Tamoyos, who
dwelt in the present provinces of Rio de Janeiro and Minas-
Geraes, the former were exterminated, and the latter were so
constantly harassed and defeated in war by the colonists, that,
though for a long time wanting unanimity, they finally were persuaded by the eloquence of an influential and eminent chief (Jappy
Assú,—a second Orgetorix) to emigrate to the distant North,—

then more than three thousand miles from their former home,—and they settled upon the southern bank of the Amazon, from its confluence with the Madeira, at various points, down to the island of Marajo. Their descendants are found this day in the country between the Tapajoz and the Madeira, among the lakes and channels of the great island of the Tupinambás. They are now called the Mandrucús,—the most warlike Indians of South America. They live in villages, in each of which is a fortress where all the men sleep at night. This building is adorned within by the dried heads of their enemies decked with feathers. These ghastly ornaments have the features and hair very well preserved.

The existing tribes, in their manners and customs, are closely allied to our North American Indians, with this exception :—that the savages south of the equator have all been found to be exceedingly deficient in any religious idea. None of them, when first visited, seemed to have the faintest conception of the Great Spirit which so strikingly characterized the simple theology of the aborigines of the Mississippi and the St. Lawrence. Attempts to civilize them have proved abortive except when they are held in a state of pupilage, as they were by the Jesuits, or under the rigid discipline of the Brazilian army.

The curious ethnologist will find in the tribes of the Upper Amazonian waters the red man who has been untouched by civilization. Mr. Wallace—who roamed for some years among these sons of the wilderness—has given us much information in regard to them, and says that one of the singular facts connected with these Indians is the resemblance which exists between some of their customs and those of nations most remote from them. Thus, the *gravatána* or blowpipe reappears in the *sumpitan* of Borneo ; the great houses of the Uaupés and Mandrucús closely resemble those of the Dyaks of the same country ; while many small baskets

and bamboo boxes from Borneo and New Guinea are so similar in.
their form and construction to those of the Amazonian Indians
that they might be supposed to belong to adjoining tribes.   Then,
again, the Mandrucús, like the Dyaks, take the heads of their
enemies, smoke-dry them with equal care, preserving the skin and
hair entire, and hang them up around their houses.   In Australia,
the throwing-stick is used; and on a remote branch of the Amazon
(the Purus) we see a tribe of Indians (the Purupurús) differing
from all around them in substituting for the bow a weapon only
found in such a remote portion of the earth, among a people so
distinct from them in almost every physical characteristic.

The aboriginal population is unknown, and there are only about
nineteen thousand catechized or Christian Indians reported by the
Minister of the Empire.

On the ocean-route from Rio to Bahia there are four small islands,
called the *Abrolhos*, ("Open your eyes,") which are dangerous pro-
jections from a bank of rocks that exhibits itself occasionally
between the seventeenth and twenty-fifth degrees of south latitude, at a distance of from two to ten leagues from the mainland.
Besides these, there is also a regular reef of rocks running quite
near the shore, and generally parallel with it, the whole distance
from Cape Frio to Maranham.   Espirito Santo, Porto Seguro,
Ilheos, and, in fact, nearly all the ports along the entire coast, are
formed by openings through this reef.

After three or four days' steaming, the lower extremity of the
island of Itaparica, with its numerous palm-trees, looms up in the
horizon, and but a short time elapses before the eye catches the
outline of tne white domes and towers of Bahia San Salvador, the
second city of the Empire.

When the steamer arrived, I was, through the kindness of Sr.
Nobre, the *guarda mor*, immediately transferred to the shore in his
Government-barge.   The walls of a circular fort rising from the
bosom of the water, built by the Dutch, frown upon the shipping;
while the fortresses on the hills command the harbor and the
entire city.

Landing at the Custom-House, I passed into the lower town,
with its narrow streets (in some places there is but one) running
parallel to the water's edge.

VIEW OF A PORTION OF BAHIA.

Along the Rua da Praya are located the Alfandega and the Consulado, through the latter of which all home-productions must pass
preliminary to exportation. Some of the *trapiches* (warehouses)
near by are of immense extent, and are said to be among the
largest in the world.

Around the landing-places cluster hundreds of canoes, launches,
and various other small craft, discharging their loads of fruit and
produce. On one part of the praya is a wide opening, which is
used as a market-place. Near this a beautiful spacious modern
building has been constructed for an exchange. It is well supplied
with newspapers from all parts of the world, and is in a cool and
airy situation. The principal commercial houses are situated on
the Rua Nova do Commercio, and these compose the finest blocks
of buildings in Brazil,— perhaps in all South America. These
edifices would adorn the business-portions of London, Paris, or
New York.

The lower town is not calculated to make a favorable impression
upon the stranger. The lofty buildings are nearly all old, although
generally of a cheerful exterior. The streets in this vicinity are
very narrow, uneven, and wretchedly paved, and at times as filthy
as those of New York. At the same time it is crowded with pedlars
and carriers of every description. You here become acquainted
with one peculiarity of Bahia. Owing to the irregularities of its
surface and the steepness of the ascent which separates the upper
town from the lower, it does not admit the use of wheel-carriages.
Not even a cart or truck is to be seen for the purpose of removing
burdens from one place to another. Whatever requires change of
place in all the commerce and ordinary business of this seaport—
and it is second in size and importance to but one other in South
America—must pass on the heads and shoulders of men. Burdens
are here more frequently carried upon the shoulders, since, the
principal exports of the city being sugar in cases and cotton in
bales, it is impossible that they should be borne on the head like
bags of coffee.

Immense numbers of tall, athletic negroes are seen moving in
pairs or gangs of four, six, or eight, with their loads suspended
between them on heavy poles. Numbers more of their fellows are
seen sitting upon their poles, braiding straw, or lying about the

alleys and corners of the streets asleep, reminding one of black
snakes coiled up in the sunshine. The sleepers generally have
some sentinel ready to call them when they are wanted for business, and at the given signal they rouse up, like the elephant to his
burden. Like the coffee-carriers of Rio, they often sing and shout
as they go; but their gait is necessarily slow and measured, resembling a dead-march rather than the double-quick step of their
Fluminensian colleagues. Another class of negroes are devoted to
carrying passengers in a species of sedan-chair called cadeiras.

PORTERS OF BAHIA.

It is indeed a toilsome and often a dangerous task for a white
person to ascend on foot the bluffs on which stands the *cidade alta*,
particularly when the powerful rays of the sun are pouring, without mitigation, upon the head. No omnibus or cab can be found
to do him service. In accordance with this state of things, he
finds near every corner or place of public resort a long row of cur-
tained cadeiras, the bearers of which, hat in hand, crowd around
him with all the eagerness, though not with the impudence, of
carriage-drivers in North America, saying, " *Quer cadeira, Senhor?* "
(" Will you have a chair, sir?") When he has made his selection,
and seated himself to his liking, the bearers elevate their load and
march along, apparently as much pleased with the opportunity of

carrying a passenger as he is with the chance of being carried. To keep a cadeira or two, and negroes to bear them, is as necessary for a family in Bahia as the keeping of carriages and horses is elsewhere. The livery of the carriers, and the expensiveness of the curtaining and ornaments of the cadeira, indicate the rank and style which the family maintains.

Occasionally you will meet a proud creole Mina negress, who rejoices in the name *par excellence* of *the Bahiana.* Her turban, her shawl, her ornaments, and her elastic step in the heeled slipper, display a native grace unattainable by modern fashion.

I regret that I have no sketch of Bahia taken from the water, — for from that point the city seems truly magnificent in its proportions; but the large cut, from a daguerreotype, gives a view of the religious metropolis of Brazil, stretching on its terraced hills around to

THE BAHIANA NEGRESS.

Montserrat. The steep ascent on which we see the *cadeira*-carriers is the same up which Henry Martyn climbed in 1805, so graphically described in the journal incorporated in the pages of his biography. The lower city, with the exception of the Rua Nova do Commercio, has been very little changed since the visit of that devoted missionary.

Some of the streets between the upper and lower towns wind by a zigzag course along ravines; others slant across an almost perpendicular bluff, to avoid, as much as possible, its steepness.

Nor is the surface level when you have ascended to the summit. Not even Rome can boast of so many hills as are here clustered together, forming the site of Bahia. Its extent between its extreme limits—Rio Vermelho and Montserrat—is about six miles. The city is nowhere wide, and for the most part is composed of only two or three principal streets. The direction of these changes with the various curves and angles necessary to preserve the summit of the promontory. Frequent openings between the houses built along the summit exhibit the most picturesque views of the bay on the one hand and of the country on the other. The aspect of the city is antique. Great sums have been expended in the construction of its pavements,—more, however, with a view to preserve the streets from injury by rain than to furnish roads for any kind of carriages. Here and there may be seen an ancient fountain of stonework, placed in a valley of greater or less depth, to serve as a rendezvous for some stream that trickles down the hill above; but nowhere is there any important aqueduct, though recent water-works, with steam-engines manufactured in France, have been lately erected east of the Noviciado, which will furnish a bountiful supply of the potable element to the city.

In contemplating Bahia from the theatre (the large building on the high terrace) we are carried back to the earliest periods of the colonial history of Brazil. The old round fort in the midst of the waves is an episode of the brief·power of Holland in this portion of America, upon which Time has made no perceptible change.

Bahia de Todos os Santos, the Bay of All Saints, was discovered in 1503 by Americus Vespucius, who was then voyaging under the patronage of the King of Portugal, Dom Manoel. In 1510, a vessel under the command of Diogo Alvares Corrêa was wrecked near the entrance of this bay. The Tupinambás, inhabiting the coast, fell upon and destroyed all who survived this shipwreck, except the captain of the vessel. The Indians spared Diogo,— probably, as some supposed, on account of his activity in assisting them to save articles from the wreck. He had the good fortune to obtain a musket and some barrels of powder and ball. He early took occasion to shoot a bird, and the Indians, terrified by the explosion no less than by its effects, called him from that moment *Caramurú,* "the man of fire."

He then conciliated their favor by assuring them that, although he was a terror to his enemies, he could be a valuable auxiliary to his friends. He accordingly accompanied the Tupinambás on an expedition against a neighboring tribe with whom they were at war. The first discharge of Caramurú's musket gained him possession of the field, his frightened adversaries scampering for their lives.

Little more was necessary to secure him a perfect supremacy among the aboriginals. As a proof of this, he was soon complimented with proposals from various chiefs, who offered him their daughters in marriage. Diogo made choice of Paraguassú, daughter of the head-chief Itaparica, whose name is perpetuated as the designation of the large island in front of the city, while that of Paraguassú, the bride, is applied to one of the rivers emptying into the bay. He built a hamlet which he denominated S. Salvador,* in gratitude for his escape from the shipwreck. This settlement was located in a place denominated Graça, on the Victoria Hill, a suburb of the city, still occasionally called *Villa Velha*, (old town.)

After the lapse of some years, a ship from Normandy anchored in front of Caramurú's town and opened communications with the shore. Diogo now determined to return to Europe; and, having supplied the vessel with a cargo, he embarked for Dieppe, accompanied by Paraguassú. He intended, if he arrived safely, to go from Dieppe to Lisbon. The French, however, would not permit this, but preferred to make him a lion in their own capital. Paraguassú was the first Indian female who had ever appeared in Paris. A splendid fête was given at her baptism, when she was christened Catharine Alvares, after the Queen Catharine de Medicis. King Henry II., accompanying his royal spouse, officiated on the occasion as godfather and sponsor.

---

* In successive editions of the narrative of the " United States Exploring Expedition" we find the following:—"The city of San Salvador, better known as Rio de Janeiro,"—which is comparable for accuracy to McCulloch's Geographical Dictionary, making the mountainous province of Rio de Janeiro to consist "mostly of plains." *San Salvador* is eight hundred miles north of Rio de Janeiro, and *San Sebastian*—the old name of Rio—has about as much similarity to San Salvador as New Orleans has to New York.

The French Government contracted with Caramurú to send out vessels which should carry him to his adopted country, and return with brazil-wood and other articles, which should be given in exchange for goods and trinkets. In the mean time, true to his original intent, he contrived to inform Dom John III., of Portugal, of the importance of colonizing Bahia. A young Portuguese, who had just finished his studies in Paris and was returning to Portugal, was the bearer of this message. This young man (Pedro Fernandez Sardinha) afterward became Bishop of Bahia.

The natives rejoiced at Caramurú's return, and his colony now increased rapidly and extended its influence in every direction.

At this period the King of Portugal, in order to secure the settlement of Brazil, divided the country into twelve captaincies, each of fifty leagues' extent on the coast, and boundless toward the interior. Each captaincy was conceded to a Donatary, whose power and authority were absolute. Francisco Pereira Coutinho, who came to take possession of Bahia, was a man rash and arbitrary in the extreme. He became jealous of the influence of Diogo Alvares, and commenced to persecute and oppress him, and finally sent him on board a ship as a prisoner.

This course exasperated the Indians, who determined on revenge. They attacked the settlement and killed Coutinho. Diogo Alvares was again restored to his original supremacy.

The growing importance of the country, together with rumors of violence practised by the Donataries, induced Dom John III. to appoint a Governor-General of Brazil, to reside at S. Salvador and to have jurisdiction over all the Donataries.

In 1549, Thomé de Souza, the first Governor-General, landed with military ceremonies at Villa Velha, but in the course of a month proceeded to choose another location for the commencement of his operations. It was that of the present Cathedral, Government Palace, and other public buildings.

Caramurú was now an old man, but was of great service to the Governor-General in consummating with the natives a treaty of peace. In four months a hundred houses were built, and various sugar-plantations were laid out in the vicinity.

From this period the city of S. Salvador, having been constituted the capital of Portuguese America, and remaining under the direct patronage of the mother-country, rapidly increased in size and importance.

The year 1624 witnessed the first depredations of the Dutcn upon the then quiet and prosperous city of Bahia. Without the least notice or provocation, a fleet from Holland entered the harbor, attacked the city, burned the shipping, and debarked men to seize the fortress of S. Antonio, and, after some fighting, gained possession of the town. This they sacked, not even sparing the churches. The captors immediately erected additional fortifications and built many new houses. They made prizes of all the Portuguese and Spanish ships that came into the harbor not knowing that the town had changed masters.

Portugal was at this time tributary to Spain. The news of the loss of Bahia caused great consternation at Madrid, and the more since it had been rumored that the English were to unite their forces with the Dutch and establish the Elector-Palatine King of Brazil. The Spanish court adopted measures worthy of its superstition and its power. Instructions were despatched to the Governors of Portugal, requiring them to examine into the crimes which had provoked this visitation of the divine vengeance, and to punish them forthwith. Novenas were appointed throughout the whole kingdom; and a litany and prayers, framed for the occasion, were to be said after the mass. On one of the nine days there was to be a solemn procession of the people in every town and village, and of the monks in every cloister. The sacrament was exposed in all the churches of Lisbon, and a hundred thousand crowns were contributed in that city to aid the Government in recovering S. Salvador.

A great ocean-fleet of forty sail, carrying eight thousand soldiers, sailed under D. Fadrique de Toledo and D. Manoel de Menezes, which in March, 1625, appeared off the bay; and after some delay, the object of which was to learn if the Hollanders had received reinforcements, D. Fadrique, satisfied that they had not, entered the harbor with trumpets sounding, colors flying, and the ships ready for action. The Dutch vessels also, and the walls and forts, were dressed out, with their banners and streamers hoisted, either to welcome friends or defy enemies, whichever these new-comers might prove to be. The city had been fortified with great care, according to the best principles of engineering,—a science in which no people had at that time such experience as the Dutch. It was defended by ninety-two pieces of artillery, and from the new fort upon the beach they fired red-hot shot.

After some severe skirmishing, the Dutch, having waited in vain for the fleet from Holland, proposed a capitulation, which was acceded to.

The Hollanders attempted to retake the city in 1638, under Mauritz, Count of Nassau, who was then in possession of Pernambuco and a large portion of the adjoining coast. They were repeatedly defeated at Bahia, but succeeded for a time at other points.

The original attack, on the part of the Dutch, grew out of purely mercenary motives. It was planned and executed under the auspices of the celebrated West India Company. Proving successful at first, the Hollanders did not content themselves with plundering the inhabitants, but determined to make the very soil their own. Their inroads were manfully resisted by the Portuguese, and the war, at different times, extended along the whole coast from Bahia to Maranham.

In 1636, Mauritz, Count of Nassau, was sent out to take command of the troops and to govern the new Empire. Under his direction active measures were set on foot; forts, cities, and palaces were built, and the country was explored in search of mines. Agriculture was undertaken with a strong hand, and it is easy to imagine what changes would have been introduced into those fertile regions by the industrious Hollanders, had not the fate of war decided against them. In the low ground, the marshes and the streams that surround the city of Pernambuco, they would have especially gloried.

But the Brazilians, under their vigilant leaders, Camarão, Henrique Diaz, (the former an Indian, the latter a negro,) Souto, and Vieyra, kept up such incessant attacks upon the Hollanders, that at last, in 1654, they were expelled from Pernambuco, and in 1661 they abandoned, by negotiation, all claim to Brazil.

It is interesting to think that, whatever motives may have urged the commercial Hollanders to attack Brazil, the Christians of that brave little Protestant country were not slow to follow up the settlements; and hence, in Pernambuco and vicinity, faithful missionary stations were established, and, when the Dutch were finally driven from the country, some of the clergymen came to New Amsterdam, and one of them was the first pastor of the Dutch Reformed Church founded at Flatbush, Long Island.

From this time the Hollanders ceased their attacks on Bahia, that city advanced in wealth and prosperity, and was the seat of the Viceroyalty until 1763, when it was transferred to Rio de Janeiro.

The position of Bahia, opposite the coast of Africa, caused it to be, from early times, an important rendezvous for those engaged in the African slave-trade. The offensive ideas now associated with that traffic among all enlightened nations are strangely in contrast with the semblance of philanthropy under which it was originally carried on. What a worthy enterprise, to send vessels to *ransom* those poor pagan captives and bring them where they could be Christianized by baptism, and at the same time lend a helping hand to those who had been so kind as to purchase them out of heathen bondage and bring them to a Christian country! Expressive of such ideas, the bland title by which the buying and selling of human beings was known during the seventeenth and eighteenth centuries, was the *"commerce for the ransom of slaves."*

Bahia increased in population and wealth, and in 1808 its prosperity was still more augmented by the Carta Regia which opened the ports of Brazil to the world.

This city was the last that remained faithful to Portugal; for, though the independence of the Empire was declared in September, 1822, it was not until July, 1823, and after severe suffering, that the Portuguese army evacuated Bahia de San Salvador. The rebellion of 1837 was frightful in the extreme; but the Imperial Government finally obtained the mastery, and from that day to this Bahia has continued quiet, and has made rapid strides of improvement.

I do not think that there is any city in Brazil that so interests the foreigner as Bahia. It is the spiritual capital of the country, being the residence of the archbishop. The churches, the convents, and other public buildings, are upon a large scale, and have no provincialism in their appearance. The people are gay and

social, and in my extended travels throughout the Empire I have
nowhere found a society equal to that of Bahia. At the house of
Mr. Gillmer, the American Consul, one is always sure to meet the
most refined and well-educated Brazilians. This gentleman is one
of the few American consuls who, by knowledge of the language
of the land where they reside, by sociability of character and ease
of manners, and by pride of country, justly represent a great
nation. Mr. Gillmer has long resided at Bahia, and by his many
excellent qualities has won the hearts of the Brazilians. The
weeks spent in his agreeable family gave me an opportunity for
making many acquaintances among the citizens of Bahia and the
foreigners resident in that city. The residence of Mr. Gillmer is in
a delightful portion of the city, where verdure and bloom abound.
Each night the breezes were laden with sweet odors, and every
morning the sun seemed to reveal new beauties of opening buds
and brilliant flowers. The house of Senhor Nobre was surrounded
by shade and fruit trees, and his large
salon was weekly filled by amateur
and professional musicians, who gave
the most charming *soirées musicales*.

THE LONG TAILED MALE
HUMMING-BIRD.

Early one morning I looked from
a window of the Consul's house,
and saw, upon the branch of a
bread-fruit-tree beneath me, a humming-bird sitting quietly upon her
tiny nest. In the midst of the foliage she appeared like a piece of
lapis lazuli surrounded by emeralds;
for her back was of the deepest blue.
Everywhere throughout Brazil this
little winged gem, in many varieties,
abounds, while in North America,
from Mexico to the fifty-seventh degree of latitude, it is said that there
is but one species of the humming-
bird. Mr. Gosse calls the long-tailed kind (*Trochilus polytmus*) the
gem of American ornithology; and well it deserves the title, if
we consider the flashes of rich golden green, purplish black, deep-

bluish gloss, and gorgeous emerald green, which irradiate from this winged jewel.

The males are among the most belligerent of creatures,—rarely meeting without having terrible combats

The city is not, however, so much distinguished for its frequentation by humming-birds as its bay is celebrated as a "whaling-ground." To "fish for whales" is a regular business at Bahia, and nearly every week, from the numerous terraces, admiring thousands can gaze upon the stirring excitement of capturing these monsters of the deep. Why they frequent this port I do not know, unless their peculiar food abound in its waters. If we descend through lime-tree hedges to the Rio Vermelho, we may have an opportunity (besides seeing the fixtures for extracting oil) of witnessing the triumphant arrival of the dead leviathan. Hundreds of people—the colored especially—throng

TROCHILUS POLYTMUS.

around to witness the monster's dying struggles, and to procure portions of his flesh, which they cook and eat. Vast quantities of this flesh are cooked in the streets and sold by quitandeiras. Numbers of swine also feast upon the carcass of the whale; and all who are not specially discriminating in their selection of pork in the market, during the season of these fisheries, are liable (*nolens volens*) to get a taste of something "very like a whale." This whale-fishery was once the greatest in the world. At the close of the seventeenth century, it was rented by the Crown for thirty thousand dollars annually.

From the Rio Vermelho we ascend by a winding path to the Victoria Hill, passing *en route* the English and American cemeteries. The latter is the only burial-ground in Brazil belonging to the citizens of the Union, and our country has long been greatly indebted to the courtesy of English consuls for suitable places of interment for natives of the United States. This cemetery is the result of private generosity, and especially of the energy and liberal subscriptions of Mr. Gillmer. It is, however, neither just

nor reasonable that he should bear the whole burden. In vain has
he appealed to our Government for aid in keeping up this resting-
place for our country's dead; and the result is, that, no allowance
being granted, the cemetery is in a sad condition. The policy of
Great Britain is noble in this respect. Everywhere she erects
chapels and provides cemeteries for her subjects; and, though
necessarily the United States cannot recognise any connection
between Church and State, yet a decent place for the burial of the
dead in foreign countries is a matter of common humanity, which
demands immediate attention from Government. I have known
parents in the United States who would have given thousands if
they could only know the spot where rested the remains of beloved
sons who, dying in hospitals, were thrust into the common receptacle
for those whose country had not made provision of a cemetery.

On the Victoria Hill may be found the finest gardens that Bahia
affords, the most enchanting walks, and the most ample shade.
Here, too, are the best houses, the best air, the best water, and the
best society. The walls of two ancient and extensive forts also
add much to the romance and historical interest of the place.
With its magnificent prospect of blue water and verdant isles, it
is a spot that combines an external beauty of the rarest quality.
It was here that Henry Martyn, who incidentally touched at this
port on his passage to India more than half a century ago, sighed
and sung,—

> " O'er the gloomy hills of darkness
> Look, my soul; be still, and gaze."

That the moral aspect of the place has not undergone any very
great change (unless it be in diminished bigotry and greater indifference) is not to be presumed, as no causes have been at work
that contemplated such a change. Everywhere there are still
evidences which give point to the remark of Martyn:—" Crosses
there are in abundance; but when shall the DOCTRINES of the
cross be held up?"

I looked upon no portion of Brazil with greater interest than
those walks, gardens, chapels, and convents visited by Henry
Martyn. The Hospital for Lepers, and the chapel where he gently
and lovingly, yet firmly, uttered his protest against corrupt religion,

are still standing: the latter, however, is no longer in use. The
pepper-plantation is torn up, but the clove-trees of which he speaks
are still flourishing. Some of the convents which he entered are
now tenantless of their monkish dwellers; for in some respects a
better day has dawned upon Brazil, and many of these huge buildings, once given up to thriftless, indolent, and vicious orders, are
now used for colleges, lyceums, libraries, and hospitals. The convent where the future missionary to Persia alone, as the sun was
setting and the cloisters were darkened, taught, with Vulgate in

A CHAPEL VISITED BY HENRY MARTYN.

hand, "the faith once delivered to the saints" to the curious and
benighted friars, still lifts its whitened walls,—walls which heard
his teachings and the prayers which he whispered for the blessing
of a pure gospel to descend upon Brazil. Have Henry Martyn's
prayers been forgotten before the Lord of Hosts? We love to
regard the petitions of the early Huguenots at Rio de Janeiro,
those of the faithful missionaries of the Reformed Church of Holland at Pernambuco, and the prayers of Henry Martyn at Bahia,
as not lost, but as having already descended, and as still to descend,
in rich blessings upon Brazil.

My intercourse with Rev. Mr. Edge, the English chaplain, was
exceedingly pleasant. He was a Cambridge man, and one of enlarged and catholic views. The chapel was better filled on the

Sabbath than any other that I saw in Brazil. In a ramble with him, I sketched, under a burning sun, the chapel above, which was near the country-seat mentioned by Martyn where he first saw the clove and the pepper. That first visit of Henry Martyn in the country, away from the house of Antonio José Correa, I believe to have been where the Hospital of Montserrat is now situated.

The day was beautifully clear, and we rode over a long, well-paved street called the Calçado, which reaches quite into the country. In the outer suburbs the cocoanut-palm grows in great profusion, and the jaca-tree waves its green, glistening foliage above the infinite variety of vegetation which adorns this Southern land. We passed the Carmelite Convent and went as far as the

N. SENHORA DE MONTSERRAT.

road which leads to the Fever Hospital: here we descended and walked to the tongue of land called Montserrat, upon which are picturesque fortifications, a row of summer-houses,—that of Mr.

Gillmer distinguished by the American flag,—and on the extreme point a small Roman Catholic chapel, more than two hundred years old, above the doorway of which I deciphered this inscription:— "*A Virgem foi concebida sem peccado original.*" Why Romanists should cling with such tenacity to the dogma of the immaculate conception, which contains nothing essential to salvation, I could never understand.

We visited the well-appointed hospital near by, which is intended particularly for those who have been smitten with the yellow fever; but its attacks have been very light for the last few years, though the cholera, in 1855, was quite fatal to the blacks and to the mixed population generally. Yet, when we consider that, out of a population of nearly a million in the province, but nine thousand fell before the cholera, the percentage is small compared with that of New York in 1833, and almost nothing when compared with the ravages of the same disease at St. Louis in 1849 and '50. In the spring of 1857, the journals of the United States teemed with the accounts of the fell swoop of the yellow fever at Rio de Janeiro, where for a short time twenty-five persons *per diem* died. It can be proved by actual statistics that no city of equal population in the United States has so good a sanitary condition as Rio de Janeiro.

The view of Bahia from Montserrat is truly magnificent. The curving lines of whitened buildings—the one upon the heights, the other upon the water's edge—everywhere separated by a broad, rich belt of green, itself here and there dotted with houses,—the fortress, the shipping, the white-capped waves, over which the whale-boats are pursuing their gigantic sport,—the distant isle of Itaparica and the blue ocean beyond,—all form a picture which at the time fills one with exhilarating delight, and ever after dwells in the cabinet of memory a choice and beautiful picture. There are few cities that can present a single view of more imposing beauty than does Bahia to a person beholding it from a suitable distance on the water. Even Rio de Janeiro can hardly be cited for such a comparison. The capital excels in the endless variety of its beautiful suburbs; but in the Archiepiscopal City beauty is con- centrated and presented at one view. In Rio, for pleasant abodes, one section competes with another, and each offers some ground of preference; but in Bahia, the superiorities seem all to be united in one section, leaving the foreigner no room for doubt that the focus is the Victoria Hill.

Beneath its brow, just on the edge of the bay, is a stately residence embowered with cool fruit and flowering trees, where fountains sweetly murmur in cadence with the musical rippling of the waters which break upon the neighboring beach. It may, however, distress some of my readers to know that this beautiful place is a snuff-factory, where the celebrated *aréa preta* is made which enjoys a monopoly in Brazil. Snuff-making and snuff-taking were found among the aborigines; but this particular snuff was the invention of a Swiss from Neufchâtel, and from which he acquired a large fortune. By his will, after enriching his relatives, he left liberal sums for the endowment of hospitals for his native canton, and also for benevolent purposes in Bahia. The main establishment (there are branches in Rio and Pernambuco) is under the superintendence of M. Barrelet, of Neufchâtel, in whose agreeable family I had that intercourse so sweet to the Christian in a foreign land.

Common-school education at Bahia is upon the best footing in the Empire, and the Bahians take great pride in showing the statistics of their various institutions. Young Dr. Fairbanks accompanied me one morning through the chief hospital and the medical college. In the latter I found that there were nearly three hundred students attending the lectures. Some of the professors— both natives and foreigners—are men of talent and erudition, and the course of instruction is probably equal to that of any medical school on the Western continent. In the library connected with the institution I saw some very large and very costly volumes on anatomy in the Russian language. They had been recently sent out from St. Petersburg, and were in every respect very finely gotten up.

Near by is the old Cathedral, an immense edifice, which has been constructed with great expense, and is superior to any church in Brazil, unless it may be the unfinished Candalaria of Rio. In a wing of this building, from which may be enjoyed a very commanding view of the harbor, is located the public library. It contains many thousand volumes, a large portion of which are in French; and it also possesses some most valuable manuscripts.

The librarian is the Hon. Chevalier de Lisboa, the accomplished

scholar and gentleman, who, as Minister-Plenipotentiary, represented Brazil at Washington in 1845. I was deeply interested in
a large and well-illustrated volume shown me by the Chevalier,
which was an account of the "Dutch in Brazil" and was published
at Amsterdam before the middle of the seventeenth century.

In the immediate neighborhood of the Cathedral are the archiepiscopal palace and seminary, and the old Jesuit College, now
used as a military hospital. The latter building, together with
the Church of Nossa Senhora da Conceição, (its steeples are seen on
the right of the large view of Bahia,) on the Praya, may almost be
said to have been built in Europe: at least, the principal stonework for them was cut, fitted, and numbered on the other side of
the Atlantic, and imported ready for immediate erection. The
President's palace is also but a short distance from this locality.
It is a substantial building, of ancient date, located upon one side
of an open square.

The Presidents of provinces are appointed by the Emperor, and
his choice is by no means confined to the particular province to be
governed. Hence Brazilian statesmen are liable to many changes
of residence: but it may be that there is wisdom in this, for it has
been said that the selections are thus made of strangers to the province so "that the influence of family connections and personal
friendships may not prove temptations to partiality in the distribution of gifts and emoluments under their control." The President
is, in fact, a Viceroy with a body-guard; and it seems to me that
the appointing-power by which he is elevated to office is one of
the most conservative elements in the Brazilian Constitution.

My colleague was at Bahia on the anniversary of the Emperor's
birth, and his felicitous description of that scene will convey an
idea of similar celebrations throughout the whole Empire:—

"The Bahians were preparing to celebrate the birthday of their youthful Emperor, the 2d of December. This anniversary is, throughout the nation, a favorite
one among the several *dias de grande gala*, or political holidays. Of these the Brazilians celebrate six. The 1st of January heads the list with New Year's compliments to His Majesty. The 25th of March commemorates the adoption of the
Constitution. The 7th of April is the anniversary of the Emperor's accession to
the throne. The 3d of May is the day for opening the sessions of the National
Assembly. The 7th of September is the anniversary of the Declaration of the
national Independence; while the last in the catalogue is the 2d of December, the
Emperor's birthday. On all these days, except the 3d of May, His Majesty holds

court in the palace at Rio. Presidents of provinces, as the special representatives of the Crown, follow the example of their sovereign, by holding levée in the several provincial capitals; but they do not presume to receive Imperial honors in their own person. The place of honor in their *sala de cortejo* is always allotted to the portrait of His Majesty. Near by, as the special representative of the throne, the President takes his place, accompanied perchance by the bishop. Before these, in measured step, pass the dignitaries invited, in the order of their rank and distinction, paying their obeisance severally to the Imperial portrait. After this ceremony, mutual compliments are exchanged by the individuals present, and the company breaks up.

"It was no ordinary celebration that was to take place at this time. During the recent session of the National Assembly at Rio de Janeiro, it had been more than intimated that the Bahians generally were doubtful in their loyalty. Not relishing such insinuations, they had resolved to make a display on this occasion which, from its unexampled magnificence, should not only demonstrate their fidelity to the throne, but should throw even the metropolis into the shade. In addition to the usual cortejo, there were to be ceremonies for three successive days and illuminations for as many nights. On the first day there was to be a grand *Te Deum*, with a sermon; on the second, a military ball at the palace; and on the third, an unrivalled exhibition of fireworks, on Victoria Hill, at the Campo de S. Pedro.

"The 2d of December came. It was not clad in the frosty robes of a Northern winter, with whistling winds and drifted snow at its heels. Nay, the North is not farther from the South than is the idea many a reader has pictured in his imagination at the bare mention of December, from the reality of the day in question. Preceded by but a brief interval of twilight, the sun threw upward his mellowest rays, burnishing the wreathed clouds of the eastern horizon. Presently from his bed of ocean he rose majestic on his vertical pathway, looking down on one of the fairest scenes nature ever presented to the eye of man. The boundless expanse of the Atlantic on the east,—the broad and beautiful bay on the south and west, with its palm-crested islands and circling mountains,—were but an appropriate foreground to the lovely picture of the city herself, reposing like a queen of beauty amid the embowering groves of the proud eminences over which her mansions, her temples, and her lordly domes were scattered.

"The day was ushered in by the roar of cannon from the several batteries and the vessels-of-war. From that moment might be seen the shipping of every nation in the harbor, gayly decked with flags, signals, and streamers of unnumbered hues.

"Being much occupied in the morning, I did not reach the Cathedral in time to listen to the discourse which preceded the Te Deum, which terminated at three o'clock P.M. At this moment there was a discharge of rockets in front of the Cathedral and a general salute of artillery from the guns of the forts and shipping. The scene was now transferred to the Government Palace, the old residence of the Viceroys, where the cortejo took place. At the same time, the troops of the city, to the number of two thousand five hundred, were paraded in the Palace Square and in the streets leading from the Cathedral to that place. These, together with all the other principal streets, had been adorned by silk and damask hangings from the windows,—the national colors, yellow and green, being most frequent and most admired. The illumination at night throughout the city, but specially at the Passeio Publico, was, of all other parts of the celebration, most interesting to me.

"This public promenade of Bahia is located on the boldest and most commanding

height of the whole town.   One of its sides looks toward the ocean, and another
up the bay, while nothing but an iron railing guards the visitor against the danger
of falling over the steep precipice by which its whole front is bordered.   For
airiness, this locality is not even surpassed by the Battery of New York, while its
sublime elevation throws the last-mentioned place into an unfavorable contrast.
The space allotted to the Battery is greater, but the variety and richness of the
trees and flowers of the Passeio Publico of Bahia fully compensate for its deficiency
in this respect.   Here it was, under the dark, dense foliage of the mangueiras, the
lime-trees, the bread-fruit, the cashew, and countless other trees of tropical
growth, that thousands of lights were blazing.   Most of these hung in long lines
of transparent globes,—so constructed as to radiate severally the principal hues of
the rainbow,—and waved gracefully in the evening breeze as it swept along, laden
with the fragrance of opening flowers.

"The calmness of a summer evening always throws an enchantment over the
feelings; but there was a peculiar richness in this scene.   Not only was the observer delighted with the varied and skilful exhibitions of artificial light around
him, but, lifting his eyes above them to the vaulted empyrean, he might there gaze
upon the handiwork of the Almighty, so gloriously displayed in the bright constellations of the Southern sky.

"The wealth, fashion, and beauty of the Bahians never boasted a more felicitous
display than was mutually furnished and witnessed by the thousands that thronged
this scene.   What an occasion was here offered to the mind disposed to philosophize
on man!   From hoary age to playful youth, no condition of life or style of
character was unrepresented.   The warrior and the civilian, the man of title, the
millionnaire and the slave, all mingled in the common rejoicings.   Never, especially, had the presence of females in such numbers been observed to grace a scene
of public festivity.   Mothers, daughters, wives, and sisters, who seldom were permitted to leave the domestic circle, except in their visits to the morning mass, hung
upon the arms of their several protectors, and gazed with undissembled wonder at
the seemingly magic enchantments before and around them.   The dark and flowing
tresses, the darker and flashing eyes, of a Brazilian belle, together with her sometimes darkly-shaded cheek, show off with greater charms from not being hidden
under the arches of a fashionable bonnet.   The graceful folds of her mantilla, or
of the rich gossamer veil which is sometimes its substitute, wreathed in some indescribable manner over the broad, high, and fancy-wrought shell that adorns her
head, can scarcely be improved by any imitation of foreign fashions.   Nevertheless,
the *forte* of a Brazilian lady is in her guitar, and the soft modinhas she sings in
accompaniment to its tones.

"On the marble monument erected in memory of Dom John's visit to Bahia
illuminated forms were fitted, and, on this occasion, displayed, in large and brilliant letters, extravagant praise to D. Pedro II.

"In another quarter, upon a high parapet overlooking the sea and bay, had
been constructed a fancy pavilion, in the style of an Athenian temple.   In front
of this, supported by the central columns, had been placed a full-length portrait
of His Majesty.   In the saloons of this palacete were stationed bands of music,
surrounded by ladies and dignitaries of the province.   The portrait of the
Emperor was concealed by a curtain until a given hour of the evening, when the
President made his appearance, and, suddenly drawing it up, gave successive
vivas to His Majesty, the Imperial family, the Brazilian nation, and the people
of Bahia,—all of which were responded to with deafening acclamations from the

multitude around, while the heavens above were resplendent with the discharge of a thousand rockets.

"On Wednesday, the festivities of the great national anniversary terminated with a pyrotechnic display. The Passeio Publico was illuminated more brilliantly than before, and all the gardens surrounding the Campo de San Pedro were lighted up with torches and bonfires. A large platform had been erected in the centre of this square, upon which the Emperor's portrait was again exhibited,—the Archbishop assisting the President to roll up the curtain from before it at the appointed hour. The concourse of people was vastly greater than it had been on any previous evening. The weather was without interruption serene and beautiful, but neither the plan nor execution of the fireworks deserved high commendation. Yet all the bustle and crowd passed away, as on the previous nights, without the slightest disturbance. This fact was certainly a happy comment upon the orderly disposition of the people. I witnessed no *funcção* in Brazil which was, on the whole, more interesting to me than this. Its superiority over the exhibitions of the usual religious festivals was manifest. In fact, the simple circumstance that it was a civic celebration, and destitute of any religious pretensions, went far to commend it to the admiration of any one who had often been shocked by those incongruous medleys of the solemn and ridiculous which are by many thought essential to the 'pomp and splendor' of religious anniversaries."

Away from the pretty Victoria Hill, in a portion of the lower town, the stranger, among other curiosities, may see what is called by its right name,—a *fabrica de imagens*, (image-factory.) It is not my intention to enlarge on worship in this city, for it is the same as throughout the Empire. Saints, crucifixes, and every species of the ghostly paraphernalia of Romanism, are here exhibited in the shops, with a profusion which I nowhere else saw, indicating that the traffic in these articles is more flourishing than in other parts. It is not in name only that Bahia enjoys the ecclesiastical supremacy of Brazil. It is the see of the only archbishop in the Empire. Its churches exceed in number and in sumptuousness those of any other city; and its convents are said to contain more friars and more nuns than those of all the Empire beside.

But I cannot pass over this subject without referring to Saint Antonio de Argoim, who seems to be the favorite patron of the calendar in Brazil. His image is in the Franciscan Convent, and his history is as follows:—

In 1595, a fleet, under the direction of some Lutherans, sailed from France, with the intention of capturing Bahia. On their way they attacked Argoim, a small island on the coast of Africa belonging to the Portuguese, and, after having committed various depredations, carried off, among other sacred things, an image of St. Anthony.

Once more at sea, the fleet was attacked with storms, which sunk several of the vessels. Those that escaped this fate were assaulted with a pestilence, during which, through pure spite toward the Roman Catholic religion, the aforesaid image was thrown overboard, having been first hacked with cutlasses. The vessel that carried it put into a port of Sergipe, and all on board were taken prisoners. These men were sent to Bahia, and the first object they saw on the praia was the very same image they had so maltreated. It had been cast up by the waters to confront them!

A worthy citizen obtained the image and placed it in his private chapel; but when the Franciscans learned what a miracle had happened, they demanded the image, and carried it in solemn procession to their convent. So great was its fame now, that King Philip ordered the establishment of a grand procession in memory of these events. And, strange to tell, popularity did for the image what the bitter hostility of the heretics could not do. Its friends, the friars, became ashamed of its old and ugly appearance, and laid it aside to make room for a more gaudy and fashionable one, which was christened in its name and presumed to be the inheritor of its virtues. Having thus been introduced to the citizens of Bahia, St. Anthony was now enlisted as a soldier in the fortress near the barra bearing his name.

In this capacity he received regular pay until he was promoted to the rank of captain by the Governor, Rodrigo da Costa. The order for his promotion lies before me, and is so curious that I give the concluding portion. After referring to a vow by the *camara municipal*, which had been unfulfilled, the Governor says,—

"Wherefore, and because we now more than ever need the favors of the afore-mentioned saint, both on account of the present wars in Portugal, and of those which may yet happen in Bahia, the said Chamber has besought me, in commemoration of the afore-mentioned vow, to assign to the said glorious St. Anthony the rank and pay of a captain in the fortress, where he has hitherto only received the pay of a common soldier.

"In obedience to this request, and subject to the approval of the King, I therefore assign to the glorious St. Anthony the rank of captain in the said fortress, and order that the solicitor of the *Franciscan Convent* be authorized to draw, in his behalf, the regular amount of a captain's pay.

"RODRIGO DA COSTA.

"BAHIA, July 16, 1705."

Now, the miracle of S. Antonio was truly notable. But the investigations of modern science, and a little more experience, have cleared up the mystery. While conversing with a gentleman, not a Romanist, at Bahia, about S. Antonio's singular voyage to the coast of Brazil, he gravely, to my surprise, stated that it was without doubt a *bona fide* account that the hacked image had floated to the Western world: all could be explained by natural laws. A few days afterward he gave me the following, which will doubtless be a novel confirmation of Lieutenant Maury's theories in regard to ocean winds and currents.

"It is not at all surprising that, in those days of gross credulity and ignorance, the appearance of the image of Santo Antonio on this coast should have been considered as a miracle, performed expressly for the purpose of bringing to condign punishment the 'pirates' for the sacrilegious act they had committed.   Of the appearance of the image on the beach, and its having floated from Africa, no reasonable doubt can be entertained; and, in proof of its entire probability, the following remarkable coincidence may be presented:—

"About fifteen years ago, the late Visconde do Rio Vermelho, a gentleman of the utmost veracity, and owner of an extensive fishery on this coast a few miles to the north of the harbor of Bahia, near Itapican, declared to the writer of the present lines that the figure-head of a vessel, somewhat injured by fire, was brought to his residence from the beach (where it had been stranded) and placed on his grounds. Shortly after, a painter from the city, engaged in painting the house, on seeing the figure immediately recognised it as one he had painted, some months previously, for a vessel which had afterward sailed for the coast of Africa, and of whose safety great fears were entertained, no news having been received from her.   It was subsequently ascertained that the vessel in question had been burned to the water's edge, on the coast of Africa,—the figure-head, singularly enough, having brought the first tidings of the disaster.

"It is likely that the figure-head, being of light cedar, and the pedestal to which it was attached, of hard wood with bolts and fastenings of iron, may have floated in a nearly upright position, thus presenting a broader surface for the action of the northeast trade-winds, and materially accelerating its passage across the Atlantic."

At Rio de Janeiro S. Antonio has long enjoyed the position and received the pay of a colonel in the regular army. How he can appropriate his salary to himself is difficult for us to understand; but it may throw some light on the subject to state that it passes through the hands of his terrestrial delegates,—the Franciscan monks,—and by a proper application you may see the accounts and receipts for his saintship's washing, clothing, &c.

Traditions respecting St. Thomas's visit to Brazil are very common in different parts of the country. Many of them were coined by the Jesuits, and they have passed currently among a credulous people. Observe the logic with which the renowned Simon de Vasconcellos proves that Saint Thomas, certainly, must have been in South America.

"With what show of reason," says the Jesuit, "could the American Indian be damned, if the gospel had never been preached to him? He who sent his apostles into all the world could not mean to leave America—which is nearly half of it—out of the question. The gospel, therefore, must have been preached there in obedience to this command. But by whom was it preached? It could not have been by either of the other apostles, Paul, Peter, John, &c. St. Thomas, therefore, must have been the man!"

No wonder the Jesuits were able to map out his travels from Brazil to Peru, to find traces of his pastoral staff, crosses erected by him, and inscriptions in Greek and Hebrew written by his hand. They even brought his sandals and mantle unconsumed out of the volcano of Arequipa. I suppose it was either in going or returning that he visited England and preached under the Glastonbury Thorn.

The commerce of Bahia suffered to some extent at the suppression of the slave-trade; but it is slowly advancing in legitimate channels. The culture of tobacco and of coffee are both increasing. Railways are projected into the interior, and steamers (not to mention the Government lines) run to the coast-towns in Sergipe and Alagoas on the north, and nearly to Espirito Santo on the south.

DARING NAVIGATION.

Sr. Martin, former President of the province, deserves great credit for his advancement of agriculture, while Senhor Lacerda, co-operating with Messrs. Carson & Gillmer, has done much toward advancing the manufacturing-interest. The finest factory in all Brazil—perhaps South America—was erected according to the plans and under the superintendence of Colonel Carson, an American of daring energy and genius. During my stay in the province of Bahia, one of the pleasantest excursions was my visit to Valença, the seat of the factory.

It was a cheerful party that accompanied Mr. and Mrs. Gillmer; and the day was so bright that our trip was most agreeable over the bay through a fleet of little whale-boats that were in hot pursuit

of their spouting game. There were a number of Brazilian gentlemen on board, who, finding the American Consul making an
excursion, came and placed their houses at the disposition of himself and companions. About noon we passed the light-house on the
Moro de S. Paulo,—a beautiful structure, built under the superintendence of Colonel Carson. We steamed up the river Una to
Valença, where the colonel joined us, and we then re-embarked in
long "dug-outs" in order to ascend the stream to the *fabrica*.

In a few moments we were at the foot of roaring rapids, upon
the borders of which the genius of this enterprising American had
erected a saw-mill, a window-sash factory, and a planing-machine;
in addition to which he had constructed a lock,—the first in Brazil,
—through which our canoes passed. In the sash-factory we saw
the chief workman, Mr. Foster, from Worcester, Massachusetts.
This establishment belonged to Dr. Bernardini, a Brazilian LL.D.,
who left the judge's bench to enjoy the more lucrative position of
a manufacturer. At Dr. B.'s order, a slave brought down, with
capital skill, several saw-logs from above the falls. The expertness
with which he balanced himself and guided in perfect safety his
clumsy craft was truly admirable, and called forth from our party
loud huzzas. The manner in which he managed the log illustrates
the descent of the rapids of the Upper Amazonian affluents.

We resumed our route, passing up the narrow stream. Upon
the banks were numerous negresses and mulatresses engaged in
washing. In looking upon them I thought, for the first time in
my life, of the nuisance of clothing in matters of manual labor.
The women (whose glistening rounded limbs were as smooth as
those of the Greek Slave) were naked to the waist, and the children—some not far from their teens—were *in puris naturalibus*.

We arrived at the factory, or, rather, at the factories; for, clustering around the large *fabrica*, whose white walls stand out in bold
relief from its background of green, are machine-shops, foundries,
&c. &c. The rattle of the looms, the cheerful smile of the merry
girls, and the indescribable din and buzz of a factory, made me
almost imagine myself near Lowell. The operatives, men and
women, are mostly from the orphan-asylum and foundling-hos-
pitals. They are under good discipline, and compare in morals
very favorably with those of the best-conducted factories in our

own land. In the foundry I saw the whole operation of modelling, moulding, and finishing, performed by negroes. The foreman of the foundry is a Brazilian negro, trained by Mr. Carson, and the most intricate machinery is here manufactured.

Extensive buildings were still going up to facilitate the manufacture of cotton cloths, which are of finer quality than those turned out at St. Alexio; and it is gratifying to state that this factory can scarcely meet the demand, and, doubtless, in a few years Messrs. Lacerda & Co. will be amply rewarded for their immense outlay. I here found a millwright (Mr. R. A. Randall) from Scituate, R.I.*

THE VALENÇA FACTORY.

After a sumptuous and truly tropical dinner, the gentleman-portion of our party sallied forth for an excursion, the end of which was to find a suitable place to sketch the immense factory.

---

* It seemed truly out of place, in this distant corner of the world, to read the names of machinists of the United States, whose workmanship was here benefiting a people speaking another tongue. The following are some of the names which I copied from inscriptions on the machinery:—C. Lewis, New York, drilling-lathe; D. Dicks, Hadley Falls, Mass., antifriction press or punch; S. Jones, Boston, improved shears; C. F. Pike, Providence, R.I., iron-planer; J. & S. W. Putnam & Co., Fitchburg, Mass., bolt-cutter. There were other machines, by J. Peck, Coventry Factory, (Anthony's,) R.I., and by Thayer, Houghton & Co., Worcester, Mass.

The *point de vue* was well chosen; but each of us carried away a piece of the foreground, in an innumerable quantity of *garapatos*, which small insects—resembling very diminutive spiders—clung to our garments with a most tenacious hold. Each one of these little fellows produces a boil; and, in some parts of Brazil, cattle in a long dry season—the insect cannot survive a drenching—have sometimes perished by the sores thus created. I hastened to the house, plunged into a bath of hot water, and then was rubbed down with a pint of rum,—more of the article, by three gills, than ever before had been applied to my *physique*, either externally or internally. This effectually stopped the depredations which had begun.

Early the next morning, Mr. Randall and I went to the spot where two of our countrymen were buried. Three Americans came out together, and he alone was left. He feelingly recounted to me the circumstances of their death as we passed up a narrow path to their resting-place. The graves were under the deep shade of two jaca-trees, and over them small obelisks had been erected. It was to me a solemn scene in that early morning hour.

After breakfast, Mr. Gillmer, Mr. Pointdexter, a young Pole, and myself, went up the river to see an upper waterfall. The shrubs, the dead stumps, and the lofty trees on the banks seemed blooming with orchidaceous plants. Rich cabinet-woods also abound in the forest. At Bahia, the Visconde Fiaz and Senhor Viana (brother of the chief collector of customs at Rio) showed me, at their residences, some of the finest specimens of furniture, made from native woods, that I ever saw. We finally reached the fall, which resembles a miniature Niagara. The river Una here pours over a ledge of rocks in such volume that it has been computed there is enough water-power to drive one hundred factories of five thousand spindles each.

On our return from our visit to the *fabrica*, we accepted the hospitality of Senhor Bernardini, who gave us a splendid dinner.

We were accompanied to the city by Colonel Carson, whom I found a most interesting man of intelligence and common sense. His life had been a wandering one. He came out to Brazil to die; but the delicious climate made him a new man, and he had truly "gone ahead,"—building saw-mills, light-houses, factories, and had been abroad, for the Provincial Government, to investigate the sugar plantations of the West Indies and the States on the Mexican Gulf, for the purpose of promoting the growth of sugar in Bahia. He gave me much information concerning the trade that *might* be between the United States and Bahia. In that second port of Brazil we have been annually losing ground. But many articles—for instance, cottons, hardware, leather, soaps, &c. &c.—might be introduced with advantage. The specimens of leather from J. Chadwick, Esq., of Newark,—the same found in the shoes of Mr. Boynton,—and the samples of cutlery and carving sent out by Mr. Garside, also of Newark, attracted, by the excellence of their quality, much attention at Rio; and the same may be said of the rope and rope-yarn manufactured at the Excelsior Works by Mr. H. Webber & Co. All of these articles, and many others, if properly managed, might be exported to Brazil, whose trade would really be worth as much as all the remainder of South America if we only had it in our possession. Formerly, large quantities of common drillings were exported from the United States to Bahia, from the York Mills, Saco, Maine, and were held in great favor by the Brazilians. This article was actually imitated at Manchester, England, and sent out to Bahia with the stamp, "York Mills, Saco, Maine," and sold as such. But, though well sized and fair-looking, it soon proved worthless and fell into disrepute, and the Brazilians to this day believe that the Yankees cheated them. In England, common cottons cannot be made equal to those manufactured in the United States, because the price of the raw article is too high, and the best cotton is consumed for fine goods, and only the "waste" for the coarser; whereas, in the American factories as good a raw article is used for the coarse cloth as for the finer textures.

Brazil annually consumes many million yards of cotton cloths, both plain and printed. She only produces about three million yards: the rest must be supplied from abroad. We honor fair and honorable competition; we admire the perseverance of John Bull in all that is good, and would have our own merchants imitate the latter quality and that only, and endeavor to have at least a fair share in the trade with Brazil, so that we may not annually have a cash-bill of fifteen millions of dollars against us when our productions are needed by the growing Empire of the South. Let our far-seeing commercial men turn their attention in this direction, and, by judicious measures, secure a foothold.

Just after nightfall our little steamer was again at the wharf, and all returned home, delighted with the excursion to Valença.

Before leaving the subject of Bahia, it becomes me to mention—without entering into particulars—that my Bible-labors there, as elsewhere throughout the Empire, were prospered; and I pray that the seed sown, where were Henry Martyn's first missionary efforts on foreign ground, may be prospered by Him who openeth and no man shutteth, and who takes care of His own word.

Note for 1866.—I profit by this space to mention that some of the most important explorations have been made since 1855,—as that of the province of Ceará, under the direction of Srs. Frei Allamão, Capanema, Lagos, and others, constituting a scientific commission. Several important rivers have been explored and mapped. The Purús, that large Amazonian affluent, is perhaps more unknown than the river Nile. Sr. Herculano Ferreira e Penha, when President of Pará, called particular attention to this river in one of his annual messages, (since translated by Dr. Spruce, the Upper-Amazonian traveller, and published by the Royal Geographical Society.) In 1862, Major J. M. da S. Coutinho, in the steamer *Pirajá*, ascended this river, taking soundings, &c., for seven hundred miles. This is remarkable; for the affluents of the Amazon are generally interrupted in their navigation at a comparatively short distance from their embouchures. It is believed that the Purús is the "Madre de Dios" of the old Spaniards, and that by this river Brazilians can go to the borders of Bolivia. An English explorer, Mr. Chandless, June 15, 1865, reached even a higher point on the Purús than Coutinho attained. I regret that want of space prevents my giving an extended account of other labors by Coutinho, Halfeld, (whose survey of the S. Francisco is a *magnum opus*,) and other river-explorers like Dr. Couto de Magalhães, who in 1863 descended the Araguaya from near Goyaz to Pará.

In Bahia is a quartette of scientific men to whom the *savants* of England, France, and America are greatly indebted; viz. Dr. Wucherer, Professor in the Medical College of Bahia, who has made many contributions to the British Museum, department of Natural History; M. Brunet, a Frenchman, who has explored the Amazon, the San Francisco, and the Parnahiba; Rev. Dr. Nicholay, the physical geographer, now rector of the English Chapel at Bahia, who has amassed important geographical *data* in this portion of Brazil; and Sr. Antonio Lacerda, Jr., a native Brazilian, educated in the United States and France, an enthusiastic lover of natural science in general, and well known in Europe and America as a most accomplished entomologist.

CHAPTER XXV.

DEPARTURE FROM BAHIA—THE VAMPIRE-BAT—HIS MANNER OF ATTACK—THE
BITTEN NEGRO—ANNOYANCES MAGNIFIED—ANACONDAS—ONE THAT SWALLOWED
A HORSE—THE MARMOSET—PROVINCE OF ALAGOAZ—THE REPUBLIC OF PAL-
MARES—PERNAMBUCO—THE AMENITIES OF QUARANTINE-LIFE—IMPROVEMENTS
AT THE RECIFE—PECULIARITIES OF PERNAMBUCAN HOUSES—BEAUTIFUL PANO-
RAMA—VARIOUS DISTRICTS OF THE CITY—A BIBLE-CHRISTIAN—EXTRAORDINARY
FANATICISM OF THE SEBASTIANISTS—COMMERCE OF PERNAMBUCO—THE POPULA-
TION OF THE INTERIOR—THE SERTANEJO AND MARKET-SCENE—THE SUGAR AND
COTTON MART—THE JANGADA—PARAHIBA DO NORTE—NATAL—CEARÁ—THE
PAVIOLA—TEMPERATURE AND PERIODICAL RAINS—THE CITY OF MARANHAM—
JUDGE PETIT'S DESCRIPTION—THE MONTARIA—DEPARTURE.

O the North! Leav-
ing the pleasant city of Bahia, we again
turn our faces toward the Amazon. Our
steamer glides rapidly over a summer sea, and, though we visit
province after province, we cannot dwell long upon their scenery
and condition, for in both they are very similar to some of the
lesser divisions of the Empire which we have already considered.
The monotony of the voyage is broken up by tinkling guitars,
merry singing, and eloquent speaking. We have embryo statesmen on board; military officers with fierce moustaches and high-

sounding titles; medical students returning to Sergipe, Alagoaz, Pernambuco, and Parahiba; witty, sallow, dirty sertanejos; black-eyed senhoras; and two or three tonsured, gambling padres. All form a fit audience; and the vociferous *apoiados, apoiadidissimos,* encourage the maiden efforts of the orators, and beguile the time as we steam along the low coqueiro-lined coast.

A hazy bank of fog hanging in the distant horizon indicates the mouth of the great Rio San Francisco, and the boundary-line between the provinces of Sergipe and Alagoaz. Sergipe is thinly populated: but in the eastern portion a considerable quantity of sugar and tobacco is cultivated; while the western districts are devoted chiefly to the rearing of cattle.

In another chapter I have spoken of the annoyances to which herds are sometimes subject from the little chigoes. The younger portions of the herds have in some places a more formidable enemy

THE VAMPIRE-BAT.

in the huge vampire-bat. The owner of large possessions in the northwestern part of Goyaz said he could not rear cattle with any success or profit, from the havoc committed among his calves by the winged demons the vampires. I have often had my own horses and mules bled and sucked by these sanguinary *phyllostomina.* They abound from Paraguay to the Isthmus of Darien; and the reports of early travellers and the figurative language of poets, so long discredited, are found to be much nearer the truth than the world has believed. Morning after morning have I seen beasts of burden, once strong, go staggering, from loss of blood drawn during the night by these hideous monsters. In almost every instance they had taken the life-current from between the shoulders, and, when they had finished their murderous work, the stream had for some time continued to flow. The extremities, however, are the usual points of attack; and the ears of a horse, the toes of a man, and the comb of a cock, are choice morceaux for the display of the vampire's phlebotomizing propensities.

The exact manner by which this bat manages to make an incision has long been a matter of conjecture and dispute. The tongue, which is capable of considerable extension, is furnished at its extremity with a number of papillæ, which appear to be so arranged as to form an organ of suction, and their lips have also tubercles symmetrically arranged. These are the organs by which it is certain the bat draws the life-blood from man and beast, and some have contended that the rough tongue is the instrument employed for abrading the skin, so as to enable it the more readily to draw its sustenance from the living animal. Others have supposed that the vampire used one of its long, sharp, canine teeth to make the incision, which is as small as that made by a fine needle. Mr. Wallace says that he was twice bitten, once on the toe, and a second time on the tip of the nose. "In neither case," writes that explorer, "did I feel any thing, but awoke after the operation was completed. The wound is a small round hole, the bleeding of which it is very difficult to stop. It can hardly be a bite, as that would awake the sleeper: it seems most probable that it is either a succession of gentle scratches with the sharp edge of the teeth, gradually wearing away the skin, or a triturating with the point of the tongue till the same effect is produced. My brother was frequently bitten by them; and his opi-

HEAD OF THE VAMPIRE-BAT, SIZE OF LIFE.

nion was that the bat applied one of its long canine teeth to the part, and then flew round and round on that as a centre, till the tooth, acting as an awl, bored a small hole,—the wings of the bat serving at the same time to fan the patient into a deeper slumber. He several times awoke while the bat was at work, and, though of course the creature immediately flew away, it was his impression that the operation was conducted in the manner above described." There is much in the dental arrangement of these *phyllostoma* to make this seem plausible. The molar teeth of the true vampire or spectre bat, are of the most carnivorous character,—the first being short and almost plain, and the others sharp and cutting and terminating in three and four points. Notwithstanding this, that most accurate naturalist and observer—Dr. Gardner—is of the opinion that it wounds its victim in a manner entirely different from the foregoing description. He says that,

"Having carefully examined, in many cases, the wounds thus made in horses, mules, pigs, and other animals,—observations that have been confirmed by information received from the inhabitants of the northern part of Brazil,—I am led to believe that the puncture which the vampire makes in the skin of animals is effected by the sharp, hooked nail of its thumb, and that from the wound thus made it abstracts the blood by the suctorial powers of its lips and tongue."

Some of these bats measure two feet between the tips of their wings. There are some persons whom a vampire will not touch, while others are constantly victimized. The alligator-riding Waterton states that for eleven months he slept alone in the loft of a wood-cutter's abandoned house in the forest, and, though the vampires came in and out every night, and hovered over his hammock, yet he could never have the pleasure of being bitten, —which amusement he doubtless would have foregone if he had had the experience of Mr. Wallace, who says that a wound on the tip of the toe is very painful, rendering a shoe unbearable for several days, and "forces one to the conclusion that, after the first time for the curiosity of the thing, to be bitten by a bat is very disagreeable."

There are instances in Northern Brazil where individuals for whom the bat entertained a great predilection had to be removed to a different portion of the country, where the bloodthirsty animals did not abound. One of Mr. Wallace's party—an old negro— was constantly annoyed with them. He was bitten almost every night; and, though there were frequently half a dozen persons in the room, he would be the party favored by their attentions. "Once," Mr. Wallace writes, "he came to us with a doleful countenance, telling us he thought the bats meant to eat him up quite, for, having covered up his hands and feet in a blanket, they had descended beneath his hammock of open network, and, attacking the most prominent part of his person, had bitten him through a hole in his trousers!"

While enumerating the various insects, reptiles, and vicious animals of Brazil, the reader who has not visited that land would be led to the belief that it is impossible to stir a foot without being affectionately entwined by a serpent, sprung upon by a jaguar, or bitten by a rattlesnake. In your fancy every bush swarms with chigoes ready to engraft their stock upon your legs, every cranny contains a scorpion waiting to ensconce himself in your pantaloons, and every pool is filled with electric eels prepared to give you a shocking reception. I can only say that, when travelling on the sea-coast

THE ELECTRIC EEL.

and in the interior, I never was more annoyed by insects than I had been in the southwestern portion of the United States; and that, with a moderate degree of care, you may journey fifty days without experiencing any thing more deadly than the bite of a mosquito. The sand-flies call forth more complaints from naturalists and travellers than do either serpents, scorpions, or centipedes; and yet all of these are more or less found throughout the interior. But difficulties only seem insurmountable in the distance: they disappear when looked boldly in the face, and do not affect the tourist and the naturalist one-tenth as much in reality as in anticipation.

THE SCORPION.

In this connection a few words may be devoted to the anaconda, the largest of the ophidian family. I confess myself to have been incredulous in regard to the powers and capacities of this huge reptile until I went to Brazil, and I have no doubt that I shall, in the opinion of some, add a few pages to the innumerable "snake-stories."

The enormous anaconda, (*Eunectes murinus*,) or *sucurujú* of the natives, (a portrait of which forms the initial letter of this chapter,) inhabits Tropical America, and particularly haunts the dense forests near the margin of rivers. The boa-constrictor, the *jiboa* of the Indians, is smaller and more terrestrial. The first of these creatures which I saw was a young one belonging to a gentleman in the province of S. Paulo. I afterward saw one in the province of Rio de Janeiro that measured twenty-five feet. Mr. Nesbitt, the engineer who took the Peruvian Government steamers to the upper affluents of the Amazon, informed me that he shot, on the banks of the Huallaga, an anaconda which measured twenty-six feet seven inches. An Italian physician at Campinas (S. Paulo) gave me an account of the manner in which the sucurujú, or anaconda, took his prey.

The giant ophidian lies in wait by the river-side, where quadrupeds of all kinds are likely to frequent to quench their thirst. He patiently waits until some animal draws within reach, when, with a rapidity almost incredible, the monster fastens himself to the neck of his victim, coils round it, and crushes it to death. After the unfortunate animal has been reduced to a shapeless mass by the pressure of the snake, its destroyer prepares to swallow it by sliming it over with a viscid secretion. When the anaconda has gulped down a heifer (by commencing with the tail and hind-feet brought together) he lies torpid for a month, until his enormous meal is digested, and then sallies forth for another. The doctor added that the sucurujú does not attempt the deglutition and digestion of the horns, but that he lets them protrude from his mouth until they fall off by decay. It had been said by some casual observers that the anaconda dies after swallowing a large animal, that the buzzards seen near him eat him up; but the doctor added that close observation showed that this statement was entirely erroneous. However, the vultures were always the close attendants of the sucurujú, to aid him in the delivering of his fæces. As to the amount of credence due to the statements of Dr. B., relative to the horns of the swallowed animal and the peculiar midwifery of the buzzards, I leave the reader to form his own opinion; but the facts are incontrovertible in regard to the capacity of the anaconda to swallow animals whose diameter is many times greater than its own. Of all the travellers and explorers whose writings I have read, Wallace and Gardner are the most moderate in their accounts as eye-witnesses, and are most particular to record nothing of which they were not fully persuaded after patient and careful investigation. Mr. Wallace says "it is an undisputed fact that they devour cattle and horses." In the province of Goyaz, Dr. Gardner came to the fazenda of Sapê, situated at the foot of the Serra de Santa Brida, near the entrance to a small valley. This plantation belonged to Lieutenant Lagoeira. Dr. G. remarks that in this valley and throughout this province the anaconda attains an enormous size, sometimes reaching forty feet in length: the largest which he saw measured thirty-seven feet, but was not alive. It had been taken under the following circumstances:—

"Some weeks before our arrival at Sapê," writes Dr. G, "the favorite riding-horse of Senhor Lagoeira, which had been put out to pasture not far from the house, could not be found, although strict search was made for it all over the fazenda. Shortly after this one of his *vaqueiros*, (herdsmen,) in going through the wood by the side of a small stream, saw an enormous sucurujú suspended in the fork of a tree which hung over the water. It was dead, but had evidently been floated down alive by a recent flood, and, being in an inert state, it had not been able to extricate itself from the fork before the waters fell. It was dragged out to the open country by two horses, and was found to measure thirty-seven feet in length. On opening it, the bones of a horse in a somewhat broken condition, and the flesh in a half-digested state, were found within it: the bones of the head were uninjured. From these circumstances we concluded that the boa had swallowed the horse entire. In all kinds of snakes the capacity for swallowing is prodigious. I have often seen one not thicker than my thumb swallow a frog as large as my fist; and I once killed a rattlesnake about four feet long, and of no great thickness, which had swallowed not less than three large frogs. I have also seen a very slender snake that frequents the roofs of houses swallow an entire bat three times its own thickness. If such be the case with these smaller kinds, it is not to be wondered at that one thirty-seven feet long should be able to swallow a horse, particularly when it is known that previously to doing so it breaks the bones of the animal by coiling itself round it, and afterward lubricates it with a slimy matter, which it has the power of secreting in its mouth."

Near Sapê many of the marmoset monkeys abound, and a very small species, sometimes called the ouistiti, (*Jacchus auritus,*) is exceedingly nimble, and not wanting in beauty.

The Brazilian girls are fond of pets; and, among others, a great favorite is this ouistiti, which is rarely ever seen out of Brazil, even in the best zoological collections. It has a skin like chinchilla fur, and its face presents none of the repulsive features of other monkeys. These little animals become very tame and sleep upon the lap or shoulders of their mistress. Their actions are most graceful and rapid. Two that a friend of mine embarked for the

United States could mount the ship's ropes ten times as rapidly as the nimblest sailor. If birds came on board, they hunted them from rope to rope, and passed along under the spar upon which their victim sat, and then pounced upon it with certain aim.   In their native forests they are very fond of insects, which they catch with great expertness.   They are excessively timid when roughly handled: one of the two referred to was teased by the sailors, and in consequence died in convulsions.   It was pitiful to see the other

THE MARMOSET.

look at himself in a glass, making a plaintive noise and licking the reflection of his own face. They were so small that a square cigar-box, the length of one "Havana," contained them both. With great care the surviving ouistiti was kept alive through a Northern winter. His food was bread, sponge-biscuit, apples, and now and then a chicken's neck or a mouse. It was curious to see him devour the latter. He began at the snout and carefully pushed back the skin, eating the bones and every thing until he reached the tail, which was all that he left inside the skin. His last effort was to aid in erecting a parsonage, by being exhibited at a fair for that purpose. But his benevolence was too much for him: the little fellow pined and died, after having endured a succession of fits; and his end was doubtless hastened by the breath of his numerous visitors, and by an escape of gas in the chamber where he was kept; for the delicate monkeys in the London Zoological Gardens were all killed by being in a room with a stove. An open grate was substituted, and their successors escaped.

Next to Sergipe in our northward route is the small province of Alagoas. It derives its name from the lake—or, strictly speaking, the inlet—on which stands its old capital, the city of Alagoas. The principal seaport of the province is Maceió. Into this port we entered, after a passage of about thirty-six hours from Bahia. As we bore up to land in the morning after our second night at sea, we found the coast very flat, sometimes exhibiting a sandy beach, and anon banks of eighty or ninety feet elevation, denominated, from their prevailing color, the Red Cliffs. We approached so near these cliffs as to perceive distinctly their stratification, which resembled successive layers of brick.

The most favored island of the Southern seas can hardly present a more lovely aspect than does the harbor of Maceió. The port is formed by a reef of rocks visible at ebb-tide, which runs north and south for a sufficient distance in a right line, and seems to form an angle with an extreme point of land on the north. From the same point the beach sweeps inward in the form of a semicircle. The sand on this beach exhibits a snowy whiteness, as if bleached by the foam of the ocean-waves that unceasingly dash upon it.

A little back from the water is a single line of white houses, embowered here and there by groves of majestic coqueiros, whose noble fruit, clustered amid their branching leaves, might be thought to resemble jewels set among the plumes of a coronet. Upon a hill-side, some distance in the rear, stands the city, containing a population of about six thousand.

My visit to Maceió was most agreeable, connected as it was with the sympathizing Brazilians and others who were glad to receive the Word, and who gave me many pleasant assurances that the sojourn of my co-laborer and predecessor had not been forgotten. One old man, with tears in his eyes, referred to Dr. Kidder's visit, and aided me in the dissemination of the Truth.

Maceió is the depôt of large quantities of cotton and sugar which are brought down from the interior. Good brown sugar can be purchased at Maceió for two dollars and fifty cents per hundredweight, and the planters admit that they can raise sugar at a profit at a market-price of less than two dollars per hundredweight.

This province, fifteen years ago, was in a constant state of turmoil; but for the last ten years it has settled down into quietness, and is advancing with the general improvement of the Empire.

After leaving Maceió, we pass along a coast interesting in the history of the past. Before us we see Cape St. Augustine, which was the first portion of the New World discovered south of the equator. Our track is that over which, in early times, sailed Cavendish and Lancaster, the great English freebooters, who devastated the Brazilian coast-towns in 1591 and '93. Here, too, passed the ships of Lord Cochrane and Admirals Taylor and Jewett, two Englishmen and an American in the service of Brazil, who by their bravery and skill defeated the Portuguese fleets and did much to secure the Northern cities to the new *régime*.

In the interior, about sixty miles from Porto Calvo, there was a curious community, hidden away amid groves of palm-trees, having a regular military and priestly government, and known as the *Republic of Palmares*. It seems almost like romance to read of a settlement composed of fugitive slaves, who were perfectly organized, and from time to time went forth on predatory excursions, carrying off treasure and cattle, and taking captive the wives and daughters of the Portuguese and then exacting a heavy ransom.

They had villages and towns; and, in addition to their marauding sallies, they carried on a regular trade with some of the colonies. They flourished for sixty years; and such, at length, became their audacity that regular war was declared against them, and for months the Portuguese sustained the severest contest that they had ever been obliged to undertake west of the coast. The little State was heroically defended; but when, after it had gallantly held out against great odds, cannon were brought to the aid of the besiegers, the Republic of Palmares fell. When all hope was gone, the leader and the most resolute of his followers retired to the summit of a high rock within the enclosure, and, preferring death to slavery, threw themselves from the precipice,—men worthy of a better fate for their courage and their cause.

In its consequences to the vanquished, this victory resembled those of the inhuman wars of antiquity. The survivors of all ages and of

PERNAMBUCO.

either sex were brought away as slaves. A fifth of the men were selected for the Crown: the rest were divided among the captors as their booty, and all who were thought likely to fly, were transported to distant parts of Brazil, or to Portugal. The women and children remained in Pernambuco, being thus separated forever from their husbands and their fathers.

Twelve hours after we had left Maceió, the towers and domes of the Recife, or Pernambuco, appeared, like those of Venice, to be gradually rising from the sparkling water. Far to our right, on a bold and verdant hill, we could descry the suburb called Olinda, (translated *the beautiful*,) seeming like a rich mosaic of white towers, vermilion roofs, bright green palm-trees, and bananeiros. It is, however, in this case distance that lends enchantment to the view; for Olinda, whose inhabitants once looked down in contempt

THE JANGADA, AND THE ENTRANCE TO PERNAMBUCO.

upon their commercial neighbors of the Recife, is now in decay. The law-school, with its three hundred students, has been transferred to Pernambuco, and this once valiant capital of the equatorial colonies of Portugal is now going rapidly to decay.

Olinda deserves to be regarded as S. Vincente, and the two places may be considered as exhibiting the classic remains of the

colonial system of Portugal.   Olinda, however, reminds us nearly
as much of the Dutch as it does of the Portuguese, being known
in the annals of Holland as the ancient ·Mauricius, upon which the
ambitious Count of Nassau staked his fortune and his fame.

As we drew near to Pernambuco, the warehouses and the shipping presented the features of a large commercial town, and the
resemblance between it and the silent Queen of the Adriatic no
longer forced itself upon the beholder.   The waves outside of the
curious reef, (*recife,*) or natural breakwater, were dotted with
lateen-sailed jangadas or catamarans, and the proprietors of these
dancing rigged rafts seemed literally at sea "on a log."

Our steamer came proudly up to the fierce little fort and white
pharo that (so low is the reef) appeared to rise from the water.
We anchored on the seaward side of the fortress and awaited with
anxious expectation the visit of the health-boat.   Every passenger,
from the wild *matuto* (forester) and sertanejo to the dignified
medico and the vain officer of the Imperial army, was rejoicing at
his approaching liberation. The health-boat came bobbing around
the fort, and we had the satisfaction of hearing that we should be
quarantined for ten days on an island four miles west of the city.
There was really no necessity for this, for our health-bill from
Maceió was immaculate.   It is needless to narrate our adventures
in getting to the quarantine; our navigation on a jangada; how
fifty persons were quartered in four rooms' (comfortable for eight
individuals) which would have been unbearable except for the
capital ventilation through the arched tiles; how merry we were,
and contented, under the circumstances; how we were refreshed
by cocoanut-milk and bracing breezes; how I had opportunities
for doing good; how we were all liberated and a hundred more
put into our places; and how kind was my reception (when I was
permitted inside of Pernambuco) by Mr. Samuel Johnson and Mr.
Hitch, (the heads of two houses, English and American.)   All this
must be unwritten history.   As has been said of a traveller's delay
in Italy, it may be said of this detention at Pernambuco, in logical
language there was no *causa causans;* but the *causa sine quá non*
was that we were in Brazil, where the "brief authority" of officials
is sometimes notoriously overbearing.

Pernambuco is the third city of Brazil, and is the greatest sugar-mart in the Empire.   Its population is variously estimated at eighty thousand and one hundred thousand.   In all respects Pernambuco is a thriving and a progressive city.   Those who remember its former unpaved streets and its other inconveniences for comfort and conveyance would now be surprised at the various changes and improvements.   Water-works have been constructed, good bridges erected, and extensive quays have been formed on the margins of the rivers that would serve, according to Mr. Hadfield, as models for the conservators of "Father Thames."   Printingpresses send forth dailies and weeklies, besides from time to time respectable-sized books and Government documents.   Education is looking up, whether we consider the common schools, the collegios, or the flourishing institution for legal instruction, which rivals that of San Paulo.

The city is divided into three parishes or districts, called, severally, S. Pedro de Gonsalves or Recife, S. Antonio, and Boa Vista, which are connected by bridges and good roads.

Many of the houses of Pernambuco are built in a style unknown in other parts of Brazil.   A description of one where my predecessor was entertained by a friend may serve as a specimen of the style referred to :—

"It was six stories high.   The first or ground floor was denominated the armazem, and was occupied by male servants at night: the second furnished apartments for the counting-room, &c.; the third and fourth for parlors and lodging-rooms; the fifth for dining-rooms; and the sixth for a kitchen.   Readers of domestic habits will perceive that one special advantage of having a kitchen located in the attic arises from the upward tendency of the smoke and effluvia universally produced by culinary operations.   A disadvantage, however, inseparable from the arrangement, is the necessity of conveying various heavy articles up so many flights of stairs. Water might be mentioned, for example, which, in the absence of all mechanical contrivances for such an object, was carried up on the heads of negroes.   Any one will perceive that the liability of mistake, in endeavoring to preserve the equilibrium of each vessel of water thus transported, exposed the lower portion of the house to the danger of a flood.   Surmounting the sixth story, and constituting, in one sense, the seventh, was a splendid observatory, glazed above and on all sides.

"The prospect from this observatory was extended and interesting in the extreme.   It was just such a place as the stranger should always seek in order to receive correct impressions of the locality and environs of the city.   His gaze from such an elevation will not fail to rest with interest upon the broad bay of Pernambuco, stretching, with a moderate but regular incurvation of the coast, between the promontory of Olinda and Cape St. Augustine, thirty miles below.   This bay is generally adorned with a great number of jangadas, which, with their broad lateen sails, make no mean appearance.   Besides the commerce of the port itself, vessels often appear in the offing, bound on distant voyages, both north and south. No port is more easy of access. A vessel bound to either the Indian or the Pacific Ocean, or on her passage homeward to either the United States or Europe, may, with but a slight deviation from her best course, put into Pernambuco. She may come to an anchor in the Lameirão, or outer harbor, and hold communication with the shore, either to obtain advices or refreshments, and resume her voyage at pleasure, without becoming subject to port-charges. This is very convenient for whaling-ships and South Sea traders. In order to discharge or receive cargo, vessels are required to come within the reef and to conform to usual port-regulations.

"Men-of-war seldom remain long here. None of large draught can pass the bar, and those that can are required—probably in view of the danger of accidents when so close to the city—to deposit their powder at the fort. Few naval commanders are willing to yield to such a requirement; while, at the same time, their berth in the Lameirão cannot be relied on for either quietness or safety. The powerful winds and heavy roll of the sea are frequently sufficient to part the strongest cables. These are sufficient reasons why Pernambuco is not a favorite naval station either for Brazil or for foreign nations. The commercial shipping is under full view from the observatory, yet it is too near at hand and too densely crowded together to make an imposing appearance.

"Olinda, seen from this distance, must attract the attention and the admiration of every one. Of this city set upon a hill, one is at a loss whether to admire most the whitened houses and massive temples, or the luxuriant foliage interspersed among them, and in which those edifices on the hill-side seem to be partially buried. From this point a line of highlands sweeps inward with a tolerably regular arc, terminating at Cape St. Augustine, and forming a semilunar reconcave, analogous to that of Bahia. The entire summit of these highlands is crowned with green forests and foliage. Indeed, from the outermost range of vision to the very precincts of the city, throughout the extended plain, circumscribed by five-sixths of the imagined arc, scarcely an opening appears to the eye, although, in fact, the country overlooked is populous and cultivated. Numbers of buildings, also, within the suburbs of the city, are overtowered and wholly or partially hidden by lofty palms, mangueiras, cajueiros, and other trees. The interval between Recife and Olinda is in striking contrast to this appearance. It is a perfectly barren bank of sand, a narrow beach, upon one side of which the ocean breaks, while on the other side, only a few rods distant and nearly parallel, runs a branch of the Beberibe River.

"At a distance varying from one-fourth to half a mile from the shore runs the bank of rocks already mentioned as extending along the greater portion of the northern coast of Brazil. Its top is scarcely visible at high-tide, being covered with the surf, which dashes over it in sheets of foam. At low-water it is left dry, and stands like an artificial wall, with a surface sufficiently even to form a beautiful promenade in the very midst of the sea. This natural parapet is approached by the aid of boats. It is found to be from two to five rods in thickness. Its edges are a little worn and fractured, but both its sides are perpendicular to a great depth. The rock, in its external appearance, is of a dark-brown color, and, when broken, it is found to be composed of a very hard species of sandstone of a yellow complexion, in which numerous bivalves are embedded in a state of complete preservation. Various species of small sea-shells may be collected in the water-worn cavities of the surface. At several points deep winding fissures extend through a portion of the reef; but in general its appearance is quite regular,—much more so, doubtless, than any artificial wall could be after hundreds of years' exposure to the wearing of the ocean-waves. The abrupt opening in this reef, by which an entrance is offered to vessels, is scarcely less remarkable than the protection which is secured to them when once behind this rocky bulwark.

"Opposite the northern extremity of the city, as though a breach had been artificially cut, the rock opens, leaving a passage of sufficient depth and width to admit ships of sixteen feet draught at high-water. Great skill is requisite, however, to conduct them safely in; for no sooner have they passed the reef than it becomes necessary to tack ship and keep close under the lee of the rock, in order to avoid the danger of running aground.

"Close to this opening and on the extremity of the reef stands the fort, built at an early day by the Dutch. Its foundations were admirably laid, being composed of long blocks of stone, imported from Europe, hewed square. They were placed lengthwise to the sea, and then bound together by heavy bands of iron. A wall of the same nature extends from the base of the fortification to the body of the reef. This wall appears to have become perfectly solidified, and, in fact, augmented by a slight crust of accumulating petrifaction. This circumstance corroborates the idea that the rock, on the whole, may be increasing, like the coral reefs of the South Sea Islands.

"The district of S. Pedro—frequently called that of the Recife—is not large. Its buildings are most of them ancient in their appearance: they exhibit the old Dutch style of architecture, and many of them retain their latticed balconies or *gelouzias*. These gelouzias were common at Rio de Janeiro at the period of Dom John's arrival. But that monarch, dreading the use that might be made of them as places of concealment for assassins, ordered them to be pulled down; and they are now rarely seen in the metropolis.

"The principal street of the Recife is Rua da Cruz. At its northern extremity, toward the Arsenal da Marinha, it is wide and imposing in its aspect. Toward the other end, although flanked by high houses, it becomes very narrow, like most of the other streets by which it is intersected. A single bridge connects this portion of the city with S. Antonio, the middle district.

"S. Antonio is the finest part of Pernambuco when considered as a city. It contains the palace and military arsenal, in front of which a wall has recently been extended along the river's bank. Just above the water's edge has been placed a row of green-painted seats for the accommodation of the public. These are inviting, mornings and evenings, although, in the absence of shade-trees, the rays of the sun, pouring upon the turfless sand, render the heat intolerable throughout the day.

"The principal streets of this section of the city, together with an open square used as a market-place, are spacious and elegant. The bridge crossing the other river is longer and more expensive than the one just described, although the depth of the stream beneath is not so great. On the southern or southwestern bank of this river stands the British Chapel, in a very suitable and convenient location. That edifice is built in modern style, and generally well attended by the English residents, on Sabbath-days, both morning and evening. Boa Vista is very extensive, and is chiefly occupied by residences and country-seats. A few large buildings stand near the river, and, like most of those in the other sections of the town, are devoted in part to commercial purposes. Beyond these, the houses are generally low, but large upon the ground, and surrounded by gardens, here denominated *sitios*. The streets here were formerly unpaved, and unhappily suffered to remain in a most wretched condition. Sand, dry and wonderfully comminuted, abounds on all sides, unless variegated by filthy pools of standing-water.

"The hedges in the environs of Pernambuco are similar to those of Rio, although generally more rank in growth. Many of the houses exhibit an expensive and at the same time tasteful style of construction. I was pointed to one in the veranda of which was arranged a collection of statues. The owner being a wealthy and notorious slave-dealer, some wag, a few years since, thinking either to oblige or to vex him, crept in by night and supplied him with a cargo of new negroes, by painting all the marble faces black."

Pernambuco has ever manifested more activity than any other of the Northern provinces. It was the first to declare against the Portuguese Government, and several times there have been commotions that threatened for a time the dismemberment of the State; but at the present time there is no province more faithful. An outbreak occurred in 1848, in consequence of a band of miscreants from the interior joining with a few disaffected in the city; but their leaders were summarily dealt with, and since that time the province has remained perfectly tranquil.

The state of religion at Pernambuco is not obviously different from that in other parts of the Empire. The monasteries are in low repute, having at present but few inmates. The hospicio of the Barbadinhos, or Italian Capuchins, has been converted into a foundling-hospital. None of the churches are remarkable for their beauty, or splendor of construction. That of Nossa Senhora da Conceição dos Militares is distinguished for a singular painting upon its walls, designed to represent the battle of the Gararápes, and to commemorate the victory which was then obtained over the heretical Hollanders.

I followed up the Bible-labors of my predecessor, and found some unexpected openings for sowing the good seed. There never was a more favorable opportunity than the present for the introduction of truth and of a pure worship into this portion of Brazil. What is most needed in view of this object is a number of fearless and faithful Brazilian preachers.

Through the English chaplain, Dr. Kidder was made acquainted with a priest who had already become convinced of the necessity of some new measures for enlightening the people, and who had recently taken an active part in circulating Bibles and tracts. He thus records his interview with this Bible-Christian:—

"I met with this padre a few days after my arrival in the city. He came into the house of a friend with whom I was dining, and, happening to lay his hand upor some of the new tracts which I brought along, he broke forth in expressions of delight, saying that he had use for a quantity of these publications. In addition to their subject-matter, he was particularly pleased with their severally bearing the imprint of Rio de Janeiro, a circumstance that indicated the radiation of light from that important point. This individual was a man fifty years old, as much like the ex-Regent Feijo in his appearance as any other Brazilian I ever saw. Part of his education he had received in Portugal, part in Brazil. He had once been chaplain to the prison-island of Fernando de Noronha. Owing to his recent change of views on several important topics, he had suffered considerable persecution from his bishop and some others of the clergy, but he seemed in no way disheartened by this.

"His opinion was, that the silent distribution of tracts and Scriptures among those persons and families disposed to read and prize them was the best method of doing good in the country at present. And most faithfully did he pursue that method, calling on me every few days for a fresh supply of evangelical publications.

"I one day returned his visit, and found him surrounded with quite a library, among which his Bible attracted my attention, as having been for a year or two past his one book. Almost every page in it was marked as containing something of very especial interest. I could but wish that all with whom the Bible is not a rare book prized it as highly as did this padre, who, after having spent the greater portion of his life as a minister of religion according to the best of his previous knowledge, now in his declining years had found the word of God to be 'a light to his feet and a lamp to his path.'"

In 1838, there occurred in this province one of the most extraordinary scenes of fanaticism which is a melancholy proof that the boast of the Romish Church is in vain that such extravagances are confined to Protestant countries. The following narrative, condensed from the official documents before me, may challenge a parallel in either history or mythology. In order that the reader may fully understand it, I will remind him that there prevails in Portugal, and to some extent in Brazil, a sect called Sebastianists. The distinguishing tenet of this sect is the belief that Dom Sebastian, the King of Portugal who, in 1577, undertook an expedition against the Moors in Africa, and who, having been defeated, never returned, is still alive, and is destined yet to make his reappearance on earth, when all that the most enthusiastic Millerarian ever anticipated will be realized. Numberless dreams and prophecies, together with the interpretation of marvellous portents confirming this idea, have been circulated with so much of clerical sanction, that many have believed the senseless whim. Nor have there been lacking persons, at various periods, who have undertaken to fulfil the prophecies, and to prove themselves the veritable Dom Sebastian.

The prime point of faith is, that he will yet come, and that too, as each believer has it, in his own lifetime. The Portuguese look for his appearance at Lisbon, but the Brazilians generally think it most likely that he will first revisit his own city, St. Sebastian.

It appears that a reckless villain, named João Antonio, fixed upon a remote part of the province of Pernambuco, near Piancó, in the Comarca de Flores, for the appearance of the said Dom Sebastian. The place designated was a dense forest, near which were known to be two acroceraunian caverns. This spot the impostor said was an enchanted kingdom, which was about to be disenchanted, whereupon Dom Sebastian would immediately appear at the head of a great army, with glory, and with power to confer wealth and happiness upon all who should anticipate his coming by associating themselves with said João Antonio.

As might be expected, he found followers, who, after a while, learned that the imaginary kingdom was to be disenchanted by having its soil sprinkled with the blood of one hundred innocent children! In default of a sufficient number of children, men and women were to be immolated, but in a few days they would all rise again and become possessed of the riches of the world. The prophet appears to have lacked the courage necessary to carry out his bloody scheme; but he delegated power to an accomplice, named João Ferreira, who assumed the title of "His Holiness," put a wreath of rushes upon his head, and required the proselytes to kiss his toe, on pain of instant death. The official letter to Sr. Francisco Rego Barras, at that time President of Pernambuco, states that "he also married every man to two or three women with superstitious rites in accordance with his otherwise immoral conduct." After other deeds, too horrible to describe, he commenced the slaughter of human beings. Each parent was required to bring forward one or two of his children to be offered. In vain did the prattling babes shriek and beg that they might not be murdered. The unnatural parent would reply, "No, my child; there is no remedy," and forcibly offer them. In the course of two days he had thus, in cold blood, slain twenty-one adults and twenty children, when a brother of the prophet, becoming jealous of "His Holiness," thrust him through and assumed his power. At this juncture some one ran away, and apprized the civil authorities of the dreadful tragedy.

Troops were called out, who hastened to the spot; but the infatuated Sebastianists had been taught not to fear any thing, but that should an attack be made upon them it would be the signal for the restoration of the kingdom, the resurrection of their dead, and the destruction of their enemies. Wherefore, on seeing the troops approach they rushed upon them, uttering cries of defiance, attacking those who had come to their rescue, and actually killing five, and wounding others, before they could be restrained. Nor did they submit until twenty-nine of their number, including three women, had actually been killed. Women, seeing their husbands dying at their feet, would not attempt to escape, but shouted, "The time is come! Viva! viva! the time is come!" Of those that survived a few escaped into the woods, the rest were taken prisoners. It was found that the victims of this horrid delusion had not even buried the bodies of their murdered offspring and kinsmen, so confident were they of their immediate restoration.

Pernambuco lies on the great eastern shoulder of the South American continent, where it pushes farthest into the ocean. Its present great commercial importance is largely owing to this fortuitous position. The city does not depend for its large exports on the fruitfulness or plenty of the region immediately surrounding it.

This region is the *sertão*, ("the wilderness, or desert,")—a term applied to much of the great promontory on which the province lies. It is a continued plain, of but little elevation above the sea, of a surface undulating to a small degree, occupied by a crisp, thin herbage on a baked ferrugineous clay, or patched over with dwarfed forests, is irregularly supplied with rain, and is very sparsely populated.

Pernambuco sends out annually four millions of dollars of exports past the angry little fort at the end of the *Recife*. A half-million reaches the United States. But its abundant beef and hides are gathered from the fat but untamed herds that riot among the sedgy meadows of the far-off San Francisco; while a portion of the cotton and sugar are harvested three hundred miles away, around the *Villa das Flores* and among the foot-hills of Santa Barbaretta,—the first mountain-chain that arrests the trade-wind as it sweeps westward, laden with rain, which pours down the little valleys that furrow the serra and fill the region below with plenty.

There are also an immense number of sugar-plantations on the

proposed railway from Pernambuco to Joazeiro. From the Recife
to the river Una—a distance of seventy-five miles—there are no
less than three hundred sugar-estates on the sections of the railway
already under contract.

The distant population of this province is as untamed as the
wilderness in which it exists. Law is worn very loosely. Society
is patriarchal rather than civil. The proprietor of a sugar or cattle
estate is, practically, an absolute lord. The community that lives
in the shadow of so great a man is his feudal retinue; and, by the
conspiracy of a few such men, who are thus able to bring scores of
lieges and partisans into the field, the quiet of the province was
formerly more than once disturbed by revolts, which gave the
Government much trouble.

Revenue, accordingly, can only be collected by import and export duties. Taxation is impossible, because there is no system
of tax-gathering vigorous enough to collect it. A few years ago
an excise was put on the herds of cattle, and the exciseman went
into the sertão for the Emperor's money. He was caught, stripped,
and imprisoned in the trunk of a dead bullock, with his head stick-
ing out. "If the Emperor wants beef," the sertanejos said, "let
his exciseman take it along."

The provincial of Pernambuco, as he enters the city from the
sertão to do his semi-annual marketing, is worthy of such an exploit, and is a notable. The highway to the city lies through
Cachingá,—a neat little hamlet two or three leagues from the
Recife. The village is hidden from the observer as he approaches
by a long valley of orange and banana trees. This is the sertanejo's
last night's halt before getting to market. He has already ridden
for twelve days, perched upon a couple of oblong cotton-bags
strapped parallel to his horse's sides, followed by his train of a
dozen horses or mules, loaded, in the same way, with cotton or
sugar. A monkey, with a clog tied to his waist, surmounts one in
place of the driver; parrot and his wife another; and a large brass-
throated macaw with a stiff blue coat of feathers another. A raw
hide protects his wares from the rain. Night after night he has
slept on the earth, or has been suspended in his inseparable hammock, slung between two trees, with only the generous, starry sky
for a covering.

Cachingá, quiet and silent by day, is boisterous by night; for, during its watches, the sertanejos accumulate about the vendas by hundreds.   The first streaking of the morning witnesses a miscellaneous distribution, over the earth, of men, jaded horses, mules, monkeys, parroquetas, and sugar and cotton bags.   The caravan is at once put in motion.   Each individual sertanejo stirs his beasts, packs their loads, goes behind the riding-horse, seizes hold of the tail, puts a foot on the hock-joint, and leaps up on the back

SERTANEJOS.

as if ascending a flight of stairs.   This is a summons to every horse of his troop—already educated to it—to take his place in the train. In an instant the motley cavalcade is rolling down the valley of the Capibaribe before the sun has absorbed the dew-drops, which are like pendent jewelry on the rank leaves of the thick orchards that overhang the road.   The sertanejo passes on, only pausing to uncover before the patron saint of all cavaliers, (who is shut up in a wooden case at the gateway of the bridge of San Antonio,) and he finally halts with his various merchandise, living and dead, in the street Trapixe.

The individuality of the sertanejo is now manifest.   On his head he wears a pindova hat, after the pattern of a sugar-loaf, attempered by experience to every condition of weather. Under it is an affluent "shock" of hair, in the midst of which, in a doubtful state of light and eclipse, is a thin, bronze face, of Portuguese configuration, with eyes significant of divided curiosity and suspicion. He is attired in a cotton shirt and unmentionables, the one scant to the elbows and unbuttoned at the throat, leaving his tanned bosom bare, and the other rolled up to the knees. His feet are all unlearned in such commercial literature as the statistics of boots and shoes.

Early morning is the busy hour of Pernambuco. The sugar-streets are thronged with a wonderful miscellany of horses, mules, asses, and sugar-bags; sugar-merchants delicately holding samples; cotton-bales, goats with their families on a morning promenade; and quitandeiras eloquently passing panegyrics on cakes, comfits, and oranges. And still the tide of loaded horses and asses pours into the Trapixe. The horses lie down to rest, and the sertanejo, fatigued with the riot of the night, and anticipating the noontide siesta, pillows himself to slumber on the neck of his steed. A wood-dealer, with twin-bundles of fagots strapped on the side of his donkey, attempts to force a way. He is followed by a poultry-dealer mounted on an ass, with an immense hamper of fowls, advertised by a dozen chicken-necks thrust at full length through the lattices. Macaws and parrots make the tenor of the busy occasion; while the ambitious trumpets of a half-dozen donkeys lend their bass semitones. In the midst of this Babel of sounds, the *sabia*—sweetest of the Southern feathered tribes of song and peer of the Northern thrush and the mocking-bird— pours out his hearty, mellow praises from a lady's window by the side of a whitewashed church.

No market-scene can anywhere be more various, checkered, and interesting than at Pernambuco, in the busy sugar-season. Before meridian, the actors have changed, and others have taken their places. The black *ganhadores*, naked to the waist, with sugar-bags on their heads, hurry from the sugar-warehouses to the lighters, at full trot, in exact pace to their own boisterous music.

Nearly the whole of Brazil is adapted to the cultivation of sugar; but it is on the sea-coast from Campos to the sixth degree of south latitude that it is produced in the greatest abundance. The export of sugar from Pernambuco is annually increasing, and its production is flourishing under the improved machinery introduced by the brothers De Mornay. In 1821 this province produced 20,000,000 pounds; in 1853 the total was 140,000,000 pounds. The whole number of pounds exported from Brazil in 1855 was 254,765,504, of which we purchased to the amount of more than a million of dollars.

The ordinary price at Pernambuco is about three cents per pound for brown and five cents for pure white sugar. The clayed or white sugars are exported to Sweden and the United States: much of the brown is sent to the Mediterranean: the consignments to England are generally put up for "Cowes and a market."

Pernambuco also exported, in 1866, 32,159,040 pounds of cotton to Liverpool. This cotton is of a good quality, and brings a higher price than the generality of that exported from the United States. To the Quakers of England this Brazilian article has the preference, because it is mostly, according to Friends Candler and Burgess, raised by the free half-breeds of the interior; and I believe that there is very little of it which has to do with slave-labor. Great Britain imported from Brazil, in 1856, 21,830,000 pounds of cotton; but, as we have seen, Pernambuco alone in 1866 exported nearly fifty per cent more. In 1854 the export of cotton from Pernambuco was not quite three million pounds. The fibre is inferior only to that of sea island.

But the Brazilian Mail-steamer awaits us. We bid farewell to our friends, and soon pass on one side the little fort at the end of the reef, and on the other the rusty cannons of old Fort do Brum, and are at once on the ocean. At the same time a hundred *jangadas*, or *catamarans*, sally out for the fishing-grounds at some indefinite distance from land,—ten, fifteen, twenty, or forty miles. These curious crafts are each composed of four logs of cork-palm, eight inches in diameter, pinned together, with a plank thrust down between them for keel and rudder, and a broad, brown lateen sail, made from fibrils, affixed to a rude mast. The catamaran flies like the wind, and the clipper—swift courser of the sea— cannot outstrip it. The fisherman, with breeches rolled up to his thigh, (for every wave submerges his palm-logs,) sits securely on a pegged stool: occasionally he dips up the brine with a calabash and dashes it over his sail. Have no fear for this frail ship-carpentry. The catamaran will re-enter the harbor to-morrow morning, or, at furthest, the next day after, laden with a cargo of most extraordinary fish,—pink-eyed, ox-eyed, and four-eyed, round-shouldered, Roman-nosed, scaly and unscaled; and among them are some wearing a quantity of tails, hairy and tufted, like a buffalo-bull's. Only once, the story goes, a catamaran was run down at night: the picked-up owner was carried to Baltimore, to return at length and find his inconsolable widow solaced by a new marriage, and some young birds in the family nest not yet old enough to fly.

Dr. Kidder once performed a voyage in a jangada to the beautiful island of Itamaracá, and his experience shows that they are breezy, watery, and safe.

A minute after passing Fortaleza de Brum, a last sight is taken of a couple of Hollandish-looking windmills; and, as we glide away we have a glimpse of Cocoanut Island, lifting up its forest of green feathers against the clear sunset-sky, and finally nothing remains but the rocky pyramid of Olinda, crowned with a cross-bearing church, and, beyond, the low shores that stretch away toward Parahiba do Norte.

There is an utter dissimilarity in the geological position of the provincial capitals of Northern Brazil. But there is a striking resemblance in the heavy stone-masonry of the houses, in the tones of the families of bells that inhabit every church-turret, in the profound sand that fills the streets, and in the twinkle of the eyes and the thin sallow faces of the male inhabitants.

The little island of Itamaracá, which, under the old Dutch Government, was the most spirited and affluent along the whole coast, has now been almost lost sight of in geography, and has been degraded from a first commercial consequence into a lean and beggared colony of fishermen and fruit-raisers. Parahiba, the capital of Parahiba do Norte, with a population of ten thousand, is situated upon the Parahiba River, some ten miles from the sea. The greenery of both shores overhangs the narrow river so closely that it seems to be approached through a cavern of verdure. Red crabs doze on the muddy beaches, and countless tribes of waders industriously pick up a living at every retreat of the tide. At the end of this arched avenue of trees, and on the hill-side of a narrow valley, whitewashed Parahiba appears, and, as our steamer draws near, the bells of a cathedral that rises above it summon the priests to perform the solemn offices for the dead.

Natal, or Rio Grande do Norte, is, on the other hand, built on low lands near the sea. The steamer does not enter it, but lies off at an anchorage two or three miles from the shore. Passengers, with their luggage, are delivered, for want of boats, on board of a vivacious raft of palm-logs that goes hobbling round at the mercy of the sea. Each wave sweeps its whole length and breadth. *En route* to his post is a military *commandant*, just assorted and dis-

PADIOLA.

charged from the ruder human clay of the steamer, and he stands erect on the float, brilliant in attire and trappings, and made more magnificent by his top-boots, which, at every plunge, fill up with water from the briny deep.

Ceará can hardly be said to have a harbor: it is only a road-stead. This city is on ground comparatively level, and but few feet higher than the ocean. The bluff, tall mountains of Ibiapaba, four or five leagues distant, picturesque as the shores of the Hudson, and visible from the sea for a hundred miles, (though not marked

on the maps,) form a beautiful background. Their sides are fretted
with coffee-plantations, and, under the glass, their profile is serrated with feathery palm-woods. Here the style of landing is
very different from that at Natal. A boat transports the passengers to the verge of the surf that always breaks on the shore.
A municipal chair, (*padiola*,) large enough for the accommodation
of a couple of beef-fed aldermen, is borne on the backs of four
stout slaves, until the water reaches their chins, and the surf, as
they advance, passes over and around them. In the swift drift of
water that precedes the breakers, the chair receives the precious
freight of human life and treasure, and is carried at once, through
the surf, to the shore.

Aracati, in the province of Ceará, and Parnahiba, in that of
Piauhy, are principally cattle-marts. There is an equally striking
difference in the productions of the different provinces. Pernambuco and Aracati are sugar-dealers; Parahiba exports cotton principally. Ceará mingles sugar and coffee, and is eminently reputable for its beef. Parahiba and Piauhy have a ruder civilization,
and accumulate hides, tallow, and beef, and gather rice on the low
plains along the rivers. Maranham, in addition to its large
exports of cotton, rice, and salt, is a druggist, collecting many
species of invigorating roots, barks, and balsams in its woods.
Pará is gratefully known to the world for its cacáo and caoutchouc.

There is a difference, too, in the appearance of the coasts. After
leaving Olinda, no highlands are seen, except the mountains behind
Ceará, until the bluff sand-hill of San Marcos is turned on entering
Maranham. After leaving Parahiba do Norte, the eye tires of the
dreary shores and hillocks of white sand, herbless and treeless,
save here and there a riband of green cocoanuts in the little
valleys, or columnar cacti that from time to time shoot up out
of the unrelieved desert as if to keep note of its utter desolation. Though, as has been observed, there is no Sahara in
Brazil, there has often been much suffering from drought in
this portion of the Empire. As seen from the deck, glistening
sand frequently stretches away beyond the reach of sight. Such
is the character of the country for hundreds of miles. This
is slowly modified as the voyage extends farther north. The white
sand-drifts are, at long intervals, striped with vegetation; then it

VIEW IN THE PROVINCE OF PIAUHY.

becomes more interspersed, until at Maranham the whole shore is clothed with the beauty, brilliancy, and luxuriance of tropical growth.

The sea-built masonry of the reef of Pernambuco appears at frequent intervals along the coast, at distances varying from one

THE CACÁO.

hundred to one thousand yards from shore. At Ceará alone it seems to pass under the land, through the sandy point of Mucoripe. The ocean, with its low, hoarse voice of habitual sorrow, often breaks over it.

Petitinga—a triangle of green in the midst of a wide desolation of sand-hillocks—is famous for the tortoise-shell (second only to that of the South Sea) gathered among these disrupted rocks.

34

But the morality of the hamlet is like that of the Bedouins. Legitimate trade is sometimes suspended to plunder a flour-vessel which has been driven ashore by a storm and the currents. Then the whole population turn salvors, and salvage covers the cargo.

The point of the coast about Cape S. Roque is dangerous to vessels making their way close to the shore, in consequence of sunken reefs and the strong current, at the rate of three or four miles an hour, that, having already swept across the ocean from the African coast, impinges on Brazil not far from Bahia, and is then deflected northwardly till it passes the mouth of the Amazon, after which it continues until it becomes known to us as the Gulf Stream. This is a serious obstacle to attempting a landing north of Cape S. Roque, because then, with an adversity both of wind and current, it is difficult to turn the cape without standing far out to sea. Before the introduction of steamers, news from Northern Brazil was sometimes received at Rio de Janeiro *viá* Europe. Mr. Southey mentions the case of a vessel sent eastward from Maranham in 1656, having troops on board for some special emergency; which, after having been out fifty days,—a time long enough to exhaust her provisions,—found it necessary to put back, and in twelve hours reached the port she had left.

Eight degrees of latitude and more than fifteen hundred miles of coast are comprehended between Pernambuco and Pará on the Amazon. The climate of all is much alike, and without any appreciable differences on account of seasons. The range of the thermometer in the shade is from 82° to 90°, scarcely ever indicating a change of more than five degrees. So equable, indeed, is the temperature of the northern coast, that one cannot but be astonished at witnessing it advance slowly, during six months of the year, from 82° to the maximum, then, turning and tracing its way back, to the minimum with equal decorum. But the quantity and distribution of rain are very unequal, and its seasons vary at different points along the coast. At Pernambuco the rain continues about three months only, and falls in inconsiderable quantities, while at Pará, by exact observation, less than sixty days of the year are without rain. But the reader must not imagine a continuous state of overhanging clouds: the sun is seen as often as at New York. The rainy season at Pernambuco is nearly

ended when that at Maranham begins. At this latter point the
tropical rain, though less continuous than at Pará, is established
in full vigor. Light occasional showers inaugurate its approach.
Every day invigorates it, till, at the height of the season, in a
bright sky, black clouds rush up suddenly from every point of
the horizon to the zenith, bring their stores together in an angry
shock, accompanied by violent lightning and thunder, and pour
them down in a deluge on the earth. At this time, although
the rain sometimes con-
tinues incessantly dur-
ing the day, there is a
usual periodicity of the
showers, at ten o'clock
in the morning and
three in the afternoon,
—lasting a couple of
hours, and with bright
skies between. So great
is their precision that
all the appointments
of the day are made
with reference to these
short times of tempest.
The rainy season of
Maranham continues
about six months, and
during this time no less
than two hundred and
thirty inches of rain
falls! So says a British
resident. What author-
ity he has for his data
I know not. The re-

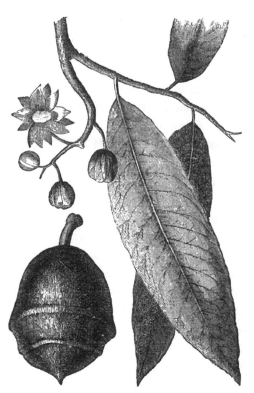

THE SAPUCAYA NUT.

mainder of the year is rainless. Still, vegetation does not droop.
Plants have in themselves the power of adaptation to great differences of seasons, and borrow and absorb the transparent moisture
which the trade-wind brings from the sea, thus maintaining their
usual rankness of growth.

And now, turning from the weather to something more stable,
we observe that the city of San Luiz de Maranham ranks as the
fourth in the Empire, and is the capital of the rich and important
province of the same name. The estuary upon which it stands
was discovered by Pinzon in 1500. Though Maranham was made a
captaincy as early as 1530, the French, in 1612, were the first to
form a permanent settlement, and, in compliment to the patron
saint and the royal family of France, named the town St. Louis and
the bay St. Mary.

The territory of the province is rather uneven in its surface,
although it has not a single range of mountains. It is watered by
a large number of rivers, both great and small. It remains to a
great extent covered with forests, in which valuable woods and
precious drugs are abundant. The soil is peculiarly adapted to the
cultivation of rice, which it produces in vast quantities. Cotton
thrives much more than the sugarcane. The indigenous fruits are
numerous and rich, and in the distant interior are many edible
nuts, among which none is more curious than the three-cornered
Brazil-nut (*Bertholetia excelsa*) and the sapucaya, (*Lecythis ollaria*.)
The latter is a capsule or nut as large as an infant's head, filled with
small, oily, eatable grains. With this capsule pretty vases and
sugar-bowls are often made. The pineapples and bananas, of
several species, deserve mention for especial excellence. Mineral
riches have not been withheld from this portion of the globe. Fine
strata of old red sandstone furnish an excellent and common
material for building; while iron and lead ores and antimony have
been discovered, although they have not yet been turned to public
advantage. Fish abound in the waters of the province; and herds
of sheep, cattle, and horses multiply rapidly on the plantations of
the interior.

San Luiz de Maranham is believed to be better built, as a whole,
than any other city of Brazil. It exhibits a general neatness and
an air of enterprise which rarely appears in the other towns of the
Empire. There are, moreover, within its bounds but few huts
and indifferent houses. None of the churches appear unusually
large or sumptuous, but many of the private dwellings are of a
superior order. The style of construction is at once elegant and
durable. The walls are massive, being composed of stone broken

fine and laid in cement. Although the town does not occupy a large extent of ground, the surface it covers is very unequal. Its site extends over two hills, and, consequently, a valley. The rise and descent in the streets are in many places very abrupt. Few carriages are in use, and, in accordance with this circumstance, there is only one good carriage-road in the entire vicinity. That road leads a short distance out of town. The cadeira is but little known here as a means of conveyance. The *rede*, or hammock, is generally used as a means of easy locomotion. It is very common, both in Maranham and Pará, to see ladies in this manner taking their passeio, or promenade. Gentlemen do not often make their appearance in public in this style,

A REDE.

although it is generally conceded that they are quite fond of swinging in their hammocks at home.

Hon. John U. Petit, who resided for a number of years at Maranham, has kindly furnished me a few of his full notes; and his descriptions of Maranham are so fresh, graphic, and full of life that I give them entire:—

"The lateral streets, crossing the two principal thoroughfares, descend rapidly to the estuaries on each side. The heavy rains dash their torrents along down their pavements and cleanse the whole city. Filth is thus made impossible. *Quebracosta* or Breakback Street deserves its name, for it drops down abruptly like a declivity.

"My first landing was made at evening, and at the end of the outpouring of the diurnal rains. Already the sun was out, and the clouds were half dispersed from the sky, except here and there a few remaining fugitives, fantastically arranged, now in crags and mountain-steeps, now in distant harvest-landscapes, now in long, blue lakes, with sloping shores of green and orange.

"But the prevailing and superabundant humidity at this season, though unfelt and obviously unseen, is yet seen in its effects. Every thing that is touched is clammy. The wet season is the green age of mould. And yet it is not so much wet as musty. Mould grows on every thing that gives it a place for rest. A greasespot on a coat, or a soiled coat-collar, becomes verdant after a night's exposure. Albino wakes you to take a cup of coffee, and you sip the liquid swinging in your hammock, just as the morning is peeping, and the velvet-breasted wren is singing from the tall crown of a bread-fruit-tree or early humming-birds are sucking nectar from the very throats of the red pomegranate-flower. Albino then improvises a lustre on your boots. But you have hardly sunk down in your hammock and waked up again, when—*presto*—your boots are grown over with a green vegetable nap, an antiquity-looking mildew. The old black, revered, neat's-leather trunk, fellow-visitor of many States, and the acquaintance of many custom-house explorers,—now standing modestly back by the wall with its lid uplifted, as though it wished everybody to look in and see its very heart,—under the novel influence, is first white, then brown, then yellowish, and, at last, green in an apparent old age. But, if this attract remark, it is only for a moment; for the mould perishes at the first hot breath of old Sol,—suddenly as the ephemera that lives a whole life and dies in crossing a sunbeam.

"Maranham, in its principal streets, is built of compacted stone-masonry. Houses are usually of two, three, or four stories, with walls of two and a half or three feet in thickness, the better to resist attacks of external heat. Maranham is nearly a finished city; but a house was erected, not long since, in the Street St. John. A train of asses and mules brought the red, ferrugineous sand-stone—just landed from Bom-Fim—up the Palace Square in panniers,—a reluctant slave compelling them from behind. The lime was carried in baskets, on the heads of slaves, from the opposite sea-shore; while, in order to mix the mortar, women marched up, loaded with water-jars, from the abundant fountain behind Praia Cujú.

"The population is affluent. The residents of the city are the proprietors of the plantations and of the numerous slaves dwelling on the fazendas of the mainland. Factors supervise them there, and the annual rents are paid without giving the masters any trouble in going after them, and the money is soon wasted in the abundance—and, sometimes, the dissipation—of the city.

"With such ample means, the children of its burghers are very well educated in the more brilliant and showy and less practical attainments of knowledge,—sometimes at home, less often abroad. Ladies more frequently than gentlemen are met with who have learned the arts of pleasing and conquest at Lisbon, Madrid, and Paris. This superior class constitutes a social realm where Roger de Coverley might live happy.

* * * * * * * * * *

"Before midnight, the streets are quiet as churchyards, and it is only the late walker who is met by the patrol with a musket on his shoulder and a bayonet at the end of it, and required to give the countersign; and, answering, it is likely, with a very difficult utterance, *Amigo*, which means that he is a particular friend of the Emperor's, is then directed to move on.

"Below the class of opulent citizens, who dwell in large stone houses having balconies at all their windows and verandas above, that shut out the invasion of the sun, first in rank is the large class of shopkeepers and artisans. For these, several schools exist. The city, too, abounds in charities. It has its home of orphans, its house of foundlings, a house of lepers, hospitals for the sick, and *misericordias*, with open doors, embracing all the children of distress.

"The Portuguese make an important element of the population in all the cities. They are spirited, ambitious, self-reliant, and money-making. They do not create wealth, but acquire it. The *Brazileiro* looks on them with habitual aversion. This had its origin in the time of the colonial dependence on Portugal, when home-bred courtiers of the monarch crowded all the walks of ambition in Church and

State, to the exclusion of the natives of the colony. The Government then was terribly unjust and oppressive. The Portuguese appointees were generally in circumstances of decayed fortune, which they went abroad to repair; and the history of the *capitanias* is only a repetition of the old story of the outrages and rapacity of the Roman proconsuls. To this deep cause of hatred another is added, in the steady flow of Portuguese colonization into the Empire, monopolizing, by vigor and ingenuity, the shopkeeping and the more skilful mechanical employments, in which a Brazilian rarely appears. Most of them come as adventurers and obtain competence, many of them affluence.

"A vessel touches in Brazil, loaded with Portuguese lads bent on making fortunes. Each has a large chest, capable of holding a whole family, At a custom-house inspection, two of the boys lift up the huge lid. In the immense cavern to which it opens are seen dispersed a shirt, 'a pair of socks,' needles and thread, and, in addition, the adventurer's stock in trade,—two or three strings of Spanish onions. In ten or twelve years the boy has become a man, and embarks his chest again to return to Portugal. But now he has it strapped with ropes to keep down the cover. Small boxes and carpet-bags cluster around it, as if they were the old chest's children; and the old chest, having no wings, but feeling maternal, hovers over them with its shadow. And, before embarking, the indefatigable Portuguese has paid duty on a considerable amount of specie. Such is the facetious and somewhat overdrawn picture by which the *Brazileiros*, the lineal descendants of a common ancestry, solace themselves over their deadly enemies the Portuguese.

"The class of Brazilians proper—the offspring of the old Portuguese emigrant, —embracing the civil functionary, the army and navy officer, the priest, and the gentleman of the city and the country—forms about one-third of the population. The Portuguese population, in number, is about one-sixth. Below these are the varieties,—making about one-half the census,—the negro, mulatto, mestizo, and Indian. The wants of the latter are few and cheap:—a house floored on the naked earth, palm-thatched at the sides and overhead, with hammocks slung diagonally across it for sitting and sleeping, and with attire exceeding Eve's garden-dress merely by a shirt or pantaloons; besides these, the sea and earth, equally bountiful, spread their tables with plenty. But individuals of one class easily shift into another. Genteel persons sometimes get out of their places and become vagabonds; while, overcoming the slightest possible obstacle on account of color, exchanges in society are made, as everywhere else, by some in subordinate ranks forcing themselves out of their positions upward.

"A musical furor rages like the dog-star. Piano and harp are vocal in the parlors and saloons. But the guitar—as in the vine-covered cottages of Portugal —is a joy forever in all the households of the poor; while its humbler types—the banjo and marimba—are an equally universal property of the black and all his derivatives. The slave that goes bareheaded, barefooted, and unshirted vexes it (the marimba,—that primitive guitar) in the soft moonlight, before his master's door, in the presence of a bevy of loitering wenches, on whose hearts, as a second instrument, he plays,—taking them captive by the sorcery of his art. The melodies of the North American plantations (the African-born airs of Virginia and Tennessee, long since threadbare in the United States) are, like the smallpox, contagious through all ranks of society. A dozen negroes, carrying a large crockery-hogshead slung over their shoulders on bamboos, are mourning, in *minor* melody, the fate of 'Poor Old Ned.' In the Street Sant' Anna, from behind a latticed door, one hears a musical voice telling Susannah not to cry.* Aristocratic pianos are loud with 'Rosa d'Alabama' and 'Senhoritas de Buffalo,' with much more music than prosody.

THE MARIMBA.

"Outside and inside, S. Luiz is a very lovable city. Good-temper, courtesy, and kindness are almost universal. This is confined to no position of life. A ready, overflowing hospitality welcomes the stranger at every door.

"It is very pleasant to draw a picture of Maranham by memory, with the bay, dotted over with little islands of verdure broad enough in some places not to permit you to see the opposite shores, folding it in the embrace of its two large estuaries; strange fishermen's craft, picturesque *montarias* and canoes, lying along the praias; dainty, tall cocoanuts fringing the profile of the city, as it seems to be thrown carelessly over the sharp ridge that advances into the bay; groves of bananas and oranges clinging on its steep sides; a redolence of sweets from native flowers filling the air; occasional *mirantes* pretentiously stretching up above the general perspective of red tiles; and the tall tower of the cathedral and the populous turrets of scores of churches pushing their rounded pinnacles into the sky.

" 'Swallows,' says Dr. Johnson, 'certainly sleep all winter. A number of them

---

* The wide diffusion of the so-called "Ethiopian Melodies" of the United States is almost incredible. In 1849, at one o'clock in the morning, I was riding from Charing Cross to the Surrey side of London, and heard a party of young Englishmen singing, at the top of their voices, "Oh, Susannah!" &c. Once, in passing over the Gloria Hill, at Rio de Janeiro, I caught the notes of the same tune, which was being performed by one of the inmates of a Brazilian cottage. But the most unexpected treat, in this particular, I experienced in 1850, at Terracina,—the ancient Anxur, and not far from the Three Taverns mentioned in Acts xxviii. 15. It was an Italian midnight; and, while I was listening to the sound of the Mediterranean wave, as it broke upon the decaying quays of Terracina, and thinking of the long past of old Rome, I was startled by a clear voice (which made the ruins around us ring) sending forth upon the night-air "Old Uncle Ned." It suddenly dashed away every thought of Italy and Rome and carried me most hastily over the ocean. I afterward discovered that the serenader was a Boston Yankee, who had wandered to this quiet nook, and who had been so singularly affected by the sacred and classic associations that he gave vent to the "Ancient Uncle Edward," as most in accordance with emotions called forth by the antiquity—classic and sacred—of Terracina.—J. C. F.

conglobulate together by flying round and round, and then, all in a heap, throw themselves under water and lie in the bed of a river.' The first greeting at Maranham to the April visitor is the dear old friend the swallow. He builds his house under the tiled eaves. It haunts church-spires in myriads, as though a religious bird. As the sun goes down and shines with diminished beams, and until he finally sinks to rest, far up in the sky little flocks of swallows are seen wheeling in giant circumferences. Sometimes their enemy the vulture, at the same hour of the evening, is up there with his family, airing, after a day spent shamefully among carcasses. Then squadrons of swallows muster and drive him from those azure fields. Now they disport themselves along the earth, now flit on lazy wing above the house-tops, or pick a zigzag way along the airy avenues, among the groves of palm and figs and oranges, or dart away, swift and unerring as an arrow, after some gay butterfly, from which—as riches cannot shield from death—his velvet bosom and painted wings cannot buy him escape. A half-dozen weeks hence, the swallow that sits at the margin of that red tile, teaching her young, with affectionate art, to fly, may, under Northern skies, at home, skim above the fragrant clover-meadows or yellow harvests, or through the blossoming orchard or butternut-clump, or lave her white bosom in the little lake, or sweep along the hill, chasing the shadow of a lazy cloud. Thus are the swallows delightfully occupied during our cold winter, and when the time to migrate arrives they gather in countless hosts on all the house-tops, preparatory to their long journey, to proclaim, with other harbingers, to Northern lands, still brown with the hues of annual death, that light-footed Spring is coming with a power of resurrection. Choicest of the gifts with which man mitigates his lot is the physical charm of all beauteous nature, its mute yet divinely-speaking flowers, and its happy birds, harmonious with more than choral sweetness.

"The sight of the pretty white village of Alcantara, of five or six thousand inhabitants, a half-dozen miles distant across the bay, makes one wish to visit the mainland. Alcantara is noted for the production of salt, gathered, as in some of the West India Islands, from natural pools supplied with water from the ocean at the recurrence of the spring-tides. A few miles farther up the coast is the village of Guimaraens, in the midst of a region abounding in cotton, rice, and mandioca.

"The twin-bays of San Marks and San José, immediately behind the island of Maranham, are reached from the interior of the province by several rivers—the Pindaré, the Mearim, and the Itapicurú—hardly more considerable than the Mohawk or the Upper Wabash. As Alcantara invites you to its shores, these rivers tempt you to ascend their mangrove-lined banks to their sources.

"The mangrove-tree is present along all the tide-water of Northern Brazil, and at high-water is standing in it at mid-waist, only its branches, sea-green leaves, and a few white blossoms above it. Behind it, on the high shore, are lines of towering palms. Vegetable propriety is outraged in the manner in which the mangrove grows. From its shaft, a half-dozen inches in diameter and a half-dozen feet high, it puts forth horizontal branches. These, in turn, drop down suckers, that become rooted into the mud and soon attain the size of the parent stem; and these, in turn, send out other branches and drop other stems, till the tree has grown into a large framework, and so strengthens itself against the tempests. In its deep shadows, where no human foot intrudes, the *sericoria* —the woodcock of the tropics—fearlessly leads abroad its young. Upon the roots oysters cling, and, at low-water, present the curious spectacle of bivalves growing on trees. The mangrove contains, in great abundance, the principle of tannin, which, in the form of a concocted extract, may become a valuable article of commerce."

The *montaria* referred to is thus described by Dr. Kidder :—

"In the river, in front of the Varadoura, a respectable collection of merchant-vessels may generally be seen at anchor. None of the water-craft, however, appear more picturesque than does the montaria,—a species of flat-boat used much on

THE MONTARIA.

these waters. In the first one which I saw, I counted ten Indians paddling it rapidly against the tide. They each held a paddle, about the size and shape of an oval spade, perpendicularly in both hands, and, all striking at once into the water, gave the boat great momentum."

We now bid adieu to the clean, the gay, the hospitable city of San Luiz, and steam for Pará.

Note for 1866.—Since this chapter was written, J. C. F. has visited the whole coast from Rio de Janeiro to Pará, and the cities of Bahia and Pernambuco four times in as many years. He would be glad to enumerate the many improvements which have taken place, in railways, &c.; but want of space forbids. He cannot, however, forget the many warm receptions, particularly at Pernambuco, from Messrs. Swift, Hitch & Rollins, (the various partners of Henry Forster & Co.,) from S. P. Johnson, from the Baron of Livramento, a "live" Brazilian, from Sr. Tasso, and from the Sá e Albuquerque family at Gararápes ; and neither is he unmindful of the kindness of two eminent Brazilian statesmen, the Visconde de Camarigibe and the Visconde de Boa Vista, as also of Dr. Vasconcellos, editor of the *Jornal do Recife*.

CHAPTER XXVI.

MAGNIFICENCE OF NATURE IN THE BRAZILIAN NORTH — THE CITY OF PARÁ — THE
ENTRANCE OF THE AMAZON — THE FIRST PROTESTANT SERMON ON THESE WATERS
— PARALLEL TO THE BLACK-HOLE OF CALCUTTA — EFFECTS OF STEAM-NAVIGATION
— IMPROVEMENTS IN PARÁ — THE CANOA — BATHING AND MARKET SCENES —
PRODUCE OF PARÁ — INDIA-RUBBER — PARÁ SHOES — THE AMAZON RIVER — MR.
WALLACE'S EXPLORATIONS — THE VACA MARINA — CETACEA OF THE AMAZON —
TURTLE-EGG BUTTER — INDIAN ARCHERY — BRAZILIAN BIRDS AND INSECTS —
VISIT TO RICE-MILLS NEAR PARÁ — JOURNEY THROUGH THE FOREST — THE
PARANESE BISHOP'S SUSPICIONS OF DR. KIDDER — STATE OF RELIGION AT
PARÁ.

WE rapidly steam over the four hundred miles between Maranham and Pará, and we have reached the eastern edge of the Brazilian North,—the maritime border of that vast basin which
contains an area equal to that of two-thirds of Europe. We are
about entering upon a region the most wonderful in its nature,—
where every object is upon the grandest scale. The mightiest
river of the world rises in the loftiest mountains of the Western
continent and flows for thousands of miles through forests unparalleled in beauty, extent, and productiveness. Here the *Victoria
Regia*, the giant of Flora's kingdom, nestles on the bosom of the
shady pools, or reposes on the still waters that are shielded by some
verdant peninsula from the rushing waves of the never-ceasing
flood that pours from the Andes. Millions of the most brilliant-
plumaged birds and insects, curious quadrupeds and reptiles, inhabit this almost *terra incognita*. Perhaps no region of our globe
possessing such wonders has been so easy of access and so little
explored. We are, however, on the eve of a great change: steam
is doing its legitimate work, and the present generation may not
live to see the Valley of the Amazon, like that of the Mississippi,
teeming with millions, but there will be a thorough knowledge of
its vast resources. Much that is visionary has been written con-

cerning the "mighty Orellana;" and those who are expecting to behold its fertile shores a half-century hence filled with a thrifty population and smiling under civilization are doubtless doomed to disappointment. And, while Southern Brazil will ever be the fit field of enterprise for the European and North American, still, there is no reason to doubt that the statement of Mr. Wallace—the most thorough explorer of the Amazon Valley—is strictly true when he says, "For richness of vegetable production and fertility of soil it is unequalled on the globe, and offers to our notice a natural region capable of supporting a greater population and supplying it more completely with the necessaries and luxuries of life than others of equal extent."

Amazonia should have a volume to itself; but this work would be incomplete without some notices of this portion of the Empire of Brazil, which has always excited a deep interest on both continents.

The city of Belem, or Pará, is usually the point of departure for those visiting the Amazonian region from the East. There was formerly a land and water route from Maranham to Pará, which has now been abandoned: according to Mr. Southey, it used to be performed by canoes passing through the continent, and coasting around not less than thirty-two bays, many of them so large that sight cannot span them. These bays are connected by a labyrinth of streams and waters, so that the voyage may be greatly shortened by ascending one river with the flow, crossing to another, and descending with the ebb. The distance thus circuitously measured is about three hundred leagues, and may be traversed in thirty days. Dr. Kidder says,—

"I met with one individual who had in early life passed through this inland passage in a much more direct course, his voyage occupying only fourteen days. It was at that golden era when Indian labor was plenty and could be secured at four cents per day. Some years after, the same individual wished to perform this voyage, but was forced to abandon it, from the difficulty of finding canoe-men to serve him even at fifty cents per day. He entertained the most delightful recollections of the route, exhibiting as it did the glories of nature in all their pristine loveliness. Nothing interrupted the security of the traveller, and nothing disturbed the silence of those sylvan retreats save the chattering of monkeys or the carolling of birds. The silver expanse of waters, and the magnificent foliage of tropical forests, taller than the world elsewhere contains, and so dense as almost to exclude the light of the sun, combined to impress the mind with inexpressible grandeur.

"The canoes were drawn up on shore every night when refreshment and repose

PARA.

were des.red, and the skilful Indians, in a few moments, could secure sufficient game for the subsistence of the party. Thus the voyage was prosecuted with little fatigue and with every diversion."

In some portions of Brazil where there are so many streams to be crossed, ferry-boats, on some occasions, were formerly extemporized. An ox-hide was the principal material for the construction, and a slave was the means of propulsion.

NOVEL FERRY-BOAT.

Pará is situated on the river of the same name, which, some contend, is but a continuation of the Tocantins, and not one of the mouths of the Amazon. Mr. Wallace inclines to the former, but general belief to the latter, opinion.

During the prevalence of certain winds, and owing to the strong currents, which force the fresh water far out to sea, the entrance of the Pará River is sometimes both difficult and dangerous. My colleague thus describes his experience:—

"We entered this mouth of the Amazon at a fortunate juncture. The weather was so clear that we distinctly saw the breakers on both the Tigoca and Braganza banks, and the tide had just commenced flowing upward. For nearly an hour we could observe, just ahead, the conflict of the ascending and descending waters. Finally, the mighty force of the ocean predominated, and the current of the river seemed to recoil before it.

"This phenomenon is called, from its aboriginal name, *pororoca*, and gives character to the navigation of the Amazon for hundreds of miles. No sailing-craft can descend the river while the tide is running up. Hence, both in ascending and descending, distances are measured by tides. For instance, Pará is three tides from the ocean, and a small vessel entering with the flood must lie at anchor during two ebb-tides before she can reach the city. Canoes are sometimes endangered in the commotion caused by the pororoca, and hence they generally, in anticipation, lie to in certain places called *esperas* or resting-places, where the water is known to be but little agitated. Most of the vessels used in the commerce of the Upper Amazon are constructed with reference to this peculiarity of the navigation, being designed for floating on the current rather than for sailing before the wind, although their sails may often be made serviceable.

"The ebb and flow of the tides in the Amazon are observed with regularity five hundred miles above the mouth, at the town of Obidos. The pororoca is much more violent on the northern side of the island of Marajó, where the mouth is wider and the current becomes more shallow.

"As we passed up the great river, the color of the water changed from the dark hue of the ocean we had'left to a light green, and afterward, by degrees, to a muddy yellow. We were barely in sight of the southeastern bank of the river; and, after we had ascended more than forty miles, the island of Marajó began to be visible on the opposite side. In the course of the day we approached nearer the continent, and the shore was seen to be uniformly level and densely covered with mangrove-thickets. The only village distinctly seen was Collares, which our commander, Captain Hayden, had captured during the revolution. The whole day we were borne along by the combined force of steam and wind, but the tide was part of the time against us. At evening a clear full moon shed down from an unclouded sky new splendor upon a scene already sublime. A most fragrant breeze from the land became more and more perceptible as the river narrowed. Two boats were the only craft we saw during the whole ascent. Finally, we came alongside the Forte da Barra, two miles distant from the city of Belem, and were hailed as we passed. The lights of the town, and of vessels in front of it, then became visible. We described a semicircle around the harbor, passing between two vessels-of-war, and came to an anchor at ten o'clock.

"The towers of the cathedral, of the palace, and of several churches, were distinctly visible in the moonlight.

"The second day after our arrival was the Sabbath, and through the courtesy of Captain H. it was arranged that I should hold a Bethel service on board the Maranhense steamer. Some American seamen were present, and several persons came from the shore. These, together with the ship's company, formed an audience to whom I announced the tidings of the kingdom of God. Making allowance for the circumstance of a public packet just clear of her passengers and the same night going to sea with another supply, the occasion was very favorable for divine service, and I felt truly grateful for the opportunity—probably the first ever enjoyed by any Protestant minister—of attempting to preach Jesus and the resurrection upon the wide waters of the Amazon. I held similar services at Pará on seven succeeding Sabbaths,—once on board an American vessel in port, and at other times in the private house of a friend.

"The location of Pará, or the city of Belem, is in 1° 28′ S. latitude and 48° 28′ W. longitude. Its site occupies an elevated point of land on the southeastern bank of the Pará River, the most important mouth of the Amazon. This city is eighty miles from the ocean, and may be seen from a long distance down the river. It has a very imposing appearance when approached from that direction. Its anchorage is very good, formed by an abrupt curve in the stream, and admits vessels of the largest draft. The great island of Marajó forms the opposite bank, twenty miles distant, but is wholly obscured from sight by intervening and smaller islands.

"The general appearance of Pará corresponds to that of most Brazilian towns, presenting an array of whitened walls and red-tiled roofs. The plan on which it is laid out is not deficient in either regularity or taste. It possesses a number of public squares, and the streets, though not wide, are well paved, or rather macadamized. The proportion of large, well-built houses is respectable, although the back-streets are mostly filled with those that are diminutive in size and indifferent in construction.

"The style of dwelling-houses is peculiar, but well adapted to the climate. A wide veranda is an essential portion of every habitation. It sometimes extends quite around the outside of the building, while a similar construction prevails on at least three sides of a spacious area within. A part of the inner veranda, or a room connected with it, serves as the dining-room, and is almost invariably airy and pleasant. The front-rooms only are ceiled, save in the highest and most expensive edifices. Latticed windows are more common than glass, but some houses are furnished with both, although preference is always given to the former in the dry season. Instead of small, dark, and unventilated alcoves and sweltering beds for sleeping, they have suspension-hooks arranged for swinging hammocks across the corners of all the large rooms, and transversely along the entire sweep of the verandas. Some dwellings contain fixtures of this sort for swinging up fifty or sixty persons every night with the least possible inconvenience.

"The effects of the revolution of 1835 are still very apparent in Pará. Almost every street shows a greater or less number of houses battered with bullets or cannon-shot. Some were but slightly defaced, others were nearly destroyed. Of the latter, some have been repaired, others abandoned. The S. Antonio Convent was much exposed to the cannonading, and bears many marks of shot in its walls. One of the missiles was so unlucky as to destroy an image perched in a lofty niche on the front of the convent."

This revolution was one of the most successful on record, where the aborigines, guided by white leaders, nearly regained their power, and for a time held in subjection the European descendants. Pará, though now prosperous, has been singularly unfortunate in the check to its progress which has been the heritage of many revolts.

The traveller, on entering this city, is struck with the peculiar appearance of the people. The regularly-descended Portuguese and Africans do not, indeed, differ from their brethren in other parts; but they are comparatively few here, while the Indian race predominates. The aboriginals of Brazil may here be seen both in pure blood and in every possible degree of intermixture with both blacks and whites. They occupy every station in society, and may be seen as the merchant, the tradesman, the sailor, the soldier, the priest, and the slave. In the last-named condition they excited most my attention and sympathy. The thought of slavery is always revolting to an ingenuous mind, whether it be considered as forced upon the black, the white, or the red man. But there has been a fatality connected with the enslavement of the Indians, extending both to their captors and to themselves, which invests their servitude with peculiar horrors.

Nearly all the revolutions that have occurred at Pará are directly or indirectly traceable to the spirit of revenge with which the bloody expeditions of the early slave-hunters are associated in the minds of the natives and mixed bloods throughout the country. The Brazilian revolution in this part of the Empire was attended with greater horrors than in any other province.

When the independence of the country was declared, Pará was for a time held by the Portuguese authorities. On the arrival of Lord Cochrane at Maranham, he despatched one of his officers, (Captain Grenfell,) with a brig-of-war, to take possession of Pará. This officer had recourse to a stratagem which, although successful, was little more creditable to his bravery than his integrity.

Having arrived near the city, he summoned the place to surrender, asserting that Lord Cochrane was at anchor below, and, in case of opposition, would enforce his authority with a vengeance. Intimidated by this threat, the city hastened to swear allegiance to the throne of Dom Pedro I., and Grenfell managed to have obnoxious individuals expelled before his deceit was found out. Opposition, however, soon sprang up: a party was organized with the intent of deposing the provincial junta. The latter, of course, claimed the protection of Grenfell. He immediately landed with his men, and, joining the troops of the authorities, easily succeeded in quelling the insurrection. A large number of prisoners were taken, and five ringleaders in the revolt were shot in the public square. Thence returning on board, he received, the same evening, an order from the president of the junta to prepare a vessel large enough to hold two hundred prisoners. A ship of six hundred tons' burden was accordingly selected. It afterward appeared that the number of prisoners actually sent on board by the president was two hundred and fifty-three. These men, in the absence of Captain Grenfell, were forced into the small hold of the prison-ship, and placed under a guard of fifteen Brazilian soldiers.

"Crowded until almost unable to breathe, and suffering alike from heat and thirst, the poor wretches attempted to force their way on deck, but were repulsed by the guard, who, after firing upon them and fastening down the hatchway, threw a piece of ordnance across it and effectually debarred all egress. The stifling sensation caused by this exclusion of air drove the suffering crowd to utter madness, and many are said to have lacerated and mangled each other in the most horrible manner. Suffocation, with all its agonies, succeeded. The aged and the young, the strong and feeble, the assailant and his antagonist, all sank down exhausted and in the agonies of death. In the hope of alleviating their sufferings, a stream of water was at length directed into the hold, and toward morning the tumult abated, but from a cause which had not been anticipated. Of all the two hundred and fifty-three, four only were found alive, who had escaped destruction by concealing themselves behind a water-butt."—*Armitage*, vol. ii. p. 108.

This dreadful scene is perhaps unparalleled in history, or finds its parallel alone in the black-hole of Calcutta. Its only mitigation consisted in its having been caused by carelessness and ignorance, without "intent to kill." It has, however, but too much affinity with the treatment of the prisoners taken and confined at the same place in the subsequent civil revolutions. Vast numbers of these unhappy men were crowded into the prison of the city and of the fort, where they were kept, without hope of release, until death set them free. Besides, a prison-ship, called the Xin Xin, was filled to its utmost capacity. Dr. Kidder has estimated that not less than *three thousand* had died on board that one vessel in the course of five or six years. My colleague thus speaks of the last great revolt at Pará :—

"The disorders that broke out at Pará in 1835 were disastrous in the extreme. They first commenced among the troops. The soldiers on guard at the palace seized an opportunity favorable to their designs, and on the 7th of January simultaneously assassinated the president of the province, the commander-at-arms, and the port-captain. A sergeant, by the name of Gomez, assumed the command, and commenced an indiscriminate slaughter of the Portuguese residents. After twenty or thirty reputable shopkeepers had been killed, these insurgents proceeded to liberate about fifty prisoners, among whom was Felix Antonio Clemento Malcher, an individual who had been elected a member of the provisional junta at the time of Grenfell's invasion, but who was subsequently arrested as the instigator of a rebellion at the Rio Acará. This Malcher was now proclaimed president, and a declaration against receiving any president from Rio until the majority of Dom Pedro II. was formally made.

"No houses were broken open on this occasion. Order was soon restored, and things remained quiet till the 19th of February. At this time, Francisco Pedro Vinagre, the new commander-at-arms, having heard that he was to be arrested for some cause, called out the soldiers and populace to attack the president. Malcher shut himself up in the Castello fort and attempted to defend himself. In the course of two or three days two hundred men were killed and the president captured.

He was sent to the fort at the Barra, below the city, as if to be imprisoned, but was murdered on the way, undoubtedly by the orders of Vinagre, who was now supreme.

"On the 12th of May an attempt was made, under the constitutional vice-president, Senhor Corrêa, to take possession of the town, by landing troops from a squadron of thirteen vessels-of-war. This attempt was repulsed, and the vessels dropped down the river. Soon after, a new president (Senhor Rodriguez) arrived from Rio. On the 24th of June he landed with a body of two hundred and fifty troops, the insurgents having retired toward the interior. Disorders still continued in the province, and, on the 14th of August, a body of Indians, led on by Vinagre and others, suddenly descended upon the capital. They obtained possession of the city and commenced an indiscriminate massacre of the whites. The citizens were obliged to defend themselves as they best could. Vinagre fell in the midst of a street-skirmish. An English and a French vessel-of-war, lying in the harbor, sent on shore a body of marines, but soon withdrew them on account of the pusillanimous conduct of the president.

"The Indians commenced firing upon the palace from the highest houses of which they could get possession, and artillery from the palace attempted to return the fire. The president, however, soon withdrew and abandoned the city to destruction. Many families succeeded in escaping on board vessels in the harbor, but many others fell victims to rapine and murder. Edurado, the principal leader after the death of Vinagre, endeavored to protect the property of foreigners, and, to some extent, succeeded: nevertheless, as fast as possible, the foreign residents withdrew from the city, and thought themselves fortunate to escape with their lives. The period that ensued might with propriety be called the reign of terror. But it was not long a quiet reign. Disorders broke out among the rebels, and mutual assassinations became common. Business was effectually broken up, and the city was as fast as possible reverting to a wilderness. Tall grass grew up in the streets, and the houses rapidly decayed. The state of the entire province became similar. Anarchy prevailed throughout its vast domains. Only a single town of the Upper Amazon maintained its integrity to the Empire. Lawlessness and violence became the order of the day. Plantations were burned, the slaves and the cattle were killed, and in some large districts not a white person was allowed to survive.

"In May of the following year, General Andréa arrived as a new president from the Imperial Government and forced his way into the capital. He proclaimed martial law, and, by means of great firmness and severity, succeeded in restoring order to the province. It was, however, at the cost of much blood and many lives. He was accused of tyranny and inhumanity in his course toward the rebels and prisoners; but the exigencies of the case were great, and furnished apologies. One of the most disgraceful things charged upon him and his officers was the abuse made of their authority in plundering innocent citizens, and also in voluntarily protracting the war so that their selfish ends might be advanced. Certain it is that the waste of life, the ruin of property, and the declension of morals, were all combined and lamentably continued; and yet in this state of things we see nothing but the fruits of that violence and injury which, from the first colonization of Pará by the Portuguese, had been practised against the despised Indians.

"In addition to the more direct consequences of the disorders, the salubrity of the country and of the city itself fearfully deteriorated. The rapid growth and the equally rapid decay of vegetable matter on the spots from which years of cultivation had banished it brought on epidemics and other fatal diseases, which swept off hundreds of the people that survived the wars. Thus, one of the richest and fairest portions of the earth was nearly desolated.

"Until 1848 it was only by slow degrees that Pará recovered. Nothing, indeed, but the extraordinary and spontaneous fertility of the whole region has enabled the province, in any considerable degree, to reclaim its business-relations. Notwithstanding all the natural beauties so profusely exhibited at Pará,—reminding one, at every step and at every glance, of the glorious munificence of the Creator, —there are but few places which suggest sadder reflections upon the wickedness and misery of man. Until within a few years, we can scarcely point to a bright spot in its history. During the early periods that succeeded its settlement by Europeans, a continual crusade was carried on against the aboriginals of the soil, for the purpose of reducing them to a state of servitude. In vain were the reasoning and power of the Jesuits arrayed in opposition to this course. In vain was African slavery introduced as its substitute. The cruel and sanguinary purposes of the Portuguese were persevered in. An innocent and inoffensive people were pursued and hunted down in their own forests like beasts of prey. Thus, iniquity triumphed; but a terrible retribution followed. The foul passions which had been nurtured in the persecution of the Indians were equally malevolent when excited against each other by the common jealousies and differences of life. For a long time previous to the outbreak of 1835, assassinations had been the order of the day. Scarcely a night passed without the occurrence of more or less. No man's life was secure. Revenge rioted in blood. This was too much the case in other parts of the country at the same period, but at Pará worse than elsewhere. Then followed the dreadful scenes already described, in which the long-degraded and down-trodden Indians, headed by factious and intriguing men, gained the ascendency in turn and drove the white population into exile."

It is a singular fact that Brazil was the first country of South America, and perhaps, for an Empire so vast, the first in the world, to bind her provinces together by steam-navigation. Pará is now reaping the fruits of this wise measure. The great old Convent of S. Antonio has but few monks, and recently the greater portion of its spacious grounds has been sold to the *Amazon Navigation Company*, (a Brazilian association.) This company is now erecting on or near these grounds the large workshops, coal-depôts, wharves, &c. so essential to the proper prosecution of their various and extended steam-interests. The Custom-House was formerly a huge ecclesiastical building, and the barracks of the standing army once belonged to the order of Carmelites. A great number of new houses have been recently erected from the Custom-House to the Castello fort, and an extensive pier has been constructed where formerly there were no facilities for landing except that which the beach afforded. The streets were, a few years since, in a wretched state; but from the date of the regular steamers on the Amazon (1853) there has been a vast improvement. Nearly all are macadamized, and are well lighted by camphene. Formerly the *rede* and the most antiquated Portuguese vehicles were the only means of land-conveyance in Pará. Mr. Henderson (to whom I am indebted for recent information) says that there are now nearly fifty coaches, (of Newark and Boston manufacture,) which are at the command of citizens or visitors; and on Sunday particularly are they most busily occupied in plying between Pará and Nazaré at the modest rate of twenty-five cents each passenger. The ladies formerly made their calls and visits by being carried in a hammock: they now ride behind a pair of handsome grays. A few years only have elapsed since nearly all the water was carried in truly Oriental style, and the following beautiful description of Dr. Kidder is still most accurate so far as nature is concerned; but in regard to the water-carriers the *picturesque* is diminishing, while the *convenient* is gaining:—

"The evening and morning scenes that may be enjoyed at Pará are indescribably beautiful. At night all is still, save the occasional rustling of a balmy breeze; and the imagination must be vivid that can picture to itself more loveliness than is exhibited when the moon walks forth in her splendor. The dark luxuriant foliage, crowning hundreds of spreading trees, is burnished with a mellow lustre too exquisite for words to portray; while the waving plumes of numerous palm-trees, glancing their reflections downward upon the beholder, add to the charms of the scenery. The opening blossoms of many fruit-trees and humbler flowers load the air with a fragrance which is none the less grateful from not being mingled, as in some of the larger towns, with offensive effluvia. The blandness of the evening air is in delightful contrast to the rigors of the noonday sun, and an occasional breeze invigorates the system after either the confinement or the exposure of the day. Although in the course of the night there falls a copious dew, yet so balmy and healthful is the atmosphere that there is no dread of exposing to it the most delicate constitution. This is the climate that of all others I would seek as a relief to enfeebled health, and especially for pectoral affections.

"A morning scene is scarcely inferior in effect. I sometimes went out to enjoy it long before the mild radiance of the moon was lost in the more powerful beams of the king of day, who at his appointed time rose through a brief twilight and hastened on his effulgent course through the cloudless ether. The Brazilians are generally early risers, and it may be remarked that in their towns generally the foreign houses are those latest opened for business. Nevertheless, there are few who walk abroad for the pleasure or exercise of walking. Almost the only persons met in my morning walks at Pará were the negroes and Indians, in countless numbers, going with earthen jars upon their heads for water.

"There is no artificial fountain in the whole city. The only source of drinking-water is a spring on the eastern side of the town. Jars of this water are sometimes carried around on horseback for sale, to accommodate those who do not keep a large supply of servants. A few wells in the suburbs, together with the current of the river, furnish water for washing and similar purposes."

Though a few tottering and almost skeleton horses may still be seen staggering under the load of four water-jars, a better day has dawned upon Pará. The introduction of more than two hundred water-carts, drawn each by a single ox, is an event to be chronicled as an advance in civilization, and shows as much improvement as macadamized streets and modern carriages. The Brazilian is far more flexible than the Portuguese. A few years ago, a benevolent citizen of the United States endeavored, at his own cost, to furnish the peasantry of some of the Portuguese islands with suitable and civilized carts instead of the inconvenient clumsy vehicles which they and their fathers before them had been using for centuries. His benevolent enterprise was entirely frustrated, for they would not give up their antiquated ox-killing carts. In 1856, Portugal was the only division of Europe, excepting Turkey, that did not possess a railway. The water-carts of Pará are similar in shape to that depicted on page 175.

While the city fronts upon the river, its rear is skirted by a shaded walk whose equal would be difficult to find in Brazil. The Estrada das Mangubeiras is a highway extending from near the Marine Arsenal on the river side to the Largo da Polvora on the eastern extremity of the city. It is intersected by avenues leading from the Palace Square and the Largo do Quartel. Its name is derived from the mangabeira-trees with which it is densely shaded on either side. The bark of these shade-trees is of a light grayish color, regularly striped with green; their product is a coarse cotton that may be used for several purposes: their appearance is at once neat and majestic.

On the grounds of the old Convent—now the Hospital—of S. José, a botanical garden was laid out in 1797; but it was neglected, and finally abandoned during the troublous times of 1823 and '35.

In 1854, during the presidency of the distinguished and talented Sebastião do Rego Barros, formerly Minister of War, the site for a new botanical garden was laid out farther from the city and on a far more extensive scale. He sent to Europe and procured five or six skilful professional gardeners, who designed a handsome plan for the new works, which will doubtless soon be prosecuted to completion.

Beyond the actual precincts of the city, one may instantly bury himself in a dense forest and become shut out from every indication of the near residence of man.

The coolness of these silent shades is always inviting, but the stranger must beware lest he loses his way and thus be subjected to many annoyances and difficulties. Formerly there were many stories told of persons who became bewildered in the mazes of these thickets, and, though but a short distance off, were utterly unable to find their way back to town. Several persons are believed to have perished in this manner.

All important posts throughout the town are regularly guarded, and whoever approaches after eight o'clock at night is hailed with a harsh, indistinct call:—" *Quem vai lá?*" (Who goes there?) The proper answer is, "*Amigo,*" (A friend,)—which many contract to a swinish grunt. To this the condescending permission, "*Passa largo!*" is generally retorted by the soldier, and the person goes by.

My colleague, in giving his experience at Pará, thus writes:—

"As my lodgings were opposite the *trem*, or military arsenal, my ears became very familiar with these exclamations, which were vociferated the whole night long. Not only these, but the piercing scream, '*As armas!*' which resounded every hour when guard was relieved, and the blowing of a horn at frequent intervals,—as, for example, at Ave Maria, when all the soldiers doff their caps in honor of the Virgin,—formed no small annoyance, at least during hours allotted to repose. Another peculiar custom of Pará is the ringing of bells and the discharge of rockets at a very early hour of the morning. I sometimes heard it at four o'clock, and with much regularity at five. [In 1862, J. C. F. occupied the same room.]

"Few objects at Pará attract more attention from the stranger than the fashionable craft of the river. Vessels of all sizes—from that of a sloop down to a shallop—are called *canoas*. Few canoes proper, however, are in use. The *montaria*, seen and described at Maranham, is very common in the harbor.

"The large *canoas*, made for freighting on the river, appear constructed for any thing else rather than water-craft. Both stem and stern are square. The hull towers up out of the water like that of a Chinese junk. Over the quarter-deck is constructed a species of awning, or round-house, generally made of thatch, to protect the navigator against the sun by day and the dew by night, and, it also may be added, against the moon; for the Paraenses are very superstitious in regard to the silver beams of Luna. Sometimes a similar round-house is constructed over the bows, giving something like homogeneity to the appearance of the vessel. This arrangement renders it necessary to have a staging or spar-deck rigged up, on which to perform the labors of navigation. The steersman generally sits perched upon the roof of the after round-house. The idea continually disturbing my mind while beholding these *canoas* was, that, being so top-heavy, they were liable to overset, as they most inevitably would if exposed to a gale of wind. They are thought, however, to answer very well their purpose of floating upon the tide. Moreover, one special advantage of the round-house is that it furnishes room for the swinging of hammocks, and thus saves the canoe-men the trouble of going on shore to suspend them on the trees. Mr. Mawe says that, in descending the Amazon, he passed a man who had moored his canoe while he fastened his bed upon some branches of a tree overhanging the water and took a nap!

AMAZONIAN CANOA.

"The street running parallel to the river and connecting with the several landings is that in which the commercial business of the place is principally transacted. At certain hours of the day it presents a very lively appearance.

"Various objects and customs are observed at Pará that appear altogether peculiar to the place. In one section of the city, when animals are slaughtered for market, vast numbers of vultures are observed perched upon the trees or wheeling lazily through the air. Along the margin of the river, both morning and evening, great numbers of people may be seen bathing. No ceremonies are observed at these very necessary, and no doubt very agreeable, ablutions. Men, women, and children—belonging to the lower classes as a matter of course—may be seen at the same moment diving, plunging, and swimming in different directions.

"There is generally a crowd of canoes around Ponta da Pedra, the principal landing-place. These, together with the crowd of Indians busily hurrying to and fro, conversing in the mingled dialects of the Amazon, are peculiar to Pará. Here may be seen cargoes of Brazil-nuts, cacáo, vanilla, annatto, sarsaparilla, cinnamon, tapioca, balsam of copaiba in pots, coarse dried fish in packages, and baskets of fruits, in infinite variety, both green and dry. Here are also parrots, macaws, and some other birds of gorgeous plumage, and occasionally monkeys and serpents, together with gum-elastic shoes, which are generally brought to market suspended on long poles to prevent their coming in contact with each other. These formerly arrived in immense quantities; but now the 'India-rubber' is mostly conveyed to market in the shape of small slabs.

"The indigenous produce of the province of Pará is immense in quantity and of great value. If the people were only industrious in collecting what nature furnishes so bountifully to their hands, they could not avoid being rich. If enterprising cultivation were added to that degree of industry, there is no limit to the vegetable wealth which might be drawn from this treasure-house of nature.

"Rice, cotton, sugar, and hides are exported in small quantities, and are produced by the ordinary methods. The trade in gum-elastic, cacáo, sarsaparilla, cloves, urucú, and Brazil-nuts, is more peculiar.

"The use of the caoutchouc or gum-elastic was learned from the Omaguas,— a tribe of Brazilian Indians. These savages used it in the form of bottles and syringes: (hence the name syringe-tree.) It was their custom to present a bottle of it to every guest at the beginning of one of their feasts. The Portuguese settlers in Pará were the first who profited by turning it to other uses, converting it into shoes, boots, hats, and garments. It was found to be specially serviceable in a country so much exposed to rains and floods. But of late the improvements in its manufacture have vastly extended its uses and made it essential to the health and comfort of the whole enlightened world. The aboriginal name of this substance was *cahuchu*, the pronunciation of which is nearly preserved in the word *caoutchouc*. At Pará it is now generally called *syringa*, and sometimes *borracha*. It is the product of the *Siphilla elastica*,—a tree which grows to the height of eighty and sometimes one hundred feet. It generally runs up quite erect, forty or fifty feet, without branches. Its top is spreading, and is ornamented with a thick and glossy foliage. On the slightest incision the gum exudes, having at first the appearance of thick, yellow cream.

"The trees are generally tapped in the morning, and about a gill of the fluid is collected from one incision in the course of the day. It is caught in small cups of clay, moulded for the purpose with the hand. These are emptied, when full, into a jar. No sooner is this gum collected than it is ready for immediate use. Forms of various kinds, representing shoes, bottles, toys, &c., are in readiness, made of clay.

"When the rough shoes of Pará are manufactured, it is a matter of economy to have wooden lasts. These are first coated with clay, so as to be easily withdrawn. A handle is affixed to the last for the convenience of working. The fluid is poured over the form, and a thin coating immediately adheres to the clay. The next movement is to expose the gum to the action of smoke. The substance ignited for this purpose is the fruit of the *wassou*-palm. This fumigation serves the double purpose of drying the gum and of giving it a darker color. When one coating is sufficiently hardened, another is added and smoked in turn. Thus, any thickness can be produced. It is seldom that a shoe receives more than a dozen coats. The work, when formed, is exposed to the sun. For a day or two it remains soft enough to receive permanent impressions. During this time the shoes are figured according to the fancy of the operatives, by the use of a style or pointed stick. They retain their yellowish color for some time after the lasts are taken out and they are considered ready for market. Indeed, they are usually sold when the gum is so fresh that the pieces require to be kept apart: hence, pairs of shoes are generally tied together and suspended on long poles. They may be seen daily at Pará, suspended over the decks of the canoes that come down the river and on the shoulders of the men who deliver them to the merchants. Those who buy the shoes for exportation commonly stuff them with dried grass to preserve their extension. Various persons living in the suburbs of Pará collect the caoutchouc and manufacture it on a small scale. But it is from the surrounding forest-country, where the people are almost entirely devoted to this business, that the market is chiefly supplied. The gum may be gathered during the entire year; but it is more easily collected and more serviceable during the dry season. The months of May, June, July, and August are specially devoted to its preparation. Besides great quantities of this substance which leave Pará in other forms, there have been exported for some years past about three hundred thousand pairs of gum-elastic shoes annually. There are, however, some changes in the form of its exportation: and a few years ago a patent was taken out, by an American in Brazil, covering an invention for exporting caoutchouc in a liquid form. The Amazonian region now supplies, and probably will long continue to supply, in a great degree, the present and the rapidly-increasing demand for this material. Several other trees—most of them belonging to the tribe *Euphorbiaciæ*—produce a similar gum; but none of them is likely to enter into competition with the India-rubber tree of Pará.

MANUFACTURE OF INDIA-RUBBER SHOES.

"Another tree, not uncommon in the province, called the massárandúba, yields a white secretion, which so resembles milk that it is much prized for an aliment. It forms, when coagulated, a species of plaster, which is deemed valuable. The trees yield the fluid in great profusion. Their botanical character has never been properly investigated. It has been said that the juice of the India-rubber tree is also sometimes used as milk, and that the negroes and Indians who work in its preparation are said to be fond of drinking it; but a young lady who drank it at Pará died from the effects of the coagulation in her stomach.

"The annato or urucú is another valuable production of Pará. This is a well-known coloring-matter of an orange dye. It is a product of the tree known to botanists as the *Bixa orellana*. This tree grows ordinarily to about the size and form of the quince-tree, and exhibits clusters of red and white flowers. Its coloring-matter was extensively used by the aboriginals at the period of discovery. By means of it they formed various kinds of paint, and were fond of besmearing the whole surface of their bodies with it.

"The preparation used in commerce is the oily pulp of the seed, which is rubbed off and then left to ferment. After fermentation it is rolled into cakes weighing from two to three pounds, and in this form is exported. Cacáo—the substance from which chocolate is prepared—is a common and valuable production of Pará. It is made from the seeds of the *Theobroma cacáo*, represented on page 529.

"It would be an interesting although an almost endless task to investigate the botany of the Amazon. Laurels are yet to be won in this field of science; and it must be set down as by no means complimentary to American botanists that they have not entered it as competitors. I have often heard of Burchell as having resided some time at Pará; but I apprehend that he was, at the period of his visit, too far advanced in years to do full justice either to his own reputation or to the interminable field here spread before him."

The most thorough exploration of the Amazon has been by an Englishman,—Mr. Alfred R. Wallace, whose attention was directed to Northern Brazil by Mr. Edwards's little book, "A Voyage up the Amazon." With the enthusiasm known only to the naturalist, he entered upon this almost untrodden field in 1848, and, after devoting himself to the study of the strange and beautiful objects which abound in the remotest portions of the interior, in 1852 he gave up his wandering and romantic life among the almost unknown aborigines, and returned to England laden with Flora's richest spoils. But, alas! the burning of the ship on his homeward voyage not only caused the loss of his entire collection, but for many days his life was exposed in an open boat upon the broad Atlantic. Notwithstanding the great loss of materials,—which every naturalist and traveller can fully appreciate,—he prepared on Northern Brazil the two most interesting volumes extant. He went not to study the government and the people, but the Indians, forests, flowers, birds, and the wild beasts of Amazonia. Whoever wishes a fresh and reliable book on nature can turn to Mr. Wallace with a surety that he will find in the "Narrative of Travels on the Amazon and Rio Negro" a deeply-interesting book for general reading, and in the "Palms of the Amazon" a little volume which the naturalist will count among his best treasures.

The waters of the great river are scarcely less productive than the soil of its banks. Innumerable species of fish and amphibious animals abound in it. Several large kinds of fish are salted and dried for use. But the commerce in this article of food does not extend beyond the coast. Owing to the style of preparation, or to the coarse quality of the fish, foreigners set no value upon it. The most remarkable inhabitant of these waters is the *vaca marina,* commonly called by the Portuguese *peixe boi,* or fish-ox. This name is evidently given on account of the animal's size, rather than from any resemblance to the ox or cow other than its being mammiferous.

The *vaca marina* cannot be called amphibious, since it never leaves the water. It feeds principally upon a water-plant (*cana brava*) that floats on the borders of the stream. It often raises its head above the water to respire as well as to feed upon this vegetable. At these moments it is attacked and captured. It has only two fins, which are small and situated near its head. The udders of the female are beneath these fins. This has been pronounced the largest fish inhabiting fresh water; but, notwithstanding its

PEIXE BOI, OR VACA MARINA.

mammoth dimensions,—being, according to various accounts, from eight to seventeen feet long, and two to three feet thick at the widest part,—its eyes are extremely small, and the orifices of its ears are scarcely larger than a pin-head. Its skin is very thick and hard,—not easily penetrated by a musket-ball. The Indians used to make shields of it for their defence in war. Its fat and flesh have always been in estimation. It served the natives in place of beef. Not having salt for the purpose, they used to preserve the flesh by means of smoke.

The waters of the Amazon up to the very base of the Andes are inhabited by several species of cetacea, of which we have very scanty information. Mr. Nesbitt—who was the chief engineer on the Peruvian Government steamers built in New York and taken up the Amazon, and who spent a number of years on the King of Waters and its affluents—has kindly furnished me several items concerning the *fauna* of that region :—

"There are thousands of the regular sea-porpoise in the Amazon and its affluents, at the very foot of the Andes. Indeed, I have seen larger schools of them in the Huallaga than I ever saw in the Hudson, and of enormous dimensions. Fish of every kind is very abundant in all the rivers and lakes.

"At the Falls of the Rio Madeira the traveller will halt and gaze with wonder at the vast multitude of fish of all kinds and sizes—from the huge cow-fish to the little sardine—struggling with might and main to ascend the foaming, dashing current, without the slightest hope of success. Presently, some monster will make a dash at a school of his small congeners, when suddenly there will be a cloud of all sorts and sizes leaping in the air and trying to dodge their ravenous pursuer. All that is necessary for one wishing a fish is to take his canoe-paddle and strike right or left, when he is sure to hit: he cannot possibly miss. Here are almost always to be found great numbers of Indians collecting, salting, and drying fish. The *peixe boi* is an excellent fish for food; I would almost as soon have it for the table, in every shape, as the best veal: indeed, it might be palmed upon the unwary for that article. It is also equal to the best dried beef for chipping, in the estimation of many.

"In this connection I might mention the *Tartaruga*, or turtle of the Amazon: these are to be found by the thousand in nearly all the affluents,—especially the Madeira, Purus, Napo, Ucayali, and Huallaga. At the season for them to deposit their eggs on the '*praias*,' the streams will be fairly speckled with them, paddling their clumsy carcasses up to their native sand-bar; for it is positively asserted by the natives that the turtle will not deposit its eggs anywhere except where it was itself hatched out. They lay from eighty to one hundred and twenty eggs every other year. Of this I have been assured by persons who have artificial ponds and keep them the year round for their own table. September and October are the months for depositing their eggs."

Dr. Kidder says:—

"The *turtle-egg butter* of Amazonia (*manteiga da tartaruga*) is a substance quite peculiar to this quarter of the globe. At certain seasons of the year the turtles appear by thousands on the banks of the rivers, in order to deposit their eggs upon the sand. The noise of their shells striking against each other in the rush is said to be sometimes heard at a great distance. Their work commences at dusk and ends with the following dawn, when they retire to the water.

"During the daytime the inhabitants collect these eggs and pile them up in heaps resembling the stacks of cannon-balls seen at a navy-yard. These heaps are often twenty feet in diameter, and of a corresponding height. While yet fresh they are thrown into wooden canoes, or other large vessels, and broken with sticks and stamped fine with the feet. Water is then poured on, and the whole is exposed to the rays of the sun. The heat brings the oily matter of the eggs to the surface, from which it is skimmed off with cuyas and shells. After this it is subjected to a moderate heat until ready for use. When clarified, it has the appearance of butter that has been melted. It always retains the taste of fish-oil, but is much prized for seasoning by the Indians and those who are accustomed to its use. It is conveyed to market in earthen jars. In earlier times it was estimated that nearly two hundred and fifty millions of turtles' eggs were annually destroyed in the manufacture of this manteiga. Recently the number is less, owing to the gradual inroads made upon the turtle race, and also to the advance of civilization."

But the Government now regulates the turtle-egg harvest, so that their numbers may not be so rapidly diminished. There are some extensive beaches which yield two thousand pots of oil annually: each pot contains five gallons, and requires about twenty-five hundred eggs, which would give five million *ova* destroyed in one locality.

Indeed, it is a wonder how the turtles can ever come to maturity. As they issue from the eggs and make their way to the water, many enemies are awaiting them. Huge alligators swallow them by hundreds; the jaguars feed upon them;* eagles, buzzards, and great wood-ibises are their devourers; and, when they have escaped these land-foes, many ravenous fishes are ready to seize them in the stream. They are, however, so prolific, that it has remained for their most fatal enemy, man, to visibly diminish their number.

The Indians take the full-grown turtle in a net, or catch him with a hook, or shoot him with an arrow. The latter is a most ingenious method, and requires more skill than to shoot a bird upon the wing. The turtle never shows its back above the water, but, rising to breathe, its nostrils only are protruded above the surface: so slight, however, is the rippling that none but the Indian's keen eyes perceive it. If he shoot an arrow obliquely it would glance off the smooth shell: therefore he aims into the air, and apparently "draws a bow at a venture;" but he sends up his missile with such wonderfully accurate judgment that it describes a parabola and descends nearly vertically into the back of the turtle. (*Wallace*.) The arrow-head fits loosely to the shaft, and is attached to it by a long fine cord carefully wound around the wood, so that when the turtle dives the barb descends, the string unwinds, and the light shaft forms a float or buoy, which the Indian secures, and by the attached cord he draws the prize up into his canoe. Nearly all the turtles sold in market are taken in this manner, and the little

---

* "The jaguar, say the Indians, is the most cunning animal in the forest: he can imitate the voice of almost every bird and animal so exactly as to draw them toward him: he fishes in the rivers, lashing the water with his tail to imitate falling fruit, and, when the fish approach, hooks them up with his claws. He catches and eats turtles, and I have myself found the unbroken shells, which he has completely cleaned out with his paws: he even attacks the cow-fish in its own element, and an eye-witness assured me that he had watched one dragging out of the water this bulky animal, weighing as much as a large ox."— *Wallace*

square vertical hole made by the arrow-head may generally be seen in the shell.

In connection with this might be mentioned the archery of some of the civilized Indians in various portions of the Empire. A large and strong bow is bent by their legs. In this way they are able to shoot game at a great distance.

CABOCLO ARCHERS.

As to the birds of the Amazon, they are everywhere brilliant beyond birds in any other portion of the world. Some, like the dancing cock of the rock, and the curious and little-known umbrella-bird, are very difficult to obtain. I can only mention the latter.

This singular bird is about the size of a raven, and is of a similar color; but its feathers have a more scaly appearance, from being margined with a different shade of glossy blue. On its head it bears a crest different from that of any other bird. It is formed of feathers more than two inches long, very thickly set, and with hairy plumes curving over at the end. These can be laid back so as to be hardly visible, or can be erected and spread out on every side, forming, as has been remarked, "a hemispherical, or rather a hemi-ellipsoidal, dome, completely covering the head, and even reaching beyond the point of the beak." It inhabits the flooded islands of the Rio Negro and the Solimões, never appearing on the mainland. It feeds on fruits, and utters a loud, hoarse cry, like some deep musical instrument,—whence its Indian name, Uera-mimbé, "trumpet-bird."

And what can be said of the countless tribes of insects that swarm in the Amazonian forests? My first acquaintance with the rich living gems of Brazil was made at the retired residence of Mr. G., in the lovely Larangeiras at Rio de Janeiro, and afterward in various parts of the Empire. I did not cease to wonder at the innumerable and brilliant hosts of Lepidoptera, Coleoptera, Heli-coniidæ, &c. &c. It would require volumes to note them. In the vicinity of Pará itself there is ample opportunity for the study of nature.

THE UMBRELLA-BIRD.

Dr. Kidder visited the American rice-mills situated twelve miles distant from the city, and thus describes the excursion :—

"Our way led through a deep, unbroken forest, of a density and a magnitude of which I had, before penetrating it, but a faint conception. Notwithstanding this is one of the most public roads leading to or from the city, yet it is only for a short distance passable for carriages. Indeed, the branches of trees are not unfrequently in the way of the rider on horseback. A negro is sent through the path periodically with a sabre to clip the increasing foliage and branches before they become too formidable. Thus the road is kept open and pleasant. Notwithstanding the heat of the sun in these regions at noonday, and the danger of too much exposure to its rays, an agreeable coolness always pervades those retreats of an Amazonian forest, whose lofty and umbrageous canopy is almost impenetrable. The brilliancy of the sun's glare is mellowed by innumerable reflections upon the polished surface of the leaves. Many of the trees are remarkably straight and very tall. Some of them are decked from top to bottom with splendid flowers and parasites, while the trunks and boughs of nearly all are interlaced with innumerable runners and creeping vines.

"These plants form a singular feature of the more fertile regions of Brazil. But it is on the borders of the Amazon that they appear in their greatest strength and luxuriance. They twist around the trees, climbing up to their tops, then grow down to the ground, and, taking root, spring up again and cross from bough to bough and from tree to tree, wherever the wind carries their limber shoots, till the whole woods are hung with their garlanding. This vegetable cordage is sometimes so closely interwoven that it has the appearance of network, which neither birds nor beasts can easily pass through. Some of the stems are as thick as a man's arm. They are round or square, and sometimes triangular, and even pentangular. They grow in knots and screws, and, indeed, in every possible contortion to which they may be bent. To break them is impossible. Sometimes they kill the tree which supports them, and occasionally remain standing erect, like a twisted column, after the trunk which they have strangled has mouldered within their involutions. Monkeys delight to play their gambols upon this wild rigging; but they are now scarce in the neighborhood of Pará. Occasionally their chatter is heard at a distance, mingled with the shrill cries of birds; but generally a deep stillness prevails, adding grandeur to the native majesty of these forests.

*   *   *   *   *   *   *   *   *   *

"On our route to Maguary, I was surprised to see lands which ten or twelve years ago had been planted with sugarcane now entirely overgrown with trees of no insignificant dimensions. Only a few acres immediately around the engenho had been kept free from these encroachments. Here was located the first mill for cleaning rice ever built in the vicinity of Pará. It was established by North American enterprise. A small water-power existed on the site; but, after the mills were constructed, it was found that this power was insufficient in the dry season: consequently, a steam-engine of sixteen horse-power was imported from the United States, and has been made to do good service. The steam-power was kept in action constantly, and, at proper seasons, the water-power also. Both were inadequate to the amount of business that offered. Several American mechanics were employed at this establishment, which, small as it is, compares favorably with any mechanical establishment in the whole country. A stream connects this engenho with the great river, and thus furnishes cheap conveyance for cargoes to and from the city."

My colleague also had some experience at Pará not quite so agreeable as riding through Amazonian forests :—

"Soon after my arrival, in company of the United States Consul, I waited on Senhor Franco, the president of the province, to whom I bore a letter of commendation. This individual had formerly been clerk in one of the English mercantile houses in Pará, and was subsequently educated as a beneficiary of the province, of which he had now become the chief magistrate. He received us with civility, and in person conducted us through the palace. I found that building one of the best of the kind in the Empire. It was built, together with the cathedral and some of the churches, in the days of that talented but ambitious prime minister of Portugal, the Marquis of Pombal, who cherished the splendid idea of having the throne of Portugal and all her dominions transferred from the banks of the Tagus to those of the Amazon. This circumstance accounts for the ample size and magnificent structure of these buildings in a town of moderate extent.

"At a proper time I waited on the juiz de direito,—the chief officer of the police, —to exhibit my passport and obtain a license of residence in the very loyal and heroic city of Pará and the province of which it was the capital. No embarrassments were put in my way, and no detention occurred. I obtained the requisite license, and kept it until I had occasion to obtain a new passport on my departure. Nevertheless, it appeared at one period that my unmolested residence in the city was very much in jeopardy.

"The old Bishop of Pará seemed to have caught the contagion of alarm from his colleague in Maranham; and both these prelates—yielding more than their sober judgment should have allowed them to certain unfounded and malicious representations sent them from some quarter—wrote to Senhor Franco concerning me as a very dangerous person, who ought not to be suffered to land in the province. The president probably satisfied himself on that point during my visit to him; and although he owed his political elevation very much to his ecclesiastical patrons, yet he managed to satisfy their apprehensions by a very short and formal correspondence with the American Consul. No person interfered with me or any of my pursuits from first to last."

The see of Pará is certainly still very much endangered by the Bible, if we may judge from the "pastoral" issued in the *Diario do Commercio* (of the 8th of April, 1857) by Dom José Affonso de Moraes Torres, "by the grace of God and of the Holy Apostolical See, Bishop of Grão Pará." The good bishop seems to be terribly exercised by what he terms *uma Sociedade Biblica ultimamente creada com o noma de Alliança Christa*, (a Bible Society lately created under the name of the Christian Alliance.) He says that its emissaries circulate books, one of which—a catechism—he has read, and that in it he "encounters a doctrine entirely opposed to the belief of the Church of Jesus Christ." That which particularly stirs up his ire is that the little book teaches that the worship of images is idolatry. He then insists that such worship is altogether right, only that the internal operation of the mind is not exactly the same as when worshipping God. He not only hurls his invectives at the little book and at heretics, but proves from Scripture that we can be doing God's service in adoring his creatures. He adduces, with decided emphasis, that Abraham worshipped the angels and *adored* the sons of Heth (!) [*adorou os filhos de Heth*, Gen. xxiii. 7.]

The true head of offence in the little book is that it contains unmutilated the Ten Commandments. I have in my possession the Ten Commandments as they are printed in all the books of religious instruction in Portugal and in some parts of Brazil, and the second commandment is entirely omitted; and, in order to make up the

36

Decalogue, the tenth commandment is thus divided. "Thou shalt not covet thy neighbor's house" figures as the ninth, and "Thou shalt not covet thy neighbor's wife," &c. &c., "nor any thing that is thy neighbor's," is the tenth.

The state of religion at Pará is by no means flattering, and the heart is as far from being reached by empty forms and gorgeous pageants on the Amazon as it is on the Tiber or the Danube. The grand annual festival of Nazaré always attracts from the city an immense crowd, who go not for religious edification, but for the nine days' feasting, dancing, fireworks, and gaming.

General reflections upon the character and tendency of such a scene of festivities—so absorbing to a whole community and so long continued—seem unnecessary. If it had no religious pretensions it would be less exceptionable; but for a people to be made to think themselves doing God's service while mingling in such amusements and follies is painfully lamentable.

Note for 1866.—The city of Pará has returned to its former size, the population now being as great (if not greater) as it was before the disastrous days of 1835–38. In 1862 the junior author could see but few traces of the rebellion in the condition of the buildings, and, though the elderly people had ineffaceable recollections of the revolt and scenes of bloodshed, the great majority of the population have grown up without sad *souvenirs.* Many improvements have taken place. Some of the most important in a material point of view are those which have been brought about by Sr. Pimento Bueno, the *gerente* of the Amazonian Navigation Company. The houses of James Bishop & Co. (J. C. Bond), H. K. Corning & Co. (Mr. Moran), and Burdett & Everett (Mr. Pond), are energetic representatives of American interests at Pará. President Brusque, who was President of the province of Pará in 1861, '62, and '63, took the deepest interest in publishing the material resources of the province of Pará, and his *Relatorios* of 1862–63 are full of the most valuable information. The latest and most reliable English book on the Amazon is the "Naturalist on the Amazon," by Henry Bates, Esq., London, 1863. This is a most charming and valuable work. Mr. B. passed nearly ten years in that equatorial region, and has given the world many important facts concerning the great valley, aside from information in regard to its natural history. Only one drawback to many is to be found, in his "Darwinian" views; but they are "put" so modestly, and his investigations are so much better than his theory, that one becomes only interested in the great theme of his book, "the King of Waters."

CHAPTER XXVII.

AMAZONAS—ITS DISCOVERY—EL DORADO—GONÇALO PIZARRO—HIS EXPEDITION—
CRUELTIES — SUFFERINGS — DESERTION OF ORELLANA — HIS DESCENT OF THE
RIVER—FABLE OF THE AMAZONS — FATE OF THE ADVENTURER — NAME OF THE
RIVER—SETTLEMENT OF THE COUNTRY—SUCCESSIVE EXPEDITIONS UP AND DOWN
THE AMAZON — SUFFERINGS OF MADAME GODIN — PRESENT STATE — VICTORIA
REGIA — STEAM-NAVIGATION — EFFECTS OF HERNDON AND GIBBON'S REPORT —
PERUVIAN STEAMERS—THE FUTURE PROSPECTS OF THE AMAZON.

AMAZONAS (or Alto Amazonas) is the most northern province of
Brazil. My colleague thus writes in regard to the history of this
vast and almost-unknown division of the Empire:—

"No portion of the earth involves a greater degree of physical interest. Its
central position upon the equator, its vast extent, its unlimited resources, its mammoth rivers, and the romance that still lingers in its name and history, are all
peculiar. Three hundred years have elapsed since this region was discovered; but
down to the present day two-thirds of it remains uncivilized and almost unexplored.

"Indeed, few persons, save the Indians, and the slave-hunters who once pursued
them, have even penetrated its remote sections, or seen any parts of it save the
banks of navigable rivers. The circumstances of its discovery will ever be considered remarkable. It was about the middle of the sixteenth century when the
fable of El Dorado filled the public mind of Europe. The existence of a New
World was then fully demonstrated, and the leaven of desire for its undeveloped
treasures had spread from court to camp, from princes to beggars, until the whole
mass of society was in a ferment. Avarice, personified under the garb of adventure, bestrode the ocean. Scarcely did her footsteps touch the shores of the New
World, ere they were bathed in blood. She commenced her work of desolation
in the fair islands of the Caribbean. She caused the din of arms to resound in
the primeval forests and aboriginal cities of the continent. She scaled the
Cordilleras, and laid waste savannahs upon both the Atlantic and the Pacific
shores.

"Among the bloodthirsty and cruel men who stood forth as leaders in the work
of conquest and plunder, Gonçalo Pizarro, the brother and associate of the con-
queror of Peru, was second to few, if any. His talents may have been less, but
his daring and cruelty were greater. In 1541, this adventurer set out from Quito,
with an army of three hundred soldiers, and four thousand Indians to serve them
as bearers of burdens, with the design of discovering the land of gold. This was

563

an imaginary kingdom, shaped out of the half-comprehended tales of the persecuted Indians and exaggerated by the most extravagant fancies.

"This fabulous kingdom received a name from the fashion of its monarch, who was said, in order to wear a more magnificent attire than any other potentate in the world, to put on a daily coating of gold-dust. His body was anointed every morning with a costly and fragrant gum, to which the gold-dust adhered when blown over him by a tube. In this barbaric attire the Spaniards denominated him EL DORADO,—the GILDED KING. No fictions concerning this monarch or his kingdom were too extravagant for credence. He was generally located in the grand city of Manoa, in which no fewer than three thousand workmen were employed in the silversmiths' street. The columns of his palace were described as of porphyry and alabaster: the throne was ivory, and the steps leading to it were of gold. Others built the palace of white stone, and ornamented it with golden suns and moons of silver, while living lions, fastened by chains of gold, guarded its entrance. With day-dreams like these dancing before the minds of commanders and soldiers, the army of Pizarro set out, cherishing the highest anticipations.

" In proceeding eastward from Quito, they were obliged to cut their way through forests, to climb mountains, and to contend with hostile tribes of Indians. Every tribe with which they met was interrogated about El Dorado, and when unable to give any intelligence of it they were put to torture: some were even burned alive, and others were torn to pieces by bloodhounds, which the Spaniards had trained to feed on human flesh.

" The effects of this dreadful cruelty returned upon the heads of its perpetrators with a terrible vengeance. As the tidings of their approach spread from tribe to tribe, the poor natives learned to flatter their hopes and send them along. The rains came on, and, lasting for months, rotted the garments from the bodies of the soldiers, who could neither make nor find a shelter. At length their provisions were exhausted, and they began to feed upon their dogs. The sick multiplied, so that they were obliged to build a brigantine in which to carry them. This was a herculean task for soldiers to perform, especially without the requisite implements. Before it was accomplished they had to slaughter their horses for food. Their troubles continued and even increased: still, with death staring them in the face, Pizarro continued to seize prisoners, and put them in irons when he supposed they desired to escape. When they at length stood upon the banks of the river Napo, not less than one thousand of the Peruvians had perished.

" The commander now heard of a larger river into which this emptied, and was told that the country surrounding the junction was fertile and abounding in provisions. He therefore determined to despatch the vessel with fifty men to procure supplies for the rest. Francisco de Orellana, a knight of Truxillo, was put in command of this expedition. The stream carried them rapidly downward through an uninhabited and desert country. When they had descended about three hundred miles, the question was started whether they should not abandon the idea of returning. They had not found food sufficient for themselves; and how could they succor the army? Besides, how could they ascend against the current in their enfeebled state? It would only be to perish with the rest. They might as well continue their descent, for 'rivers to the ocean run,' and there was some chance that they might in this way not only save their lives but also immortalize their names by new discoveries. Orellana urged these considerations with so much plausibility, that all consented save two,—a Dominican friar and a young knight of Badajoz, who contended against the plan as treacherous and cruel. Orellana disposed of this objection by setting the knight on shore, to perish or return to the army as he best could. The friar became an easy convert to the new scheme, and thenceforward took a prominent part in it. Orellana renounced the commission he had received from Pizarro, and received an election from his men as their commander, so that he might make discoveries in his own name, and not under delegated authority in the name of another.

"It was on the last day of December, 1541, that this adventurous voyage was commenced, after mass had been said by the Dominican. Their prospects were gloomy enough. Their stock of provisions was wholly exhausted, and they were forced to boil the soles of their shoes and their leathern girdles, in hope of deriving nourishment from them.

"It also became necessary to build a better vessel. This being accomplished with great difficulty and delay, they resumed their voyage. Sometimes they met with a kind reception from the Indians, but more generally they had to fight their way with great losses and imminent danger of complete destruction.

"It was in the month of June that, during a battle with a hostile tribe, they discovered what they reported to be Amazons. Friar Gaspar, the Dominican, affirms that ten or twelve of these women fought at the head of the tribe which was subject to their authority. He described them as very tall and large-limbed, having a white complexion, and long hair plaited and banded around their head. Their only article of dress was a cincture, but they were armed with bows and arrows. The men fought desperately, because, if they deserted, they would be beaten to death by these female tyrants; but, when the Spaniards had slain some seven or eight of the latter, the Indians fled. These stories were generally believed to have been deliberate falsehoods fabricated with the idea of giving consequence to the voyage. The existence, however, of a powerful tribe of Amazons in that portion of South America was a subject of deliberate inquiry and grave discussion for at least two centuries. Condamine and others favored the opinion that there had been such a people, of which some remnants remained till about the time of Orellana, soon after which they became extinct by amalgamation with surrounding tribes. The Spanish historian Herrera has given detailed accounts of the adventures of Orellana, compiled from his own statements, endorsed by his veracious chronicler, Friar Gaspar. They contain, however, but little authentic information. But, strange as it may seem, modern investigation (as will be seen hereafter) has proved that the veracious frade apparently spoke the truth.

"In the course of seven months they reached the ocean. After some repairs made upon their vessels, they sailed out of the great river during the month of August, and on the 11th of September they made the island of Cubagua. Orellana proceeded thence to Spain, to give an account of his discoveries in person.

"The excuse he presented for deserting Pizarro was accepted, and, on solicitation, he received a grant of the conquest of the regions he had discovered. He had but little difficulty in raising funds or enlisting adventurers for his expedition. It, however, proved disastrous. His fleet arrived out in 1544, but, amid the labyrinth of channels at the mouth of the river, it was impossible to find the main branch. After a month or two spent in beating about, without being able to ascend the river or to accomplish any important object, Orellana succumbed to his misfortunes, and, like many of his men, sickened and died. He was the first to descend the embouchment of the Amazon; but Pinzon is said to have discovered the mighty current in 1500.

"Mr. Southey had so much respect for his memory, that he made an effort in his history to restore the name of Orellana to the great river. He discarded Maranon,

as having too much resemblance to Maranham,* and Amazon, as being founded
upon fiction and at the same time inconvenient. Accordingly, in his map, and in
all his references to the great river, he denominates it Orellana.

"This decision of the poet-laureate of Great Britain has not proved authoritative
in Brazil. *O Amazonas* is the universal appellation of the great river among those
who float upon its waters and who live upon its banks, and is now given to the new
province whose capital is the Barra do Rio Negro.

"Pará, the aboriginal name of this river, was more appropriate than any other.
It signifies 'the father of waters.' The term 'Pará River' designates the southern,
in opposition to the northern, principal mouth of the Amazon, and also the province
through which the mighty river finds the ocean."

The name *Amazonas* has been stated by some to be derived from
the Indian word *Amassona*,—a term, it is pretended, applied to the
wonderful phenomenon of a high tide of these rivers two days
before and two days after full-moon, which extends to the very
confluence of the Madeira. As this tide is very destructive to
small craft, the natives called it *Amassona*, ("boat-breaker.") This
story, it seems to me, has no foundation whatever. I do not believe
Amassona to be an aboriginal term; for the Portuguese substantive
amás means "a heap," and the simple verb *amassar* means "to
knead," "to bruise," &c.; while the reflex verb *amassar-se* means
"to heap up itself."

The origin of the name and the mystery concerning the female
warriors, I think, has been solved, within the last few years, by the
intrepid Mr. Wallace, who left the beaten track,—the bed of the
great river,—and in the remotest haunts of the wild man, by his
persevering patience and his knowledge of the *Lingoa Geral*, has
given much information to the world concerning the little-known
interior.

I believe it will now be found that, although the early monkish
chroniclers of the New World often used their imaginations instead
of being content with facts, they were in this case not so culpable
as many have supposed. They really believed that they had fought
with female warriors, and certainly *appearances* were in favor of
their truthfulness. Mr. Wallace, I think, conclusively shows that
Friar Gaspar and his companions saw Indian male warriors who
were attired in habiliments such as Europeans would attribute

---

* Both words have evidently a common origin, being derived from the Portuguese
mare, "the sea," and *nao*, "not,"—*not the sea*, as a great river near its mouth
appears to be.

to woman.   Mr. Wallace visited numerous tribes on the upper
affluents of the Amazon, and, in speaking of their language, habits
of dress, and other characteristics, he says,—

"The use of ornaments and trinkets of various kinds is almost confined to the
men.   The women wear a bracelet on the wrists, but no necklace, or any comb in
the hair: they have a garter below the knee, worn tight from infancy, for the purpose of swelling out the calf, which they consider a great beauty.   While dancing
in their festivals, the women wear a small *tanga*, or apron, made of beads prettily
arranged: it is never worn at any other time, and immediately the dance is over
it is taken off.

"The men, on the other hand, have the hair carefully parted and combed on
each side and tied in a queue behind.   In the young men it hangs in long locks
down their necks, and, with the comb, which is invariably carried stuck on the top
of the head, gives to them a most feminine appearance: this is increased by the
large necklaces and bracelets of beads and the careful extirpation of every symptom
of beard.   Taking these circumstances into consideration, I am strongly of opinion
that the story of the Amazons has arisen from these feminine-looking warriors encountered by the early voyagers.   I am inclined to this belief from the effect they
first produced on myself, when it was only by close examination that I found they
were men; and, were the front parts of their bodies and their breasts covered with
shields such as they always use, I am convinced any person seeing them for the
first time would conclude they were women.   We have only, therefore, to suppose
that tribes having similar customs to those now existing on the river Uaupes inhabited the regions where the Amazons were reported to have been seen, and we
have a rational explanation of what has so much puzzled all geographers.   The
only objection to this explanation is, that traditions are said to exist among the
natives, of 'a nation of women without husbands.'   Of this tradition I was myself
unable to obtain any trace, and I can easily imagine it entirely to have arisen from
the suggestions and inquiries of Europeans themselves.   When the story of the
Amazons was first made known, it became, of course, a point with all future travellers to verify it, or, if possible, to get a glimpse of these warlike ladies.   The
Indians must no doubt have been overwhelmed with questions and suggestions
about them, and they, thinking that the white men must know best, would transmit
to their descendants and families the idea that such a nation did exist in some distant part of the country.   Succeeding travellers, finding traces of this idea among
the Indians, would take it as a *proof* of the existence of the Amazons, instead of
being merely the *effect* of a mistake at first, which had been unknowingly spread
by preceding travellers seeking to obtain some information on the subject.

"In my communications and inquiries among the Indians on various matters, I
have always found the greatest caution necessary to prevent one's arriving at wrong
conclusions.   They are always apt to affirm that which they see you wish to believe, and, when they do not at all comprehend your question, will unhesitatingly
answer, 'Yes.'"

Having thus explained the origin of the word *Amazonas*, we will
again turn to the historic sketch of Dr. Kidder:—

"About seventy years after the events (the voyage of Orellana) above narrated,
the Portuguese began to settle in Pará, advancing from Maranham.   In 1616, Francisco Cadeira, the first chief-captain, laid the foundations of the present city of Pará.

under the protection of Nossa Senhora de Belem. In 1637, another party descended the Amazon from Quito. It was composed of two Franciscan friars and six soldiers, who had been sent on a mission to the Indians upon the frontiers of Peru. The mission proved unsuccessful. Some of the missionaries grew weary and returned ; others persisted until the savages attacked and murdered the commander of their escort of soldiers, when all dispersed. Those who were disheartened at the prospect of the dreadful journey back to Quito committed themselves to the waters, as Orellana had done nearly a century before. They reached Belem in safety, but so stupefied with fear as to be unable to give any satisfactory account of what they had seen. It was enough for them to have escaped from the horrid cannibals through whose midst they had passed.

"In the same year, the first expedition for the ascent of the Amazon was organized. It was commanded by Pedro Teixeira, and was composed of seventy soldiers, twelve hundred native rowers and bowmen, besides females and slaves, who increased the number to about two thousand. They embarked in forty-five canoes. The strength of the opposing current and the difficulty of finding their course amid the labyrinthian channels of the river rendered their enterprise one of unparalleled toil. Many of the Indians deserted, and nothing but unwearied perseverance and great tact enabled Teixeira to keep the rest. After a voyage of eight months, he reached the extent of navigation. Leaving most of his men with his canoes at this place, he continued his journey overland to Quito, where he was received with distinguished honors. He was accompanied on his return by several friars, whose business it was to record the incidents and observations of the voyage. A considerable amount of authentic information was thus collected and published to the world. The party reached Belem in December, 1639, amid great rejoicings. After this, voyages upon the Amazon became more common.

"In 1745, M. La Condamine, a French academician, descended from Quito, and constructed a map of the river, based upon a series of astronomical observations. His memoir, read before the Royal Academy on his return, remains to this day a very interesting work. In modern times, the most celebrated voyages down the Amazon have been described at length by those who accomplished them,—e.g. Spix and Von Martius, Lister Mawe, Lieutenants Smyth, Herndon and Gibbon, and Mr. Wallace.

"The expeditions to which I have alluded have generally been prosperous, and not attended with any peculiar misfortunes. Not so with every voyage that has been undertaken upon these interminable waters. The sufferings of Madame Godin des Odonnais have hardly a parallel on record. The husband of this lady was an astronomer associated with M. Condamine. He had taken his family with him to reside in Quito, but, being ordered to Cayenne, was obliged to leave them behind. Circumstances transpired to prevent his returning for a period of sixteen years, and when finally he made the attempt to ascend the Amazon he was taken sick and could not proceed. All the messages that he attempted to send his absent wife failed of their destination. In the mean time a rumor reached her that an expedition had been despatched to meet her at some of the missions on the Upper Amazon. She immediately resolved to set out on the perilous journey. She was accompanied by her family, including three females, two children, and two or three men, one of whom was her brother. They surmounted the Andes and passed down the tributary streams of the Amazon without serious difficulties ; but the farther they entered into the measureless solitudes that lay before them, the more their troubles increased. The missions were found in a state of desolation under the ravages of the smallpox. The village where they expected to find Indians to conduct them down the river had but two inhabitants surviving: these poor creatures could not aid them, and they were left without guides or canoe-men. Ignorant of navigation, and unaccustomed to either toil or danger, their misery was now beyond description. Their canoe, in drifting on the current, filled with water, and they barely escaped with life and a few provisions. They managed to construct a raft; but this was soon torn to pieces upon a snag. The forlorn company again escape to the shore, and, as their only alternative, attempt to make their way on foot. Without map or compass, they know not whither they go. In attempting to follow the windings of the stream they become bewildered, and finally plunge into the depths of the forest. Wild fruits and succulent plants now furnish them their only food. Weakened by hunger, they soon fall victims to disease.

"In a few days Madame Godin, the sole survivor, stood surrounded by eight dead bodies! Imagine the horror that overwhelmed her as she saw one after another of her friends and family in the agonies of death! In the desperation of the hour she attempted to bury them, but found it impossible. After two days spent in mourning over the dead, she roused up with a determination to make another effort to seek her long-lost husband. She was now nearly three thousand miles from the ocean, without food, and with her delicate feet lacerated by thorns. Taking the shoes of one of the dead men, she started upon her dreary way. What phantoms now torture her imagination and people the wilderness with frightful monsters! But she wanders on. Days of wretchedness and nights of horror ensue. At length, on the ninth day, she heard the noise of a canoe, and, running to the river-side, she was taken up by a party of Indians. Suffice it to say that they conducted her to one of the missions, from which, after long delays and great exposure, she was finally conveyed down the Amazon and restored to her husband after nineteen years' separation. They returned to France together and spent the remnant of their days in retirement; but Madame G. never fully recovered from the effects of her fright and sufferings.

"Even at this day, the traveller upon the waters of the Amazon, above Pará, finds himself in a wild and uncultivated region. He will scarcely see fifty houses in three hundred miles. There are but few settlements directly on the river. Most of the villages are on the tributary streams and the *Iyuaripés*, or bayous. The houses universally have mud floors and thatched roofs; and, though the population is increasing, I fear that for a long time to come the great majority of the inhabitants in the immediate vicinity of the Lower Amazon will be such as are depicted in the engraving.

"Notwithstanding all the beautiful theories respecting steam-navigation on the waters of the Amazon and its tributaries, nothing was accomplished deserving the name until 1853. As far back as the year 1827, an association, called the South American Steamboat Company, was organized in New York, with the express design of promoting that navigation. It owed its origin to the suggestion of the Brazilian Government through its *chargé d'affaires*, Mr. Rebello, resident in the United States, who stipulated decided encouragements, and the grant of special privileges on the part of His Majesty Dom Pedro I. A steamboat was fitted out and sent to Pará, and other heavy expenses were incurred by the company; but, through a lack of co-operation on the part of Brazil, the whole enterprise proved a failure. Claims for indemnification to a large amount were for a long time pending before the Brazilian Government.

"After 1838, small Government steamers were from time to time sent up the Amazon as far as the River Negro. Such voyages were repeated at intervals, and sufficed for steam-navigation on the Amazon until 1853. The globe does not else-where present such a splendid theatre for steam-enterprise. Not only is the Amazon navigable for more than three thousand miles, but the Tocantins, the Chingú, the Tapajos, the Madeira, the Negro, and other affluents, are unitedly navigable several thousand more. All these rivers flow through the richest soil and the most luxu-rious vegetation in the world."

Near their margin is found the giant of Flora's kingdom, whose discovery a few years since is as notable a fact to the naturalist world as the regular opening of steam-navigation upon the Amazon is to the commercial world.

Of all the Nymphæaceæ, the largest, the richest, and the most beautiful is the marvellous plant which has been dedicated to the Queen of England, and which bears the name of *Victoria Regia.* It inhabits the tranquil waters of the shallow lakes formed by the widening of the Amazon and its affluents. Its leaves measure from fifteen to eighteen feet in circumference. Their upper part is of a dark, glossy green; the under portion is of a crimson red, fur-nished with large, salient veins, which are cellular and full of air, and have the stem covered with elastic prickles. The flowers lift themselves about six inches above the water, and when full blown have a circumference of from three to four feet. The petals unfold toward evening: their color, at first of the purest white, passes, in twenty-four hours, through successive hues from a tender rose-tinge to a bright red. During the first day of their bloom they exhale a delightful fragrance, and at the end of the third day the flower fades away and replunges beneath the waters, there to ripen its seeds. When matured, these fruit-seeds, rich in fecula, are gathered by the natives, who roast them, and relish them thus prepared.

The description of this magnificent plant explains the admiration experienced by naturalists when beholding it for the first time. The celebrated Haenke was travelling in a *pirogue* on the Rio Mamoré, in company with Father Lacueva, a Spanish missionary, when he discovered, in the still waters close to the shore, this gigantic

Nymphæaceæ. At the sight the botanist fell upon his knees, and—
as a not very pious French writer very Frenchily records—expressed
his religious and scientific enthusiasm by impassioned exclamations and outbursts of adoration to the Creator,—"an improvised
Te Deum which must have deeply impressed the old missionary."

THE VICTORIA REGIA AND THE BOAT-BILL.

In 1845, an English traveller, Mr. Bridges, as he was following
the wooded banks of the Yacouma, one of the tributaries of the
Mamoré, came to a lake hidden in the forest, and found upon it a
colony of *Victoria Regias*. Carried away by his admiration, he was
about to plunge into the water for the purpose of gathering some
of the flowers, when the Indians who accompanied him pointed to
the savage alligators lazily reposing upon the surface. This information made him cautious; but, without abating his ardor, he
ran to the city of Santa Anna, and soon obtained a canoe, which
was launched upon the lake which contained the objects of his
ambition. The leaves were so enormous that he could place but

two of them on the canoe, and he was obliged to make severa.
trips to complete his harvest.

Mr. Bridges soon arrived in England with the seeds, which he
had sown in moist clay. Two of these germinated in the *aquarium*
of the hothouse at Kew.   One was sent to the large hothouses of
Chatsworth: a basin was prepared to receive it, the temperature
was raised, and the plant was placed in its new resting-place on
the 10th of August, 1849.   Toward the end of September it was
necessary to enlarge the basin and to double its size, in order to
give space to the leaves, which developed with great rapidity.
So large did they become that one of them supported the weight
of a little girl in an upright position.

The first bud opened on the beginning of November. The flower
in bloom was offered by Mr. Paxton (the celebrated designer of the
London Crystal Palace) to his monarch, and the great personages
of England hastened to Windsor Castle to admire the beautiful
homonym of their gracious sovereign.

The name given to this marvellous plant by Lindley was happily
chosen; but the natives of the Amazon call it "Uapé Japona,"—the
Jacana's oven,—from the fact that the jacana is often seen upon it.

The jacana is a singular
spur-winged bird, twice
the size of a woodcock,
provided with exceedingly
long and slender toes (from
which the French term it
the surgeon-bird) which
enable it to glide over
various water-plants.   It
inhabits the marshes, and
woods near the water, and
many a time in the in-
terior I have seen it stealing over the lily-leaves on
the margin of rivers.

THE JACANA.

Returning from this di-
gression to the capabilities of the great river for steam-navigation,
we remark that the extent of the Amazon and its affluents is

immense. From four degrees north latitude to twenty degrees south, every stream that flows down the eastern slope of the Andes is a tributary of the Amazon. This is as though all the rivers from St. Petersburg to Madrid united their waters in one mighty flood.

Geographers have never fully agreed which of the upper tributaries deserves to be called the main stream of the Amazon; but the most recent explorers are decided in considering the Tanguragua or Upper Marañon as its principal source. This rises in a lake—Lauricocha—situated almost in the region of perpetual snow. Nearly all the branches of the Amazon are navigable to a great distance from their junction with the main trunk, and, collecting the whole, afford an extent of water-communication unparalleled in any other part of the globe. There is a total of ten thousand miles of steam-navigation below all falls; and, these obstructions once passed, steamers could be run for four thousand miles. The most navigable of all the branches is the Purús.

A volume of fresh water, constantly replenished by copious rains, pours forth with such impetus as to force itself—an unmixed current—into the ocean to the distance of eighty leagues. While the principal branch of the Ganges discharges 80,000 cubic feet of water per second, and the large Brahmapootra 176,200 cubic feet every sixtieth part of a minute, the Amazon sends through the narrows at Obidos 550,000 cubic feet per second. (*Von Martius.*)

This "king of waters" is remarkable for its wide-spreading tributaries. On the north side, the first from the west, below the rapids of Manseriche, is the Morona, and then come in succession the Pastaça, Tigre, Napo, Iça, Japurá, Rio Negro, and many streams of lesser note. From the south it receives—proceeding from west to east—the Huallaga, Ucayali, Yavari or Javary, Huta, Hyuruay, Teffé, Coary, Purús, Madeira, Tapajos, Chingú, and Tocantins. Most of these affluents discharge their waters into the Amazon by more than one mouth, which frequently are widely apart. Thus, the two most distant of the four mouths of the Japurá are more than two hundred miles asunder, and the outer embouchures of the Purús are about one hundred miles from each other. In the upper portion of its course the Amazon divides Equador from Peru, between which its width varies from half a mile to a mile; beyond the limits of Equador it increases to two miles; and below the Madeira—its most considerable tributary, having a course little less than two thousand miles in length—it is nearly three miles. Between Faro and Obidos—to which place the tide reaches—it decreases to less than a mile; but below Obidos it widens again, and, after the junction of the Tapajos, it is nearly seven miles across. The width of the channel of Braganza do Norte—the northern mouth of this vast river—is thirty miles opposite the island of Marajó and fifty at its embouchure; that of the Tangipurá Channel is eighteen miles at the junction of the Tocantins and thirty at its mouth.

While the whole area drained by the Mississippi and its branches is 1,200,000 square miles, the area of the Amazon and its tributaries (not including that of the Tocantins, which is larger than the Ohio Valley) is 2,330,000 square miles. This is more than a third of all South America, and equal to two-thirds of all Europe. Mr. Wallace has startled Englishmen with the fact that "all Western Europe could be placed in it without touching its boundaries, and it would even contain the whole of our Indian Empire."

In 1851–52, Lieutenants (U. S. N.) Herndon and Gibbon descended the Amazon,—one by its Peruvian and the other by its Bolivian tributaries. Their interesting reports were published by the order of Congress, and are laid before the world. Lieutenant Gibbon passed over the most unknown route, and hence his work possesses more intrinsic interest. Lieutenant Herndon's volume not only for the moment awakened the United States and England to the importance of the Amazon, but the fact of his descent of that river and his inferences—many of them totally visionary—aroused the Brazilian Government to the performance of their duty, and in 1852–53, Brazil, by treaty with Peru, engaged to run steamers, under the Brazilian flag, from Pará,—the contractors to have the monopoly of steamboat-navigation on the Amazon for thirty years, with an annual bonus of one hundred thousand dollars for the first fifteen; the voyage to be performed by two steamers,—one ascending the Amazon from Pará, the other descending it from Nauta, and meeting the up-boat at Barra.

Nauta is in Peru, on the right bank of the Amazon, forty-six leagues below the junction of the Huallaga, and has a population

.of one thousand. This company, under the leading of that enterprising Brazilian, the Baron of Mauá, immediately sent its
first steamer from Pará to Nauta. The association, in return for
privileges granted, contracted to found numerous colonies in the
provinces of Pará and Amazonas. Nearly every month colonists
under the direction of the Amazon Navigation Company arrive
from Portugal and her islands at Pará. The colonies at Obidos
and at Serpa, and another at the mouth of the Rio Negro, did not
prove successful. Although the company engaged to plant colonies above the Barra of the Rio Negro, one on the Rio Teffé, (above
V. de Ega,) three on the Madeira, at Crato and Borba, two on the
Tapajos, not far from Santarem, and three on the Tocantins, it is
doubtful if the contract be carried out.

The contract made by the company with the Portuguese emigrants was this :—

"They bind themselves to work for the company for two years at a certain compensation per diem, and to be housed and fed during that period ; and, at the end
of their apprenticeship, each person is entitled to a certain portion of open land in
fee-simple,—the heads of families to have a comfortable house on their portion, no
matter whether they were married before engaging or during their service."

I asked Mr. Nesbitt—a practical engineer who was for three
years travelling on the Amazon and some of its navigable tributaries—his opinion of the steamers employed by the company.
His reply (April, 1857) was as follows :—

"Thus far they have succeeded well. The company have fully complied with
their part of the contract both in Brazil and with Peru. There were seven steamers
in successful operation in April, 1856, and two new boats expected every week:
one of these two was the 'Bay City,' built in New York for the Sacramento and San
Francisco trade, but was so badly twisted in trying to double Cape Horn that she
put back to Rio de Janeiro for repairs, and was sold for the benefit of the under-
writers and purchased for the Amazon Company. The names of the seven steamers
that were running are the 'Tapajoz,' 'Rio Negro,' 'Marajo,' 'Monarcha,' 'Cametá,'
'Tabatinga,' and 'Solimões.' The 'Rio Negro' and 'Tapajoz' were the packets from
Pará to the Barra do Rio Negro,—making semi-monthly trips : but, after the 1st
of January, 1857, there was to be a weekly packet. The 'Marajo' ran between the
Barra and Nauta, in Peru,—making a trip every two months, and, after January,
1857, the trips were to be monthly. The 'Monarcha' was running on the Rio
Negro, rom the Barra to the mouth of the Rio Branco, and intended to go as far
as Barcellos and Moreira—still higher—whenever the water in the Rio Negro would
permit, which would be about eight months in the year. The Rio Negro, a few
leagues above the Barra, spreads out into a very wide bay of some leagues in
breadth, which renders steam-navigation more difficult than anywhere else on the
lower river, as it becomes shallower on account of the great width ; but above this

bay there is no trouble. There are several lakes adjacent to the Rio Negro, where large quantities of fish are caught, salted, and dried for market. There are a great many splendid localities for farming-purposes on the Rio Negro above the Barra. The 'Solimões' was intended for the Rio Tapajoz. The 'Cametá' was a regular packet on the Tocantins, between the city of Pará and the town of Cametá, —making monthly trips.

"All these steamers had as much business as they could well do,—those for the Barra more than they could do; hence the necessity for weekly trips.

"These steamers were fast superseding the square, stem-and-stern, crawling river-canoas; for as soon as a trader makes one trip in a steamer he begins to set some value upon time, and forsakes his three-month mode of getting up stream for a three or four days' trip. Captain Pimento Bueno, (son of the distinguished Senator,) the energetic and gentlemanly general superintending agent, told me that, with the Government bonus and the merchants' business, the steamers paid exceedingly well. They are all good boats, and most of them built of iron, as that material is decidedly the best, on account of the worms that are so destructive in the Amazon. Every town on the river furnishes wood at a fixed rate. The business of the steamers is constantly on the increase; and the industrious inhabitants of any of the villages can collect their syringa, Brazil-nuts, sarsaparilla, cacáo, &c. &c. and send them down to Pará by the steamer, and, on her return-trip, receive their money. This is creating new artificial wants, and, of course, making the people exercise more industry for the purpose of supplying their newly-awakened demands.

"These steamers certainly have done wonders in the last four years toward revolutionizing the whole business of the Amazon Valley; for, even from Moyabamba, Tarapota, and other Peruvian towns among the mountains, they now bring down their products in canoes and on bolsas (rafts) to meet the steamer at Nauta, which they never thought of doing before. Neither are the advantages of steam confined to the business-relations of life; but there is evidently an increasing desire on the part of the great mass of the people to learn more of the outside barbarians."

Mr. Nesbitt thus states the effect of the sight of a steamer on the remote population of the Upper Amazon :—

"As we would be passing a sand-bar on the upper rivers in Peru, where a steam boat had never before been heard of, and while all the fishermen and fish-driers would be standing in amazement, gazing at the 'monster of the vasty deep,'—not knowing whether it was a spirit from the diabo or some new saint sent by the Immaculate Virgin,—I would touch the steam-whistle, which would give such an unearthly screech that men, women, children, dogs, and monkeys would take to their heels and run for dear life, and would never stop to allow me to make the amende honorable."

I was desirous to obtain from this observant and practical man an opinion in regard to the views and theories of Lieutenants Maury and Herndon concerning the Amazon. In reply, he made the following statement:—

"I think that Lieutenant Maury's letters are painted rather beyond nature; but his ideas of the Amazon Valley and its capabilities are certainly, on the whole, nearer the mark than any other writer I have ever read. His theory of climate, and his reasons why the Valley of the Amazon is not like the same latitudes in Africa, &c. &c., are assuredly correct, in my humble opinion; for I was forcibly impressed with their correctness while on the spot.   The rainy season is not the incessant 'pouring down' of Africa, Central America, and the Orinoco-region.   It is more of a showery season: it is true sometimes when it rains '*it pours*,' but the showers are of short duration comparatively, and they fall at such regular intervals that one can make his calculations for business-engagements almost to a certainty.   And you will never have a day without seeing the sun more or less.

"The dry-season is not feverish and scorching; for scarcely a week—certainly not a fortnight—passes without one or more good showers.   Such a thing as crops suffering for the want of moisture is not known on the Amazon.   Although the days may be warm, the nights are always cool and pleasant, with very heavy dews.

"Lieutenant Herndon's ideas of the low banks were just such as any person would form who travelled down the river in a canoe, as it is impossible for any one thus situated to form a correct estimate of the country.   It would require years—not a few months—to learn the Valley as it ought to be learned.   There is not nearly so much land subject to inundation as Herndon estimated: notwithstanding, there are considerable portions that are overflown at high floods.   Herndon's expedition left its work unfinished; but it was of vast service to the country on the Amazon, both directly and indirectly,—as that expedition, I have not the least doubt, was the lever that moved the Brazilian Government to promote steam-navigation on the Amazon.   So *that* was the beginning; 'but the end is not yet.' "

In regard to the steamers ordered by Peru—which made the contract with Dr. Whittemore, formerly of Lima—to be built at New York and transported in pieces to Pará, to be run in connection with the steamers of the Brazilian and Amazon Navigation Company, Mr. Nesbitt gives me the following information:—

"I went out with the steamers to the Amazon, was with them while they were being reconstructed in Pará, and, after they were ready to start up the river, I took command of one of them.   Dr. Whittemore, our leader, commanded the other, and proceeded as far as the town of Obidos, where he turned them both over to me to deliver to the proper authorities, assisted by his friend, Mr. Z. B. Cavaly.   Dr. Whittemore then returned to New York.

"These steamers were not iron,—as frequently stated by newspaper paragraphs,—but were constructed of pure Georgia pine, frame, planking, and all.   The smallest one was ninety feet long, called the *Huallaga*; the other was one hundred and ten feet in length, called the *Tirado*, in honor of the then Secretary of State of Peru."

In reply to the question, How did the Peruvian steamers turn out? Mr. N. replied as follows:—

"They did not turn out so well as was anticipated, or as could have been desired for the credit of our country, whence they came.   They were built very light, and poorly finished and furnished; so much so, that the Peruvian Government officer who was appointed to receive them refused to do so, so that we were left some twenty-five hundred miles up the river from the ocean, with a couple of steamers and two American crews, without any provision being made either by the contractor or by the Peruvian Government for our support; and of the stores we had on board

37

a great portion was in a damaged state. Under these circumstances, the agents of the contractor were, from the necessity of the case, compelled to compromise with the Governor-General of Eastern Peru,—Colonel Francisco Alvarado Ortiz,—who had no authority delegated to him in the matter whatever by the Government of Peru, but who, in this disagreeable juncture, acted very fairly and was exceedingly liberal. By the compromise I had to remain in charge of the steamers until the Supreme Government would act in the matter. But the controversy is not yet finally settled, I believe, as a part of the contract-money is still due, and the Government refuses to pay it, on the ground that the contract was not complied with on the part of the contractor.

"One of them, the Huallaga, never turned a paddle-wheel after she reached the port of Nauta, but was tied up to the bank, and was rotting all the time that I was there. The other, the Tirado, made a few trips to various points above. I took her on two occasions up the Rio Huallaga almost to Chasuta, which is nearly *three thousand five hundred miles from the ocean : one of these trips was made during the lowest stage of water, and I never found less than fifteen feet water anywhere in the river-channel,*—so that a steamer of ten feet draught can pass from the *Pongo de Sal* to the Atlantic Ocean any day in the year. These steamers are at the present moment becoming more useless every day. Neither of the two boats have been run for any purpose since I left them, eighteen months ago; neither, indeed, can they be used, as the Peruvians know nothing about the management of steamboats and the engineers have all returned to the United States. The use of them has never been worth a dollar to the Government, and never will be.

"The Salt-rapid on the Huallaga, below Chasuta, is a natural curiosity. The banks of the river for more than a league are one solid mass of rock-salt, hard and clear as ice, in some places of a bluish-red color, and in others almost white, apparently the whole very pure, and in sufficient quantity to supply all South America for centuries.

"I have ascended the Huallaga, Ucayali, Pastaça, Madeira, and a short distance above the Barra do Rio Negro. The Huallaga, as before mentioned, is navigable for steamers the year round, for vessels of ten feet draught, as high as the Pongo de Sal, without the least trouble,—and to Chasuta, with ordinary caution and care,— and for canoes from Tinga Maria (only three hundred miles from Lima) to the mouth, down stream; but the ascent by canoes is very difficult. The country is excellent, being very healthy and fertile, with numerous villages all along the banks. The Pastaça is a very fine little affluent, and is navigable for steamers several hundred miles the greatest part of the year; but there are a number of tribes of hostile Indians on its lower waters. The land is most excellent, and the best Peruvian bark on the upper rivers is found on this stream. There are sometimes small quantities of gold brought down by the friendly Indians near its head-waters : I have seen some very fine specimens of it. The Ucayali can be ascended by a light-draught steamer nearly six hundred miles a part of the year, and as far as Sarayacu the whole year. The Rio Madeira is also a fine stream: it is navigable for any class of river-steamers to the Falls; but at no time can a steamer ascend these rapids. However, above the dozen rapids, there is plenty of water for several hundred miles, for a small steamer, the year round."

In 1853, a translation of Lieutenant Maury's letters was published in the widely-circulated *Correio Mercantil* of Rio de Janeiro; and I well remember the commotion his communications on the Amazon caused at the capital, in connection with a report that a "flibustier-ing" expedition was fitting out at New York to force the opening of the great river.

It is certainly a matter of deep regret that one whose writings and scientific investigations have (notwithstanding his manifold short-comings in regard to his own country) blessed and are bless-ing the world, should have permitted himself to make use of lan-guage which could only inflame a sensitive nation, and of some arguments which can only tend to "flibustiering." If Lieutenant Maury had left out the offensive language, and.a portion of his reasoning, which has been by Brazilians legitimately construed as nothing less than an advocacy of the theory that might makes right, I believe that it would have spared much unnecessary sus-picion and jealousy. Since that time a better feeling has been growing between the two countries; and we are sure that the time will come when both governments will be closely linked by com-mercial interest,* while we should receive Brazil's great staples free

---

* In the United States Commercial Relations for 1864 I find the following are the two leading staples of trade, by James Monroe, Esq., Consul at Rio de Janeiro.

The total importation of flour in Rio de Janeiro during the year 1863 amounted to 319,852 barrels, of which 241,362 barrels were from the United States. The number of barrels imported from the United States in 1862 was 261,865, and in 1861 302,061.

There were exported from this city in 1863 1,353.273 bags of coffee, against 1,487,583 bags for 1862, and 2,064,335 bags for 1861. This decrease of ex-portation has been due to a falling off in the crops. During both the years last named all the coffee raised in this province which was not required for home consumption was exported at high prices. The decrease in the amount of the crops has been due, in part, to unfavorable seasons and the ravages of an insect which attacks the tree and sometimes the flower and newly-formed fruit, but still more to defective modes of agriculture and the want of labor. The lack of laborers might be in part supplied by the introduction of suitable machinery. This has been done to a small extent; but improvements of this kind seem to spread slowly among the great plantations in the interior. The partial failure of the crops upon many old estates is no doubt owing to the continued cropping of many successive years without making the necessary returns to the soil.

While the exportation of coffee from this port to the United States in 1861 was 756,355 bags, in 1862 it was but 394,656 bags, and in 1863 only 388,875 bags. The causes of this decrease in the consumption of coffee in our country are too well understood to require explanation here. It is a striking example of the manner in which important events in countries widely separated become related to each other—events having no common origin in material or political causes, or in the plans of any human intelligence—that the falling off in the coffee crop of Brazil for the past three years has been nearly balanced by the decrease in the demand for the article in the United States.

of duty, and that which is exported by us to Brazil ought not to be heavily taxed. The property of our citizens dying intestate is administered by the Brazilian Government in a manner that never gives satisfaction. Outrages committed upon citizens of the United States in distant portions of the Empire in 1853 very tardily met with redress from the interior magistrates, whose feelings toward *Norte Americanos* were embittered by the conclusions arrived at after reading the letters of *Tenente Maury*. It was long ere we regained the sympathies which we had in 1850, when it was proposed, in case of war with England, that the whole Brazilian coast-trade should be put under the flag of the United States.

At Rio, Senhor de Angelis replied to Lieutenant Maury's "Amazon and the Atlantic Coasts of South America," (*Port. trans.,*) and his arguments, supported by Vattel and other writers on international law, are very ably stated. His volume, however, contains at its close some very pointed and plain language in regard to Texas and Greytown, which adds nothing to his argument.

We hope, however, that the judicious policy of the Union will always maintain a course which should be that of a country professing the principles of justice and liberality.

Whether the Amazon region, at least in the vicinity of the great river, can ever be thickly peopled by a more Northern race, remains to be seen. It is in one range of temperature, (not like the Mississippi, which enjoys every variety of climate,) and is as yet an almost unbroken wilderness. Some persons who have given much attention to this subject argue from the nature of the case that the provinces of Pará and Amazonas can never become flourishing rendezvous for Northerners. But, as Brazil differs from all other tropical countries, it may be that the "howling wilderness" of the Amazon will yet smile with industry and civilization. This was my conviction when in that valley in 1862.

As the case stands, Brazil certainly has the right, and the sole right, to control the rivers within her own borders, no matter if they do rise in other states; and, as previous to the treaty which gave the United States the right of descending the St. Lawrence no other country would have had the right to force England to open to the United States that river because many of her tributaries have their rise in the territory of the Union, so there is no justice in any proposition to force Brazil to concede the free navigation of the Amazon. So much for the principle involved. But we are happy to say that Brazil has waived all narrow state policy and precedent, and on the 7th of September, 1867, declared, by Imperial decree, the Amazon open to the flags of all nations, from the Atlantic to Peru. Brazil has thus applied to the Amazon the most enlarged political economy, and will in due time reap incalculable benefits in the development of her vast resources.

About one-half of Bolivia, two-thirds of Peru, three-fourths of Equador, and one-half of New Grenada, are drained by the Amazon and its tributaries. For the want of steam-communication the trade of all these parts of those countries goes west over the Andes to Callao. There it is shipped, and, after doubling Cape Horn and sailing eight or ten thousand miles, it is then only off the mouth of the Amazon, on its way to Europe or the United States; whereas, if the navigation of the Amazon were free, the produce of the interior could be landed at Pará for what it costs to convey it across the Andes to the ports of the Pacific.

Note for 1866.—In 1862, J. C. F. travelled on the Amazon from its mouth to the limits of Peru, in the new steamers Manáos, Belem, Icamyaba, and the Inca. The Amazon Navigation Company (whose President is the well-known Baron de Mauá, a Brazilian nobleman and financier at Rio de Janeiro, and whose chief director at Pará is the energetic Sr. Pimento Bueno) has conferred the greatest blessings upon that region, which is yet destined to see a greater development. The line has increased its steamers and its efficiency generally; and now the Peruvians have inaugurated a line on their share of the Amazon and the branches up the Solimões and the Huallaga. The colonies referred to on page 575 have ceased, from various causes; but the valley has gained. It is still a vast wilderness; but the volumes of Bates, ("Naturalist on the Amazon," London, 1863,) the observations of M. Brunet, a thorough but most modest explorer, the labors of Coutinho, Costa de Azavedo, and Soares Pinto in 1861, '62, and '63, and the magnificent explorations now being conducted by Professor Agassiz, will turn the attention of the world to this wonderful basin. The kind attentions of Sr. Pimento Bueno, Charles Jenks Smith, and Dr. Coutinho, at Pará; of Dr. Peixoto (Municipal Judge at Cametá), Dr. Marcos at Villa Bella, Mr. Jeffries at Obidos, and Sr. José de Freitas Guimarães, Henrique Antonii, Dr. Gustave, and Charles Collyer contributed much to procure the collections for Professor Agassiz made by me in 1862.

CONCLUSION.

THE authors, in reviewing the ground which they have gone over in this volume, only feel the imperfection of their labors and how difficult has been the task to give in so small space a just and general view of Brazil. They have compared the Empire not with England and the United States, but with other countries of the New World which have been peopled by descendants of the Latin race. This they believe to be the true mode of comparison. Many errors may thus be avoided. In the year 1857 their attention was called to an editorial in one of the most widely-circulated and influential papers of our country, in which occurs the following sentence :—

" To those who wish to know how deep human nature can sink in moral degradation and the extreme limit of monarchical imbecility, we recommend a reading of Ewbank's 'Brazil,' whose details of hopeless superstition, general ignorance, and political demoralization have no parallel."

We have already shown our appreciation of the author referred to by direct quotations from his work; and had he who penned this editorial remembered that Mr. Ewbank (more than 20 years ago) was a stranger abiding for a few months in a new and curious country, and published a journal of observations and events which he jotted down from the impressions of the moment, and makes but few generalizations, he (the editor) would not have been so sweeping in his condemnation of Brazil. He seems, however, to have entirely overlooked one of Mr. Ewbank's few general conclusions. Had he read it he would doubtless have been convinced that there was something hopeful in Brazil. As the opinions of the author in question have been often quoted to us as entirely at variance with any encouragement in regard to the Empire ruled by Dom Pedro II., we cite from his last chapter the following, which is to the point :—

" The character of the Brazilians, I should say, is that of an hospitable, affectionate, intelligent, and aspiring people. They are in advance of their Portuguese progenitors in liberality of sentiment and in enterprise. Many of their young men visit Europe, others are educated in the United States: add to this an increasing intercourse with foreigners,—the means ordained by Divine Providence for human improvement,—and who does not rejoice in their honorable ambition and in the career opened before them? It must be remembered, however, that no one people can be a standard for any other, for no two are in the same circumstances and conditions. The influence of climate, we know, is omnipotent; and, from their occupying one of the largest and finest portions of the equatorial regions, it is for them to determine how far science and the arts within the tropics can compete with thei progress in the temperate zones. As respects progress, they are, of Latin nations next to the French. In the Chambers are able and enlightened statesmen; and the representatives of the Empire abroad are conceded to rank in talent with the ambassadors of any other country. As for material elements of greatness, no people under the sun are more highly favored, and none have a higher destiny opened before them. May they have the wisdom to achieve it!"—*Ewbank's Sketches of Life in Brazil.*

It is impossible to appreciate the present condition of Brazil without taking into view the influences of the mother-country. Notwithstanding the wealth and glory of Portugal during the short period of her maritime supremacy, there are few countries in Europe less fitted to become the model of a prosperous state in modern times. In whatever light we consider Portugal or her institutions, we find them altogether behind the spirit of the age. Yet that country, as insignificent in size as it is indifferent in condition, held nearly half of South America under the iron sway of colonial bondage from the period of its discovery until 1808,—we might almost say 1822.

The short space of forty four years is all that Brazil has yet enjoyed for the great object of establishing her character as an independent nation. During that period she has had to contend with great and almost numberless difficulties. A large proportion of the inhabitants were persons born or educated in Portugal, and consequently imbued with the narrow views and the illiberal feelings so common to the Portuguese. The laws, the modes of doing business as well as of thinking and of acting, that universally prevailed, were Portuguese. All these required decided renovation in order to suit the circumstances of a new empire rising into being during the progress of the nineteenth century.

Such a renovation is not the work of a day; and if it should appear that as yet it has only properly commenced, still, the Brazilian nation will stand before the world as deserving the highest credit. She has broken off bonds that had remained riveted upon

her for ages. She has advanced from a degrading colonial servitude to a high and honorable position among the nations of the
earth. What is perhaps still better, she cherishes a desire for
improvement. She directs a vigilant eye toward other nations;
she observes the working of their different institutions, and mani-
fests a disposition to adopt those which are truly excellent, as far
and as fast as they can be adapted to her circumstances.

Her finances are in a very good condition. But she should
be ready to accept and to court a greater reciprocity among the
nations of the earth, and should abandon all narrow policy.

The revenues of the Empire are almost entirely the product of
heavy duties upon commerce. Unfortunately, the nation has but
few manufactures to call for her high tariff as a means of protection. Her duties upon imports constitute a direct tax upon internal consumption; while the duties upon exports embarrass her
trade abroad. Thus, agriculture is doubly oppressed, and it is
under the burden of great difficulties that the immense resources
of the country are to a comparatively small degree developed.

Were there no other means of providing for the expenses of
government, it would, perhaps, be idle to dwell upon this ruinous
process, unless it were to comment upon it as a necessary evil.
But is there no possibility of raising a revenue for Brazil from the
sale of public lands? Millions upon millions of acres remain as
yet unappropriated, notwithstanding the utter carelessness with
which the richest and most valuable portions of the public domain
have hitherto been yielded to the ownership of whomsoever might
incline to take possession of it. Might not Government surveys be
instituted, and the whole country brought under legal demarca-
tion? Hitherto, not one-fiftieth part of it was ever surveyed;
and even in some populous districts great uncertainty respecting
boundaries still exists. It is understood that a reform in this
direction has been begun. But what advantages could result
from these surveys, unless spontaneous foreign immigration were
encouraged?

Great things have been done in this respect, but more still remains to be accomplished. But the colonial system has not proved
the success which its friends had hoped. The popular mind is
waking up to the true mode. Intelligent emigrants are needed.

Open wide the doors, let the Government throw off all restriction of passports and every tax upon the emigrant, and the great and small proprietors will not have to resort to expensive means to induce immigration: it will flow of itself.

Education is daily exciting increased attention. In the new system of school-instruction, the French model has been generally followed. Having already described institutions of the various grades,—from the primary school to the law-university,—it will now be sufficient to remark that a great degree of improvement upon the former state of things is already manifest; but at the same time the work of educational reform has only commenced. The teachers' salaries are too low; the interest among the common people requires to be more fully excited; and a very serious obstacle is to be overcome in the want of suitable school-books.

It is sad to often find hinderances to the cause of education in the very men who ought to be leaders in the movement for the intellectual as well as the moral training of the young. A single instance and a general remark will illustrate what we mean.

A priest residing in one of the most prominent cities of the Empire, and, indeed, exercising his functions beneath the very shadow of one of the universities, was heard to say, "*Não gosto de livros; gosto mais de jogar,*" ("I have no relish for books; I like gaming better.") In corroboration of these remarks is the language of a distinguished Brazilian statesman, uttered before the Imperial Legislature :—

"A narrow strip on the coast is all that enjoys the benefits of civilization; while, in the interior, our people are still, to a great degree, enveloped in barbarism." In immediate connection with this remark, the same gentleman added, "We have been unable to do any thing, and nothing can be accomplished without the aid of a moral and intelligent clergy."

Notwithstanding the picture sketched in these brief but just intimations, there is much room to hope for Brazil on the score of education. The schoolmaster is abroad in the Empire; the press is at work; but the number of scholars has not proportionately increased with the population since 1855. Let slavery be done away, and the next decade will witness a vast improvement.

The history of Brazilian literature is brief; yet, under the circumstances in which it has sprung up, that literature must be considered creditable. Of all that has been written in the Portu-

guese language within the last hundred years, Brazil has produced
her full proportion of what is meritorious. The volumes of the
Canon Pinheiro (Rio de Janeiro) on Portuguese literature, and of
Wolf (Berlin) on Brazilian literature, sufficiently attest this.
Portugal has never produced a scientific man superior to José
Bonifacio de Andrade: indeed, for years she borrowed this distinguished Brazilian to adorn her university of Coimbra and her
medical schools of Lisbon. The only living Portuguese prose
writer who excels the Brazilian prose writers is Alexander Herculano of Lisbon. He is the present master in historic writing, and,
though differing from them both, may be compared to Lord
Macaulay or Mr. Prescott. As a prose writer, however, Torres
Homem, a Brazilian statesman tinged with as much African blood
as courses the veins of Alexander Dumas, is by the admission of
literary men at Rio their first prose writer. It is to be regretted
that he has not devoted himself more to literature. Perhaps the
most popular native writer of fiction is Sr. Alencar, author of the
Guarani. He has had the good taste and foresight to take up a
native subject. In historic writings, while there have been many
essays, the largest works are those of Varnhagen (now Brazilian
minister to Peru) and Pereira da Silva. The former has amassed a
vast amount of material for future writers of history, while the
latter is now publishing what he purposes to be an exhaustive
history of his country. The Quarterly Review of the Imperial
Geographical and Historical Institute for thirty years has been
enriched by well-written articles and essays in history and geography. In political writing the Brazilian press has abounded.
Formerly their political theories were greatly influenced by French
writers, but at the present time no foreigner so influences the minds
of the younger and middle-aged Brazilian statesmen as John Stuart
Mill. The key-note and, indeed, the burden of Sr. Zacarias' *Poder
Moderador* is John Stuart Mill on *Liberty.* In the law universities
of San Paulo and Pernambuco are many able professors and writers
on law; while the medical colleges of Rio de Janeiro and Bahia
have writers equally eminent in their department. There are
wanting lay discussions on religious subjects; but we are glad to
see an essay on religious toleration by Sr. I. B. Bareto of Pernambuco. It is, however, in poetry that, at the present time, Brazil

excels the mother-country. The names of Magalhaens and Gonçalves Dias, in poetry, stand deservedly high. Gonçalves Dias is
supreme in lyric poetry. His sad and tragic end on board a wrecked
ship in sight of his native land caused the profoundest emotion
throughout Brazil. There are many other poets of merit, but who
have only written fugitive pieces. Within the last few years the
example of Dom Pedro II. has influenced the young men to the
study of the English and American poets. Excellent translations
of the poetry of Longfellow and Whittier have, among others, been
made by the Emperor, M. M. Lisboa, Pedro Luiz, and Bittincourt
S. Paio. Porto Alegre, Macedo, Norberto, and Assis are well
known.

It may perhaps be considered by some as a misfortune, in a literary point of view, to Brazil, that her language is the Portuguese.
A prejudice against that language prevails extensively among
foreign nations; and, although that prejudice is in a great degree
unjust, it will not soon be overcome. The learned have seldom
been induced to acquire that knowledge of the language which is
essential to an appreciation of its real merits. Those who have
formed its acquaintance accord to it high praises. Mr. Southey,
for example, has declared it to be "inferior to no modern speech,"
and to contain "some of the most original and admirable works
that he had ever perused." Schlegel, in his "History of Literature," bears the very highest testimony to the beauty and copious-
ness of the Portuguese language, and cannot restrain his admiration for De Camões. Of the *Lusiad* a distinguished French writer
has said, "It is the first epic of modern times." (It must be remembered that the Latin nations have never been able to comprehend
Milton.) M. de Sismondi says, "The distinguished men whom
Portugal has produced have given to their country every branch
of literature." And again:—"Portuguese literature is complete :
we find in it every department of letters." (*De la Littérature du
Midi de l'Europe*, t. iv. p. 262.) "The Portuguese language,"
says M. Sané, "is beautiful, sonorous, and copious: it is free
from that gutturalness with which we reproach the Spanish : it
has the sweetness and flexibility of the Italian and the gravity
and descriptiveness of the Latin." (*Poesie Lyrique Portuguaise*,
p. xc Paris, 1808.) In fine, it may be remarked that no living

language—not excepting the Spanish and Italian—is so near in every respect the tongue of old Imperial Rome as that of Lusitania. If the Brazilians, possessing such a language, shall develop the genius and the application necessary to such a result, they may yet, by creating a literature worthy of themselves, secure the respect and admiration of the world.

Notwithstanding so little is known of the Portuguese language to certain classes of the literati, it prevails wherever there are or have been settlements of that nation,—not only in Brazil and the Portuguese Islands, but along the coasts of Africa and India, from Guinea to the Cape of Good Hope and from the Cape of Good Hope to the Sea of China,—extending over almost all the islands of the Malayan Archipelago.

How interesting it would be to witness light and truth radiating from Brazil and spreading their influences to each of those distant climes! Before such an event can be reasonably anticipated, how great must be the changes in the moral and religious condition of the Empire!

The ecclesiastics are notoriously corrupt. The report of a Minister of Justice not many years ago contains the following language:—

"The state of retrogression into which our clergy are falling is notorious. The necessity of adopting measures to remedy such an evil is also evident. . . . The lack of priests who will dedicate themselves to the cure of souls, or who will even offer themselves as candidates, is surprising. . . . It may be observed that the numerical ratio of those priests who die or become incompetent through age and infirmity is two to one of those who receive ordination. Even among those who are ordained, few devote themselves to the pastoral work. They either turn their attention to secular pursuits, as a means of securing greater conveniences, emoluments, and *respect*, or they look out for chaplaincies and other situations, which offer equal or superior inducements, without subjecting them to the *literary tests*, the trouble and the expense, necessary to secure an ecclesiastical benefice.

"This is not the place to investigate the causes of such a state of things; but certain it is that no persons of standing devote their sons to the priesthood. Most of those who seek the sacred office are indigent persons, who, by their poverty, are often prevented from pursuing the requisite studies. Without doubt, a principal reason why so few devote themselves to ecclesiastical pursuits is to be found in the small income allowed them. Moreover, the perquisites established as the remuneration of certain clerical services have resumed the voluntary character which they had in primitive times, and the priest who attempts to coerce his parishioners into the payment of them almost always renders himself odious, and gets little or nothing for his trouble."

At the present time Brazil is in want of nothing so much as pious, self-denying ministers of the gospel,—men who, like the

Apostle to the Gentiles, will not count their lives dear unto themselves that they may win souls to Christ. And is it too much to hope that God in His providence will raise up such men in His own way, especially when we reflect that His own Word shall not return unto Him void, and that faithful prayer shall never be forgotten before the throne of the Most High.

We might have unfolded before the reader many more incidents of labor in our Master's cause in Brazil, but have, from proper motives, withheld details: we believe that we have every encouragement to hope for Brazil in a religious as well as a political point of view.

Several things are of instant importance to the present and future welfare of Brazil. *First*, immediate legislation in regard to putting an end, by judicious measures, to slavery in the empire. The difficulties, though great, are not like those of the United States, where there is such an unreasonable prejudice against color, and where before the recent rebellion millions claimed a divine right for perpetuating the accursed institution. The Brazilians have always had a higher moral theory on the rights of the negro than the North Americans. The economical phase of the subject is unquestioned. The very best estimates are that the number of slaves from 1851 to 1861 has decreased more than one million; and the statistics in the *Relatorios* of the Ministers of Finance and of Public Works, Agriculture, and Commerce, prove that at the end of the same decade the great tropical productions of coffee, sugar, cotton, tobacco, &c. had increased more than thirty per cent. The seriousness with which her own statesmen are taking up this subject, the openly-expressed and well-known anti-slavery sentiments of the Emperor, all are so many earnests that the appeals to Brazil by such calm, true men as Laboulaye (*vide Journal de Débats* for July, 1865) and Cochin of Paris, and of devoted friends of freedom in England and the United States, will not be in vain. We sincerely hope that Brazil will not be the last nation to hold men in bondage, and that, in a generous rivalry, she may anticipate Spain in wiping off this foul blot from her national escutcheon.

Second, suitable legislation should be immediately had in regard to religious disabilities. In Appendix I we have given an article from a Roman Catholic editor at Rio on this important subject.

In January, 1866, most eloquent and forcible addresses to the same purpose were delivered in public meeting at the Rio Exchange, by Dr. Furquim d'Almeida, whose breadth of view and practical political economy give to him something of the character of the English Cobden, and of that American friend of Brazil, A. A. Low, Esq., the large-minded President of the New York Chamber of Commerce. No country can ever reach a high development, moral, material, and intellectual, unless soul-liberty in its fullest extent be incorporated in the political theory and practice of its people. The speech of Dr. Furquim d'Almeida is a just and strong corollary from Pedro Luiz's famous anti-Jesuit speech in the parliament of 1864, A. C. Tavares Bastos' *Cartas do Solitario*, and I. B. Barreto's essay on *Intolerance*.

Third, it is highly important in a material view that Brazil should remodel her laws in regard to the manner of raising her revenue. This subject is referred to on page 584. It is to be hoped that the Paraguay war will not only see that fomenter of discord, Lopez Jr., driven from South America, but that the necessity for increased revenue will cause timid financiers and those wedded to old Portuguese notions to combine with a few men of nineteenth-century notions to carry through a law for a direct and an equalized mode of taxation. In a few cities there is, to a small degree, direct taxation; but too often is it of a narrow character,—levying upon the foreigner, and having no general application. There are some men in the province of Rio de Janeiro, outside of the neutral ground of the metropolis, who are capitalists and large land-owners. The junior author recalls one, who had no family besides his wife, and who informed him that he (the capitalist) owned eight square leagues of land, in addition to his personal property, which is immense, but that he did not have to pay a penny for taxes, either on his real or personal estate. Now, a common road-side shopkeeper without children, having an income of $2000 per annum, would have to pay to the general government just as many indirect taxes for the clothes that he wore and the wine that he drank (the principal imported articles that both used) as the man worth his hundreds of thousands. By lowering the import dues, by eradicating altogether the system of export duties, and by beginning with a moderate direct impost, agriculture and commerce will flourish. — 1868. A modified direct taxation has been inaugurated.

Fourth, there should be no exclusiveness in regard to teachers and professors in the higher public institutions of learning. According to the present laws, if a gifted man of science, being a foreigner, should wish to remain in Brazil for six years to teach his particular branch in a public institution, he could not obtain a .place: it can be given to none but a native or naturalized Brazilian. We do not blame the Brazilians for cultivating a spirit of nationality, but we do find fault with any thing that will foster a spirit of narrowness. M. Agassiz was for years professor in a university under the auspices of the Prussian government; but Professor Agassiz did not lose his Swiss nationality by serving under the King of Prussia, neither was he esteemed a less competent or a less faithful teacher because he was not a Prussian. When the same *savant* came to the United States, he remained some years before becoming naturalized, and he would have been held by the public in the same estimation even if he had not become an American citizen. There is scarcely a first-class institution of learning in the United States without a foreign professor; and it has worked greatly to the advancement of education. This spirit of exclusion in Brazil is to be found in other ranks of life where the calling is much more humble than that of teaching.

Brazil can afford this, and in so doing will only be carrying out the same enlarged policy which characterized her conduct in abrogating the monopoly of the coast trade, and in opening the Amazon to the flags of all nations. As Brazil is, and will be for many years, an almost purely agricultural country, a great benefit will arise from the recent (1867) opening of the coast trade to the flags of all nations. The Amazon, which heretofore has been jealously guarded, is now open to the developments which will arise from making it a free river. We therefore hope Brazil will do away with all exclusiveness in regard to mental as well as material things.

Lastly, red tape demands the attention of the *Assembléa Geral.* Red tape, to an indefinite extent, exists in all the public offices outside of the Imperial legislature. In that body there is a great freedom from red-tapeism. If a subject is rightly presented through the usual channels, it goes through the regular parliamentary forms, and is much less impeded than it would be in London or Washington. The wording of propositions, bills, and laws is singularly free from the almost endless legal tautology in the documents of a similar character brought into the British Parliament or into the Congress of the United States. But many of the affairs in the public offices are subject to the greatest delays, from the highest to the lowest functionaries, and are nearly as much involved in red tape as they would be in Portugal and Spain, or as a case of Chancery in an English court. There are too many citizens, as well as officials, who constantly cry, *O Governo, O Governo;* and, the government being expected to do every thing, no individual activity is developed. Here is great room for improvement.

The reforms indicated are all very urgent; but the first three are of such grave importance to Brazil that the hearty prayer of every well-wisher of the country is that the Brazilians may have wisdom to achieve them in such a manner as shall redound to her highest good.

In finishing this volume, we cannot do better than to give a literal translation of extracts from the letter of welcome and of instruction to some emigrants from the South of the United States, written by Sr. Paula Souza, the Minister of Public Works, Agriculture, and Commerce, in 1865–66; and also a few extracts from the powerful speech of Dr. Furquim d'Almeida on religious toleration. We believe that this letter, as well as the speech of Furquim d'Almeida, expresses the mind of His Majesty the Emperor and of the enlightened minister; and if all Brazilian statesmen, on this and other questions, are inspired by the same enlarged sentiments, their country cannot fail to make rapid progress towards the highest civilization.

RIO DE JANEIRO, October 9, 1865.

Brazil is an immense territory, as you well know, bounded on the north in 4° north latitude by English and Dutch Guiana and Venezuela, and on the south in almost 34° south latitude by the La Platan Republics; on the east by the Atlantic Ocean, and on the west by the Republics of Peru, New Granada, &c. It contains all climates, and produces, if not naturally, at least with less labor than in any other part of the world, almost all the products of all the zones; and the fruitfulness of its soil is not inferior to the variety of its climate, and repays with usury the labor that seeks its good services. Irrigated by immense and gigantic rivers, almost all navigable, and some of which are now navigated, it offers the cheapest system of travel, and transportation for the exuberance of production which seeks other countries and exportation; immense forests and wide-spread plains lie there unprofitable, but awaiting only the man or men that solicit production from them. Its mineralogical riches are not inferior to the variety, abundance, and excellence of its woods.

From the margins of the Amazon and its confluents to the shores of the Paraguay and Paraná, you will find a soil rich from its geological composition, a salubrious climate, and a conformation of surface yielding with little trouble the ways of communication for the products, where you and your associates may fix your residence, adopting this country as your home, to go hand in hand with us, (who receive you fraternally,) in raising it, by your energy and industry, to the height of its destinies,—destinies revealed to us by the magnificence of its nature.

Our form of government differs little, at bottom, from that under which you and your companions were reared. Our President rules during life; and the presidency is transferred by inheritance; and, without criticizing what our contemporaries of North America do, I will say that we find in this advantages of order and stability which the United States of America alone, among all the republics, has been able to present. As to other differences, most are those of habits and customs. We alike adore liberty as the fecund principle for the development of man and the human family, and alike respect governmental forms as a guarantee of this liberty.

Descendants of Portuguese, and Catholics, we outwardly differ from the founders of the city of Providence, Rhode Island, and from the Puritan Dissenters of Massachusetts. We do not, however, differ in the adoration of the same principle. Come, then, to Brazil, where you will be welcome, and can live well, as is your due.

In the property that you may be able to bring there is one kind of which our legislation does not permit entry: it is that which, perhaps, may consist of slaves; I must say more,—that the importation of even the free African is forbidden by law. If, then, any of your associates possess property of this nature, they should get rid of it. This, however, is not meant to imply that, once among us, they may not employ their capital in that manner; unhappily, we still possess slaves, and trade in them within the Empire, from one province to another, is permitted.

We have, already, lands surveyed and marked off in several provinces, whither those emigrants who wish them can go immediately and establish themselves: their extent, however, is small, and does not comport with a rapid and instantaneous introduction of a great number. This will not be, however, a thing to cause difficulty, or to retard emigration, because the government is resolved to go on establishing the emigrants on lands, making, consecutively and proportionately, the surveys and limits, giving provisional title-deeds guaranteeing to the holder the definite title-deeds in the legal form. The law does not permit the gift of lands, and requires its sale; but the price is so low, and the facilities for payment so great, that negligence only, or complete idleness, will be incapable of satisfying them. The price varies from half a real to a real and a half the square braça, (11d. or 21 cents, to 2s. 8d. or 63 cents, the acre,) according to the quality of the land and its topographical situation: we have thus, then, a square league of 3000 braças square (10,764 acres) for a sum of 4500 dollars at the minimum, or 8250 dollars at the maximum price, whilst, in the pamphlet accompanying this, Sr. Sarmiento, Minister Plenipotentiary and Envoy Extraordinary of the Argentine Republic, says that the lands of the Argentine Confederation are sold at from 10,000 to 40,000 patacoons (1¼ dollars each) the Spanish league, (about 7700 acres.) [N.B. The *real* is equivalent to half a *mill* of the United States.]

With us, as there, both horned cattle and sheep produce magnificently, as like-wise all other animals which man subjects to his rule and use.   Coffee, sugar-cane, cotton, indigo, quinine, vanilla, tobacco, as also all alimentaceous articles, pro-duce wonderfully, and form the fountain of the private and public wealth of the Empire.   Foreigners obtain naturalization with facility.   The colonist is at the end of two years a *de facto* Brazilian, if he desire it.   Any foreigner as well as the colonist can, after two years, become a citizen by making his declaration before any Camera municipal.   A few days are sufficient, if naturalization is sought through our Parliament, which can consider emigrants as importers of any industry, or capital, or those worthy from their personal qualifications.   In this last category you and yours will be received, and you will be able to return in March, as Brazilian citizens, to the United States, to import your property, machines, and industry of every kind.

If our advancement is not yet great, if our development is lingering, this does not arise from political commotions, or perturbation of public order; the Bra-zilian people, as worthy and brave as any other, are more than any other sociable and affable,—of a gentleness of disposition nearly degenerating into a vice, and which has even prejudiced its good reputation abroad.   If there is some indolence in our character, it has, as compensation, a profound sentiment of duty and propriety.

The Brazilian citizen is free in the widest sense of the term: if in the great cities or towns we see in practice all our administrative system, it is not the less true that a great part of the interior lives more or less well with part of it, sus-tained by that sentiment of duty and propriety and by the tendency of its gentle and tolerant disposition.

We are Catholics; we have a religion of the State; but we force no one to follow it; the constitution merely requires that the deputies profess that religion; all religions may be practised, except that they cannot be in temples with the out-ward form.   Our municipal life has some resemblance to that of the townships of North America.   Every four years every citizen who possesses an income of 200 milreis, 100 dollars, from real estate, can vote, (if he be not guilty of an un-bailable offence,) if he be more than twenty-five years old, or than twenty-one years if a military officer, a priest in holy orders, a graduate of any academy, or married.   The voters assemble and vote for those citizens whom they desire to represent them during the period of four years.   These are justices of peace, and form the municipal chamber, that is, the municipal executive and legislative power for that quadrennial period; the police are nominated by the government of the provinces; we have the *habeas corpus* applied to all cases of offences, and guarantee-ing the liberty of the citizen; the right of complaint is sacred, and permitted even to the slave; the press is free; and the jury tries the greater part of the offences.

Whatever may be the divergence of the opinions of the political parties in the Empire, all are agreed in preserving what we possess: as in every part of the world some seek to achieve the future with greater rapidity, some with less, but all starting from the principle of the rights acquired, which no one desires to lessen.

Brazilian legislation grants certain and specific favors to emigrants, and the government of Brazil seeks eagerly to enlarge those favors; in your letter you desired to know them, and I, to furnish them complete, have ordered the legisla-tion in favor of emigrants to be compiled and delivered to you.   I would advise you to begin your journeys through Brazil by the province of S. Paulo.   Sr. Street, a naturalized Brazilian citizen, has orders to accompany you; and, as he now knows us, he will be able to furnish you all the information that you may require.

From S. Paulo go to Paraná, Santa Catharina, Rio Grande, and, on return, if you desire, you can travel through our provinces of the interior and north, where you will obtain data and information that will enable you to give a just and sure idea of us and our country to your associates, and if, then, you resolve on settling among us, your resolution will be the fruit of mature reflection and studies, which will please us more, because we open our arms to you with fraternal solicitude, without desire to attract by hyperbole of phrase, but only by the truth of fact: then we will be able to say that there are no evils that do not bring good, since the ills of our contemporaries of the North have brought us good, since they brought us the influx of North American energy, activity, and industry; and our grief at seeing them divided will be compensated for by the pleasure of the new elements of approximation and union that are offered to us.

I am, with pleasure,

Yours, &c.

ANTONIO FRANCISCO DE PAULA SOUZA.

EXTRACTS FROM A SPEECH MADE BY DR. FURQUIM D'ALMEIDA,

At the Exchange of Rio de Janeiro, on the occasion of forming the International Emigration Society, January 26, 1866.

But there are not only material embarrassments which we will have to remove to attract a great current of spontaneous emigration: the moral ones are much more important and much more difficult to combat. They are the old prejudices still encastled in our customs and in our laws, and maintained by a false patriotism and an intolerant religious spirit. Powerful enemies, everywhere opposing the most tenacious resistance to every innovation, to every idea of progress, these prejudices among us will not allow themselves to be vanquished easily: they will struggle while they have strength, and will yield only at the last extremity. We must count upon a bloody struggle, but we should not be discouraged on this account; on the contrary, we should invest ourselves with more patience and more courage to attack them and overcome them. This is the principal mission of our enterprise.

Moral embarrassments are represented by three orders of facts, civil, political, and religious; and may be translated into civil, political, and religious inequality as regards the foreigner who wishes to adopt our country as his own.

The civil inequality is quite patent. Our civil legislation prior to the law of September 11, 1861, did not acknowledge marriages not celebrated according to the prescriptions of the Catholic Church; that is, marriage purely and simply civil did not exist: consequently, marriages celebrated between Protestants or any other Dissenters, or by any other Church, were null, and for these the legitimacy of their families, the first base of every well-organized society, was wanting.

The law of September 11, 1861, wishing to satisfy in some measure the just complaints that were raised from all sides against this state of things, took a middle course, which does not satisfy the just reclamations of those who ask for civil marriage, and has much displeased the defenders of a purely religious and Catholic marriage.

This law does not establish civil registry; it contents itself with merely tolerating marriages celebrated between Dissenters according to the rites of their various faiths, and by their respective clergy. But nothing is changed as regards mixed marriages, and they, in the silence of the law, are regulated by the anterior legislation.

The inequality and injustice in this point are manifest. The Dissenter is tolerated merely to marry according to the ritual of his faith, when civil marriage should have been allowed to him as a perfect right, subject to no restriction, and to none of the many abuses to which marriage not civil can give place. Now, the marriage of Dissenters being merely tolerated, the consequence is that it is considered illegitimate in the eyes of the Catholic religion, the religion of the State, and the ecclesiastical authorities judge themselves authorized to consider it as such whenever occasion offers.

Let us suppose a case which can very easily take place. A married couple, of any dissenting faith, weary of one another, come to an understanding that they ought to separate and marry again: they address themselves to any one of our bishops, abjure their religion, adopt Catholicism, and ask license to marry whomsoever each may choose.

The bishop does not oppose the least doubt; he receives them into the bosom of the Catholic Church, and grants them license to marry a second time, it being that the Catholic Church considers as simple *concubinatio* a marriage not made before it and according to its precepts. Facts like these have already taken place among us; and their repetition must sap the basis of family, must withdraw from it all its moral strength, and implant immorality sanctioned by law.

On the other hand, the law of September 11, 1861, regulates nothing with respect to mixed marriages: consequently, they continue to be performed according to the previous legislation, that is, they are made before a Catholic priest and according to the Catholic rites and usages sanctioned by the civil law. Now, the Catholic Church does not permit marriage between a Dissenter and a Catholic, unless with the condition that the Dissenter bind himself by oath to rear and educate the children in the Catholic religion.

What injustice, what humiliation to the Dissenter who may wish to form ties with the families of the country! He has to subject himself to a hard and humiliating condition if he wish to obtain a Brazilian wife. He is obliged to stifle the cries of his conscience, which clamors that his religion is the best, and to swear that his children will be educated in the principles of that which he believes worse than his own. Gentlemen, do you know of a prescription more unjust, more intolerant, more absurd?

Beyond all, the worst is that it is contrary to our Constitution, which establishes liberty of conscience, and is useless because there is no method of enforcing it. Our Constitution guarantees to all the free exercise of his religion, with the sole restriction that the places of worship may not have the exterior form of a temple, *i.e.* with steeples and bells.

That is to say, every one may follow the creed that pleases him, and may educate his family in the same religious principles, without any authority having power to call him to account. Now, then, shall the civil legislation remain in flagrant contradiction to the Constitution in exacting that the Dissenter marrying a Catholic shall bind himself by oath to educate the children in the Catholic religion? Such a prescription is an exaction merely vexatious and humiliating, without any practical result, since our civil legislation has no penal sanction for it. What is the authority charged with its execution?

A voice.—The ecclesiastic authority.

SR. FURQUIM.—This has at its disposal neither the secular arm nor the penal sanction: it can merely lay hand on excommunication, which to-day is worth nothing. (Cries of No! No!) The Catholic himself amongst us can abjure his religion without any authority being able to call him to account, for the Constitution guarantees to all full liberty of conscience. (New cries of No! No!) Are we perchance in the Middle Ages? Can we be under the dominion of the Inquisition? So it might seem on hearing such warm and intolerant "No! No!" Happily we are in the nineteenth century, and in one of the most free and tolerant countries of modern times. I can, therefore, speak to you with all frankness and liberty. I am a Catholic, I was educated in this religion, I intend to belong to it until death; but my reason tells me that it is needful to give to all the right of adoring God according to their conscience. (Great cheers.)

By all that I have just exposed to you in relation to our legislation on marriages, you can appreciate how much it is incomplete, unjust, and unequal.

In the political part the same injustice and inequality exist: our Constitution forbids to the naturalized foreigner access to certain elevated charges of the State, such as Deputy and Minister of State. There is in this a great injustice and inequality. To invite the foreigner to form part of our nationality, abandoning all that is dear to him in his country, asking him to come with his family, his industry, his labor, his capital, enriching and aggrandizing our country,—to close on him the doors to the highest charges of the country he adopts, is an absurdity only explicable by the circumstances and the epoch in which our Constitution was promulgated.

We had just declared our independence, and the country was yet in hostilities with the mother-country. The exclusion of foreigners from certain of the higher offices of the State was established on purpose to take these offices and keep them from the Portuguese. Now it is absurd, and has no more a reason to exist. It is an odious exclusion,—above all in a new country that has need to attract emigration with all its force.

It remains to us to speak of the religious inequality in which the foreigner is placed relative to the native. This inequality transudes through every pore of our laws, beginning with the Constitution, which establishes that the religion of the State is Catholic, and considers it as a civil and political institution which has a distinct place among the various branches of our social organization.

For it are destined all the official honors; churches constructed at the cost of the State; an important place in the estimates; imposts paid by all the followers of all religions, and of which it alone has the advantage. To other faiths the Constitution merely concedes tolerance; it admits them, but with a certain distrust, with a certain reserve, in which Dissenters can discern a species of contempt. On the other hand, the Constitution exacts, for the exercising of certain offices, the oath to maintain the Catholic religion. It is a new embarrassment, a new injustice to the naturalized foreigner who belongs to a dissenting faith. Either he must be untrue to his conscience, or he has to see himself excluded forever from aspiring to the many high charges of the State.

All these embarrassments, united to those we already mentioned in the part relative to marriages, constitute the most difficult part of our programme. The religious question arouses serious difficulty on both sides. On one hand we have to overcome the prejudices of the country in that respect; on another, the sectaries of dissenting faiths show the highest repugnance to come to a country where their faith is merely tolerated, whilst marriage, which is the basis of the family and of society, does not rest upon solid and secure bases, and in which the difference of religion excludes them from certain elevated charges of the State.

They are serious obstacles; but they must be vanquished if we wish a wide current of spontaneous emigration to travel towards our country. From the countries of the Latin race and of the Catholic religion few emigrants can come to us; the Latin race has little tendency towards emigration. For a proof I will cite France, which with all her power and resources has shipwrecked in the enterprise of peopling her colonies. The tendency to emigration only exists in the Anglo-Saxon and Teutonic races. If, then, we seriously wish to people our country, we should open its gates to all races and religions, abolishing all the religious embarrassments that still exist in our laws relative to Dissenters.

By all I have just set forth to you, you must have comprehended what is our end in undertaking to found an international association of emigration, and what is the programme which it should have in view. You recognize that, to obtain a wide current of emigration to our country, it is indispensable, first of all, to treat of removing the obstacles that oppose themselves to it within the country. We see that these obstacles are material and moral; that among the material surges up the competition of slave labor, which it is needful to combat. We see that it is necessary to develop and perfect our ways of communication, to survey and mark off the public lands in localities appropriated to colonization.

As to the moral obstacles, we recognize as the principal the civil, the political, and the religious inequality, and we see that it is indispensable to reform our legislation on marriage, establishing civil marriage, admitting the naturalized foreigner to all the offices of State, and putting an end to the differences of religion in all that I said respecting the civil and political rights of the naturalized foreigner.

Our end, then, is very patent; our programme very clear. We need to employ all the means within our reach to remove all the material and moral obstacles that oppose themselves to emigration. It is in this sense that all the powers of our association should be directed. If we in heart wish that our country be enriched and aggrandized; if we wish that there travel hither a wide and vast emigration of individuals of all the advanced races of Europe and the United States, who profess all varieties of faith; if we wish them to settle and amalgamate with our population, forming a homogeneous and strong nationality, and not constituting in the bosom of our country little nationalities distinct in race, in language, in religion, in customs, enemies and rivals, without cohesion among them,—if, in fine, we wish that our country fifty years hence be a nation on the European or North American model, and not an insignificant nation on the African, the Chinese, or the Indian model, the road to follow is this that we have just traced. Let us follow it with boldness, with perseverance, with sincere patriotism. (Many cheers and shouts of "Well done.")

NOTES.

No. 1.

AMERICUS VESPUCIUS fares worse at the hands of some Portuguese authors than Pinzon. The Padre Ayres de Casal, in his *Corographia Brasilica*, urges that the Florentine "never accompanied Gonçalho Coelho or Christopher Jaques in their explorations of the coast of Brazil." Gen. J. I. d'Abreu Lima, in a note (page 8) to his *Historia do Brazil*, roundly asserts that Americus Vespucius did not accompany the two navigators mentioned above, (*todavia o que se póde negar com boas authoridades é que elle accompanhasse aos dois primeiros exploradores Portuguezes acima mentionados.*) It is true, also, that Robertson throws doubt upon some of the dates of Americus Vespucius, but more recent writers, of equal authority, give the account as stated in the text. This hesitation on the part of some Portuguese and Spanish historians, in regard to Americus, is doubtless influenced by the sentiment, on one side, that the employment of the Florentine by the King D. Manoel necessarily supposes an underrating of the Lusitanian navigators,—which does not follow, because the latter, in the expeditions referred to, appear to have had the supreme command: on the side of the Spaniards, they never could forgive Americus for having supplanted, in the New World, the name of Columbus, of whom they are as proud as if he were a Castilian.

No. 2.

It is commonly supposed that the wood yielding the red dye, *Cæsalpinia Brazilletto*, derived its common name, *Brazil-wood*, from its being principally imported from, and produced in, Brazil. This, however, is not the fact. It has been shown that woods yielding a red dye were called *Brazil-woods* long previously to the discovery of America, and that the early voyagers gave the name *Brazil* to that part of the continent, to which it is still applied, from their having ascertained that it abounded in such woods.—*Bancroft's Philosophy of Colors*, ii. 316–321.

No. 3.

The Padre Ayres Casal, in his *Corographia Brasilica*, says that the squadron "entered the Bay of *Santa Luzia*, which name was changed to that of *Rio de Janeiro*, because it was entered on the first day of the year, 1532." Any examination of the facts of the case as detailed by almost every other chronicler will not bear out the statements of Padre Ayres Casal.

No. 4.

Diario de Pedro Lopez de Souza, page 14, in which he explicitly says, "*Sabbado 30 de Abril, no quarto d'alva, eramos com a bocca do Rio de Janeiro.*"

No. 5.

The Madeira Christians were compelled to flee for refuge to the United States, in 1850; and in 1852 most intolerant acts were sanctioned by the Portuguese Government, in order to put an end to the so-called Protestant heresy in that island.

NOTES

Appendix A.

CHRONOLOGICAL SUMMARY OF THE PRINCIPAL EVENTS THAT HAVE TRANSPIRED IN THE HISTORY OF BRAZIL.

A.D. 1500. The continent of South America discovered on the 26th of January, by Vincent Yanez Pinzon, a companion of Columbus, and the first European who crossed the equator on his way to America.

" April 21, Pedro Alvarez Cabral, commander of the second Portuguese fleet that doubled the Cape of Good Hope, discovered that portion of the Brazilian coast now called Espirito Santo.

" On May 3, he landed at Porto Seguro.

1503. The Bay of All Saints discovered by Americus Vespucius.

1510. Diogo Alvarez Corrêa (Caramurú) shipwrecked at Bahia, (Bay of All Saints.)

1530. The unexplored territory of Brazil divided into captaincies by the King of Portugal.

1531. Martin Affonso de Souza entered the Bay of Nitherohy, (*Rio de Janeiro*,) previously visited by De Solis and Majellan. On the 22d of January he discovered the harbor of San Vincente, and there founded the first European colony.

1548. Numbers of Jews, having been stripped by the Inquisition of Portugal, were banished to Brazil.

1549. Thomé de Souza, the first governor-general, founded the city of San Salvador, (Bahia.)

1552. The first bishop appointed, to reside at Bahia.

1555. Villegagnon occupied the Bay of Rio de Janeiro with a colony of French Protestants, and built the fort which still bears his name, upon a small island in the harbor.

1567. The French expelled by the Portuguese and Indians.

" The city of St. Sebastian founded.

1572. The government of the colony of Brazil divided between two captains-general, resident severally at S. Salvador and Rio de Janeiro. Hence the name *Brazils*.

1576. The government again reduced to the jurisprudence of one captain-general, residing at Bahia.

1580. Brazil, in connection with Portugal, brought under the dominion of Spain.

1591. Thomas Cavendish, the English adventurer, sacked and burned S. Vincente.

1593. James Lancaster, commanding a marauding expedition, fitted out of London, captured and plundered Pernambuco.

1594. The French established a colony at Maranham.

1615. The French expelled from Maranham.

" The city of Belem (Pará) founded by Francisco Caldeira.

1624. The Dutch invaded Bahia.

1630. Second invasion of the Dutch, in which they took possession of the whole coast of Brazil, from the river of S. Francisco to Maranham. Pernambuco was their seat of government.

1637. Expedition of Pedro Teixeira, from Pará to Quito, by way of the river Amazon.

1640. Portugal and her colonies freed from the Spanish yoke.

1646. The Dutch defeated in the battle of the Guararapés, near Pernambuco; and in

1654. Finally expelled from Pernambuco.

1661. Holland abandoned, by negotiation, all claim to Brazil.

1675. The diocese of Bahia constituted an archbishopric.

1693. Regular mining for gold commenced.

1697. Settlements made in Minas-Geraes.

" Destruction of the famous Republic of the Palmares.

1710. Assault of the French upon Rio de Janeiro under Du Clerc.

1711. Capture of that city by Du Guay Trouin, and ransom by its inhabitants.

1713. Northern limits of Brazil defined by the treaty of Utrecht.

1729. Discovery of the diamond-mines in Serro Frio.

1758–60. Forcible and complete expulsion of the Jesuits from Brazil.

1763. Transfer of the capital from Bahia to Rio.

1805. Rev. Henry Martyn visited Bahia.

1808. Arrival of the royal family of Portugal.

" Publication of the Carta Regia.

" Establishment at Rio of the first printing-press in Brazil.

1811. Second printing-press established at Bahia. *Remark.*—These two were the only presses in use up to 1821.

601

1815. Brazil elevated to the rank of a Kingdom.
1817. Revolt in Pernambuco.
1818. Acclamation and Coronation of D. John VI.
1821. The Constitution of the Cortes of Portugal proclaimed and adopted at Rio.
" 24th April, D. John VI. returned to Portugal, leaving his son, Dom Pedro, Regent of Brazil.
1822. 7th September, Declaration of Independence.
" 12th October, Acclamation of D. Pedro as Emperor.
" 1st December, Coronation of D. Pedro I.
" " " Session of the Assembly convoked to draft a Constitution.
1823. Montevideo united to Brazil, under the title of the Cisplatine Province.
" The new Constitution offered to the Brazilians by the Emperor.
1824. March 25.—Sworn to, throughout the Empire.
" Revolt in Pernambuco. Confederation of the Equator proclaimed and suppressed.
1825. Independence of Brazil recognised by Portugal, August 29.
" Birth of the Imperial Prince D. Pedro II., December 2.
1826. On the death of King Dom John VI., the Emperor of Brazil, heir-presumptive to the Crown of Portugal, abdicated that crown to his eldest daughter, D. Maria II.
" Final separation of Montevideo from Brazil, that province becoming the Cisplatine Republic.
1831. Abdication of D. Pedro I., and Acclamation of D. Pedro II.
1832. War of the Panellas for the Restoration of the first Emperor.
1834. Reform of the Constitution, creating Provincial Assemblies.
1835. Revolution broke out in Pará, January 7.
" Diogo Antonio Feijo elected Regent.

1836. Donna Januaria recognised as Imperial Princess, and heiress to the throne.
1837. Feijo renounced the Regency, September 19.
" Pedro Araujo Lima appointed Regent pro tempore.
" Revolt in the city of Bahia, November 7.
1838. Restoration of Bahia, March 15.
" Death of José Bonifacio de Andrada.
" Lima elected to the Regency.
1839. First steam-voyage along the northern coast.
1840. Abolition of the Regency and Accession of Dom Pedro II. to the full exercise of his prerogative as Emperor.
1841. The Emperor's Coronation, July 18.
1843. Imperial marriages.
1844. The treaty between Brazil and England, signed in 1827, expired by limitation, November 11.
1845. Birth of the Imperial Prince D. Affonso.
1846. Birth of Donna Isabella, (heiress-apparent.)
1847. June 11, death of D. Affonso.
" July 13, Birth of Donna Leopoldina.
1849. December, First appearance of yellow fever.
1850. Suppression of the slave-trade. First steamship-line to Europe.
1852. Overthrow of the Buenos Ayrean Dictator Rosas by the aid of the Brazilian arms.
" Ground broken for the first railway.
1853. The first locomotive on the Mauá Railway, and steamers on the Amazon.
1854. Rio de Janeiro lit by gas.
1855. Surveys for various railways.
1857. The first section of the Pedro Segundo Railway finished.
1858–59. First section of Pernambuco Railway opened. Bahia Railway do.
1864. Marriage of the two princesses.
1865. War with Paraguay.
1866. Opening of S. Paulo Railroad.
1866. March 19, Birth of a son to D. Leopoldina.
1867. Dec. 7, Birth of a second son to the same.

IMPERIAL FAMILY.

The Crown of Brazil is hereditary in the line of direct succession.

EMPEROR—DOM PEDRO II. d'Alcantara, born Dec. 2, 1825. Salary, $440,000; and income from large estates.

EMPRESS—DONNA THERESA CHRISTINA, sister to the King of the Two Sicilies. Salary $54,800

IMPERIAL PRINCESSES—DONNA ISABELLA, heiress-apparent, born in 1846; DONNA LEOPOLDINA, born in 1847.

Donna Isabella married, October 16, 1864, the Count d'Eu, eldest son of the Duke de Nemours.

Donna Leopoldina, the second princess, married the Prince Auguste de Saxe-Coburg, December, 1864.

Emperor's Sisters—DONNA JANUARIA, born 1822. Married to the Prince D. Luiz Conde d'Aquilla, 1843. DONNA FRANCISCA, born in 1824. Married to the Prince de Joinville, 1843.

IN PORTUGAL.

Ex-Empress of Brazil, the Duchess of Braganza. DONNA AMELIA AUGUSTA, widow of Dom Pedro I.; born in 1812.

NOTE.—In case of the death of D. Pedro II. without issue, his sister Donna Januaria, who has three children, will succeed to the throne; and at her decease her eldest child will be the Monarch of Brazil.

Appendix B.

ABSTRACT OF THE BRAZILIAN CONSTITUTION, SWORN TO ON THE 25TH OF MARCH, 1824, AND REVISED IN 1834.

(1) BRAZIL is declared an Independent Empire, and its Government Monarchial, Constitutional, and Representative. (2) The Reigning Dynasty is to be Dom Pedro I. and his successors. (3) The Roman Catholic religion is constituted that of the State; but the exercise of all others is permitted. (4) The unrestricted communication of thought, either by means of words, writings, or the agency of the press, exempt from censure, is guaranteed: with the condition that all who abuse this privilege shall become amenable to the law. (5) A guarantee founded on the principles of the English Habeas Corpus Act. (6) The privileges of citizenship are extended to all free natives of Brazil, to all Portuguese resident there from the time of the Independence, and to all naturalized strangers. (7) The law is declared equal to all; all are liable to taxation in proportion to their possessions. (8) The highest offices of the State are all laid open to every citizen; and all privileges, excepting those of office, abolished. (9) The political powers acknowledged by the Constitution are the Legislative, the Moderative, the Executive, and the Judicial; all of which are acknowledged as delegations from the nation. (10) It is declared that the General Assembly shall consist of two chambers: the Chamber of Deputies are to hold their office for four years only; the Senators are appointed for life. (11) The especial attributes of the Assembly are to administer the oaths to the Emperor, the Imperial Prince, the Regent, or the Regency; to elect the Regent or Regency, and to fix the limits of his or their authority; to acknowledge the Imperial Prince as successor to the throne, on the first meeting after his birth; to nominate the guardian of the young Emperor in case such guardian has not been named in the parental testament; to resolve all doubts relative to the succession on the death of the Emperor or vacancy of the throne; to examine into the past administration, and to reform its abuses; to elect a new dynasty in case of the extinction of the reigning family; to pass laws, and also to interpret, suspend, and revoke them; to guard the Constitution, and to promote the welfare of the nation; to fix the public expenditure and taxes; to appoint the marine and land forces annually upon the report of the Government; to concede, or refuse, the entry of foreign forces within the Empire; to authorize the Government to contract loans to establish means for the payment of the public debt; to regulate the administration of national property and decree its alienation; to create or suppress public offices, and to fix the stipend to be allotted to them; and, lastly, to determine the weight, value, inscription, type, and denomination of the coinage.

(12) During the term of their office, the members of both Houses are alike exempted from arrest, unless by the authority of their respective Chambers, or when seized in the commission of a capital offence. For the opinions uttered during the exercise of their functions, they are inviolable. (13) All measures for the levying of imposts and military enrolment, the choice of a new dynasty in case of the extinction of the existing one, the examination of the acts of the past administration, and the accusation of Ministers, and of Councillors of State, are required to have their origin with the House of Deputies. For the indemnification of its members, it is decided that a pecuniary remuneration shall be allotted to each during the period of the sessions. (14) The number of the Senators is fixed at one-half that of the Deputies, and the members are required to be upwards of forty years of age, and to be in actual possession of an income amounting to at least eight hundred milreis per annum. (15) It is their exclusive attribute to take cognizance of the individual crimes committed by the members of the Imperial Family, Ministers, or Councillors of State, as well as of the crimes of Deputies during the period of the Legislature. Their annual stipend is fixed at fifty per cent. more than that of the Deputies.

(16) The Members of both Chambers are to be chosen by Provincial Electors, who are themselves to be elected by universal suffrage,—in which only minors, monks, domestics, and individuals not in the receipt of one hundred milreis per annum, are excluded from voting. (17) The Senators are nominated by the Provincial Electors in triple lists, from which three candidates the Emperor selects one, who holds office for life. (19) Each Chamber is qualified with powers for the proposition, opposition, and approval of projects of law. In case, however, the House of Deputies should disapprove of the *amendments or*

603

additions of the Senate, or *vice versâ*, the dissenting Chamber shall have the privilege of requiring a temporary union of the two Houses, in order that the matter in dispute may be decided in General Assembly.

(20) A *veto* is conceded to the Emperor; but it is only suspensory in its nature. In case three successive Parliaments should present the same project for the Imperial sanction, it is declared that on the third presentation it shall, under all and any circumstances, be considered that the sanction had been conceded. (21) The ordinary annual sessions of the two Houses of Legislature are limited to the period of four months.

(22) To each province of the Empire there is a legislative Assembly, for the purpose of discussion on its particular interests, and the promotion of projects of law accommodated to its localities and urgencies; but these Assemblies are not invested with any power excepting that of proposing laws of provincial interest.

(23) The attributes of the *moderative power* (which is designated the key to the entire political organization, and which is vested exclusively in the hands of the Emperor) are the nomination of Senators, according to the before-mentioned regulations; the convocation of the General Assembly whenever the good of the Empire shall require it; the sanction of the decrees or resolutions of the Assembly; the enforcement or suspension of the projects of the provincial Assemblies during the recess of the Chambers; the dissolution of the House of Deputies; the nomination of Ministers of State; the suspension of magistrates; the diminution of the penalties imposed on criminals; and the concession of amnesties.

(24) The titles acknowledged in the Constitution as appertaining to His Majesty are "Constitutional Emperor and Perpetual Defender of Brazil." His person is declared inviolable and sacred, and he himself exempt from all responsibility. He is, moreover, designated as the chief of the *executive power*, which power is to be exercised through the medium of his Ministers. Its principal functions are the convocation of a new General Assembly in the third year of each legislature, the nomination of bishops, magistrates, military and naval commanders, ambassadors, and diplomatic and commercial agents; the formation of all treaties of alliance, subsidy, and commerce; the declaration of war and peace; the granting of patents of naturalization, and the exclusive power of conferring titles, military orders, and other honorary distinctions. All acts emanating from the executive power are to be signed by the Ministers of State, before being carried into execution; and those Ministers are to be held responsible for all abuses of power, as well as for treason, falsehood, peculation, or attempts against the liberty of the subjects. (25) In addition to the *Ministry*, a Council of State is also appointed, the members of which are to hold offices for life. They are to be heard concerning all matters of serious import, and principally on all subjects relating to war and peace, negotiations with foreign States, and the exercise of the moderative power. For all counsels wilfully tending to the prejudice of the State, they are to be held responsible.

(26) The *judicial power* is declared independent, and is to consist of judges and juries for the adjudication of both civil and criminal cases, according to the disposition of future codes for this effect. The juries are to decide upon the fact, and the judges to apply the law. For all abuses of power the judges, as well as the other officers of justice, are to be held responsible. It is within the attributes of the Emperor to suspend the judges in the exercise of their functions; but they are to be dismissed from office only by a sentence of the supreme courts of appeal instituted in all the provinces.

(28) The *presidents of the provinces* are to be nominated by the Emperor; but their privileges, qualifications, and authority are to be regulated by the Assembly.

(29) If, after the expiration of four years, it should be found that any articles of the Constitution required reform, it was decreed that the proposed amendment should originate with the House of Deputies; and if, after discussion, the necessity of the reform was conceded, an act was to be passed and sanctioned by the Emperor in the usual manner, requiring the electors of the Deputies for the next Parliament to confer on their representatives especial powers regarding the proposed alteration or reform. On the assembling of the next House of Deputies, the matter in question was to be proposed and discussed, and, if passed, to be appended to the Constitution and solemnly promulgated. (The reforms were few,—the two principal being the regulation of succession in case of the death of D. Pedro II. without issue, his sister Donna Januaria, or her children, becoming heirs; and changing the provincial councils to provincial Assemblies.)

(30) Finally, civil and criminal codes are organized; the use of torture is abolished; the confiscation of property is prohibited; the custom of declaring the children and relatives of criminals infamous is abrogated, and the rights of property and the public debt are guaranteed.

Appendix C.

The following lines were composed by D. Pedro II., and written by him in the album of one of the Maids of Honor. They were doubtless never intended for the public eye, but were obtained through a member of the diplomatic corps at Rio Janeiro. Their didactive character and great compactness in the Portuguese make a poetic translation exceedingly difficult; but they have been kindly and very faithfully rendered into English verse for this volume by Mr. D. Bates, of Philadelphia, whose "Speak Gently" has become a household word.

Se fui clemente, justiceiro, e pio,
Obrei o que devia. É mui pesada
A sujeição do sceptro; e quem domina
Não tem ao seu arbitrio as leis sagradas:
Fiel executor deve cumpri-las
Mas não pode altera-las. É o throno
Cadeira da Justiça; quem se assenta
Em tão alto lugar, fica sujeito
Á mais severa lei; perde a vontade!
Qualquer descuido chega a ser enorme,
Detestavel, sacrilego delicto!
Quando no horizonte o sol espalha
Sobre a face da terra a luz do dia,
Ninguem o admira, todos o conhecem;
Mas se eclipsado acaso se perturba,
Nesse instante infeliz todos se assustão,
Todos o observão, todos o receião:
Logo se premiei sempre a virtude,
Se os vicios castiguei, nada merecei.

P. II.

Dec. 1852.

If I am pious, clement, just,
  I'm only what I ought to be:
The sceptre is a weighty trust,
  A great responsibility;
And he who rules with faithful hand,
  With depth of thought and breadth of range,
The sacred laws should understand,
  But must not, at his pleasure, change.

The chair of justice is the throne:
  Who takes it bows to higher laws;
The public good, and not his own,
  Demands his care in every cause.
Neglect of duty,—always wrong,—
  Detestable in young or old,—
By him whose place is high and strong,
  Is magnified a thousandfold.

When in the east the glorious sun
  Spreads o'er the earth the light of day,
All know the course that he will run,
  Nor wonder at his light or way:
But if, perchance, the light that blazed
  Is dimm'd by shadows lying near,
The startled world looks on amazed,
  And each one watches it with fear.

I likewise, if I always give
  To vice and virtue their rewards,
But do my duty thus to live;
  No one his thanks to me accords.
But should I fail to act my part,
  Or wrongly do, or leave undone,
Surprised, the people then would start
  With fear, as at the shadow'd sun.

APPENDIX D.

SLAVERY AND THE SLAVE-TRADE IN BRAZIL—ENGLAND AND BRAZIL.

[Translated from the Jornal do Commercio of Rio de Janeiro of May 26, 1856.]

It is impossible to undertake, with greater energy and with more honesty than our Government did, the difficult task of suppressing the slave-trade. This is a truth which cannot be contested, it being a self-evident fact.

Notwithstanding the old usages of our agricultural and manufacturing industry, which were actually based upon the slave-trade, and which must have suffered from its suppression, prejudices did not even spring out of these circumstances. Injured interests, habits broken up, did not even raise a cry : reason prevailed, and the prospect of future national welfare was acknowledged, and the whole nation and its Government did not hesitate to accept all the sacrifices of the present, in order to leave to future generations the country freed from this centennial crime, however painful may be its just expulsion.

In consequence of this change of opinion in Europe, and especially in England, toward Brazil, we should have thought that the relations between the Governments of Brazil and Great Britain had attained such a degree of brotherly esteem that it might be wished to exist between the official representatives of both nations joined by so many ties of mutual interest. We were convinced that, seeing the efforts made by the Brazilian Government properly supported by the general opinions of the people, the English Cabinet would certainly give it credit and the homage of its sympathies. But the notes addressed by the British legation to the Imperial Cabinet, when an attempt was made to land slaves near Pernambuco, and especially the last of their notes, have completely destroyed our illusions on this subject.

After having subdued the indignation caused by reading that note, considering its full extent, we said to ourselves, "What can the British Government mean when, in our present circumstances, it assails us with such a threat?" Is it the suppression of the slave-trade? Certainly not. If proper reflection could not suggest to that Government that by carrying the threatened measure into execution they would only promote and encourage that very trade which we are anxious to suppress, we would recommend to them the lessons given by the years 1830 to 1850 inclusive!

Public opinion in support of our Government has strongly sustained, and maintained with all possible watchfulness, with all the power of reason, the conviction that the suppression of the slave-trade is a true national interest: this conviction gave to our Government an incalculable strength, by which it was able to obtain the entire and immediate extinction of that trade, so that whole years have passed without any attempt being made to violate this law.

And when an attempt of this kind is occasionally made, it is always done through merchants of Lisbon and in Africa connected with North American adventurers, and carried on in vessels from the United States; and even the Brazilian Government succeeds in discovering the agents of this crime, and manages to watch and accompany them and to arrest them at the very moment when they are going to perpetrate it.

And in view of these facts the British Government, instead of congratulating our functionaries and applauding their efforts, sends us insults and threats.

In the two attempts made by Americans to establish the slave-trade, praise must be given to the Government of Brazil alone, which has so ably succeeded in defeating and repelling them. England must be conscious enough that with all her squadrons on the coast of Africa, and on the vast seas of this Empire, committing even all the silly excesses of the Aberdeen bill, it would not have effected any thing against attempts of that kind; and when our Government, by its measures and vigilance, succeeds in obtaining this admirable result, we find it difficult to explain the object of the note alluded to. But why, on this occasion, did not the British Government act as it would do if it believed that insults and threats are the best means to suppress this trade? They ought to direct their threats and insults not against us, who are innocent in this case, but against the United States.

The crime was wholly of foreign origin, and its authors were in New York and Boston. Brazil has not arms long enough to reach them ; but every thing that could be done was actually done, and, at the very moment that a North American crime was about to be perpetrated, a Brazilian authority stopped it.

606

But Albion's arms are long, and, with its diplomacy and cruisers, why does not the Government of Great Britain turn all its means of action and all its arrogant demands toward the Cabinet at Washington? Why does she not compel it to prevent such criminal enterprises at the hands of its bold adventurers and filibusters?

The following is the contract between a number of Mina blacks (who freed themselves) and the captain and consignee of the British brig Robert,—in which vessel they sailed for their native land, and arrived safely:—

"CHARTER PARTY.

"RIO DE JANEIRO.

"On the 27th of November, 1851, it is agreed between George Duck, master of the British brig called the *Robert*, A 1, shall receive in this port sixty-three free African men (women and children included in this number) and their luggage, and shall proceed to Bahia, and remain there, if required, fourteen days, and then proceed to a safe port in the Bight of Benin, on the coast of Africa not south of Badagry, (the port of destination being decided in Bahia,) and deliver the same, on being paid freight here, in this port, the sum of £800, to be paid before the sailing of the next British packet. The master binds himself to provide for the said passengers sixty pounds of jerked beef, two and a half alquieres of farinha, and one-half an alquiere of black beans, daily; a cooking-place and the necessary firewood to be furnished by the captain; half a pipe—say sixty gallons—of water to be supplied daily. The master is allowed to take any cargo or passengers and luggage that may offer at Bahia for the benefit of the ship.

"Passengers and luggage to be on board on or before the 15th of December, 1851, and disembark within forty-eight hours after the ship's arrival at the port of destination.

"Penalty for non-performance of this agreement, five hundred pounds sterling.

"GEORGE DUCK,
"RAPHAEL JOSÉ OLIVEIRA."

APPENDIX E.

TABLES OF BRAZILIAN COINS, WEIGHTS, AND MEASURES.

THE following statistics, from the consular bureau of the United States, were most carefully made out by J. S. Gillmer, Esq., American Consul at Bahia, and forwarded in his reports to the State Department at Washington. These are the most correct computations of Brazilian coins, weights, and measures, ever presented to the English and American public.

Table exhibiting the legal gold and silver coins of Brazil, with their weights in dwts. and grains Troy, fineness, and comparative value in Federal money of the United States:—

GOLD.

Denomination.	Dwts.	Grains.	Comparative Value.
Pecas	9	$5\frac{1}{3}$	$ 8.20
Moedas	5	$4\frac{1}{2}$	4.62
Soberanos (20 milreis.)	11	$12\frac{5}{8}$	10.24
Half do	5	$14\frac{1}{3}$	5.12

SILVER.

Denomination.	Dwts.	Grains.	Comparative Value.
Patacão	17	7	$ 1.00
Two patac.as	5	0	30
Two-milreis piece,	16	$9\frac{1}{2}$	94
One do	8	$4\frac{3}{4}$	47
Five hundred reis piece	4	$2\frac{1}{3}$	$23\frac{1}{2}$

COPPER COIN

is composed as follows:—

The real (pl. reis) imaginary.
Five-reis piece, (imaginary.)
Ten " " (out of use.)
Twenty-reis do. one vintem.
Forty " do. two vintems.

The latter weighs 18 dwts. 10 grains, of the nominal value of $2\frac{1}{5}$ cents. Twenty-five of these pieces make a milreis, or 1000 reis, the real being merely used as a numeral.

The above calculations are not given as absolutely correct, but, with the exception of very slight fractional differences, they are so.

PAR OF EXCHANGE.

The Brazilian "Soberano," or twenty-milreis piece of the recent coinage, being worth (according to its relative value compared with our gold coin) $10.24, it follows that the "par of exchange" between the two countries is $51\frac{1}{5}$ cents per milreis; but, the currency of Brazil being more than one-half composed of Government paper money, this standard cannot be applied to commercial transactions as a guide, and in the absence of direct exchange transactions with the United States, we must be governed by the rate of exchange on London, which either rises or falls as influenced by the commercial or other vicissitudes of the day.

The rate of exchange on London being twenty-eight pence per milreis, by taking the value of the pound sterling at $4.80 cents, the result is fifty-six cents as the value of the milreis in United States currency.

WEIGHTS AND MEASURES.

The "Marco" is divided into

8 Ounces,
64 Octaves,
192 Scruples,
4608 Grains,—which are equal to 3541½ Troy grains, or 229.460 French grammes,—83 lbs. Troy weight being equal to 135 "Marcos."

COMMERCIAL WEIGHTS.

The "Aratel," or Round, contains

2 Marcos,
4 Quartos,
16 Ounces,
128 Octaves, and
9216 Grains,—which are equal to 7082 Troy grains,—110.729 pounds being equal to 112 lbs. avoirdupois.

32 pounds = 1 Arroba = 32⅗ lbs. avoirdupois.
4 Arrobas or 128 lbs. (Portug.) = 1 Quintal = 129½ lbs. avoirdupois.
13½ Quintals or 54 Arrobas = 1 ton = 1748¼ lbs. avoirdupois.

DRY MEASURES.

The "Alqueire" of Bahia, in daily use for corn, mandioca, &c., contains 2475 cubic inches, equal to 1.15 Winchester bushels, and is divided into halves and subdivided into quarters, eighths, &c.

The "Moio" of Bahia contains 30 alqueires, or "Fangas," as they are called when used for measuring lime. The "Moio," therefore, is equal to 34.6 Winchester bushels.

The "Alqueire" of Rio de Janeiro contains 2322 cubic inches, equal to 1.08 bushels.

(The "Moio" of Lisbon is composed of 15 Fangas, and each Fanga of 4 Alqueires; the Lisbon Alqueire contains 824.832 cubic inches; the Lisbon "Moio," therefore, is equal to 23.02 bushels.)

LIQUID MEASURE.

Duties are exacted at the custom-houses of the Empire on liquids by the "Medida" of Rio de Janeiro, which contains 162.4 cubic inches, 142.241 "Medidas," being equal to 100 gallons; but in the different provinces they are sold by local measure.

In the province of Bahia, oil, rum, &c. are sold by the Canada of Bahia, which contains 435 cubic inches, equal to 1.883 gallons,—one Canada, therefore, being nearly equal to 1⅞ gallons.

The "Canada" is divided into halves and subdivided into quarters, called "Quartillos," eighths, &c.

CLOTH MEASURE.

The "Covado" and
"Vara."

The former is equal to 26.7 inches, and the latter equal to 43.3 inches: each is divided into halves, thirds, quarters, and eighths.

LONG MEASURE.

12 lines	= 1 inch.
8 inches	= 1 Palmo.
12 inches	= 1 Pé or foot.
5 Palmos	= 1 Vara.
2 Varas	= 1 Braça.
935.276 Braças	= 1 mile, (Port.)
3 miles	= 1 league.
18 leagues	= 1° of latitude.

LAND MEASURE.

Land in Brazil is bought and sold by the "Tarefa" of 900 square Braças, or 3600 square Varas, which are equivalent to 4330 (Eng.) square yards.

The "Geira" of land in Portugal is considered equivalent to 4840 square Varas, equal to 5821 square yards.

Note for 1866.—The French metrical system is now adopted by law; but the present denominations will doubtless be kept for a long time. The heavy copper money ought to be destroyed, and little coins, like the American three-cent piece, be substituted, so that the immense amount of paper money (like omnibus and ferry tickets) may be abolished. The American Bank Note Engraving Company of New York are now preparing very beautiful new postage-stamps, adorned with the head of the Emperor.

APPENDIX F.

POPULATION.

NOTHING is more difficult to ascertain with correctness than the population of Brazil. No census of the whole country has as yet been taken; and when we see it stated from "official documents," it means nothing more than conjecture and approximation. I believe that the population of Brazil is not far from nine millions. Thomas J. Adamson, Esq., United States Consul at Pernambuco, a careful collector of statistics, gives, in 1864, the latest estimated population of Brazil at more than ten millions. I think that his Brazilian authority made the estimate too high.—J. C. F.

Estimated Population of Brazil, by Thomas J. Adamson, Esq., Pernambuco.

Provinces.	Free Population.	Slave Population.	Proportion of Slaves to Free Persons.
Amazonas.	68,000	1,000	1 to 68
Pará.	300,000	20,000	1 to 15
Maranham.	330,000	70.000	1 to 4.714
Piauhy.	200,000	20,000	1 to 10
Ceará.	504,000	36,000	1 to 14
Rio Grande do Norte.	200,000	25,000	1 to 8
Parahiba.	250,000	30,000	1 to 8.137
Pernambuco.	1,040,000	260,000	1 to 4
Alagoas.	250,000	50,000	1 to 5
Sergipe.	220,000	55,000	1 to 4
Bahia.	1,100,000	300,000	1 to 3.666
Espiritu Santo	50,000	15,000	1 to 3.333
Rio de Janeiro	1,000,000	400,000	1 to 2.500
São Paulo.	700,000	80,000	1 to 8.750
Paraná.	80,000	20,000	1 to 4
St. Catharina.	135,000	15,000	1 to 9
Rio Grande do Sul.	380,000	40,000	1 to 9.500
Minas.	1,200,000	250,000	1 to 4.800
Goyaz.	205,000	15,000	1 to 13.666
Matto Grosso.	95,000	5,000	1 to 19
Total.	8,307,000	1,707,000	
Grand total.	10,014,000		

THE YELLOW FEVER OF BRAZIL.
(WRITTEN FOR "BRAZIL AND THE BRAZILIANS" BY A. R. EGBERT, M.D.)

In a publication like the present, any elaborate medical disquisition on the yellow fever of Brazil would be obviously misplaced; yet in a work upon that country a brief sketch of this disease seems necessary.

Owing to the peculiar situation of the Brazilian Empire, any one unacquainted with the country would naturally suppose that it would abound in those causes which, in all tropical countries, are so inimical to the lives of strangers. This is not the case, but exactly the reverse. Lying immediately under "the Line," Brazil is, for its situation. singularly mild and healthful. Its climate is delightful, and, along the coast especially, is tempered by a cool and never-failing breeze; while, in the interior, the elevation of the country compensates for its proximity to the Equator,—thus proving that climate must never be judged by latitude alone. All these things go to show why Brazil has been so free from the ravages of that "terrible scourge," the yellow fever.

Like all other epidemics, yellow fever hides its origin in the mists of the past. These giant devastators of nations have had no chroniclers to record their birth and early history. Some physicians imagine they can find this fever described in the writings of Hippocrates; but they forget that the peculiar symptoms on which they rely to establish the identity—black vomit and yellowness of the skin—are by no means peculiar to the disease in question. The prevalent opinion among those who have investigated the subject is that the disease is of modern origin; and some facts seem to connect it with the slave-trade. It certainly made its appearance simultaneously with that traffic, and some of our Southern physicians are convinced that it, like the blacks, was imported from Africa.

As far as our knowledge extends, Père Dutertre is the earliest writer who can be said to have alluded to this "frightful scourge of the warmer shores of the Atlantic." He saw it in 1635, in the Antilles, and expressly tells us that before that time it was unknown in those islands. In 1647 it was in Barbadoes. Père Labat found it raging at Martinique in 1649. The earliest period at which this epidemic occurred in the territory of the United States was in 1693, at Boston. Since then it has been, unfor tunately, too well known to our ancestors over the whole Atlantic coast.

It first appeared in Brazil in December, 1849, or January, 1850, and committed its greatest ravages in 1850, in the maritime provinces. It was especially violent at Pará, Bahia, and Rio de Janeiro. Pernambuco escaped. Bad as it was, the accounts of its ravages were greatly exaggerated. In the whole Empire of Brazil, the population of which is more than seven millions, there were from this disease, in 1850, in fourteen thousand deaths; and, according to the official reports, there were not quite four thousand deaths from yellow fever in the city of Rio de Janeiro,—whose population is three hundred thousand. Dr. Paulo Candido and Dr. Merrilles, who stand deservedly high in the medical profession, corroborate this statement. Dr. Lallemant, an eminent German physician of the first professional ability at Rio exaggerates, it seems to us, both the number of cases and deaths: the former he places at one hundred thousand, and the latter at ten thousand,—which seems to be utterly at variance with the statement of all the reports from other and equally credible sources. But, even admitting Dr. Lallemant's figures, we can see how much less was the mortality than at New Orleans, (a city of one-third the population of Rio,) where in the month of August, 1853, 5269 perished from this fell disease. And yet it has been represented that the capital of Brazil is the most unhealthy place in the world! According to Dr. Lallemant, 475 died at Rio in 1851; 1943 in 1852; 853 in 1853; and only four in 1854. In 1857 a few scores of cases occurred, but we have not the exact number at hand.

In 1854 the disease had entirely disappeared, and has not since shown itself until in the beginning of 1857, and in the month of March of that year it ceased.

There is little doubt that the cause of yellow fever is peculiar and specific. But great diversities of opinion exist upon the nature of this cause. Some consider it to be a living, organized, microscopic being, and others regard it as a species of ferment. Strong reasons are adduced in favor of both theories; but nothing is positively and definitely known of the nature of the cause.

As to whether the disease be contagious or not, authorities are divided. But it is now beginning to be generally conceded that it is not contagious; and the burden of proof is certainly in favor of this view of the subject.

Yellow fever exhibits a great diversity of phenomena, occasioned by a variety of influences,—assuming the particular form in accordance with the circumstances of its appearance,—scorbutic, typhous, or whatever the case may be.

[The symptoms are then described. The writer thus continues:—]

These symptoms generally last from a few hours to three days, when they subside, leaving the patient cheerful and hopeful. But this is a delusive calm, and continues from a few hours to twenty-four. Then set in debility and prostration. In severe cases the weakness is extreme: the pulse is quick, irregular, and feeble; the skin is yellow, orange, or of a bronzed aspect; the blood appears to be nearly stagnant in the capillaries, and the dependent and extreme parts of the body become of a dark purplish hue. The tongue is now often brown and dryish in the centre, or smooth, red, and chapped; and sordes occasionally collects about the gums and teeth. The stomach resumes its irritability, and the *black vomit* appears. The bowels often give way and discharge large quantities of black matter, similar to that ejected by the stomach,—and occasionally hemorrhage takes place from various parts of the body; low delirium sets in; an offensive odor sometimes exhales from the whole body; the eyes become sunken and the countenance collapsed, and death takes place, often quietly, but sometimes in the midst of convulsions.

Occasionally patients will die of yellow fever without either the black vomit, yellowness of the skin, or hemorrhage appearing.

Instead of pursuing this fatal course, the system very often reacts after the period of abatement, and a secondary fever sets in, which may be of various grades of violence. It continues a variable length of time,—sometimes speedily terminating in health, and sometimes running into a typhoid form, which may last, with various results, for two or three weeks or more. In severe cases the convalescence is always extremely tedious, and the patient is often incommoded by obstinate and unhealthy sores or abscesses in various parts of the body.

In some cases the animal functions seem to be at first almost untouched. The patient may be walking in the streets and nothing call attention to his case, unless, it may be, an unusual expression of countenance. Upon his pulse being examined, it is found to be exceedingly feeble, if not quite absent at the wrist. Black vomit and death speedily ensue. These have been called "walking cases."

The modes of treatment are many and widely different.—sometimes none of the slightest use.

[As the treatment of yellow fever in the United States is within the reach of all, it has been thought best to omit mention of it here, and only to insert Dr. Egbert's account of the Brazilian method as laid down by one of the first physicians of the Empire.—J. C. F.]

The prevention of the disease is of course even more important than its treatment. Individuals who are unable to leave the place where the disease prevails should select a residence in the highest and healthiest spots; should sleep in the highest parts of the house; should avoid the night-air; should abstain from fatiguing exercise, exposure to alternations of temperature, and excesses of all kinds; should endeavor to maintain a cheerful and confident temper; should use nutritious and wholesome but not stimulating diet; and, if compelled to enter any spot where the atmosphere is known to be infected, should take care not to do so when the stomach is empty or the body exhausted by perspiration or fatigue.

According to the best medical authorities in the United States, attempts to guard against this disease by low diet, bleeding, purging, or the use of mercury, are futile.—if not worse; for they weaken the system, and the weaker the system the less is it able to resist the entrance of the poison, or its influence when absorbed.

The following mode of treatment is that recommended and pursued by Dr. Paulo Candido, of Rio, and was under him eminently successful.

"The first step is to cleanse the digestive canal. Castor oil, in a dose of two, four, or even six ounces, must be administered without delay, whatever be the state of the patient. If he obstinately rejects this remedy, employ citrate of magnesia or neutral salts in sufficient quantity to produce eight evacuations. This effect ought to be kept up the succeeding days, but with greater moderation. Neither foreign substances nor intestinal secretions ought to be allowed to remain: they become the centres of poisonous matter. The torpor of the intestines does not allow us to trust wholly to purgatives: it is necessary to administer injections, and I make use of the following mixture:—

" ℞.—Expressed juice of Persicaria, cut up and steeped in water...... 2 lbs.
　Lemon-juice (skin and pulp cut and squeezed)........................ 4 oz.
　Sulphate of Soda.. 4 "
　Socotrine Aloes... 4 "
　Camphor, and Sulphate of Quinine, each................................ 1 drachm.
　M.—Saturate with kitchen salt.
　Q. S. for two or three enemas.

"If persicaria cannot be obtained, it may be replaced by the same quantity of infusion of chamomile, orange-leaves, or sea-water.

"These injections must be given every two hours, as hot as possible: they are rejected immediately, but are usually followed by an abundant perspiration; but the use must be continued.

"Hot sinapisms at the soles of the feet, the knees, and the thighs, ought to be employed from the first, conjointly with the above remedies, and repeated until some abatement of fever ensues.

"Friction all over the body, particularly on the abdomen, groin, armpits, arms, with the following:—

" ℞.—Camphorated Vinegar.. 1 lb.
　Sulphate of Quinine.. 2 drachms.
　Tincture of Quinine... 2 oz.
　Creosote.. 1 drachm.
　M.

"A drachm of creosote in half a pound of spirits of wine, to rub the abdomen, arms, and sides, is an excellent means of provoking perspiration and producing other effects. These frictions must be performed under the coverings of the bed, in order not to chill the patient, and must be continued for three or four hours. Besides their antiseptic action, they produce perspiration.

"A weak infusion of borage, sweetened, every hour, very hot, each infusion prepared at the time of being taken; or of hot gum-water.

"If the perspiration cannot be effected in two or three hours, we must have recourse to the tincture of aconite napel, (monk's-hood,) one drachm of, in two pounds of water, to take by spoonfuls every quarter of an hour, without interrupting the other means.

"Besides, in four hours after the evacuants have been administered, the use of interior chloride must commence:—

" ℞.—Eau de Labarraque... 2 drachms.
　Distilled water, slightly acidulated with Muriatic Acid...... ½ bottle.
　M.—

"Take three spoonfuls of this mixture in half a cup of fresh water, or simply a spoonful of Eau de Labarraque in a glass of pure water, and take a spoonful of this solution every quarter or half hour.

"Sugar must never be added to Eau de Labarraque. It must be saturated with chloride, which is easily known by the smell, and kept out of the light.

"For very delicate persons the dose must be weaker. All these means must be continuous: they do not contradict each other.

"At the end of twenty-four hours, the malady is generally subdued; but the medicaments must not cease, but the employment of them relaxed or the intervals augmented.

"Relapses, and that deceitful calm that is so often noticed preceding death, take place from the abdominal secretions having been permitted to be reabsorbed. Therefore the medicaments must be continued.

"I permit no broth, oranges, wine, or any thing else, until two days after the symptoms have disappeared and when the pulse has lowered perhaps to forty.

"I have often had recourse to sialagogues for the secretion of saliva: these are such substances as ginger, cinnamon, liquorice-root, kept in the mouth. I advise amateurs to smoke cigars.

"Tonics, especially the preparations of quinine, are very useful in small repeated doses when only weakness remains.

"I ought to add, that if the terrible symptom of suppression of urine takes place, I give to the patient a drachm of nitrate of potash dissolved in a bottle of water,—half a cupful every half or quarter of an hour; injections of an ounce of camphorated vinegar in two cupfuls of tepid water; frictions of the same vinegar or camphorated oil of almonds on the abdomen repeated at short intervals.

"I have no faith in bleeding, leeches, cupping, calomel, quinine internally, ammonia, laudanum, opium, arsenic, turpentine, nitrate of silver, ice, hot or cold baths, &c."

The treatment of Dr. Paulo Candido differs very materially from that pursued by the prominent physicians of the United States. It also differs from that pursued in the West Indies. The reason of this is, I presume, owing to the different character of the disease in Brazil. The yellow fever first appeared in Brazil on the 28th of December, 1849, and remained in the country from that time until March, 1854; in December, '57 it reappeared in a milder form, and in April disappeared.

The following is a schedule, from official records, of the number of deaths in the Empire and in the Capital, (where it was the most severe,) separately, during each year:—

	Population.	Deaths in 1850.	Deaths in 1851.	Deaths in 1852.	Deaths in 1853.	Deaths in 1854.
Empire.	7,000,000	14,000	8719	9527	8531	
Rio de Janeiro.	300,000	3827	475	1943	853	04

This table shows that the disease was comparatively light, the percentage being small.

The following is an extract from the "Report of the Minister of the Empire" for 1855.

"The yellow fever, as an epidemic, may be considered nearly extinct in this city, (Rio.) This benefit is particularly owing to the very vigilant sanitary policy that has been established. The great number of ships from all parts of the world which frequent this port has ever been the great focus of infection for this and other epidemics.

"Happily, this has been combated by the disinfecting measures that have been resorted to, and by the prompt succor that has been rendered to the afflicted crews, who, as soon as the epidemic shows itself, are conducted in the steamer (health-steamer) to the maritime hospital of Jurujuba, where they receive the most judicious and careful treatment. This hospital merits all praise. During the past year there entered 1627 patients, (not all yellow fever:) cured, 1576; died, 40. Therefore the mortality was less than 2½ per cent."

The origin of this pestilence in Brazil is a mooted point, and has given rise to the most conflicting views among the best observers: for example, Dr. Pennell, of Rio, and Dr. Patterson. of Bahia, entertain precisely opposite opinions,—the former contending for the indigenous, the latter for the foreign, origin of the disease; and both offer cogent arguments and striking facts in support of the opposite conclusions.

The scope of this paper does not admit of medical discussion; yet, as the facts observed by Dr. Pennell are highly important. and, as his conclusions entirely coincide with those of Dr. Dundas, a short sketch of them will be given.

They state that for some years the fevers of the country had been clearly changing their character, and the genuine remittent had been little seen for three years; that it was replaced in 1847, '48, and '49, by a fever of its own class, popularly known as the "Polka fever," but in reality a remittent; and that this fever was, in its turn, superseded by the yellow fever, a disease with similar features.

Coincident with these and other changes in the diseases of Brazil, the climate in its broad features had altered strangely. Thunder-storms—formerly of daily occurrence at a certain hour, so that appointments for business or pleasure were made in reference to them as to taking place "before" or "after" the shower during the summer—are now but seldom heard. There was, too. at the commencement and during the continuance of the pestilence, a stagnation and want of elasticity in the atmosphere, from the cessation to a great degree of the fresh and regular winds from the sea,—a change very perceptible and very oppressive.

The supporters of the theory of the foreign origin of yellow fever insist that it was imported by a certain ship from New Orleans to Bahia, (some say to Pernambuco,) and thence diffused throughout the Empire. Some of them urge that it was imported from Africa by slave-ships, whilst the facts adduced by Dr. Pennell go far to establish, as already stated, its indigenous parentage. Dr. Dundas says that in support of this opinion we have the strong additional fact that for the last forty years there has existed, uncontrolled by any efficient quarantine-laws, an extensive intercourse with the United States, Africa, and the West Indies,—the very hotbeds of yellow fever,—and yet up to 1849 Brazil remained perfectly healthy. Can we then in reason believe, if the disease be deemed really importable, that the maritime cities of Brazil could, under such circumstances, have escaped infection for a period of forty years? Though it is usual to say that no epidemic has visited Brazil, yet several of the older writers, as Rocha Pita in 1666, Père Labat in 1686, and Fereira da Rosa in 1694, have recorded the appearance of epidemics closely resembling the yellow fever, which, after persisting for some years, and desolating some of the large cities on the coast, finally passed away.

Drs. Pennell and Dundas conclude, from the above and other facts, that the yellow fever, which recently afflicted Brazil, is not an imported disease, but owes its origin to certain obscure atmospheric disturbances, embracing variations of temperature, hygrometric influence, electrical tension, atmospheric pressure, &c.; and, judging from the previous history of Brazil, we believe that these unfavorable conditions are but temporary: and we are rejoiced to be able to hope that the disease has nearly passed away, that Brazil will maintain its character of unparalleled salubrity among the tropical regions of the globe, and will deserve its title of "the Italy of the New World."

The following statements will show the greater healthfulness of Brazil as compared with the United States.

In 1847, in New Orleans, there were 2252 deaths from yellow fever. The population was about 90,000.

In 1853, there were, from May 26 to October 22, 8406 deaths from the yellow fever. The population of the city was more than 100,000; but, owing to so many having fled, it was estimated that not more than 50,000 people were in the city during the prevalence of the epidemic.

In 1854, there were nearly 14,000 cases of yellow fever in New Orleans; from July 14 to October 15, there were 2420 deaths from this cause. The population was about 102,000.

In Mobile, during the year 1853, there were, from August 1 to September 16, 611 deaths from yellow fever. Population of the city, 12,500.

In Natchez, in 1853, there were, from July 17 to September 20, 263 deaths from yellow fever. Population, 5000, of which only 2000 remained in the city.

In Charleston, in 1854, there were from fifteen to twenty deaths daily during the height of the disease. Population, 29,000.

In Galveston, in 1854, there were from fourteen to fifteen deaths daily. Population, 7000.

In Savannah, during the year 1854, from August 23 to October 17, there were 919 deaths from yellow fever. Population, 11,000. Three-fourths of the population fled to the country: the roads a few miles from the city were lined with the tents of the fugitives.

In general, it has been found that from one-half to two-thirds of the population flee from the cities in the United States when any severe epidemic prevails; and this must be born in mind whilst reading the above data.

In the terrible scourge at Norfolk and Portsmouth, Va., in 1855, 45 per cent. of the whole population died from yellow fever. The city was nearly deserted, there being scarcely a sufficient number to take care of the sick. The duration of the disease was one hundred and twenty-seven days.

Now, compare these data with the table before mentioned, and we immediately see the comparative immunity of Brazil from the yellow fever even during its most fatal visits. Under such circumstances further comments, so far as comparison with the United States is concerned, are useless.

It is very probable that the mildness of the climate may have exerted a greatly modifying influence upon the disease, rendering it less severe and less fatal.

In writing the above article we do not profess to have done any thing more than to have made a mere compilation from different authorities and arranged them to suit our purpose. We therefore, whatever may be the merit of the production, disclaim all originality.

The authorities we have been enabled to consult, and from which we have drawn our *matériel*, are as follows :—

Medical News and Library for 1853 and 1854.

Dr. Wood's Practice of Medicine.

New Orleans Medical and Surgical Journal for 1853.

Report of the Minister of the Empire of Brazil.

Harper's New Monthly Magazine, 1857.

Sketches of Brazil, (a medical work,) by Robert Dundas, M.D., Supt. of the British Hospital at Bahia.

Conseils contre la propagation de la fièvre jaune, by Dr. Paulo Candido, Rio de Janeiro.

And the Report of Dr. Lallemant, of Rio de Janeiro.

APPENDIX G.

The following statement shows the annual aggregate imports into Brazil from foreign countries, in contos de reis. (A conto = £112 10s. exch. 27d. per 1$000.)

Annual statement of exports from Brazil, from 1841 to 1855. (In contos de reis.)

1840–41...57,727	1845–46...52,193	1850–51...76,918
1841–42...56,040	1846–47...55,740	1851–52...92,860
1842–43...50,639	1847–48...47,349	1852–53...87,336
1843–44...55,289	1848–49...51,569	1853–54.. 84,863
1844–45...57,228	1849–50...59,165	1854–55...84,780

1840–41...41,670	1845–46...53,630	1850–51...67,788
1841–42...39,084	1846–47...52,449	1851–52...66,640
1842–43...41,039	1847–48...57,925	1852–53...73,644
1843–44...43,800	1848–49...56,789	1853–54...76,842
1844–45...47,054	1849–50...55,032	1854–55...90,570

Statement of principal exports in four periods of five years each, and in 1863–64.
The canada is nearly two gallons; the arroba, 32¼ lbs. avoirdupois.

ARTICLES.		1st Period. 1844–5 to '48–9. Average.	2d Period. 1849–50 to '53–4. Average.	3d Period. 1853–4 to '57–8. Average.	4th Period. 1858–9 to '62–3. Average.	1863–64.
Rum...............	canadas	2,709,669	2,654,820	2,847,935	2,313,782	1,784,993
Cotton...........	arrobas	714,959	956,237	979,365	846,934	1,297,228
Rice...............	"	291,262	256,865	....	....	....
Sugar	"	7,591,885	8,652,252	7,765,695	8,361,918	7,941,310
Hair...............	"	31,740	47,081	44,537	40,381	52,786
Cacáo	"	190,203	276,506	223,058	273,746	284,190
Coffee	"	7,873,952	8,850,183	11,224,544*	10,933,697*	8,183,293*
Hides, salted...	number	680,028	512,078	498,884	634,454	764,336
Hides, dry	arroba	675,283	533,653	448,498	369,748	445,625
Diamonds.......	oitavas	632	6,364	....	....	....
Tobacco	arrobas	326,343	499,204	548,504	693,126	907,218
India-Rubber..	"	38,336	105,784	143,130	164,380	232,288
Maté.............	"	254,474	404,221	461,952	549,615	719,069
Gold (bullion).	oitavas	194,808	195,756	75,401	370,586	31,898
Sarsaparilla....	arrobas	3,469	5,003			

* Average for these two periods is much affected by the partial destruction of the coffee-trees by an insect in 1861–62. In the year 1860–61 there was the greatest crop ever raised in Brazil. It amounted to 14,585,258 arrobas. In the year 1865 no less than 9,584,611 arrobas of coffee were exported from Rio and Santos alone: so that there is a great gain on 1863–64.

Statement of principal imports in four periods of five years each, and the year 1863–64.

ARTICLES.	Average. 1844–45 to 1848–49.	Average. 1849–50 to 1853–54.	Average. 1843–54 to 1857–58.	Average. 1858–59 to 1862–63.	1863–64.
	Value in Contos.	Value in Contos.	Value in Contos.	Value in Contos.	Value in Contos.
Cotton (manufactured)............................	16,781	26,445	30,350	30,501	23,970
Wool... "	2,926	4,821	6,116	4,963	4,401
Linen.. "	1,905	2,510	2,638	2,616	2,992
Silk..... "	1,287	1,892	2,730	2,865	2,350
Mixed.. "	1,571	2,222	4,127	2,670	2,735
Wines..	3,058	3,321	3,145	4,608	5,632
Flour (Wheat)	3,457	4,330	5,495	7,679	4,142
Hardware......................................	2,193	3,256	4,371	6,167	4,797
Salt Fish......................................	1,212	1,584	2,867	2,773	1,383
Crockery, porcelain, and cut glasses.........	932	1,403	1,880	1,712	1,462
Specie...	2,050	6,929	7,686	4,376	19,607
Salt	796	687	853	1,026	1,326
Butter..	1,186	1,394	1,571	2,149	1,940
Machinery....................................	213	242	277	796	621
Drugs...	467	724	1,094	1,456	1,498
Tea...	277	272	....	....	....
Copper..	398	404	....	....	....
Coal..	542	1,068	1,458	2,540	1,833
Furniture.....................................	163	115	....	....	....
Arms..	206	316	....	....	....
Boots and Shoes..............................	314	329	....	....	....
Beef and Pork	750	1,760	....	....	.
Oil..	608	566	696	1,004	1,122
Spirits, distilled..............................	400	467	890	1,661	1,665
Powder..	241	330	....	....	....

The importation of Brazil in three periods was made by the principal importers as follows:—

	1844–5	1854–5	1863–4.
	contos	contos	contos.
Great Britain & Possessions	30,503	45,450	64,838
France and Possessions	7,441	9,978	23,110
Portugal and Possessions	4,552	6,468	6,346
Spain and Possessions	737	1,230	2,250
United States	5,703	6,991	6,259
Hanseatic Cities	2,725	4,884	5,453
River La Plata	1,711	4,217	9,062
Belgium	868	1,671	1,805
Chile	92	1,128	146
Sardinia (Italy after 1860)	328	755	778
Austria	475	260	776
Others	2,093	1,648	2,222
	57,228	84,780	123,045

The exports of Brazil were made
To

	1844–5	1854–5	1863–4.
	contos	contos	contos.
Great Britain & Possessions	11,306	29,274	52,485
France and Possessions	2,462	8,172	17,060
Portugal and Possessions	4,216	4,649	6,662
Spain and Possessions	697	877	4,316
United States	9,210	23,807	21,666
Hanseatic Cities	4,844	6,675	1,184
River La Plata	2,427	4,175	4,014
Belgium	1,612	2,783	620
Chile	165	1,479	1,188
Sardinia (Italy after 1860)	1,072	1,217	565
Austria	3,125	1,624	764
Others	5,918	5,838	18,946
	47,054	90,570	129,470

N.B.—A conto of reis (1000$) = £112 10s.

Flour exported from the United States to Brazil.

1855–56	386,306 bbls.	1860–61	427,161 bbls.
1856–57	498,264 "	1861–62	376,315 "
1857–58	531,796 "	1862–63	410,094 "
1858–59	484,355 "	1863–64	410,862 "
1859–60	507,544 "	1864–65	362,066 "
Total	2,408,265 "	Total	1,986,498 "
Average	481,653 "	Average	397,299 "

Flour formerly paid 3 milreis per bbl. in Brazil, but since 1858 only 900 reis.

Coffee imported into the United States from Brazil.

1855–56	1,004,838 bags.	1860–61	1,137,088 bags.
1856–57	1,254,939 "	1861–62	567,146 "
1857–58	898,421 "	1862–63	366,908 "
1858–59	1,202,190 "	1863–64	574,182 "
1859–60	979,498 "	1864–65	474,843 "
Total	5,429,886 "	Total	3,120,167 "
Average	1,085,977 "	Average	624,033 "

In former times coffee was admitted free into the United States, but now pays a duty of 5 cents a pound, which is a very heavy tax upon it; and it is believed that, if the duty was reduced, coffee would be more largely imported, and the revenue would not suffer. Brazil should reciprocate such a reduction by removing her export duty.

The four principal articles of export from Brazil in twenty-four years. Arroba = 32 lbs.

	Cotton. (arrobas.)	Coffee. (arrobas.)	Sugar. (arrobas.)
1840–41	691,875	5,059,223	6,698,391
1841–42	639,580	5,565,325	4,817,577
1842–43	685,149	5,897,555	5,209,721
1843–44	814,255	6,294,281	5,682,980
1844–45	826,445	6,229,277	7,476,286
1845–46	645,345	7,034,582	7,110,804
1846–47	606,882	7,947,753	6,963,960
1847–48	639,288	9,307,292	7,409,349
1848–49	849,416	8,354,840	8,801,616
1849–50	1,109,314	5,935,817	7,993,986
1850–51	883,440	10,148,268	9,907,860
1851–52	898,250	9,544,858	7,480,099
1852–53	997,908	9,923,982	10,681,344
1853–54	892,273	8,698,036	8,258,378
1854–55	869,960	13,027,526	7,951,422
1855–56	1,024,801	11,651,806	7,448,582
1856–57	1,088,025	13,026,299	7,670,430 ·
1857–58	1,014,550	9,719,054	7,257,758
1858–59	751,348	11,168,110	10,506,245
1859–60	854,624	10,307,293	5,735,070
1860–61	670,860	14,585,258	†4,451,188
1861–62	872,210	9,880,924	10,571,970
1862–63	1,085,628	8,716,836	9,345,371
1863–64	1,297,228	8,183,293	7,794,310
1864–65	*655,374		

* First six months.

† One portion of the crop in Bahia (one of the largest producers) for 1860–61 has not been rendered. The demand for cotton since 1862 has diminished the production of sugar.

Estimates of expenditures for 1866–67.

Ministry of Empire	5,100	contos.
" Justice	3,139	"
" Navy	7,975	"
" War	14,583	"
" Foreign Affairs	848	"
" Finance	18,042	"
" Agriculture, Public Works, & Commerce.	9,185	"
Total	58,875	"
Estimated receipts	55,000	"
Deficit	3,875	"

The internal funded debt up to March, 1865, was as follows:—

Contos	78,419	at 6 per cent.
"	1,837	at 5 "
"	119	at 4 "
Total	80,376	
Debt not converted	466	
Treasury notes in circulation	12,400	
Total	93,242	
External debt	70,640	
Grand total	163,882	= £18,337,500

The Government paper money circulating in Brazil, which in 1855 was 45,000 contos, has been reduced to 29,094 contos.

[The various tables, with a single exception, of Appendix G were prepared with great care by M. le Chevalier d'Aguiar, Brazilian Consul at New York.]

The following on the Internal and Foreign Debt of Brazil was published in January, 1866, by Henry Nathans, Esq., Broker, at Rio de Janeiro.

INTERNAL DEBT OF BRAZIL.

Total emission of 6 per cent. Stock..	86,752:400$	
5 per cent. "	1,384:400$	
4 per cent. "	119:600$	
		88,256:400$

Redeemed.

Of six per cent. emission...3,672:000$		
" five " " "	161:200$	
" four " " "	$	3,833:200$

Total emission in circulation... 84,423:200$

In 80,793 Bonds of 1.000$ each 6 per cent....................	80,793:000$	
" 885 " " 800$ " 6 per cent....................	708:000$	
" 1,577 " " 600$ " 6 per cent....................	946:200$	
" 1,583 " " 400$ " 6 per cent....................	633:200$	
		83,080:400$
" 691 " " 1:000$ " 5 per cent....................	691:000$	
" 487 " " 600$ " 5 per cent....................	292:000$	
" 600 " " 400$ " 5 per cent....................	240:000$	
		1,223:200$
" 113 " " 1:000$ " 4 per cent....................	113:000$	
" 11 " " 600$ " 4 per cent....................	6:600$	119:600$
		84,423:200$

Of this amount there are possessed by religious charitable establishments, life-insurance companies, &c............................ 17,048:200$
which, with the exception of 3 and 4,000:000$, will finally revert to the country: the remainder will only find its way on the market if extraordinary mortality should occur.
The banks hold either for their own account or hypothecation................................. 851:400$

FOREIGN DEBT OF BRAZIL.

	Real Value.	Nominal Value.	Redeemed.	Nominal Value in Circulation.
Loan of 1839	£312,512	£411,200	£94,400	£316,800
" 1852	954,250	1,040,600	170,700	866,900
" 1859	508,000	508,000	89,900	418,100
" 1863	3,300,000	3,855,300	37,800	3,817,500
May 19, 1858	1,425,000	1,526,500	248,800	1,277,700
Mar. 15, 1860				
Account of União and Industria Road............	475,000			
Account of Pernambuco Railroad..................	400,000	} 1,373,000	125,900	1,247,100
Account of Mucury Colony...........................	135,000			
	7,709,762	8,714,600	767,500	7,947,100
October 19, 1865..	5,000,000	6,943,613	....	6,963,613
Total foreign debt..............................	£12,709,762	£15,678,213	767,500	*£14,910,713

Loan of 1839 in 1869 emitted at 5 per cent.
 1852 in 1882 " 4½ "
(a) 1859 in 1879 " 5 "
(b) 1858 in 1888 " 4½ "
(c) 1860 in 1890 " 4½ "
 1863 in 1893 " 4½ "
 1865 in 1902 " 5 "

(a) In 1859 the loan of 1829 matured, and, the Bonds being then at par, new Bonds were given in exchange at same price and interest as the old ones.

(b) and (c) These two loans were raised for the benefit of Dom Pedro II. Railway, Pernambuco Railway, Carriage Road of União e Industria and Mucury Colony. All of these companies, with the exception of the Pernambuco line, are now Government property; and her quota of these loans is £400,000.

The amortization of loans herein given is calculated till 1865, and in March, 1866, only, can the exact amount be ascertained. The amortization is as follows:—

1 per cent. for loans of 1839, 1852, 1859, 1865.
2 " " " 1858 and 1860.
1 65-100 " " 1863.

* It will be seen that Mr. Nathans makes the grand total of Brazil's indebtedness over six million pounds sterling more than Chevalier d'Aguiar, who puts it at £18,337,500. This apparent discrepancy is reconciled by adding the new loan of £5,000,000, made October 19, 1865.—J. C. F.

APPENDIX H.

RECENT DISCOVERIES OF COAL IN BRAZIL.

(From "The Anglo-Brazilian Times," of July 8, 1865.)

LETTER FROM PROFESSOR AGASSIZ.

OUR readers are aware that since this illustrious stranger arrived in Rio he has not been for a moment at rest. Whilst his assistants have been, each in his special department, working towards the attainment of the object which they have in view in Tropical America, the professor himself has been the most active of the party, trying to win from Nature the secrets which she holds; and we are informed that many new and interesting facts have been added to the domain of science.

We have from the outset looked upon this expedition with great interest, in so far as we have seen in it a value lying beyond the fields of pure science. The speculations of the philosopher of to-day may to-morrow become the established facts of commerce, and it will be impossible for the investigations of Agassiz to leave behind them only barren results. His labors may in the end yield us a harvest of material wealth: indeed, we have before us at this moment one very pertinent illustration of this fact, which we may assume is but the forerunner of many others of the same kind.

Our readers have for a long time heard of the famous coal-beds of Candiota, in the province of Rio Grande do Sul. The expectations of many are turned in that direction, as the most valued instance of the hidden wealth of Brazil. Mr. Plant has so far awakened or revived an interest in these things, that from time to time the topic has been made a public one, has been regarded as a question for commercial action, and has been debated each time with growing interest in the Legislature. We are not to-day talking of the value of this matter in the abstract: our minds have been long made up on this subject: we only wish to show how the opinion of a man like Agassiz at once settles the whole question, and leaves only to commerce the practical development of plans for making available this most important element in a nation's wealth and power.

Mr. Plant, as a geologist, submitted to the examination of the professor such fossils and geological illustrations of the province of Rio Grande do Sul as he supposed would be of interest and would help to complete the collections which are being made for the United States. The importance of these fossils, and the sure deductions which science draws from them, appear to have startled and delighted the great savant; and a few days since the following letter was placed in the hands of Mr. Plant. We print it *verbatim*, as it is of such a nature as to become at once important, and will show the Government of Brazil that if it only follows up the path opened up by science, the results as a source of wealth cannot be doubted.

"RIO, June 18, 1865.

"DEAR SIR:—I have not yet returned my thanks for the fine specimens you have presented to me, though ever since I saw them I have looked for a moment's leisure to do so.

"However, this gives me an opportunity of expressing a more mature opinion concerning their geological age, which I am glad to have an opportunity of recording, especially since the examination I have made of them has satisfied me of the correctness of some views concerning the fossils of the oldest geological formation, in which I had little confidence. That these organic remains all belong to the Carboniferous period is unquestionable; and it is the close affinity with the characteristic fossils of Europe which particularly interests and in a measure surprises me. Had the whole collection been made in Pennsylvania, I would not more decidedly have recognized its Carboniferous characteristics, down to the rocks underlying and overlying the fossiliferous beds; and the photographs you have shown me of the localities leave no doubt of the great extent and value of the coal-beds proper of the river Candiota, whilst the coal itself may fairly be compared to the best in the market, judging from the specimens you have shown me and those I owe to your kindness.

"With my best wishes for the further success of your geological explorations, in which I hope you may hereafter also include the Drift and erratics, now that you are satisfied of their existence in Brazil,

"I remain,

"Yours, very truly,

"L. AGASSIZ.

"N. PLANT, Esq."

We think our subscribers will join with us in our opinion that we have much to look for from this expedition, fitted out by the munificence of Nathaniel Thayer, Esq., of Boston. It may be that this expedition has been projected with a conviction of the value of science as an agent in commerce. We are sure the indirect results will, for Brazil, be very important. One of the world's greatest minds is breaking in upon a region almost unknown. His aim is, we know, to extend the empire of mind and to storm the stronghold of nature, making her subservient to the wants of man; and in parting with Professor Agassiz and his co-laborers for a season, we can but give him our benediction, with the hope that in a few months he will be among us again, rich with treasures from the Amazon.

We are also at liberty to state that Sr. Capanema, whose abilities as a geologist are too well known to need comment, has seen Mr. Plant's collection of fossils from the Candiota coal-mines, and has arrived at the same conclusion as Professor Agassiz in respect to the coal-beds belonging to the Carboniferous period.

[So interesting to science and to commerce are these coal discoveries in Rio Grande do Sul, that I asked Mr. Plant to give me full information on this subject. Under date of July 24, 1865, he has forwarded me that which constitutes the remainder of Appendix H.—J. C. F.]

THE COAL-FIELDS OF THE RIVER JAGUARÃO, AND ITS TRIBUTA-
RIES THE RIVERS CANDIOTA AND JAGUARÃO-CHICO, IN THE PRO-
VINCE OF RIO GRANDE DO SUL.

THE coal-basin of the river Jaguarão is situated in the southern part of the province of Rio Grande
do Sul, between lat. 31° and 32° South, and long. 324° and 325°, (French Meridian,) in the valley of
the Jaguarão and its tributaries the rivers Candiota and Jaguarão-chico. It covers an area of about
fifty miles by thirty, its greatest diameter being from north to south. The coal strata, which the
geological section illustrates, and from whence the accompanying specimens have been obtained and
the thickness of the beds determined, are exposed in an elevated escarpment on the bank of the
river Candiota, at a place called "Serra Partida," where they appear in the following order of
superposition:—

The uppermost bed (No. 1) is composed of sandstone of a highly ferruginous nature, resembling
in its appearance the "Gres Bizarre" of Europe. It contains nodules of a silicious peroxide of iron,
yielding from 25 to 35 per cent. of metal. It varies considerably in its thickness, in some places
being completely worn away, and in others attaining a depth of upwards of 200 feet. Immediately
below this occurs a bed (No. 2) of coal-shale, very argillaceous, and perhaps unfit for fuel: it possesses
a thickness of 9 feet, and can be seen cropping out wherever the superincumbent bed has been denuded: it rests upon a bed (No. 3) of sandy ochreous shale, containing spetaria of an ochreous oxide
of iron, which, together with the iron-stone found in the sandstone, will, in all probability, be turned
to profitable account when the coal-beds are worked. Underneath this is a bed (No. 4) of bituminous
coal, 3 feet thick. The mineral, although it leaves a high percentage of ash, will be found useful
in melting the iron-ores from the interstratifying beds; and there is every reason for supposing that
it will be found of a better quality when the bed is fairly worked. The samples tested were taken
from very near the surface, which may in some measure account for its apparent impurity: it rests
on a bed (No. 5) of white clay, or schist, containing innumerable impressions of fossil plants (perhaps
aquatic), the general appearance of which would lead one to conclude that these Carboniferous deposits belong to a later period than that assigned to the coal-measures of England and the United
States, were such a conclusion not confuted by the fossil ferns found in the other interstratifying
shales: it has a thickness of 5 feet, and overlies a bed (No. 6) of good coal, 11 feet thick. This coal
resembles very much in its appearance the Newcastle, and may be traced for many miles along the
banks of the river Candiota, sometimes forming the bed of that river, and of the small streams falling
into it; it is separated from another seam by a thin parting of blue clay (No. 7). The coal of the
lower bed (No. 8) appears to be even of a better quality than No. 6: it has a clean, shining fracture,
and in some places thin seams of pure cannel coal may be traced along the bed. It is highly in-
flammable, boiling up like oil during combustion. This coal has been used as fuel in various ways
with marked success. It has been tried on the steamers navigating the "Lagoa dos Patos" in the
province of Rio Grande, and although it left a greater portion of ash than the Cardiff coal, it was
found to be a good caking coal, and served every purpose of a steam fuel. Below this is another bed
(No. 9) of blue clay, containing vestiges of fossil plants. In every thing else it is similar to the upper
bed of the same mineral, and has a thickness of 9 feet. It reposes on the thickest seam (No. 10) of
coal exposed in the escarpment at the "Serra Partida." This is the lowest bed of coal exposed in any
part of the coal-field of Candiota; but in all probability other beds will be found nearer to the centre
of the basin, or this, as well as the incumbent beds, may become thicker, judging from the fact that
all the beds appear to thicken as they approach the middle of the valley of the river Jaguarão. The
great thickness (25 feet) and the good and homogeneous character of the seam are important features
in this coal-field. The mineral (although taken from near the decomposed face of the cliff on the
river Candiota) was found to leave even less ash than that from the seam above. It has frequently
been used on steamers with the same success as Newcastle coal. The coke obtained from this coal
by Mr. W. G. Ginty, of the Rio Gas Works, (see Mr. Ginty's report,) was even better than that derived
from Newcastle coal. It overlies a bed (No. 11) of ironstone shale, which, in a scientific point of
view, is the most important deposit of the coal-measures of the Jaguarão, from the fact of its containing impressions of organic remains, by which the geological age of the coal-beds can be determined;
the fossil plants found imbedded in this shale all belong to the same genera as those which charac-
terize the coal-fields of Great Britain and the United States,—the most abundant belonging to the
genera "Lepidodendron" and "Glossopteris;" others have been recognized as being similar to the
ferns found in the very oldest secondary rocks, thus leaving no uncertainty as to the true Carboniferous character of the coal-measures of the river Candiota. This seam is very prolific of fossils; and
there can be no doubt that when these immense beds of mineral treasure are worked, many new and
interesting forms of vegetable life will be brought to light to enrich our knowledge of the coal-fields
of the Southern hemisphere. The iron-stone shale is very rich in metal, and will, doubtless, be
worked as an iron ore, when the mines are opened. Below this there occurs another bed (No. 12) of
sandstone, similar in all respects to the uppermost bed, after which is a bed (No. 13) of fine crystal-

GEOLOGICAL SECTION OF STRATA SHOWN ALONG THE VALLEY OF
THE CANDIOTA, RIO GRANDE DO SUL.

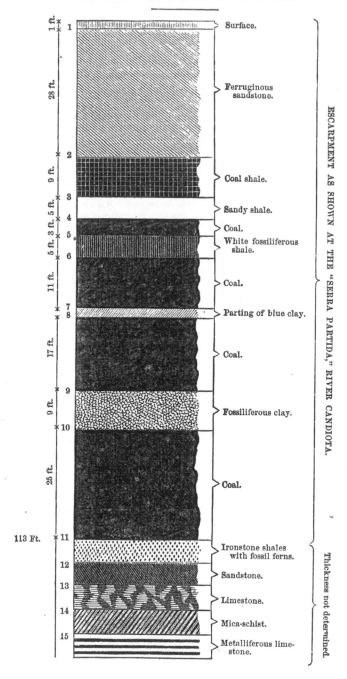

line lin.estone, containing small fragments of graphite disseminated throughout the mass: it is tra. versed also by veins of a very pure carbonate of lime in the form of double-refracting spar, which in some places attain a considerable thickness. This limestone will not only be of immense value as a calcined lime, but also as a flux in smelting the iron ores. The three things essential for the erection of smelting works are thus found on the same district interstratifying each other,—the ore, the fuel, and the flux, all of the very first quality,—a combination of mineral riches (only waiting for the hand of man to realize them) scarcely to be found together in one spot in any other part of the globe. Evidently, the two lowest beds of these coal-measures are mica-schist (No. 14), and another limestone rock (No. 15), of a very dark and compact nature. It is scarcely possible to determine which is the lowermost, as in some places the mica-schist is seen lying on the sienite which surrounds the coal-basin, and in others the limestone: the name of "Metalliferous limestone" has been given it, owing to the innumerable crystals and thin veins of sulphuret of iron which appear in it. In all probability, other metalliferous veins will be found in this limestone.

Nearly the whole of the coal-basin of the valley of the Jaguarão is enclosed by sienitic hills of from 200 to 300 feet high; the sides towards the coal-field slope gently downwards till they disappear under the sandstone overlying the coal: on the other side, the sienite, after presenting an uneven and undulating aspect for some three or four leagues, gradually subsides into an even country, which continues on almost perfectly plain till the seaport city of Rio Grande do Sul (S. Pedro) is reached. So that the company already formed for making the survey for a railway to carry the mineral riches of the valley of the Jaguarão down to a seaport, where the coal can be shipped to the different ports along the coast of Brazil and to the river Plate, will find no difficulty in discovering a route along which a cheap line of rails can be laid down.

The engraving opposite p. 347 (from a photographic view of the different escarpments in which the coal-beds are shown along the river Candiota) will show the great facilities afforded for working the coal in almost any part of the basin, by open cuttings. Tram-ways can be laid down branching off in different directions from the main trunk line, along which the coal-wagons can be run right into the seams of coal, thereby rendering the sinking of expensive shafts quite unnecessary.

The general dip of the beds is from 5° to 10° S. W., and in no place are there signs of subsequent upheavals or dislocations of strata visible, so that very little obstruction will be met with in carrying the tram-ways along the seams as the working of them goes forward.

It is almost unnecessary to dwell upon the immense value of these coal-deposits as a commercial enterprise, when it has been already ascertained, by a "running survey" of the country between the seaport of Rio Grande do Sul (S. Pedro) and the coal-mines of Candiota, that in all probability the coal will be delivered on board vessels lying in the port of Rio Grande at perhaps less than Rs. 7$000 per ton, where it is at the present moment being sold at Rs. 24$000, and as soon as a bill is passed allowing vessels of all nations to trade between the Brazilian ports, there will be no lack of enterprising ship-owners to carry the Rio Grande coal to Rio de Janeiro, in which port alone the enormous amount of 180,000 tons of coal are annually imported for a price which will enable the coal-mining company to sell the Candiota coal, in the market of the capital of the Brazilian Empire, for about Rs. 15$000 per ton, a price which will annihilate any competition from foreign markets, seeing that the foreign coal is seldom sold for less than Rs. 22$000 per ton.

The consumption of coal in the river Plate is perhaps as great as that of Rio de Janeiro, and the facilities for supplying the markets of Buenos Ayres and Montevideo from the coal-mines of the river Candiota are still greater than those for supplying Rio. The coal can be sent from the mines, put on board coll.ers, and delivered in Montevideo, in three or four days, at about half the cost of de-livering the same article in Rio, and in a market where coal is never less than fifteen dollars per ton, or Rs. 30$000. The consumption of coal along the Brazilian coast and in the river Plate increases yearly, and in all probability it will be found, after the coal-mines of Candiota have been opened for a few years, that a single line of railway will not be found sufficient to carry the supply of coal to meet the increasing demands.

Rio de Janeiro, 20th July, 1865.   Nathaniel Plant.

The Brazilian Coal-Fields, by Edward Hull, B.A., F.G.S.

(Note from the Quarterly Journal [England] of Science, No. II. April, 1864.)

The immense empire of Brazil, occupying one-third of the continent of South America, with an area of upwards of 3,000,000 of square miles; considerably larger than Russia in Europe; watered by the largest river in the world, which, with its tributaries, is navigable for many hundred miles from its mouth; its western bounds stretching to the spires of the Andes, and its eastern washed by the waves of two oceans,—such a country as this would appear fitted to occupy the foremost rank amongst the nations of the Western hemisphere, provided its boundless resources were turned to account by an intelligent people, and civilization were advanced by wise laws.

It is satisfactory to reflect, that while most of the surrounding republics—the shattered limbs of

Spanish America—are tossed on the waves of anarchy, Brazil enjoys a peaceful government, unde. a constitutional monarchy; personal freedom, with political security; monarchical principles combined with popular rights. We notice these points in the government of Brazil because they afford the highest guarantee of national progress and development of industrial pursuits. Nor are the raw materials necessary for the attainment of a high position among the manufacturing communities of the world absent from the soil of Brazil.

The northern half of the empire is physically not unlike the plain of Northern Italy on a large scale. Covered with forests springing from a rich alluvial soil, and watered by the Amazon and its giant branches, it is prodigiously fertile. The southern half is hilly, and sometimes mountainous, and gives birth to the Rio de la Plata. One of the peaks of the Organ Range rises behind the harbor of Rio de Janeiro to an altitude of 7,500 feet. It was once supposed that this great empire—rich in precious stones and nearly all the metals from gold to iron inclusive—was devoid of one natural product, useful, if not absolutely essential to the full utilization of the other mineral treasures, namely, coal; but such a supposition was altogether erroneous, as recent investigations have fully shown.

A writer in a recent number of the Quarterly Review for 1860 mentions [in a Review of Brazil and the Brazilians] the existence of a coal-field about forty miles from the sea [in the province of Rio Grande do Sul]. This is all that was known on the subject on this side of the Atlantic, till very recently.

To a countryman of our own, Mr. Nathaniel Plant, we are indebted for a full account of the position and resources of three distinct coal-fields which he has recently explored in the southern part of the empire. The largest presents some features of peculiar interest, which we proceed briefly to lay before our readers.

The first notice of these minerals seems to have been taken by a Mr. Guilherme Bouleich, in the province of Rio Grande do Sul. This appears to have been in the year 1859.

The matter, however, seems to have been lost sight of until the end of 1861, when Mr. N. Plant, who for several years had been examining the mineral districts of Rio Grande do Sul, and other parts of South America, determined to make a fuller exploration of the coal districts; and he has now sent to this country an account of the very remarkable deposits of mineral fuel to be met with, together with those unbiassed witnesses,—photographic views and rock specimens.*

The Candiota coal-field is the largest of the three which have yet been discovered. It lies between lat. 31° and 32° S., and is thus at the extremity of the province of Rio Grande do Sul. It is traversed by the river Jaguarão and several of its tributaries, along whose banks the seams of coal crop out. There are two great seams of bituminous coal, the lower being 25 feet in thickness, and separated by only a very few feet of shale from the upper bed, (or series of beds, which is 40 feet in thickness.) In some places the intermediate bands of shale which separate the mineral into distinct layers, thin away, in which case a solid seam of no less than 65 feet is formed, unsurpassed, wo believe, in vertical dimensions by any similar formation yet discovered. We have handled specimens of the coal; and, though taken from the outcrop, it is scarcely distinguishable, except by a slight brownish hue, from the ordinary coal of our own country.

The coal strata reposes on a series of shales, sandstones, and crystalline limestones, the whole of which are supported by mica-schist, and finally by sienite.

Iron is also present, as in the coal-formation of Britain, both in the form of bands of clay-ironstone and as a roof for the seams of coal. At the top of the cliff formed by the outcrop of the coal-seams there occurs a mass of silicious iron-ore several yards thick, a sheet casting from which, taken by Mr. N. Plant, was exhibited at the late Industrial Exhibition among other Brazilian products. Thus there occur in close proximity to each other the ore, the fuel, the flux and clay necessary for the establishment of iron-furnaces.

The several minerals thus united rise in the form of an elevated escarpment, which may be traced for several leagues, affording the utmost facility for working by open works, or tunnels driven into the sides of the hills.

From its base stretches a gently sloping plain of basalt, over which a railway to a port on the Rio Gonsalo might be laid down at a very moderate cost.   *   *   *   *   *   *   *   *   *
*   *   *   *   *   *   *   *   *   *   *   *   *   *   *   *   *   *   *   *

After an inspection of the fossil plants which have been sent over to this country, there can be no doubt, we think, that these beds belong to the Carboniferous age. Mr. Plant has sent over several pieces of ironstone on which are imprinted very distinct specimens of Lepidodendron, and several ferns not unlike those of the coal measures of Britain. A gentleman, also, who has studied the coal measures of Nova Scotia, which are of the same age as those of Britain, refers in a letter. which we have seen, to fine specimens of Sigillaria and Stigmaria, both of which are characteristic of this period. Specimens of these, however, we have not seen in the collection we have examined; but, nothing can be more distinct than the fronds of Lepidodendron already referred to. While on this

---

* These have been laid before the Geological Society, Manchester, by his brother, Mr. S. Plant, Curator of the Royal Museum, Salford.

subject, we may be allowed to remark that although, on the authority of Professor M'Coy, the age of the Australian coal-fields was for some time considered to be Jurassic, the recent investigations of the Rev. W. B. Clarke go to establish the Carboniferous age of these beds. Mr. Clarke has sent to England a collection of fossils from the New South Wales coal-fields, containing specimens of Lepidodendron and Spirifer; and thus it would appear that during the same great epoch, so pre-eminently Carboniferous, deposits of coal were being elaborated over both sides of the equator,—a marvellous instance of the uniformity of nature's operations in early geologic times.

The importance of these great deposits of coal to the commerce of the eastern seaboard of South America need not be dwelt upon. At the present time nearly 200,000 tons of coal are annually imported into Rio de Janeiro alone, at a cost of forty-nine shillings per ton, and from this depot other coast towns are supplied. When once the coal-fields of Candiota are opened up, the Brazilian Government may be supplied at nearly half the price, and our own little island be spared the doubtful honor of providing fuel for a continent on the other side of the globe.

(Signed)   EDWARD HULL.

REPORT ON THE CANDIOTA COAL, BY W. G. GINTY, ENGINEER-IN-CHIEF OF THE RIO DE JANEIRO GAS WORKS.

MR. NATHANIEL PLANT:

DEAR SIR :—I have received and examined your samples of Brazilian coal from Candiota with great interest, and I am glad to be able to congratulate you on its really good quality.

The samples you sent me were too small for complete and satisfactory analysis in the apparatus at my disposal. I found also the samples varied a good deal in appearance and quality. This has arisen, no doubt, from their having been obtained from various positions on the nearly perpendicular face of the immense stratum, and from variable periods of exposure, as, owing to the crumbling away or disintegration of pieces under the incessant action of the weather, these samples may have been exposed for periods varying from each other as seconds do from centuries.

The Candiota coal resembles the Newcastle steam-coal (which comes to this market, at least) very much in structure, cleavage, and general appearances; nor does it differ very much from Newcastle coal in its useful properties, except that it contains more than double the quantity of ash, which is detrimental to its heating powers; but this objection is likely enough to disappear altogether in samples from the deeper parts of the mine.

The coke from the Candiota coal is, however, very different in appearance from that of the Newcastle coal, and resembles the coke of (what is sold here as) Cardiff coal in its silvery-colored laminations.

Some of this Candiota coal, however, especially that of the lower seam, is very friable, and is evidently what is called caking coal (that is, it boils or becomes molten during the process of carbonization): however all the qualities of the coke from the Candiota coal are very good.

As you say the dip or inclination of the seams or strata of this Candiota coal is 5° from the plane of the horizon, I think it most reasonable to presume that a much finer, more compact, and equable quality of coal may be calculated upon at lower depths. 5° is a gradient of 1 in 11.4, or 8.77 per cent., or 462 feet per mile. Thus, in such an immense field as you have described to me, there is ample margin for obtaining other than surface coal, which for obvious reasons, in Brazil as elsewhere, cannot be as pure, as compact, or as uniform in quality as that obtained at great depths. I shall watch the prosecution of your explorations in this direction with great interest.

The following are the results of my examinations (as far as they went) on the Candiota coal,—the samples of Newcastle, Cardiff, and Wigan Cannel, with which it is compared below, having been tried at the same time in the same apparatus:—

	Specific Gravity Water, 1.000	Per-cent. of Coke.	Cubic Feet of Gas per Ton.	Illuminating Power in Standard Candles.
Candiota coal (mean of three qualities)	1.240	63	6,900	5.00
Do. do. lower seam	1.230	60	8,198	5.80
Newcastle	1.250	62	——	——
Cardiff	1.275	80	——	——
Gas, or Cannel coal (Case and Morris)	1.240	62	9,600	20.50

From the appearance of the lower seam, I do not despair of your finding a good gas coal for us in the Candiota district, and thus freeing the Brazilian Gas Companies from the fearful tax they have to pay in the shape of freights from England, amounting to from 200 to 300 per cent. on the value of the materic prima. I send you labelled samples of the different qualities of coke above referred to.

I remain your obedient servant,

(Signed)   W. G. GINTY, Mem. Inst. C. E.,
                                   Engineer Rio de Janeiro Gas Company

THE GOLD-MINES OF NORTHERN BRAZIL.

GOLD is plentifully diffused in veins, lodes, and deposits of auriferous earth throughout the primitive mountain valleys of Northern Brazil. Rivers and streams, laden with water-worn particles of the "precious metal," testify the fact. But, in the absence of capital and skilled labor for its extraction, gold in Northern Brazil is confined to the acquisition of private adventurers or to aboriginal tribes. In South Brazil, in the province of Minas-Geraes, in the vicinity of São João del Rey, the gold mines worked by English companies have proved up to this time the most remunerative in South America. In 1865, that energetic Brazilian Sr. Jaccomo Tasso, of Pernambuco, called the attention of English capitalists to the gold-regions of Parahyba do Norte, and soon after a company was formed under the title of the "Tasso Brazilian Gold-Mining Company (limited)," having a capital of £200,000 (with power to increase), at £5 per share. The officers of the association, who have long been in commercial and other relations with Brazil, are as follows:—CHARLES CAPPER, Esq., Merchant, London. CHARLES SAUNDERS, Esq., Merchant of Liverpool and Pernambuco. CHARLES BARBER, Esq., London. WILLIAM CREMER, Esq., London. EDWARD JOHNSTON, Esq., Merchant of London, Liverpool, and Rio. SEBASTIAN PINTO LEITE, Merchant of London, Manchester, and Liverpool. BONAMY PRICE, Esq., London, Director of the St. John del Rey Gold Company.

This company availed itself of the local knowledge and territorial possessions of Sr. Tasso, which are situated in the heart of the gold district of Parahyba, determined to undertake the work of gold-mining in Northern Brazil on a scale and with such approved machinery as shall render the enterprise one of immediate and great productiveness. In pursuance of this resolution, the company entered into a contract for the purchase of the estates of Sr. Tasso, at Pianco, in the province of Parahyba, Brazil, (on which estates eight gold-bearing lodes have already been discovered,) and also for exercising the rights of exploration and pre-emption, within the provinces of Parahyba and Pernambuco, now conceded to that gentleman by the Brazilian Government. An imperial concession for these purposes includes the unexpired right for four years to explore the interior of both provinces for minerals, and to appropriate and work to the extent of 150 "datas," or about 25,000 acres, eligible either for gold or for other mining enterprise.

When the concession was obtained, Mr. William Reay and Mr. Thomas Andrew, surveyors of practical experience in Brazilian gold-mining, were sent from England to Sr. Tasso's estate in the district of Pianco, through which flows one of the affluents of the river Piranhas. Several rich lodes of gold were traced by them, and portions of the ore assayed with satisfactory results. Notwithstanding the necessary imperfection of such methods as could be adopted on the spot, the result of twenty-six assays gave an average of about 16 dwts. 5¼ grains of fine gold per ton of ore, as taken in each case from the surface of the several lodes then traced. The chief of these assays yielded on the average 1 oz. 9 dwts. 23 grs. per ton of ore, and five samples obtained from different parts of the "Boa Esperança" lode yielded 2 oz. 9 dwts. and 15 grs. of gold per ton. Later "extracts" from the lodes gave even more extraordinary results. Mr. Charles Martin, London, reporting on them, says:—"I send you herewith copy of the assays I have had made of your quartz, which are surprisingly rich, and which would pay enormously if the bulk should prove any thing like the sample."

	oz.	dwt.	grs.			oz.	dwt.	grs.	
No. 5 Gold.........	3	10	12	per ton of 20 cwt.	No. 6 Gold.........	12	5	15	per ton of 20 cwt.
Silver.......	0	18	0	" "	Silver.......	3	15	0	" "

Another sample assayed by Messrs. Johnson, Matthay & Co., of London, yielded:—

Gold.........6·350 oz. per ton of 20 cwt. | Silver.........4·350 oz. per ton of 20 cwt.

In addition to the "eight lodes" mentioned, two lodes, of a still richer description, are believed to lie near the centre of the estate. The grounds on which investments were recommended were— 1st. That the provinces of Parahyba and Pernambuco, over which the rights of the company will extend, contain the richest gold-mines in the empire. 2d. That, besides the lodes of gold, the locality contains an enormous quantity of auriferous ore detached from the lodes, available at comparatively little expense. 3d. That the surface of the land already explored at Pianco is remarkably favorable for mining operations. 4th. That the neighborhood being populous and fertile, labor is cheap, and supplies of cattle, corn, and other produce are easily obtainable. 5th. That a main road, along which a great part of the cotton exported from Pernambuco is now conveyed, runs through the site of the concession at Pianco. 6th. That Sr. Tasso transfers to the company not only his right of exploration and pre-emption throughout the provinces of Parahyba and Pernambuco, and the concession of thirty-six "datas" (equal to about 6000 acres) already obtained at Pianco, but also the absolute ownership of all his estate in that vicinity, estimated at about 12,000 acres, together with buildings, stock, timber, and all his rights appertaining thereto. 7th. And, finally, that a moderate outlay appears sufficient to insure very large returns. If only fifty tons of ore per day were reduced, it is estimated that a year's operations, at that rate, would yield a profit on the cost of working of not less than £46,500.

APPENDIX I.

(From "The Anglo-Brazilian Times," of October 24, 1865 (edited by a Roman Catholic), Rio de Janeiro.

RELIGIOUS DISABILITIES.

HOWEVER diminished the influence religious bigotry may now openly exert upon the conduct of civilized peoples, it is not by any means a powerless element of mischief, even among those nations who put forward the strongest pretensions to the perfect establishment of religious toleration, and who allow to the dictates of conscience the utmost latitude of thought and action.

Nations, even more than individuals, with difficulty loosen and divest themselves from those trammels which hereditary reverence throws around them,— those trammels which arise from the "straining at a gnat and swallowing a camel" scruples of doctrinal education,—from that Phariseeism that lies at the root of all intolerance, (whose life and nature are egotism and self-conceit,) which bestows opprobrious terms upon its brother man, and says to him, "Stand aside: I am holier than thou;" "Thy faith is not my faith: therefore thou art a fool, and a godless man, and unfit to rule thyself." Religious "cant," as it has been termed, is still rampant and destructive among some of the freest and most progressive nations; and the cry of "Religion and the Church in danger!" still proves sufficient, at times, to drown the dictates of conscience and even-handed justice.

What wonder, then, when commercial, Protestant England and polished and tolerant Papal France still furnish occasional examples of outbursts of religious intolerance, that Brazil, a land lately emerged from her colonial chrysalis, barely entering into contact with the outer world, (her liberal constitution and laws rather the work of enlightened monarchs and progressive statesmen, acquiesced in by a gentle people, than the expression of the inner feeling of the advanced masses,) should still maintain upon her statute rolls legislative provisions contrary to the liberality and spirit of that constitution? If religious disabilities are hurtful in old and densely-populated countries, they are a hundredfold more mischievous in a new and thinly-settled one like Brazil; for they tend to repel instead of attracting immigration, and they deprive those who are subjected to them of that feeling of equality of rights and interests which it should be the great endeavor of a people to inculcate among those citizens of other lands who come among them to be of them and with them.

Brazilian legislation disables any but Roman Catholics from becoming Deputies, and, constructively, from becoming "electors,"—the "delegates" of the United States. Whence arises danger to the "Good Estate" if his fellow-citizens give their suffrages to a "heretic" whom they trust and wish to honor? Do the Papist and Jewish legislators of Great Britain and Prussia advocate insurrection and immorality? Does Louis Napoleon find his Jewish senators and his Minister of Finance less faithful than his Roman Catholic ones? Do conscience and honor make their resting-place only in the bosom of members of a "state religion"? And are not the property, the interests, the intelligence of the dissenting portion of a population as worthy of, and as much entitled to, representation, as the property and interests of members of the State Church?

Does the legal restriction of a cross or steeple render the heretical house of prayer less a temple of that God whose worship was first carried on in upper rooms, in caves, in groves? Will the sight of heretical peristyles and decorations uproot from the minds of the faithful the precepts of a Church whose claim it is to possess the keys of heaven and hell, to loose and unloose? Can a legislature which enacts such stringent laws of mortmain, that provides for the extinguishment of the confraternities, and for schools to educate and enlighten the rising generation,—can they, can the ministers of the state religion themselves, believe that the "outward signs of temple" will undermine and overthrow a worship bequeathed to them by the martyrs that carried it triumphantly through the opprobrium and the bloody scenes of Jewish and Roman persecution? No! it cannot be that they seriously believe it!—to do so would be to doubt themselves and their religion; but a few intolerant minds raise the war-cry of "The Church in danger!" and "cant" carries the day against the secret wishes of the tolerant majority.

But perhaps nothing is likely to operate more injuriously on the successful issue of immigration than the legislative intolerance of "civil" marriages. These have been permitted by the laws of several nations which furnish emigrants, not merely by such Protestant countries as Great Britain and the United States, but likewise by Roman Catholic France and Italy. Great numbers of their marriages have been celebrated according to the simple form and regulations of their permissive laws of civil matrimony: yet Brazil refuses to allow them force or validity within her territories! What will be the effect of such intolerance upon the proud emigrants from the Southern States, whom Brazilians are so wisely endeavoring to attract, in whose country a simple rite and registry before a country justice is a usual form of marriage, yet is a marriage and an evidence unrecognized and invalid in Brazilian courts of law! Will these Americans, so rightly jealous of their own and their families' reputation, willingly seek Brazil while she stamps upon their children the brand of illegitimacy? Is not this also intolerance rampant behind the mask of "danger to the Church"?

These are matters well and deeply to be pondered by Brazilian legislators, and on which some Brazilian O'Connell can win reputation and the homage of men.

Let him boldly couch the spear and overthrow the dragon of intolerance at the next session of the legislature, and he will find the demon "cant," like Apollyon, needs but a vigorous thrust to send him howling from the presence of the assembled wisdom of Brazil.

Immigration is yet a tender exotic in this country, that calls for gentle handling and careful culture. The weakling plant is promising fairly now, and now is the time to nourish it, to cull away every weed that threatens injury, that the little upshoot, gaining strength and vigor day by day, deep-rooting and wide-branching, may spread around through all the land, bringing prosperity and happiness to millions, and riches and power to Brazil.

NOTE.—*Dissenters in the new House of Commons.*—There are in the House of Commons, as representatives of English constituencies, thirteen Independents, twelve Unitarians, five Jews, three Catholics, three Quakers, one Baptist, and one Wesleyan; as representatives of Irish constituencies, thirty-one Catholics, one Quaker, and one Independent; as representatives of Scotch constituencies, three United Presbyterians, two Free Churchmen, one Independent, and one Unitarian: making the total number of Dissenters in the new House of Commons forty-four, and the number of Catholics thirty-four: gross total, seventy-eight.

APPENDIX J.

PROFESSOR AGASSIZ'S LABORS ON THE AMAZON.

THE wonderful discoveries made by Professor Agassiz in the *fauna* of the Amazon have attracted the attention of the scientific world. In due time these great results will doubtless be published by the professor himself. But so great is the interest manifested in these explorations that I avail myself of two letters written by Dr. J. M. da Silva Coutinho, (the Brazilian explorer of the Purús,) who accompanied the professor on the Amazon.

The first letter of Dr. Coutinho from Manáos, under date of November 7, 1865, gives the following account of the professor's researches:—

"In the beginning of September I wrote my first letter from this capital, giving a slight notice of my labors so far. We had then more than 300 species collected in Pará, Tajipurú, Gurupá, Porto de Moz, Monte Alegre, Villa Bella, and Serpa. In Santarém we collected only some four species. We spent fifteen days on the passage from Pará.

"From Manáos we started on the 10th of September, on board the *Icamiaba*, bound for Tabatinga, intending to pursue our voyage in the Peruvian steamers thence to the village of Jurimaguas, and in canoe and on foot from this landing to the eastern side of the Andes.

"Several of our travelling companions were to stay at Tabatinga, S. Pagés, Nauta, and Laguna, to make collections in the Maranhão, Uallaga, Ucayalle, Napo, Hyauary, and other affluents of the Solimões or Upper Amazon.

"We planned this journey while awaiting the state of the Rio Negro, still at the beginning of the ebb, as we could make valuable collections only forty days later.

"In Teffé, during the stay of the steamer, we collected some specimens of the *Acará petroina*, with its eggs in its mouth, and Professor Agassiz had then an opportunity of studying this curious phenomenon, of such great scientific interest.

"We found the ebb far advanced there, and the people of the place said that there was already abundance of fish.

"We started the same day from Teffé [where Bates collected in 1857–59].

"In Fonte Boa, Tunantins, and S. Paulo, we found the river lower; and some of the inhabitants said that the freshet could not delay long.

"This circumstance caused us to alter the plan of the voyage, and Professor Agassiz resolved to return from Tabatinga to Teffé, to remain at work there, profiting by the best time for the fisheries of the Solimões, whilst Dr. Coutinho and another would go to the side of the Andes to study the geologic formations and the vestiges of the ancient glaciers.

"When we came to Tabatinga, this plan was further altered in view of the news from Peru. The civil war had invaded the districts of Caxamarca and

Chachapoyas, through which we had to pass, and there was neither safety on the way nor means of effecting the trip. Besides this, as the excursion to Peru could be made at any time, and the ichthyological study of the Solimões only when the river is lowest, almost all the fishes disappearing as soon as the freshet begins, we agreed to leave the work of the Andes to a later period, and to use the time of low water in the Solimões.

"In Tabatinga we found the remnant of the Spanish commission, which had descended by the Napo, having traversed the republic of Ecuador. One of the members was very sick. [Spanish should doubtless read "Peruvian."—J. C. F.]

"There we left Mr. Bourget to make collections from the Hyauary, and Mr. James and another to explore the Iça, Hyutay, and Hyuruá.

"On the 24th of September we were in Teffé. The first fishing that we did was on the beaches of *Nogueira*, five miles off, opposite the city. The enthusiasm of Professor Agassiz seemed to reach ecstasy on seeing the great number of species collected in only three casts of the net. 'This success is so great that I feel my head splitting,' said he, contemplating the fishes on the beach.

"We continued with much profit our labors in the basin of Teffé, passing afterwards to the left side of the Solimões in company with Major Estulano, who furnished us the occasion and means of making a fine collection. The best result we obtained was in Lake Boto, which is one of the deposits of water so curious in the islands of the Amazon.

"The *Paraná-Mirim*, the channel which separates two islands, is obstructed at the higher part, either because of a bank formed before it turning aside the current, or of the increase of the beaches at the upper parts of the islands. The Paraná-Mirim passes then to the condition of a gulf or bay. During the low stage of the river, these deposits contribute with their contingent, and so the sands little by little advance towards the mouth. When the river rises there is not the least emptying of the gulfs: on the contrary, they receive a part of the water of the river. Finally, the mouth dries in the summer, plants grow, the sediment increases the soil rapidly, and the gulf is transformed into a lake.

"As I said, the labors in Lake Boto were very satisfactory. The professor had here another opportunity of verifying the principle established by him many years ago upon the resemblance of the adults and young of diverse genera of the same family.

"In the Teffé he discovered a new genus of the family of the *Scomberesoces*, which he has named *Limnobelone*. This genus is distinguished from the others by having the dorsal and anal fins larger, and the caudal rounded. The maxillars are like those of the genus *Belone*.

"In Lake Boto we caught a young fish of a new genus, having the inferior maxillar much larger than the superior, entirely different from the adult, and, under this point of view, perfectly resembling another genus of the same family, the *Hemiramphus Braziliensis*, which is found in the Atlantic Ocean and is common at Rio de Janeiro.

"Not less important was the discovery made by the professor in some fishes of the family of the *Siluroides*, of having only two bones in the opercular apparatus, when, until now, it was believed to have three.

"In Teffé great assistance was rendered by Dr. Romauldo, the Juiz de Direito, Captain João da Cunha, and Lieutenant Pedro Mendes. The old fisherman Vicente Marquez gave us important information on the habits of the fishes, according to which we could settle on secure bases for the distribution into species.

"On the 18th, our companion Mr. James arrived, having visited the Iça and Hyutahy, but not having had time to examine the Hyuruá.* On the 21st, the *Icamiaba* anchored in the port on her return from Tabatinga, bringing Mr. Bourget. Both brought upwards of two hundred species. We embarked on the same day, and came to this capital on the 23d of October.

"From want of health and of alcohol, we have not done here as much as we desired. In all we collected seventy-six species, almost all new, during the three days we passed at Lake Hyanuary. The most notable discovery was a new genus of the family of the *Chromides*, which has the caudal fin shaped like a lance, to which Professor Agassiz gave the name of Dr. Coutinho.

"The President of the Province accompanied us to Hyanuary, and furnished us all requisites. Dr. Tavares Bastos and other gentlemen also went with us.

"Up to this time we have collected 776 species, of which 650 are new.

"The professor said, before coming to the Amazon, that he would be well satisfied if he collected 250 new species. The result, then, has been extraordinary; and the professor says that it is a true *revelation* for science.

"We supposed that there would be diversity of the species in the black and white water, in the lakes and rivers, in the upper part and the mouths; but no one had imagined that it would extend to the same region where all circumstances are identical.

"The species of Pará are entirely different from those of Tajapurú, these from those of Gurupá, these from those of Monte Alegre, and so on. Even between near places there is a notable difference, as we observed in the lakes of José-Assu and Maximo, which are not four miles apart, and lie on the same side of the Tupynambaranas.

"The Amazon, then, comprehends a great number of ichthyologic faunæ, or provinces occupied by distinct species.

"The knowledge of this fact opens new horizons to scientific researches, and is the surest base for the study of the distribution of species.

"Having established, therefore, the great principle, it remains to know the number of the ichthyologic provinces, the extent of their limits or the situation of the points of contact, and the causes which determine the differences. All these questions exact long labors and study, but their result must be extraordinary and perhaps one of the finest results yet obtained in the study of nature.

"The surprise increases when we reflect that the climate does not vary throughout a great extent of the Amazon.

"The same phenomenon that is observed in the main stream takes place in the tributaries; and, as our labors were made in a few places of the Amazon, and merely, besides, in the Tapajos, Hyauary, Iça, and Teffé, some leagues from their mouths, an idea may be formed of the great result of a complete exploration that would embrace the courses and all the tributaries. There is no exaggeration in supposing the existence of between two and three thousand species in the valley of the Amazon. Until now only a little more than 100 were known. Wallace collected 205 in the Rio Negro; his collection, however, was in most part lost.

"With our labors in the Negro, Madeira, and in Maués and other parts of the province of Pará, we hope to find, perhaps, 300 species more, reaching thus a

---

* Dr. Coutinhos writes with an initial H all those names of the affluents of the Amazon which commence with a J in the map of Brazil affixed to this work.

number exceeding 1000, which are as many as are known at present in the Mediterranean.

"When Linnæus published the 6th edition of his System of Nature, a little more than a century ago, the number of known species on all the globe did not exceed 300. Now, however, the labors of only three months give the knowledge of almost 800 in the Amazon." [April, 1866.—The final result of five months is said to be 1300 species.—J. C. F.]

SECOND LETTER.

"Manáos, November 24, 1865.

"As 1 said before, we resolved to put off the trip to Peru in consequence of the news received at Tabatinga of affairs in that republic. The low stage of water, also, was nearly at its close, and thus but little time remained to make useful collections. As soon as the rise commences, the beaches become covered and the margins inundated, while the greater number of the species of fish leave the rivers for the *Igapó*, the forest that borders the river and remains flooded during winter. The fishes that remain seek the deepest places, and the use of the net or hook becomes almost impossible. The only recourse to the Indian is the arrow and harpoon, but fishing then becomes very slow, and not rarely the fisher has to return empty-handed.

"The collection that we made in the Teffé and its neighborhood was magnificent. Besides numerous species, Professor Agassiz discovered many new genera, acquiring knowledge, likewise, of some important laws to which certain species are subject in their development, ignorance of which has caused mistakes in classification.

"Schomburg, for example, established as a special characteristic the adporous protuberance found in the head of some fishes of the genus *Cychla:* we, on the contrary, find that this is accidental. We had occasion to examine the *Tucunaré* (the indigenous name) without the protuberance, with it beginning to develop, and after it had attained its growth.

"This curious phenomenon takes place in the commencement of winter. The fishermen account for it by saying that the fish, while entering the 'igapó,' strikes its head against the trees, and therefore the inflammation appears. The true reason, however, is the critical state in which the 'Tucunaré' is at the beginning of winter, which is when it spawns.

"Through the profusion of some genera of the family of *Chromids*, permitting the comparative study of the young and the adults, the professor was enabled to lay down the following law:—that the same species presents different radicals according to age.

"In some new species of the *Siluroids* the professor found, likewise, only two bones in the opercular apparatus, whereas all the species previously known had three.

"He verified again the principle established by him many years ago on the resemblance of adult and young individuals of different genera of the same family. A new genus was found belonging to the family of the *Scomberosoces*, which he called Lymnobelona, and which was distinguished from the others by having the dorsal and anal fins greater, and the caudal rounded. As in the genus *Belona*, the maxillars are equal.

"Some time afterwards we caught a young one of the new genus, having the lower maxillar much greater than the upper one, and in this point of view somewhat resembling the other genus of the same family, the *Hemiramphus Braziliensis*, inhabiting the sea, and found on almost all the coast of Brazil.

" At Manáos we collected 150 more species, almost all in the Lake Hyanuary, opposite the city.

" To-day there arrived in the *Icamiaba* a fine collection brought by the two parties sent to Lakes Manacápurú and Cudayás. In two bottles alone the professor found sixty-eight species; and we hope to find a greater number in the eight barrels not yet opened.

" The collection made up to now consists of 970 species, of which more than 700 are new. Imagine the surprise and pleasure of Professor Agassiz in view of so great a result, who when entering the Amazon considered himself fortunate should he find 250 species.

" The results that we obtained at the first points made us believe that the great river possessed many distinct faunæ of the ichthyological provinces: however, when at our last two we explored some *igarapés* (smaller rivers) and lakes, distant from one another not more than 1200 yards, we saw then that the number was extraordinary. The collection of so great a number of species in a few places, and during only three months, surprises the more when we reflect that, in 1840, Captain Wilkes collected only 600 in a voyage round the globe, with three ships, in an expedition lasting four years.

" As yet we cannot determine what number of species a single region contains, on account of the rarity, and of the little time at command. At present we are treating of this important question in Lake Hyanuary.

" The family of *Chromids* is one of those which show the greatest sensibility to changes of their mode of living, and therefore the most important discoveries made by us appertain to them. It comprehends, in the Amazon, the fishes known as the *Tucunaré*, Jacundás, and Acarás.

" Fifteen new genera have now been established by the professor, embracing the fishes that live by preference in the igarapés, and he thinks of separating two of these genera into families, in consequence of their general characters, some belonging to the *Gobioides* and others to *Cyprinodontes*. Those fishes are known by the name of *Amoré*, and are found only at the west part of the island of Marajó, at the place Taypurú. This circumstance makes us believe that the *Amorés* determine approximately the point to which the waters of the ocean reach in the Amazon.

" The most interesting genus, however, in the opinion of the professor, is one discovered by us in Lake Hyanuary, opposite this city. It came, so to say, to strengthen the union of the families, serving as intermediate between others rather separated, as those of the *Jacundás* (*Crenicychla*), and divers *Acarás* (*Satanoperea, Hygrogonus*). The structure of the caudal fin of the new genus is an exaggeration of the *Jacundá* type, in consequence of the prolongation of its medial rays, whilst the dorsal and anal fins are equally elongated in the posterior part, as in the true *Acarás*. The body resembles the *Jacundá*.

" According to the professor, the name *Polymorphes* suits the *Chromids* well. Its genera resemble the greater part of those of the families inhabiting the ocean, many rivers of the East Indies, and other points of the globe. In the Amazon, then, representatives of a great proportion of the inhabitants of the waters are found. This family is one of those most widely spread, although the greater number of its species are found in South America. Represented in the whole of tropical Asia, it extends to Africa on the western coast to the Cape of

Good Hope. In America it is met with in all the streams of the continent, from Patagonia to the Gulf of Mexico; being substituted in the North by the *Elychtids* (*Centrarchids*), which might well be joined with the true *Chromids*.

" The professor has paid particular attention to the study of the *Chromids*, on account of the difficulties presented in the knowledge of the species.

" It is almost impossible to determine with precision the special characteristics of these fishes without examining a great number of specimens, because the young and adult differ considerably in some genera, and there is a notable difference between the two sexes. In some genera the young have a more elongated form than the adults; the contrary happening in others. It is by the color, however, that the adults are most distinguishable from the young. The *Tucunarés*, for example, when completely developed are remarkable for their brilliant colors, transversal stripes, and a beautiful dark-blue spot on the tail, having a yellowish or pink fringe. The young, on the other hand, are pale, and have only a longitudinal stripe. In the new *Pleoropo*, the longitudinal stripes that the young have are substituted in adolescence by some black spots on the sides and tail. The contrary is observed in the genus *Mesonauta*. A line of black dots, which the young have, disposed diagonally on the sides, is transformed by age into a continuous streak; and in other genera modifications more or less remarkable, according to age, are observed.

" It is therefore evident that, to appreciate these great differences, the examination of one individual is insufficient. For this reason the descriptions published up to the present time cannot give an exact idea of these fishes, and it is very probable that some fishes, considered as different species, are only the same at different ages. Still more: the adults vary, likewise, according to the season, and at the time of spawning, during which they have the most brilliant colors. On this point, however, the inquiries require much time. The difference of character between the sexes does not appear to be the same in all the genera. The most considerable we found among the *Cychla* and *Geophagus* (*Tucunaré* and *Acaráyné*). The male possesses at the milting season an adporous protuberance in the head, as I have already mentioned, and which Schomburg gives as special characteristic of the species *Cychla trifasciata* and *Cychla nigromaculata*.

" The habits of the *Chromids* are very variable. Whilst some swim at the surface, as the *Lepterophilum* (*Acará-pena*) and the *Mesonauta* (*Acará Meréré*), others descend a little below. Of this kind is the genus *Cychla;* but the *Hygrogonus* does not leave the bottom, and, buried in the mud, often escapes the net. This genus, which comprehends the Acará-assu, is one of the most beautiful, from the carmine spots that the fishes have on the tail, upon the dorsal and side fins. Its habit is to leave eggs in the orifices found in the banks, and to remain there until the young can accompany it.

" The fishes of genera *Geophagus* and *Satanoperea* (*Acaráyané* and *Acaratinga*) keep the eggs in a pouch formed by the superior pharyngeal bones, which curve upon the branchial arches.

" The professor had an opportunity of studying the complete development of the eggs, (observing the newly-born in the branchial pouch up to their state of swimming freely,) in the species to which he gave the name of the Emperor, *Geophagus Pedroinus*, and of which he made a complete study.*

---

* This fish was first discovered on the 22d of November, 1862, by Sr. Henrique Antonii and J. C. F., while collecting specimens for Professor Agassiz, in an *igarapé* on the island of Papagayos, opposite the Manáos.

" The configuration of the head is very curious in these species that preserve the eggs in the branchial pouch. They have a nervous protuberance, resembling the electric lobule of the *Malapterurus*, in the part posterior to the cerebellum, serving as a root to the nerve that is prolonged hence to the inferior branchial arch, forming, evidently, such a centre of special action as is shown in the marsupial pouch. For this reason the protuberance well merits the name of *'genetic lobule.'*

" About thirty years ago the family of the *Chromids* was established, almost at the same time, by Heckel and Müller, with some genera of the *Labroides* and *Scienoides* of Cuvier. The number of its species was then very limited. In the British Museum catalogue published in 1862, which gives the last and most complete enumeration of this family, the number of its species in all the world is 110, distributed in 19 genera. Of these species, only 12 belonged to the Amazon. Now, however, we have here 120 species, almost all new,—that is, a greater number than were known in 1862 in all points of the globe.

" At another opportunity I will speak of the families of the Siluroids and Characins.

" In the next fortnight we purpose setting out for Mauós, and thence to Parñ."

A VOLCANO IN SOUTHERN BRAZIL.

Captain Burton, F.R.G.S., English Consul at Santos, writes a brief but interesting letter to " The Anglo-Brazilian Times," in regard to the discovery of a volcano in Southern Brazil, about half-way between S. Paulo and Paranaguá.

" Mr. Editor:—I was canoeing down the river of Iguapé,—in this consular district,—it is called, ridiculously enough, the Ribeira, or rivulet,—when, calling on the excellent vigario (vicar) of Xiririca, M. J. Gabriel da Silva Cardoso, and looking over his Livro do Tombo, (Parish Register,) I was struck by the name of a place, in the Tupi or Lingua Geral 'Vutupoca,' translated *Morro que rebenta*, 'hill that explodes.' On the other side of the river, bearing southwest from the vicarage, rose the morro, clothed with trees *cap-à-pié*, an isolated gradual cone, with a distinctly volcanic outline. Its northeastern face is, I was told, a perpendicular rock.

" The fearful rains of January, 1866, prevented my ascending the Exploding Hill. But the result of many local inquiries was that as lately as fifteen years ago flame has been seen rising from the hill, and the phenomenon was accompanied by rumblings and explosions which extended across the river to the opposite range of Bananal Pequeño.

" You will, I hope, hear from me again. Should this report of a dormant volcano in Southern Brazil be confirmed by absolute exploration, the discovery will be of no little value in a geographical point of view. And these lines may perhaps, should I be unable to carry out my project, induce another and a better man to undertake the task. It is not, you will remember, half a century ago when the scientific men of Europe declared that no volcanic formations, and certainly no volcanoes, could be found in this magnificent empire.

" I am, sir,
" Your obedient servant,
RICHARD F. BURTON, F.R.G.S.
" Hotel Milton, Santos, Brazil."

APPENDIX K.

THERMOMETRICAL OBSERVATIONS AT RIO DE JANEIRO IN 1864.

Months.	Centigrade.			The same reduced to Fahrenheit.			
	7 A.M.	1 P.M.	5 P.M.	7 A.M.	1 P.M.	5 P.M.	Average.
January	25.350	27.250	26.719	77.630	81.050	80.094	79.591
February	25.399	27.820	27.040	77.718	82.076	80.672	80.155
March	24.970	27.155	28.052	76.946	80.879	82.494	80.107
April	24.240	26.762	26.081	75.632	80.172	78.946	78.250
May	21.920	23.986	23.484	71.456	75.175	74.271	73.634
June	20.029	22.806	22.392	68.052	73.051	72.306	71.136
July	19.229	22.221	21.855	66.612	71.998	71.339	69.984
August	20.130	22.781	22.017	68.234	73.006	71.631	70.957
September	20.700	22.838	22.414	69.260	73.110	72.345	71.571
October	21.844	23.951	23.127	71.319	75.112	73.629	73.353
November	23.321	24.815	24.672	73.978	76.667	76.410	75.684
December	24.839	26.519	25.233	76.710	79.734	77.419	77.954

Meteorological and other observations by Lieutenants (Brazilian Navy) José da Costa Azevedo and João Soares Pinto, of the Commission for settling the limits between Brazil and Peru.

Average Temperature, each month for six months, at Pará, (lat. S. 1° 27' 06",) from observations in the street of·S. Jeronimo, from November to April inclusive, 1861–62.

	Reaumur.			The same by therm. Fahrenreit.		
	7 A.M.	1 P.M.	5 P.M.	7 A.M.	1 P.M.	5 P.M.
November	20.0	22.2	21.7	77	82	80.8
December	20.1	22.6	21.5	77.2	82.8	80.3
January	20.4	22.6	21.4	78	82.8	80.1
February	19.5	22.8	21.8	75.8	83.3	81
March	19.3	22.5	22.1	75.4	82.6	81.7
April	19.5	23.5	21.8	75.8	84.8	81
Average for six months	19.6	22.8	21.7	76	83.3	80.8

Average Temperature each month for six months in 1862, at Manáos (Barra), from observations on the grounds for the Cathedral [Igrega Matriz].

	Reaumur.			The same by therm. Fahrenheit.		
	7 A.M.	1 P.M.	5 P.M.	7 A.M.	1 P.M.	5 P.M.
May	20.9	22.2	21.8	79	82	81
June	20.5	22.2	22.5	78.1	82	
July	20.9	23.3	23.4	79	84.4	84.6
August	20.8	23.9	23.1	78.8	85.7	84
September	20.3	24.6	23.7	77.6	87.3	85.3
October	21.8	24.9	24	81	88	86
Average for six months	20.9	23.5	23.2	79	84.8	84.2

Note.—The lowest average at Pará recorded by Srs. Soares and Pinto was that of December, at 5 A.M., when it was Reaumur 18.7, (Fahrenheit 74.7;) but there is no 5 A.M. record for the four succeeding months. The lowest average at Manáos in the six hottest months of the year was in June, at 10 P.M., Reaumur 19, (Fahrenheit 74.7,) and at 3 A.M., Reaumur 19.2, (Fahrenheit 74.9.) Pará is dryer and a little cooler than Manáos.

	Latitude South.	Longitude West of Greenwich.	Ordinary Level of River above that of the Ocean.	Declivity of the Amazon current, in English feet.
			Metres.	Feet. Inches.
Pará...............................	1° 27′ 06″	48° 26′ 17″	10.71	35 1.6
Breves*..............................	1 43 08	50 25 39	12.49	40 11.7
Gurupá..............................	1 24 57	51 34 47	13.09	42 11.3
Prainha..............................	1 49 00	53 25 54	14.62	47 11.6
Santarem............................	2 24 50	54 39 14	15.38	50 5.5
Obidos†............................	1 55 03	55 26 29	17.70	58 0.8
Villa Bella‡.........................	2 37 25	56 40 58	24.23	79 5.9
Serpa................................	3 08 05	58 22 24	25.26	84 2.2
Manáos (or the Barra)§.........	3 08 04	59 57 03	28.19	92 5.8
Coary................................			35.09	115 1.5
Teffé (or Ega)......................			36.79	120 8.4
Fonte Boa...........................			37.34	123 9.8
Tunantins...........................			38.03	124 9.2
Villa de S. P. d'Olivenca........			38.26	125 6.3
Tabatinga...........................	4 15 00	69 52 13	45.99	150 10.6

* On the island of Marajo, (southwest portion,) 131 geographical miles from Pará.

† The tide reaches Obidos during the lowest stage of water.

‡ Sometimes called Villa Nova.

§ Before the careful observations of Srs. J. da Costa Azevedo and Soares Pinto, the estimates of the elevation of the river above the level of the ocean were mere guesses: *e. g.* the level of the river at Manáos was placed by Spix and Martias at 169 (metres) 57cms. (English feet 556 4 inches) above the ocean, Castelnau at 62m. 48cms. (204 feet 11¼ inches), and by Herndon, 1475 feet(!); whereas the real level above the ocean is 92 feet 5¼ inches. Can there be found elsewhere in this world such a channel for internal navigation?

THE DECLIVITY OF THE AMAZON PER LEAGUE* (PORTUGUESE) FROM TABATINGA TO PARÁ.

	Inches.
From Tabatinga to Villa S. P. d'Olivença...........................	2.720
" Villa d'Olivença to Teffé (Ega)...............................	2.400
" Teffé to Coary..	2.870
" Coary to Manáos......................................	2.970
" Manáos to Serpa (129 geographical miles)..........	2.940
" Serpa to Villa Bella (159 do.)..........	3.200
" Villa Bella to Obidos (105 do.)..........	3.690
" Obidos to Santarem (73 do.)..........	3.860
" Santarem to Prainha (100 do.)..........	4.140
" Prainha to Gurupá (143 do.)..........	4.910
" Gurupá to Breves (119 do.)..........	5.900
" Breves to Pará (131 do.)..........	7.810
" Para to mouth of the Amazon...........................	5.847
Average declivity per league..............................	4.090, or 1 inch per mile.

Count de Castelnau's observations in 1844 make the declivity per league 4.14 inches.

* The ordinary Portuguese league is about four English miles.

Note.—I observed the tide at Obidos in November, 1862; Mr. Bates, in 1855, observed it on the Tapajos, a distance of 530 miles from the ocean. This tide is of fresh water banked up or driven inward by the regular ocean tide.—J. F. C.

STEAM

ROUTES

TO

BRAZIL.

TIME TABLE OF THE UNITED STATES AND BRAZIL LINE.

(From the agent, W. R. Garrison, Esq., 5 Bowling-Green, New York.)

Passage from New York to Rio de Janeiro.	Stay in Port. hours.	*Passage from Rio de Janeiro to New York.*	Stay in Port. hours.
Leave N. Y 22d of each mo. of 30 days, 3 P.M.		Leave Rio the 25th of each month, at 2 P.M.	
Arrive at St. Thomas (1400 miles) the 29th	12	Arrive at Bahia (725 miles) the 29th	12
" Pará (1610 ") the 7th	22	" Pernambuco(375 ")the 1st	13
" Pernambuco (1090 ") the 15th	11	" Pará (1090 ") the 6th	34
" Bahia (375 ") the 17th	12	" St. Thomas (1610 ") the 14th	12
" Rio Janeiro (725 ") the 20th		" New York (1400 ") the 21st	

Total . . . . 5200 miles run.
Total running time 25 days, 15 hours.
Four calls 2 " 9 "

28 days.

N. B.—In months having 31 days the steamers leave New York the 23d.

Total running time, 22 days 16 hours.
Four calls, 2 " 23 "

25 days 15 hours.

From New York to Rio and back 58 days 15 hours.

N. B.—The steamers will leave Rio the 26th of months having 31 days.

The United States and Brazil Mail Steamship Company corresponds, first, at St. Thomas with the English, French, and Spanish steamers which run to 43 ports in the West Indies, Mexico, Central America, New Grenada, Venezuela, and the Guianas; with the English and French lines to Europe; 2d, at Pará it corresponds with the Amazon Navigation (Brazilian) Company's steamboats, which run up as far as Peru and are in connection with Peruvian steamboats on the upper Amazon, and with Brazilian coast steamers for Maranham, Ceará, &c.; 3d, at Rio with the French and English steamers which go to Montevideo and Buenos Ayres, (the French line leaving Rio on the 22d of each month.) We understand that the time is to be shortened. Mr. A. Arango, of New York, was the successful advocate of the company at Rio in 1865. Mr. J. F. Navarro was the New York agent until June, 1866.

INDEX.

Abdication of Dom Pedro I., 1, 83.
Aboriginal names, 306.
Aborigines, 466.
Academy of Fine Arts, 261.
Academy of Laws, 369.
Acclamation of Dom Pedro II., 223.
Administration of Brazilian law, 253.
Advertisements, 254.
Agassiz, 210, 250, and Appendix J, 627.
Alagoaz, province of, 511.
Alto-Amazonas, province of, 563.
Amazon River, 539, 551.
    canoa, 551.
    Cetacea of, 555.
    discovery of, 563.
    entrance to, 541.
    expeditions, 563, 568, 574, 581.
    explorations, 554.
    first Protestant sermon on the, 542.
    future of, 581.
    Navigation Company, 547.
    origin of name, 566.
    source, 573.
    steamers, 575.
    tributaries, 573.
    Valley (area of), 574.
Amazons, tribe of, 576.
Amenities of quarantine life, 514.
American Cemetery, 485.
    factory, 274.
    machinery, 501.
    Seaman's Friend Society, 200.
    sheep, 431.
Americus Vespucius in Brazil, 49.
Anacondas, 508.
    one that swallowed a horse, 509.
Andrada, Antonio C., 215, 218, 222, 373, 377.
    death of, 383.
    José Bonifacio, 72, 73, 83, 215, 224, 272, 323.
    Martin Francisco, 73, 220, 224, 376, 383.
Annoyances magnified, 507.
Ant-hills, 359.
Araguya explorations, 463.
Armadillo, 193.
Astronomy under difficulties, 429.
Asylums, 107-123.
Aymores, ferocity of, 471.

Bahia, city of, 475, 476.
    history of, 478, 479.
    recaptured from Hollanders, 482.
    sociability of, 484.

Bahia, view of, 489.
Baronial style, 440.
Bastos, A. C. Tavares, 139, 186, 197.
Bates, 581.
Bay of Rio de Janeiro, 14.
Beautiful panorama, 516.
    scenery, 345.
Bees, indigenous, 455.
Bennett's, 205, 210.
Bible Christian, 519.
    distribution, 256, 306, 336.
Bird Colony, 405.
Bishop Moura, 379.
Boat-bill, 571.
Boa Vista, 279.
Botanical gardens, 208.
Brazil, discovery of, by Pinzon, 46.
    Cabral, 47.
    first governor, 50.
    independence, 71.
    origin of name, 49.
    revolution in, 72.
Brazilian Constitution, 76, and Appendix B, 603.
    dinner, 310.
    funerals, 203.
    Historical and Geographical Institute, 261.
    Jupiter Pluvius, 272.
    literature, 251, 586.
    writers, 586.
Bridge of novel construction, 411.
Brotero of San Paulo, 426.
Brotherhoods, 107-123.
Burial of the Innocents, 343.
Burton's discovery of a volcano, Appendix J, 633.

Cadeiras, 476, 477.
Campinas, 400.
Campos, 465.
Campo Santa Anna, 211.
Candiota coal, Appendix H, 617
Canoe voyage, 328.
Canta Gallo, 292.
Cape Frio, 464.
Captain Foster, 235.
Carumarú, 478.
Cascades, 206.
Ceará, 527.
    exploration of, 502.
Charlatanism, 342.
Chinese tea, its culture in Brazil, 419, 420.
City of Pittsburgh, 235.
Climate of Brazil, 269, and Appendix K, 634.

Coal-mines, 347, and Appendix H, 617.
Coffee, its history and culture, 449.
Colleges, 178.
Colonia Donna Francisca, 334.
    Joinville, 332.
Commerce of Brazil, 194, Appendix G, 614.
Commerce with United States, 501.
Constancia, 283.
Constitution of Brazil, 76, and Appendix B, 603.
Convents, 107–123.
Cool resorts, 270.
Cormorant and slavers, 315.
Coronation of D. Pedro II., 225.
Cotton-factory, 493.
Council of State, 227.
Course of law study, 371.
Curious items of trade, 360.
Curious trial, 265.

Daring navigation, 497.
Deceived custom-house officials, 421.
De la Condamine, 568.
Desterro, 344.
Diamond- and gold-mines, 448, 462, 463, and Appendix H, 624.
Difficulties overcome, 241.
Discoveries of new species of fish by Agassiz, Appendix J, 627.
Diseases in Brazil, 417, and Appendix F, 609.
Distinguished men, 373.
Dom João VI., 64, 69.
Dom Pedro I., 69, 71, 73–85.
Dom Pedro II., 231–250.
Dr. Kane and Lieut. Strain, 459.

Education, 163, 176, 178.
El Dorado, 564.
Emigrants' instructions, terms, &c., 333, 592.
Emperor of Brazil, 231–250.
    his remarkable talents, 232.
    on board an American steamer, 235.
Emperor's birthday, 491.
English Cemetery, 201.
    chapel, 202.
    engineer, 318.
Enslavement of the Indians, 368.
Espiritu Santo, 465.
Events after abdication of Dom Pedro I., 213.
Excursions, 207.
Exhibition of United States manufactures, 239.
Expenses of travelling, 6, 295.
Exploration of rivers, 463, 502.
Extent of the empire, 433.
Extraordinary fanaticism of Sebastianists, 520.

Fallen forest, 338.
Falls of Itamarity, 435.
Family recreations, 175.
Feijo, Bishop, senator and regent, 216, 380.
Finest steam-voyage in the world, 198.
Fire-flies, 293.
First Protestant church in America, 54.

First steamer at Coimbra (Upper Paraguay), 459
Fish on the Amazon, Appendix J, 627.
    on the Madeira, 556.
Forest flowers and scenery, 277.
Foundling Hospital, 113.
Fourth of July in an English family, 426.
Frade Vasconcellos, 357.
French in Brazil, 54, 62.
Funerals, 203.
Furquim d'Almeida's speech, 594.

Gigantic fig-tree, 437.
Gillmer, 484.
Godin, Madame, 568.
Goyaz, province of, 453, 454.
Great ant-eater, 445.
Guaraná, Preface to 8th Edition.

Happy valley, 287.
Heath, Mr.. 284.
Heaven of the moon, 357.
Herds and herdsmen, 348.
Herndon's explorations, 574.
Historical and Geographical Institute, 261.
Historical data, 369.
Hollanders in Brazil, 481.
Home-life, 163, 169.
Homeward bound, 423.
Hospitalities of a padre, 385.
Hospitals, 107–123.
Huguenots, 54.
Humming-bird, 484.
Hunter, Hon. William, 227.

Image factory, 494.
Immediate reforms needed, 589.
Imperial Academy of Fine Arts, 261.
    marriages, 229.
Improvements of Recife, 514.
Incorrect judgments, 310.
Indian archery, 558.
    revolution at Pará, 543
Indians, 351, 470.
India rubber, 552.
In puris naturalibus, 498.

Jacaná, 572.
Jaguar, or Brazilian tiger, 445, 557.
Jangada, 525.
Journals of Rio, 253.
Journey to San Paulo, 354.
Judge Petit's description of Maranham, 533, 534

Keel-bill, 437.

Lady's impressions of travel, 273.
Lasso, 349.
Law students and convents, 361.
Limeira, 402.
Lingoa Geral, 471.
Literature, 251, 588.
Lodging and sleeping, 395.
Longfellow, Hawthorne, and Webster, 249.

Long river-route to Atlantic, 456.
Lopez and Paraguay, 352.

Machado, Alvares, 377.
Magnificence of nature, 539.
Mandioca, 189.
Maranham, city of, 533.
Marmoset, 510.
Martyn, Henry, 486, 487.
Matto Grosso, 456.
Mawe's experience, 362.
Medical schools, 180, 490.
Mexico and Brazil compared, 77.
Mines and other resources of Minas Geraes, 448.
Miracle explained, 495.
Missionary efforts in San Paulo, 386.
Montaria, 538.
Montserrat, 498.
Mulatto improvisator, 272.
Museum, 266.
Musical cart, 447.
    innkeeper, 426.

Natal, 527.
National library, 259.
Navigation on Paraguay, 459.
New disease, 416.
New York Historical Society, 237.
Night among the lowly, 398, 399.
Night travelling, 354.
Night with a boa-constrictor, 403.
Nova Friburgo, 295.

Old Congo's spurs, 397.
Omnibus, 38, 42.
Opening of Parliament, 211.
Orchidaceous plants, 341.
Organ Mountains, 277.
    their height, 282.
Origin of Indian civilization, 467.
Overthrow of Rosas, 351.

Paca, 446.
Palace of viceroys, 27.
    S. Christovão, 248.
Palm-tree and its uses, 468, 469.
Pará, 540, 551, 560.
Paraguay tea, or maté, 321.
Parahiba do Norte, 526.
Parallel to the Black Hole of Calcutta, 545.
Paraná, province of, 326.
    first president of, 321.
Paula Souza's letter to American emigrants, 592.
Paviola, 527.
Pedro II. Railway, 302.
Peixe-boi, 555.
Pennsylvanian in Brazil, 402.
Pernambuco, 515.
    commerce of, 521.
    houses of, 515.
Peter Parley in Brazil, 439.
Petropolis, 301.

Pizarro, Gonçalo, 563.
Plan of colonization, 412, 413.
Plant, Nathaniel, 347, 607, 609.
Plantation in Minas Geraes, 438.
Plantation orchestra, 441, 442.
Ponte da Area, 192.
Population of interior, 522.
Porters, 476.
Portuguese language and literature, 588.
Praia Grande, 187.
Probable extension of tea-culture in America, 422.
Produce of Amazon, 553.
Proposition to abolish celibacy, 381.
Provincial revolts, 351.
Purús River, 502.

Railroads, 299, 302, 365.
Religious disabilities, 594, and Appendix I, 625.
Republic of Palmares, 512.
Respect for San Paulo, 364.
Reverses of Jesuits, 368.
Rio de Janeiro, 21.
    Bay of, 15, 51.
    capital of Brazil, 187.
    City Improvements Company, 89.
    Custom-House, 30.
    early condition, 61.
    Exchange, 28.
    founded, 58.
    historic reminiscences, 14.
    journals, 251.
    libraries, 259.
    literary and scientific societies, 220.
    markets, 170.
    municipal government, 124.
    paving, 45, 87, 106.
    public promenade, 41.
        squares, 25, 38, 211.
    Rua Direita, 27.
        Ouvidor, 36.
    schools, 177.
    splendid views, 19, 22, 88, 192, 205.
    " tigers," 89.
    under the viceroys, 63.
Rio Grande do Norte, 527.
Rio Grande do Sul, 347.
Romantic life of a naturalist, 404.
Russian vessels in limbo, 317.

Sabbath observance, 188.
Saint Catherine's, 346.
Saint Vincent, 312.
San Domingo, 187.
San Francisco do Sul, 325.
San Paulo, 361.
    history of, 366.
San Sebastian, 307.
Sanitary condition of Brazil, 121.
Santo Aleixo, 271.
Santos, 309.
Schools, 177.
School-teacher, 335.
Sebastianists, 520.

Senhor José and a little difficulty, 524, 525.
Serra do Cubitão, 355.
Sertanejo and market-scene, 524.
Shells and butterflies, 346.
Sketch of the Vergueiros, 407.
Skilful negroes, 498.
Slavery, 132.
Slave-trade, 483.
Societies, 261.
Southern provinces, 303.
Speculations in town-lots, 279.
Star-Spangled Banner, 426.
Statesmen of Brazil, 186.
Stingless bees and sour honey, 455.
Sugar and cotton mart, 525.
Survey of the La Plata, 457.
Sweet lemons, 439.
Swiss bachelors, 287.
Sydney Smith's "Immortal," 273.

Tapioca, 191.
Tapir, 288.
Tasso gold-mine, Appendix H, 624.
Temperature and periodical rains, 530, 531.
Terrestrial paradise, 367.
Thermometrical tables of Rio de Janeiro and
    Pará, Appendix K, 634.
Tijuca, 205.
Todd's Students' Manual, 287.
Tolling-bell bird, 331.

Toucan, 291.
Travelled trunk, 329.
Tropeiros, 360.
Tupi Guaraní, 470.
Turtles and turtle-egg butter, 556.

Ubatuba, 305.
Umbrella-bird, 559.
United States and Brazil mail steamers, 197, and
    Appendix L, 636.

Valencia, 498.
Vampire bat, 504, 505, 506.
Vergueiro, 378.
Vesper hours, 442.
Victoria Regia, 571.
View from Inga, 192.
Village cemetery, 339.
Visionary hotel-keeper, 365.
Visit to Feijo, 380.
Volcano, Appendix J, 633.

White ants obedient to the Church, 443.
Wonderful image of St. Anthony, 494, 495.
Wreck of the frigate Thetis, 464.

Yankee Doodle, 47.
Ybecaba, 407.
Youth renewed, 239.
Ypiranga, 361.